Markov Chains with Asymptotically Zero Drift

This text examines Markov chains whose drift tends to zero at infinity, a topic sometimes labelled as "Lamperti's problem". It can be considered a subcategory of random walks, which are helpful in studying stochastic models like branching processes and queueing systems.

Drawing on Doob's h-transform and other tools, the authors present novel results and techniques, including a change-of-measure technique for near-critical Markov chains. The final chapter presents a range of applications where these special types of Markov chains occur naturally, featuring a new risk process with surplus-dependent premium rate.

This will be a valuable resource for researchers and graduate students working in probability theory and stochastic processes.

DENIS DENISOV is Reader in Probability in the Department of Mathematics at the University of Manchester. His research interests include multidimensional random walks, ordered and conditioned random walks, Markov chains, and diffusion processes.

DMITRY KORSHUNOV is Professor in the School of Mathematical Sciences at Lancaster University. He is an expert in stochastic processes, Markov chains, large-deviation theorems, heavy-tailed phenomena, and limit theorems. He previously co-authored *An Introduction to Heavy-Tailed and Subexponential Distributions* (2011, 2013).

VITALI WACHTEL is Professor for Mathematics at the University of Bielefeld. His research interests include, besides Markov chains, exit times and conditional distributions for multidimensional random walks, branching processes, and large deviations.

NEW MATHEMATICAL MONOGRAPHS

All the titles listed below can be obtained from good booksellers or from Cambridge University Press.
For a complete series listing visit www.cambridge.org/mathematics.

11. B. Bekka, P. de la Harpe, and A. Valette *Kazhdan's Property (T)*
12. J. Neisendorfer *Algebraic Methods in Unstable Homotopy Theory*
13. M. Grandis *Directed Algebraic Topology*
14. G. Michler *Theory of Finite Simple Groups II*
15. R. Schertz *Complex Multiplication*
16. S. Bloch *Lectures on Algebraic Cycles (2nd Edition)*
17. B. Conrad, O. Gabber, and G. Prasad *Pseudo-reductive Groups*
18. T. Downarowicz *Entropy in Dynamical Systems*
19. C. Simpson *Homotopy Theory of Higher Categories*
20. E. Fricain and J. Mashreghi *The Theory of H(b) Spaces I*
21. E. Fricain and J. Mashreghi *The Theory of H(b) Spaces II*
22. J. Goubault-Larrecq *Non-Hausdorff Topology and Domain Theory*
23. J. Śniatycki *Differential Geometry of Singular Spaces and Reduction of Symmetry*
24. E. Riehl *Categorical Homotopy Theory*
25. B. A. Munson and I. Volić *Cubical Homotopy Theory*
26. B. Conrad, O. Gabber, and G. Prasad *Pseudo-Reductive Groups (2nd Edition)*
27. J. Heinonen, P. Koskela, N. Shanmugalingam, and J. T. Tyson *Sobolev Spaces on Metric Measure Spaces*
28. Y.-G. Oh *Symplectic Topology and Floer Homology I*
29. Y.-G. Oh *Symplectic Topology and Floer Homology II*
30. A. Bobrowski *Convergence of One-Parameter Operator Semigroups*
31. K. Costello and O. Gwilliam *Factorization Algebras in Quantum Field Theory I*
32. J.-H. Evertse and K. Győry *Discriminant Equations in Diophantine Number Theory*
33. G. Friedman *Singular Intersection Homology*
34. S. Schwede *Global Homotopy Theory*
35. M. Dickmann, N. Schwartz, and M. Tressl *Spectral Spaces*
36. A. Baernstein II *Symmetrization in Analysis*
37. A. Defant, D. García, M. Maestre, and P. Sevilla-Peris *Dirichlet Series and Holomorphic Functions in High Dimensions*
38. N. Th. Varopoulos *Potential Theory and Geometry on Lie Groups*
39. D. Arnal and B. Currey *Representations of Solvable Lie Groups*
40. M. A. Hill, M. J. Hopkins and D. C. Ravenel *Equivariant Stable Homotopy Theory and the Kervaire Invariant Problem*
41. K. Costello and O. Gwilliam *Factorization Algebras in Quantum Field Theory II*
42. S. Kumar *Conformal Blocks, Generalized Theta Functions and the Verlinde Formula*
43. P. F. X. Müller *Hardy Martingales*
44. T. Kaletha and G. Prasad *Bruhat–Tits Theory*
45. J. Schwermer *Reduction Theory and Arithmetic Groups*
46. J. Kotus and M. Urbański *Meromorphic Dynamics I*
47. J. Kotus and M. Urbański *Meromorphic Dynamics II*
48. S. Garibaldi, H. P. Petersson, and M. L. Racine *Albert Algebras over Commutative Rings*
49. D. Pellicer *Abstract Chiral Polytopes*
50. C. Xu *K-Stability of Fano Varieties*
51. D. Denisov, D. Korshunov, and V. Wachtel *Markov Chains with Asymptotically Zero Drift*

Markov Chains with Asymptotically Zero Drift

Lamperti's Problem

DENIS DENISOV
University of Manchester

DMITRY KORSHUNOV
Lancaster University

VITALI WACHTEL
Universität Bielefeld, Germany

CAMBRIDGE
UNIVERSITY PRESS

CAMBRIDGE
UNIVERSITY PRESS

Shaftesbury Road, Cambridge CB2 8EA, United Kingdom

One Liberty Plaza, 20th Floor, New York, NY 10006, USA

477 Williamstown Road, Port Melbourne, VIC 3207, Australia

314–321, 3rd Floor, Plot 3, Splendor Forum, Jasola District Centre, New Delhi – 110025, India

103 Penang Road, #05–06/07, Visioncrest Commercial, Singapore 238467

Cambridge University Press is part of Cambridge University Press & Assessment, a department of the University of Cambridge.

We share the University's mission to contribute to society through the pursuit of education, learning and research at the highest international levels of excellence.

www.cambridge.org
Information on this title: www.cambridge.org/9781009554220
DOI: 10.1017/9781009554237

When citing this work, please include a reference to the DOI 10.1017/9781009554237

First published 2025

A catalogue record for this publication is available from the British Library

Library of Congress Cataloging-in-Publication Data
Names: Denisov, Denis author | Korshunov, Dmitry, 1966– author — Wachtel, Vitali author
Title: Markov chains with asymptotically zero drift : Lamperti's problem / Denis Denisov (University of Manchester), Dmitry Korshunov (Lancaster University), Vitali Wachtel (Universität Bielefeld, Germany).
Description: Cambridge, United Kingdom ; New York, NY : Cambridge University Press, 2025. | Includes bibliographical references and index.
Identifiers: LCCN 2024058145 (print) | LCCN 2024058146 (ebook) | ISBN 9781009554220 hardback | ISBN 9781009554237 epub
Subjects: LCSH: Random walks (Mathematics) | Asymptotic distribution (Probability theory) | Markov processes
Classification: LCC QA274.73 .D46 2025 (print) | LCC QA274.73 (ebook) | DDC 519.2/33–dc23/eng/20250214
LC record available at https://lccn.loc.gov/2024058145
LC ebook record available at https://lccn.loc.gov/2024058146

ISBN 978-1-009-55422-0 Hardback

Contents

Preface

The main goal of this text is a comprehensive study of time-homogeneous Markov chains on the real line whose drift tends to zero at infinity. We call such processes Markov chains with asymptotically zero drift. Traditionally this topic is referred to as Lamperti's problem.

Time-homogeneous Markov chains with asymptotically zero drift may be viewed as a subclass of random walks that are perturbed in space. The latter are of basic importance in the study of various applied stochastic models, among them branching and risk processes, queueing systems, etc. Random walks generated by sums of independent and identically distributed random variables have been well studied; see e.g. the classic textbooks by W. Feller [63], V. V. Petrov [132], and F. Spitzer [143]; for a recent development of the theory of random walks we refer to A. A. Borovkov and K. A. Borovkov [22]. There are many monographs devoted to various applications where random walks play a crucial rôle. We just mention the books on ruin and queueing processes by S. Asmussen [8, 7]; on insurance and finance by P. Embrechts, C. Klüppelberg, and T. Mikosch [57] and by T. Rolski, H. Schmidli, V. Schmidt, and J. Teugels [137]; and on stochastic difference equations by D. Buraczewski, E. Damek, and T. Mikosch [31].

In applied stochastic models, if one allows the process considered to be dependent on its current state, we often get a Markov chain which has asymptotically zero drift; we demonstrate this in the Chapter 11, where we discuss in particular branching and risk processes, stochastic difference equations, and the ALOHA network.

The study of processes with asymptotically zero drift was initiated by J. Lamperti in the 1960s in a series of papers. In particular, he classified such Markov chains in [111, 113], where the conditions for positive recurrence, recurrence, and transience were derived via a martingale technique. In [112], Lamperti revealed a new class of limit theorems for transient Markov chains, including theorems for the weak convergence of a properly normalised square of a Markov chain to a Γ-distribution; the proof is based on the method of moments.

Since then the martingale approach for the study of Markov chains with asymptotically zero drift has been further developed. In each particular problem the main point has been to construct an appropriate test (Lyapunov) function such that when applied to a Markov chain it produces a sub- or supermartingale. Modern state-of-the-art research in this direction can be found in the recent monograph of M. Menshikov, S. Popov, and A. Wade [121]; this was preceded of the monographs by G. Fayolle, V. Malyshev, and M. Menshikov [61] and of A. A. Borovkov [23]. We have been influenced by these books and by further contacts with their authors.

The main advantage of the martingale approach is that the test functions considered are mostly elementary, which on one hand simplifies calculations while on the other hand allows us to derive deep results.

However, it is clear that elementary test functions do not allow us to track the subtle asymptotic behaviour of Markov chains when we are interested in the precise asymptotics of, say, the tail-invariant measure. For that reason, there is a need for a novel approach to this kind of problem. Our approach developed in this book includes many novel elements, and much of the material presents original research. The two main ingredients are as follows.

(i) To study the tails of recurrence times and the tails of invariant measures of recurrent chains we follow Cramér's approach, which is based on an appropriate change of measure. More precisely, we apply a kind of Doob's h-transform to the transition kernel of a chain killed upon entering an appropriately chosen set. This approach differs from the method of Lyapunov test functions, where one considers functions of Markov chains. The main advantage of Cramér's approach consists in the fact that it allows us to work with a new Markov chain whose jumps are stochastically bounded, like the original jumps. This is in contrast with an approach based on the consideration of a function of a Markov chain where – in the case of functions growing faster than linear – the jumps are not usually stochastically bounded: they blow up at infinity.

In order to perform a Doob's h-transform of a substochastic transition kernel one needs a positive harmonic function for that kernel. By definition, every harmonic function is a solution to a certain equation. Thus, the analytical properties of the solutions are a priori unclear and have to be studied. This problem is very hard in general. In order to overcome this difficulty we suggest the following modification of Doob's transform: instead of using harmonic functions with unclear properties we perform a change of measure to a superharmonic function which is chosen to be sufficiently close with a harmonic function while having the analytical properties needed for our analysis. The resulting kernel is then substochastic, where the total transition probability mass can be less than 1 at some or all states. However, the loss of mass can be controlled effectively.

(ii) We develop an approach that allows us to construct the superharmonic functions needed for (i) – starting from the ratio of the drift to the second moment of the jumps – such that after a change of measure based on a superharmonic test function we get a transition kernel which is almost stochastic far from the origin. It turns out that the same approach can be used to construct Lyapunov test functions for the classification of Markov chains. Of course, the test functions constructed in this way are not as elementary as in the martingale approach; however, using them we can derive better criteria for transience, recurrence and positive recurrence. We can also derive precise asymptotics for various characteristics of Markov chains, and that is our main contribution.

In Chapter 2 we provide a basic classification of Markov chains, with many improvements on the results known in the literature. In Chapter 3 we are interested in the down-crossing probabilities for transient Markov chains. Chapters 4 and 5 deal comprehensively with limit theorems for transient Markov chains, including convergence to Γ- and normal distributions, while Chapter 6 deals with the corresponding renewal measures. Chapter 7 explains how we can apply Doob's h-transform to Markov chains. Chapters 8 and 9 develop techniques needed for deriving precise tail asymptotics of the power and Weibullian types respectively. In Chapter 10 we demonstrate how powerful this approach is by studying Markov chains with asymptotically constant negative drift. Finally, Chapter 11 presents various applied stochastic models in which Markov chains with asymptotically zero drift arise naturally, and hence the above results for Markov chains are applicable to those models, which leads to novel results.

As discussed in Section 1.3 for random walks delayed at zero and in [24] for Markov chains, the invariant measure of a Markov chain with negative drift bounded away from zero far from the origin is heavy-tailed – all positive exponential moments are infinite – if and only if the jumps are heavy-tailed. As we discuss in this book, Markov chains with asymptotically zero drift give rise to heavy-tailed invariant measure, whatever the distribution of jumps, even if they are bounded random variables. So, stationary Markov chains with asymptotically zero drift provide an important example of a stochastic model where a light-tailed input produces a heavy-tailed output.

Most of this research monograph is based on novel results obtained following the approach described above. This book should be of interest for PhD students and researchers in the area of Markov chains and their applications.

We are thankful to many colleagues for helpful discussions, contributions, and bibliographical comments, notably to D. Buraczewski, S. Foss, M. V. Menshikov, and S. Popov.

The authors started to work on the topic of this book while participating in the programme on Stochastic Processes in Communication Sciences held at the Isaac Newton Institute for Mathematical Sciences in the University of Cambridge, in 2010. Later, the book was mostly written while the authors worked, together or individually, at Augsburg, Bielefeld, Lancaster, Ludwig-Maximilian, and Manchester Universities and the Sobolev Institute of Mathematics in Novosibirsk; we thank our home institutions for their hospitality.

Manchester — Denis Denisov
Lancaster — Dmitry Korshunov
Bielefeld — Vitali Wachtel
March 2024

Notation and Conventions

Intervals	(x,y), an open interval; $[x,y]$ a closed interval; half-open intervals are denoted by $(x,y]$ and $[x,y)$
Integrals	$\int_x^y$, integral over the interval $(x,y]$
$\mathbb{R}$, $\mathbb{R}^+$, $\mathbb{R}^d$	the real line, the positive real half-line $[0,\infty)$, and d-dimensional Cartesian space
$\mathbb{Z}$, $\mathbb{Z}^+$	the set of integers; the set $\{0,1,2,\ldots\}$
$C^\gamma(\mathbb{R})$	the class of γ times continuously differentiable functions
$\mathbb{I}(A)$	the indicator function of A, that is, $\mathbb{I}(A)=1$ if A holds and $\mathbb{I}(A)=0$ otherwise
$\to$, $\downarrow$, $\uparrow$	convergence, monotone decreasing convergence, and monotone increasing convergence
O, o, and $\sim$	Let u and v depend on a parameter x which tends, say, to infinity. Assuming that v is positive we write $u(x)=O(v(x))$ if $\limsup_{x\to\infty}\|u(x)\|/v(x)<\infty$; $u(x)=o(v(x))$ if $u(x)/v(x)\to 0$ as $x\to\infty$; $u(x)\sim v(x)$ if $u(x)/v(x)\to 1$ as $x\to\infty$; $u_n(x)=o(v_n(x))$ uniformly for all n if $\sup_n\left\|\frac{u_n(x)}{v_n(x)}\right\|\to 0$ as $x\to\infty$.
$a\wedge b$, $a\vee b$	$\min(a,b)$, $\max(a,b)$ respectively.
a^+, a^-	the positive part of a, $a^+=\max(a,0)$, and negative part of a, $a^-=\max(-a,0)$
$\sigma(X_\theta,\theta\in\Theta)$	the σ-algebra generated by the family of random variables $\{X_\theta,\theta\in\Theta\}$.
$\mathcal{B}(S)$	the Borel σ-algebra in the space S

$\mathbb{P}\{B\}$	the probability (on some appropriate space) of the event B
$\mathbb{P}\{B \mid A\}$	the conditional probability of B given A
$\mathbb{E}\xi$	the mean of the random variable ξ
$\mathbb{E}\{\xi; B\}$	the mean of ξ over the event B, $\mathbb{E}\xi\mathbb{I}(B)$
$F * G$	the convolution of distributions F and G
$:=$ $(=:)$	the quantity on the left (right) is defined to be equal to the quantity on the right (left)
$\le_{\mathrm{st}}$ $(\ge_{\mathrm{st}})$	the random variable on the left is stochastically not greater (not less) than the random variable on the right
$=_{\mathrm{st}}$	equality in distribution
$\Rightarrow$	weak convergence of random variables to a random variable or distribution
$\xrightarrow{p}$	indicates convergence in probability
$\square$	indicates the end of a proof
$\{X_n\}$	a Markov chain
$P(x, B)$	the transition probabilities of a chain $\{X_n\}$, that is, $\mathbb{P}\{X_{n+1} \in B \mid X_n = x\}$
$P_x\{\cdot\}$	the distribution given $X_0 = x$
$\xi(x)$	the jump of $\{X_n\}$ from x
$m_k(x)$	the kth moment of the jump $\xi(x)$, $\mathbb{E}\xi^k(x)$
$m_k^{[s]}(x)$	the s-truncated kth moment of the jump $\xi(x)$, that is, $\mathbb{E}\{\xi^k(x);\ \lvert\xi(x)\rvert \le s\}$
τ_B	the time of the first entry of X_n to a Borel set B, that is, $\min\{n \ge 1 : X_n \in B\}$
$H(B)$, $H_x(B)$, $H_x^X(B)$	the renewal measure of a Borel set B generated by X_n, that is, $\sum_{n=0}^{\infty} \mathbb{P}\{X_n \in B\}$, $\sum_{n=0}^{\infty} \mathbb{P}_x\{X_n \in B\}$
$r(x)$	a reference function which describes the asymptotic behaviour of the ratio $-2m_1^{[s(x)]}(x)/m_2^{[s(x)]}(x)$ in the case of a recurrent chain or $2m_1^{[s(x)]}(x)/m_2^{[s(x)]}(x)$ in the case of a transient chain
$R(x)$	the integral of a function $r(x)$, $\int_0^x r(y)dy$
$U(x)$	either $\int_0^x e^{R(y)}dy$ or $\int_x^\infty e^{-R(y)}dy$ depending on whether a recurrent or transient chain is considered
$\Gamma_{k,\theta}$	stands for a Γ-distribution with shape parameter k and scale parameter θ, that is, a distribution with probability density function $\frac{1}{\Gamma(k)\theta^k} x^{k-1} e^{-x/\theta}$, $x \ge 0$; the expectation is $k\theta$ and the variance $k\theta^2$

N_{a,σ^2}	a normal distribution with expectation a and variance σ^2
$\Phi(x)$	the standard normal cumulative distribution function
$\log_{(m)} x$	the mth iteration of the logarithm of x: $\log_{(m)} x = \log \log_{(m-1)} x$
$e^{(m)}$	a solution to the equation $\log_{(m)} x = 1$

1

Introduction

In this chapter we introduce basic notions needed in what follows. We also discuss nearest neighbour Markov chains and diffusion processes, which represent the two classes of Markov processes whose invariant measure, in the case of positive recurrence, or Green function, in the case of transience, is available in closed form. A closed form makes possible the direct analysis of such Markov processes: their classification and the tail asymptotics of the invariant probabilities or Green function. This discussion sheds some light on what we may expect for general Markov chains.

1.1 Countable Markov Chains

Let us start with a simple process, a *countable time-homogeneous Markov chain* $X = \{X_n, n \geq 0\}$, which is a stochastic process with a countable *state space*, which can always be reduced to $S = \mathbb{Z}^+$. The process is determined by an initial distribution X_0 and a collection of *transition probabilities* $p_{xy} \geq 0$, $x, y \in S$, such that $\sum_{y \in S} p_{xy} = 1$ for all $x \in S$ and

$$\begin{aligned}\mathbb{P}\{X_{n+1} = x_{n+1} \mid X_n = x_n, X_{n-1} = x_{n-1}, \ldots, X_0 = x_0\} \\ = \mathbb{P}\{X_{n+1} = x_{n+1} \mid X_n = x_n\} = p_{x_n x_{n+1}},\end{aligned} \tag{1.1}$$

for all time epochs n and all sequences of states $x_{n+1}, x_n, \ldots, x_0$ in S. In words, the probability of moving from one state to another does not depend on the trajectory that determined how X arrived in its current state. This memoryless property can be equivalently defined as independence of the future and the past given the current state, that is,

$$\mathbb{P}\{BA \mid X_n = x_n\} = \mathbb{P}\{B \mid X_n = x_n\}\mathbb{P}\{A \mid X_n = x_n\}$$

for any $n \geq 1$ and events $B \in \sigma(X_{n+1}, X_{n+2}, \ldots)$ and $A \in \sigma(X_0, \ldots, X_{n-1})$.

Definition 1.1 A random variable T taking non-negative integer values which is possibly improper, is called the *stopping time* if, for all $n \in \mathbb{Z}^+$, the event $\{T \le n\}$ belongs to the σ-algebra $\sigma(X_0, X_1, \dots, X_n)$.

The Markov property (1.1) can be extended to stopping times as follows. If T is a stopping time then the process $\{X_{T+n}\}_{n\ge 0}$ is also a Markov chain with initial distribution X_T. Moreover, for any $x \in S$, this chain is independent of X_0, X_1, ..., X_{T-1} given $X_T = x$. This property is called the *strong Markov property*.

For any state $x \in S$, denote by τ_x the first hitting time of x,

$$\tau_x := \inf\{n \ge 1 : X_n = x\},$$

with standard convention $\inf \emptyset = \infty$. For all x, τ_x is a stopping time.

Definition 1.2 A state x is called *positive recurrent* if $\mathbb{E}_x \tau_x < \infty$.

Definition 1.3 A state x is called *non-positive* if it is not positive recurrent, more precisely, if either $\mathbb{P}_x\{\tau_x = \infty\} > 0$, or $\mathbb{P}_x\{\tau_x < \infty\} = 1$ and $\mathbb{E}_x \tau_x = \infty$.

Definition 1.4 A state x is called *recurrent* (*persistent*) if $\mathbb{P}_x\{\tau_x < \infty\} = 1$.

Definition 1.5 A state x is called *null recurrent* if $\mathbb{P}_x\{\tau_x < \infty\} = 1$ while $\mathbb{E}_x \tau_x = \infty$.

Definition 1.6 A state x is called *transient* if $\mathbb{P}_x\{\tau_x < \infty\} < 1$.

By the strong Markov property, the time lengths between consecutive visits of the chain to a fixed state x are independent and identically distributed. Therefore, a state x is transient if and only if $\mathbb{P}_x\{\tau_x < \infty\} < 1$, which is equivalent to the convergence of the following series (*a Green function*):

$$\sum_{n=0}^{\infty} \mathbb{P}_x\{X_n = x\} = \mathbb{E}_x \sum_{n=0}^{\infty} \mathbb{I}\{X_n = x\} < \infty.$$

Definition 1.7 The *period* of state x is defined as

$$d_x := \gcd\{n \ge 1 : \mathbb{P}_x\{X_n = x\}\}.$$

A state x is called *aperiodic* if $d_x = 1$.

Definition 1.8 A Markov chain X_n is called *irreducible* if, for all x and y, $\mathbb{P}_x\{X_n = y\} > 0$ for some n.

Notice that, for an irreducible countable Markov chain, the following solidarity property holds true: the positive recurrence, non-positivity, recurrence, null recurrence, transience, or aperiodicity of any state implies the same property for all other states.

Definition 1.9 A measure $\{\pi_x\}_{x\in S}$ is called *invariant* (or *stationary*) for a countable Markov chain $\{X_n\}$ if

$$\pi(y) = \sum_{j\in S} \pi(x) p_{xy} \quad \text{for all } y \in S.$$

Definition 1.10 A probability distribution $\{\pi_x\}_{x\in S}$ is called *asymptotic* (or *limiting*) for a countable Markov chain $\{X_n\}$ if

$$\mathbb{P}_x\{X_n = y\} \to \pi(y) \quad \text{as } n \to \infty \quad \text{for any } x \in S.$$

An asymptotic distribution – if exists – is necessarily an invariant probability measure; however, this does not hold vice versa. For Markov chains with finitely many states, the following ergodic theorem is a major result.

Theorem 1.11 *Any finite irreducible aperiodic Markov chain possesses an asymptotic distribution.*

For a Markov chain with infinitely many states the last result may fail in general. For example, a simple random walk with transition probabilities $p_{x,x+1} = p > 1/2$ and $p_{x,x-1} = 1-p < 1/2$ is irreducible; however, there is no convergence to an asymptotic distribution. This Markov chain is transient which is only possible due to the infinite number of states.

Theorem 1.12 *Let $\{X_n\}$ be a countable irreducible Markov chain. Fix some $x \in S$. If $\{X_n\}$ is recurrent then a measure π defined by*

$$\mu(y) := \mathbb{E}_x \sum_{n=1}^{\tau_x} \mathbb{I}\{X_n = y\} = \sum_{n=1}^{\infty} \mathbb{P}_x\{X_n = y, n \le \tau_x\}, \quad y \in S, \quad (1.2)$$

is a σ-finite invariant measure for $\{X_n\}$.

Proof Let us first check that $\mu(y) < \infty$ for all $y \in S$. By its definition, $\mu(x) = 1$. Since $\{X_n\}$ is irreducible, there exists a state y such that $p_{yx} > 0$. Then the random variable

$$\mathbb{E}_x \sum_{n=1}^{\tau_x} \mathbb{I}\{X_n = y\}$$

is stochastically bounded by a geometric distribution with success probability $p_{yx} > 0$, hence $\mu(y) < \infty$. Then, by the solidarity property, $\mu(z) < \infty$ for all $z \neq x$.

Now let us show that μ is invariant. Indeed, for $z = x$,

$$\begin{aligned}\sum_{y\in S}\mu(y)p_{yx} &= p_{xx} + \sum_{y\neq x}\sum_{n=1}^{\infty}\mathbb{P}_x\{X_n = y, n \le \tau_x\}p_{yx}\\ &= p_{xx} + \sum_{n=1}^{\infty}\sum_{y\neq x}\mathbb{P}_x\{X_n = y, n < \tau_x\}p_{yx}\\ &= p_{xx} + \sum_{n=1}^{\infty}\mathbb{P}_x\{\tau_x = n+1\}\\ &= \mathbb{P}_x\{\tau_x < \infty\} \;=\; 1 \;=\; \mu(x),\end{aligned}$$

because $\{X_n\}$ is recurrent. For any $z \neq x$,

$$\begin{aligned}\sum_{y\in S}\mu(y)p_{yz} &= p_{xz} + \sum_{n=1}^{\infty}\sum_{y\neq x}\mathbb{P}_x\{X_n = y, n < \tau_x\}p_{yz}\\ &= p_{xz} + \sum_{n=1}^{\infty}\mathbb{P}_x\{X_{n+1} = z, n+1 < \tau_x\}\\ &= \sum_{n=1}^{\infty}\mathbb{P}_x\{X_n = z, n < \tau_x\}\\ &= \mu(z),\end{aligned}$$

by the definition of $\mu(z)$ for $z \neq x$. □

So, any irreducible recurrent Markov chain possesses a σ-finite invariant distribution. However, the existence of a σ-finite invariant distribution does not guarantee recurrence, as the following example demonstrates. For a simple random walk on $\mathbb{Z}$, the Haar measure assigning $\mu(x) = 1$ for all $x \in \mathbb{Z}$ is invariant whatever the success probability p.

For positive recurrence there is a criterion in terms of an invariant measure, as follows.

Theorem 1.13 *For a countable irreducible Markov chain $\{X_n\}$, the following are equivalent:*

(i) *some state is positive recurrent;*
(ii) *all states are positive recurrent;*
(iii) *the measure μ defined in (1.2) is finite;*
(iv) *there exists a probability invariant measure π.*

Then

$$\pi(y) \;=\; \frac{1}{\mathbb{E}_y\tau_y} \quad \textit{for all } y \in S.$$

Proof The equivalence of (i) or (ii) to (iii) is immediate from the definition (1.2), because

$$\sum_{y\in S}\pi(y) = \sum_{y\in S}\mathbb{E}_x\sum_{n=1}^{\tau_x}\mathbb{I}\{X_n = y\}$$
$$= \mathbb{E}_x\sum_{n=1}^{\tau_x}\sum_{y\in S}\mathbb{I}\{X_n = y\} = \mathbb{E}_x\tau_x,$$

which is only finite if $\{X_n\}$ is positive recurrent.

The most difficult implication is (iv)→(iii). It follows from the observation that any invariant measure π satisfies the equalities

$$\pi(y) := \pi(x)\mathbb{E}_x\sum_{n=1}^{\tau_x}\mathbb{I}\{X_n = y\} = \pi(x)\sum_{n=1}^{\infty}\mathbb{P}_x\{X_n = y, n \le \tau_x\}, \quad y \in S.$$

For a proof, see e.g. Meyn and Tweedie [126, Theorem 10.4.9]. □

1.2 Real-Valued Markov Chains

Now let us proceed with a *time-homogeneous Markov chain* $X = \{X_n, n \ge 0\}$, whose state space is a Borel subset S of $\mathbb{R}$, that is, for all $x \in S$ and Borel sets $B_0, \ldots, B_{n-1}, B_{n+1} \in \mathcal{B}(S)$,

$$\mathbb{P}\{X_{n+1} \in B_{n+1} \mid X_0 \in B_0, \ldots, X_{n-1} \in B_{n-1}, X_n = x\}$$
$$= \mathbb{P}\{X_{n+1} \in B_{n+1} \mid X_n = x\}.$$

We usually simply say that X_n takes values in $\mathbb{R}$, keeping in mind that the corresponding transition probabilities may be defined only on some subset S of the real line.

Denote by $P(\cdot,\cdot) : S \times \mathcal{B}(S) \to [0,1]$ the *transition probabilities* of $\{X_n\}$:

$$P(x, B) = \mathbb{P}\{X_{n+1} \in B \mid X_n = x\};$$

this function is measurable in x for each fixed B and is a probability measure for each fixed x, that is, it is a stochastic transition kernel. Then, for all n and B,

$$\mathbb{P}\{X_{n+1} \in B\} = \int_S P(y, B)\mathbb{P}\{X_n \in dy\}.$$

Let $\mathbb{P}_x\{\cdot\} = \mathbb{P}\{\cdot \mid X_0 = x\}$ and $\mathbb{E}_x\{\cdot\} = \mathbb{E}\{\cdot \mid X_0 = x\}$.

Denote by $\xi(x)$, $x \in S$, a random variable corresponding to the *jump* of the chain at point $x \in S$, that is, a random variable with distribution

$$\begin{aligned}\mathbb{P}\{\xi(x) \in B\} &= \mathbb{P}\{X_{n+1} - X_n \in B \mid X_n = x\} \\ &= \mathbb{P}_x\{X_1 \in x + B\}, \quad B \in \mathcal{B}(\mathbb{R}).\end{aligned}$$

In the sequel we will always assume that S is a right unbounded set. Furthermore, for ease of notation, we assume that $P(x, B)$ is defined for all $x \in \mathbb{R}$.

Denote the kth moment of the jump at point x by

$$m_k(x) \; := \; \mathbb{E}\xi^k(x).$$

Definition 1.14 We say that a Markov chain $\{X_n\}$ has an *asymptotically zero drift* if $m_1(x) = \mathbb{E}\xi(x) \to 0$ as $x \to \infty$.

The study of processes with asymptotically zero drift was initiated by Lamperti in a series of papers [111, 112, 113].

The first topic of basic importance is the classification of Markov chains, which is discussed in detail in Chapter 2. For any Borel set $B \subset \mathbb{R}$ denote by τ_B the time of the first entry of $\{X_n\}$ to B,

$$\tau_B := \inf\{n \geq 1 : X_n \in B\}.$$

If B is a singleton then we can repeat the classification of B as in the previous section. However, this does not work well for Markov chains that are truly real-valued, as it could happen that then $\mathbb{P}_B\{\tau_B < \infty\} = 0$. For that reason we introduce a classification of a general Borel set B with respect to X_n that reduces to the classification presented in the last section if B is a singleton.

Definition 1.15 A set B is called *positive recurrent* if $\mathbb{E}_x\tau_B < \infty$ for all $x \in B$.

Definition 1.16 A set B is called *non-positive* if it is not positive recurrent; more precisely, if either $\mathbb{P}_x\{\tau_B = \infty\} > 0$, or $\mathbb{P}_x\{\tau_B < \infty\} = 1$ and $\mathbb{E}_x\tau_B = \infty$ for some $x \in B$.

Definition 1.17 A set B is called *recurrent* if τ_B is finite a.s. for all initial states $x \in B$.

Definition 1.18 A set B is called *null recurrent* if τ_B is finite a.s. and $\mathbb{E}_x\tau_B = \infty$ for all initial states $x \in B$.

Definition 1.19 A set B is called *transient* if $\mathbb{P}_x\{\tau_B < \infty\} < 1$ for all initial states $x \in B$.

Definition 1.20 A measure π is called *invariant* for $\{X_n\}$ if

$$\pi(B) = \int_S P(x, B)\pi(dx) \quad \text{for all } B \in \mathcal{B}(S).$$

In [111] Lamperti showed that if $S = \mathbb{R}^+$, $\limsup X_n = \infty$, and $\mathbb{E}|\xi(x)|^{2+\delta}$ is bounded for some $\delta > 0$ then

- $2xm_1(x) \le m_2(x) + O(x^{-\delta})$ implies that some neighborhood of zero is recurrent,
- $2xm_1(x) \ge (1+\varepsilon)m_2(x)$, for some $\varepsilon > 0$ and all sufficiently large x, implies that any compact set is transient.

In [113] Lamperti proved that $2xm_1(x) + m_2(x) \le -\varepsilon$ is sufficient for the positive recurrence of any compact set and that $2xm_1(x) + m_2(x) \ge \varepsilon$ implies the non-positivity of any compact set (for the case of either null recurrence or transience). These criteria were improved later by Menshikov et al. [124]. Instead of the existence of moments of order $2 + \delta$ they assumed that $\mathbb{E}\xi^2(x)\log^{2+\delta}(1 + |\xi(x)|)$ is bounded. Moreover, they obtained a more precise classification of positive recurrence, null recurrence, and transience which involves iterated logarithms.

In the next section we discuss classical random walks to show the difference between them and Lamperti processes. That is followed by a couple of sections devoted to two types of specific processes – nearest neighbour Markov chains and diffusion processes – where many characteristics of interest may be computed in closed form by following quite elementary calculations; this provides the basic intuition needed to approach general Markov chains with asymptotically zero drift.

In Section 1.6 we describe our approach to general Markov chains with asymptotically zero drift.

1.3 Random Walks

Let us consider a fundamental example of Markov chains, random walks. We start by recalling some important asymptotic results, which will be extended to Lamperti's Markov chains later.

Definition 1.21 *A random walk* with initial state x is a sequence of partial sums, $S_0 = x$,

$$S_n := S_{n-1} + \xi_n = x + \xi_1 + \cdots + \xi_n, \quad n \ge 1,$$

where the ξ_n are independent and identically distributed random variables.

Any random walk is a Markov chain with transition kernel

$$P(x,B) = \mathbb{P}\{\xi_1 \in B - x\}, \quad x \in \mathbb{R}, \quad B \in \mathcal{B}(\mathbb{R}).$$

It is a *space-homogeneous* Markov chain because all its jumps $\xi(x)$, $x \in \mathbb{R}$, are distributed as ξ_1. Roughly speaking, it is a process with continuous statistics in the sense that there are no boundary effects in this model.

If $\mathbb{E}|\xi_1| < \infty$ then the strong law of large numbers holds, that is,

$$S_n/n \to \mathbb{E}\xi_1 \quad \text{a.s. as } n \to \infty.$$

This implies, in particular, that if $\mathbb{E}\xi_1 > 0$ then the set $(-\infty, \widehat{x}\,]$ is transient for all $\widehat{x} \in \mathbb{R}$. If $\mathbb{E}\xi_1 < 0$ then the set $(-\infty, \widehat{x}\,]$ is positive recurrent. It is also well known that in the case $\mathbb{E}\xi_1 = 0$ the random walk S_n is null recurrent, that is, any bounded set is null recurrent.

In addition, if $\mathbb{E}\xi_1^2 < \infty$ then the central limit theorem holds, that is,

$$\frac{S_n - n\mathbb{E}\xi_1}{\sqrt{n\mathbb{V}\text{ar}\,\xi_1}} \Rightarrow N_{0,1} \quad \text{as } n \to \infty.$$

The simplest process with discontinuous statistics – that is, with boundary effects – is a random walk delayed at zero, which is defined next.

Definition 1.22 *A random walk delayed at zero (a Lindley recursion)* is a stochastic process $W = \{W_n, n \geq 0\}$ such that, for all $n \geq 1$,

$$W_n = (W_{n-1} + \xi_n)^+ := \max(0, W_{n-1} + \xi_n),$$

where the ξ_n are independent and identically distributed random variables that are independent of $W_0 \geq 0$.

The process W is a Markov chain with transition kernel

$$P(x,B) = \mathbb{P}\{(x+\xi_1)^+ \in B\}, \quad x \in \mathbb{R}^+, \quad B \in \mathcal{B}(\mathbb{R});$$

this is a particular example of a Markov chain that is asymptotically homogeneous in space, defined below, because its jumps satisfy the following weak (and in total-variation distance) convergence:

$$\xi(x) =_{\text{st}} (x+\xi_1)^+ - x \Rightarrow \xi_1 \quad \text{as } x \to \infty.$$

Definition 1.23 We say that a Markov chain $\{X_n\}$ is *asymptotically homogeneous in space* if

$$\xi(x) \Rightarrow \xi \quad \text{as } x \to \infty, \tag{1.3}$$

for some random variable ξ. Equivalently, $P(x, x+\cdot) \Rightarrow \mathbb{P}\{\xi \in \cdot\}$.

Let $W_0 = 0$. Then

$$W_n = \max(0, \xi_n, \xi_n + \xi_{n-1}, \xi_n + \xi_{n-1} + \xi_{n-2}, \ldots, \xi_n + \xi_{n-1} + \cdots + \xi_1);$$

hence, for all n, W_n is equal in distribution to the maximum

$$\begin{aligned} M_n &:= \max(0, \xi_1, \xi_1 + \xi_2, \xi_1 + \xi_2 + \xi_3, \ldots, \xi_1 + \xi_2 + \cdots + \xi_n) \\ &= \max_{0 \le k \le n} S_k, \quad \text{where } S_0 = 0. \end{aligned}$$

An application of the Lindley recursion $\{W_n\}$ is the waiting time process in the single-server queue system with $\xi = \sigma - \tau$, where σ represents the typical service time and τ the typical inter-arrival time. Among applications of the process of maxima M_n is the collective risk process, with $\xi = X - c\tau$, where X represents the typical claim size, τ the typical inter-arrival time, and c the premium rate; here $\mathbb{P}\{M_\infty > x\}$ represents the ruin probability given the initial reserve $x > 0$.

If $\mathbb{E}\xi_1 > 0$ then $\{W_n\}$ is a transient Markov chain (any bounded set is transient), which satisfies the central limit theorem provided that $\mathbb{E}\xi_1^2 < \infty$:

$$\frac{W_n - n\mathbb{E}\xi_1}{\sqrt{n\mathbb{V}\text{ar}\,\xi_1}} \Rightarrow N_{0,1} \quad \text{as } n \to \infty.$$

If $\mathbb{E}\xi_1 = 0$ then $\{W_n\}$ is null recurrent (any bounded set is null recurrent), and, by the functional central limit theorem (Donsker's theorem),

$$\frac{W_n}{\sqrt{n\mathbb{V}\text{ar}\,\xi_1}} \Rightarrow \sup_{t \le 1} B(t) \quad \text{as } n \to \infty,$$

where $B(t)$ is a Brownian motion; see, e.g. Billingsley [16, Section 10].

If $\mathbb{E}\xi_1 < 0$ then $\{W_n\}$ is positive recurrent (any bounded set is positive recurrent) and possesses a unique invariant probability measure, say π_W. This measure is the distribution of $M_\infty := \max_{n \ge 0} S_n$, and the distribution of W_n converges to π_W in the total-variation metric, that is,

$$\sup_{B \in \mathcal{B}(\mathbb{R})} |\mathbb{P}\{W_n \in B\} - \pi_W(B)| \to 0 \quad \text{as } n \to \infty.$$

The distribution π_W is explicitly known in few cases only. The tail behaviour of π_W is understood very well; it depends heavily on the existence of positive exponential moments of ξ_1. For that reason the following classes of distributions are introduced:

Definition 1.24 We say that a distribution F is *light-tailed* if

$$\int_{\mathbb{R}} e^{\lambda x} F(dx) < \infty \quad \text{for some } \lambda > 0.$$

A random variable ξ is called *light-tailed* if its distribution is light-tailed.

Definition 1.25 We say that a distribution F is *heavy-tailed* if

$$\int_{\mathbb{R}} e^{\lambda x} F(dx) = \infty \quad \text{for all } \lambda > 0.$$

A random variable ξ is called *heavy-tailed* if its distribution is heavy-tailed.

Definition 1.26 We say that a function $g(x)$ is *long-tailed* if, for any fixed y, $g(x+y) \sim g(x)$ as $x \to \infty$. A distribution F with right unbounded support is called *long-tailed* if $F(x,\infty)$ is a long-tailed function.

Any long-tailed distribution is necessarily heavy-tailed.

Definition 1.27 A distribution F on $\mathbb{R}^+$ is called *subexponential* if

$$(F * F)(x,\infty) \sim 2F(x,\infty) \quad \text{as } x \to \infty.$$

A distribution F of a random variable ξ is called *subexponential* if the distribution of ξ^+ is subexponential.

Any subexponential distribution is necessarily long-tailed and hence heavy-tailed; see e.g. [67, Lemma 3.2].

In order to describe the tail behaviour of π_W, let us introduce $\varphi(\lambda) = \mathbb{E}e^{\lambda\xi_1}$ and $\beta = \sup\{\lambda \ge 0 : \varphi(\lambda) \le 1\}$. Given $\mathbb{P}\{\xi_1 > 0\} > 0$, it follows that $\beta < \infty$. It turns out that the asymptotic behavior of $\mathbb{P}\{M_\infty > x\}$ depends heavily on the values of β and $\varphi(\beta)$. The following three different cases are considered:

(i) $\beta > 0$ and $\varphi(\beta) = 1$, the Cramér case;
(ii) $\beta = 0$, the heavy-tailed case where all positive exponential moments of ξ_1 are infinite;
(iii) $\beta > 0$ and $\varphi(\beta) < 1$, the intermediate case.

In the Cramér case, under the additional assumption $\varphi'(\beta - 0) < \infty$, for some $c \in (0,1)$ we have

$$\mathbb{P}\{M_\infty > x\} \sim ce^{-\beta x} \quad \text{as } x \to \infty;$$

this result goes back to H. Cramér, see e.g. [38] or [63, Chapter XII]. In Chapter 10, a similar exponential asymptotics of invariant probabilities of this type is proven for a broad class of Markov chains on $\mathbb{R}$ that are asymptotically homogeneous in space and have asymptotically negative drift.

In the heavy-tailed case, the tail asymptotics for M_∞ is only available under subexponential-type conditions, namely,

$$\mathbb{P}\{M_\infty > x\} \sim \frac{1}{|\mathbb{E}\xi_1|}\int_x^\infty \mathbb{P}\{\xi_1 > y\}dy \quad \text{as } x \to \infty$$

if and only if the integrated tail distribution F_I on $\mathbb{R}^+$ defined by its tail

$$\overline{F}_I(x) := \min\Big(1, \int_x^\infty \mathbb{P}\{\xi_1 > y\}dy\Big)$$

is subexponential; see e.g. [67, Theorem 5.12].

In the intermediate case, we have $\mathbb{E}e^{\beta M_\infty} < \infty$. In addition, if the function $e^{\beta x}\mathbb{P}\{\xi_1 > x\}$ is long-tailed then

$$\mathbb{P}\{M_\infty > x\} \sim c\mathbb{P}\{\xi_1 > x\} \quad \text{as } x \to \infty,$$

for some $c \in (0, \infty)$ (in the lattice case x must be taken as a multiple of the lattice step), if and only if the distribution of the random variable ξ_1^+ belongs to the so-called class $\mathcal{S}(\beta)$; see [14, Theorem 1] and [102, Theorem 2]. In that case $c = \mathbb{E}e^{\beta M_\infty}/(1 - \varphi(\beta))$.

So, the invariant measure of $\{W_n\}$ is light-tailed if and only if the distribution of ξ_1 is light-tailed. As we will see in what follows, for Markov chains with asymptotically zero drift the situation is very different – the invariant measure is always heavy-tailed apart from degenerate cases.

1.4 Nearest Neighbour Markov Chains

In this section we discuss nearest neighbour Markov chains, which represent one of the two classes of Markov chains for which either the invariant measure, in the case of positive recurrence, or the Green function, in the case of transience, is available in closed form. A closed form makes possible the direct analysis of such Markov chains: their classification and the tail asymptotics of their invariant probabilities or of Green function. This discussion should shed some light on what we may expect for general Markov chains. Another class is provided by diffusion processes, which are discussed in the next section.

Definition 1.28 A Markov chain $\{X_n\}$ on $\mathbb{Z}^+$ is called a *nearest neighbour (skip-free* or *continuous) Markov chain*, if $\xi(x)$ takes only the values -1, 1, or 0, with probabilities $p_-(x)$, $p_+(x)$, and $p_0(x) = 1 - p_-(x) - p_+(x)$ respectively, with $p_-(0) = 0$.

Let

$$p_+(x) = p + \varepsilon_+(x) \quad \text{and} \quad p_-(x) = p - \varepsilon_-(x), \quad p \le 1/2,$$

where the probabilities are assumed to be neither 0 nor 1, so that we have an irreducible Markov chain.

Assume that $\varepsilon_\pm(x) \to 0$ as $x \to \infty$, which corresponds to the case of asymptotically zero drift, $m_1(x) = \varepsilon_+(x) + \varepsilon_-(x) \to 0$ as $x \to \infty$. Then the second moment of the jumps is convergent: $m_2(x) \to 2p$ as $x \to \infty$.

1.4.1 Positive Recurrence

To find a sufficient condition for the positive recurrence of $\{X_n\}$, let us consider a test function $L(y) = y^2$. Its drift at all states $x \geq 1$ equals

$$\begin{aligned}\mathbb{E}L(x+\xi(x)) - L(x) &= 2x\mathbb{E}\xi(x) + \mathbb{E}\xi^2(x)\\ &= 2(\varepsilon_+(x)+\varepsilon_-(x))x + 2p + \varepsilon_+(x) - \varepsilon_-(x),\end{aligned}$$

so the chain is positive recurrent if

$$\limsup_{x\to\infty}(\varepsilon_+(x)+\varepsilon_-(x))x < -p, \tag{1.4}$$

see e.g. Lamperti [111] or Section 2.2. Let us denote the stationary probabilities of $\{X_n\}$ by $\pi(x)$, $x \in \mathbb{Z}^+$.

Proposition 1.29 *Under the condition* (1.4), *for some* $c_1 \in \mathbb{R}$,

$$\pi(x) \sim \exp\left(\frac{1}{p}\sum_{k=1}^{x}(\varepsilon_+(k)+\varepsilon_-(k)) + c_1\right) \quad \textit{as } x \to \infty, \tag{1.5}$$

provided that

$$\sum_{k=0}^{\infty}\varepsilon^2(k) < \infty, \tag{1.6}$$

where $\varepsilon(k) := \max(|\varepsilon_-(k)|, |\varepsilon_+(k)|)$.

Proof If the chain $\{X_n\}$ is positive recurrent then its stationary probabilities $\pi(x)$, $x \in \mathbb{Z}^+$, satisfy the equations

$$\begin{aligned}&\pi(0) = \pi(0)p_0(0) + \pi(1)p_-(1),\\ &\pi(x) = \pi(x-1)p_+(x-1) + \pi(x)p_0(x) + \pi(x+1)p_-(x+1), \quad x \geq 1,\end{aligned}$$

which are equivalent to

$$\begin{aligned}\pi(0)p_+(0) &= \pi(1)p_-(1),\\ \pi(x+1)p_-(x+1) - \pi(x)p_+(x) &= \pi(x)p_-(x) - \pi(x-1)p_+(x-1)\\ &\ \ \vdots\\ &= \pi(1)p_-(1) - \pi(0)p_+(0) = 0,\end{aligned}$$

which yields $\pi(x)p_-(x) = \pi(x-1)p_+(x-1)$ for all $x \geq 1$. Hence we obtain the following solution:

$$\pi(x) = \pi(0)\prod_{k=1}^{x}\frac{p_+(k-1)}{p_-(k)}, \quad x \geq 1, \tag{1.7}$$

where

$$\pi(0) = \left(1 + \sum_{x=1}^{\infty}\prod_{k=1}^{x}\frac{p_+(k-1)}{p_-(k)}\right)^{-1}.$$

So $\{X_n\}$ is positive recurrent if and only if

$$\sum_{x=1}^{\infty}\prod_{k=1}^{x}\frac{p_+(k-1)}{p_-(k)} < \infty;$$

see Harris [77] or Karlin and Taylor [87, pp. 86–87], in which these calculations are carried out for the case where $p_0(k) = 0$ for all $k \geq 1$.

Since $\varepsilon_\pm(k) \to 0$,

$$\begin{aligned}\prod_{k=1}^{x}\frac{p_+(k-1)}{p_-(k)} &= \frac{p_+(0)}{p_+(x)}\prod_{k=1}^{x}\frac{1+\varepsilon_+(k)/p}{1-\varepsilon_-(k)/p} \\ &\sim \frac{p_+(0)}{p}\prod_{k=1}^{x}\frac{1+\varepsilon_+(k)/p}{1-\varepsilon_-(k)/p} \quad \text{as } x \to \infty.\end{aligned}$$

The logarithm of the product on the right-hand side equals

$$\begin{aligned}&\sum_{k=1}^{x}\big(\log(1+\varepsilon_+(k)/p) - \log(1-\varepsilon_-(k)/p)\big) \\ &\quad = \frac{1}{p}\sum_{k=1}^{x}\big(\varepsilon_+(k)+\varepsilon_-(k)\big) + \sum_{k=1}^{x}\delta(k),\end{aligned} \tag{1.8}$$

where $\delta(k) = O(\varepsilon^2(k))$ as $k \to \infty$, for $\varepsilon(k) := \max(|\varepsilon_-(k)|, |\varepsilon_+(k)|)$. Hence, for some $c_1 \in \mathbb{R}$,

$$\pi(x) = \pi(0)\prod_{k=1}^{x}\frac{p_+(k-1)}{p_-(k)} \sim e^{\frac{1}{p}\sum_{k=1}^{x}(\varepsilon_+(k)+\varepsilon_-(k))+c_1} \quad \text{as } x \to \infty,$$

provided (1.6) holds. □

Let us consider a couple of examples with expressions for specific ε. In what follows we need the following results on the harmonic series and generalised harmonic series.

Proposition 1.30 *For the truncated harmonic series,*

$$\sum_{x=1}^{n} \frac{1}{x} = \log n + \gamma + O(1/n) \quad \text{as } n \to \infty, \tag{1.9}$$

where γ is the Euler constant.

For the truncated generalised harmonic series, for any $\alpha \in (0,1)$,

$$\sum_{x=1}^{n} \frac{1}{x^\alpha} = \frac{n^{1-\alpha}}{1-\alpha} + \gamma_\alpha + O(1/n^\alpha) \quad \text{as } n \to \infty. \tag{1.10}$$

The first example of expressions for ε concerns a drift of order $-\mu/x$.

Example 1.31 If $\varepsilon_+(x) \sim -\mu_+/x$ and $\varepsilon_-(x) \sim -\mu_-/x$ as $x \to \infty$ in such a way that

$$\sum_{x=0}^{\infty} \left| \varepsilon_+(x) + \varepsilon_-(x) + \frac{\mu_+ + \mu_-}{x} \right| < \infty$$

then (1.4) yields a positive recurrence of the chain provided that $\mu := \mu_+ + \mu_- > p$ and (1.5) implies an asymptotic equivalence, for some $c_2 \in \mathbb{R}$:

$$\pi(x) \sim e^{-(\mu/p)\log x + c_2} = \frac{e^{c_2}}{x^{\mu/p}} \quad \text{as } x \to \infty. \tag{1.11}$$

In Chapter 8 the power asymptotics of invariant probabilities of this type are extended to a broad class of Markov chains on $\mathbb{R}$ with asymptotically zero drift of order $-\mu/x$.

The second example of expressions for ε concerns a drift of order $-\mu/x^\alpha$, $\alpha \in (0,1)$.

Example 1.32 If $\varepsilon_+(x) \sim -\mu_+/x^\alpha$ and $\varepsilon_-(x) \sim -\mu_-/x^\alpha$ as $x \to \infty$, for some $\mu_+, \mu_- > 0$ and $\alpha \in (1/2, 1)$, in such a way that

$$\sum_{x=0}^{\infty} \left| \varepsilon_+(x) + \varepsilon_-(x) + \frac{\mu_+ + \mu_-}{x^\alpha} \right| < \infty$$

then the series $\sum \varepsilon^2(x)$ is again convergent and we observe a Weibullian asymptotic behaviour of the invariant probabilities,

$$\pi(x) \sim c_3 \exp\Big(-(\mu_+ + \mu_-)x^{1-\alpha}/p(1-\alpha)\Big) \quad \text{as } x \to \infty. \tag{1.12}$$

If now $\alpha \in (1/3, 1/2]$ then the series (1.6) diverges and the quadratic terms in (1.8) make a significant contribution to the asymptotic behaviour of the invariant probabilities,

$$\pi(x) \sim c_4 \exp\Bigl(-\frac{\mu_+ + \mu_-}{p(1-\alpha)} x^{1-\alpha} + \frac{\mu_-^2 - \mu_+^2}{(2\alpha-1)2p^2} x^{1-2\alpha}\Bigr) \quad \text{as } x \to \infty.$$

If $\alpha \in (1/4, 1/3]$ then we need to keep the cubic terms in the Taylor expansion of the logarithm, which adds a further correction term of order $x^{1-3\alpha}$ to the exponential function, and so on.

General Markov chains on $\mathbb{R}$ with asymptotically zero drift of order $-\mu/x^\alpha$, $\alpha \in (0,1)$, are considered in Chapter 9, where Weibullian-type asymptotics of invariant probabilities are proven.

1.4.2 Transience

Let a nearest neighbour Markov chain $\{X_n\}$ be irreducible and transient. Then $\mathbb{P}_x\{\tau_x < \infty\} < 1$ for all x and hence the renewal measure (Green function)

$$\begin{aligned} h_{x_0}(x) &:= \sum_{n=0}^{\infty} \mathbb{P}_{x_0}\{X_n = x\} \\ &= \mathbb{E}_{x_0} \sum_{n=0}^{\infty} \mathbb{I}\{X_n = x\} \end{aligned}$$

is finite for all $x_0, x \in \mathbb{Z}^+$, because

$$\begin{aligned} h_{x_0}(x) &= \mathbb{P}_{x_0}\{X_k = x \text{ for some } k\} \sum_{n=0}^{\infty} \mathbb{P}_x\{X_n = x\} \\ &= \mathbb{P}_{x_0}\{X_k = x \text{ for some } k\} \frac{1}{1 - \mathbb{P}_x\{\tau_x < \infty\}} < \infty. \end{aligned}$$

Since we are considering a Markov chain that jumps by 1 only, $h_{x_0}(x) = h_x(x)$ for all $x_0 \le x$. In the next result we find $h_{x_0}(x)$ in closed form.

Proposition 1.33 *Under the condition*

$$\sum_{u=1}^{\infty} \prod_{z=1}^{u} \frac{p_-(z)}{p_+(z)} < \infty, \tag{1.13}$$

the following representations hold true:

$$\begin{aligned} h_{x_0}(x) &= \frac{1}{p_+(x)} \sum_{u=x\vee x_0}^{\infty} \prod_{z=x+1}^{u} \frac{p_-(z)}{p_+(z)} \\ &= \frac{1}{p_-(x)} \sum_{u=x\vee x_0}^{\infty} \prod_{z=x}^{u} \frac{p_-(z)}{p_+(z)}. \end{aligned}$$

Proof We first look for a function $g(x,z) \geq 0$ such that, for all x, the process

$$Z_n = g(x, X_n) - \sum_{k=0}^{n-1} \mathbb{I}\{X_k = x\}, \quad n \geq 0, \tag{1.14}$$

is a martingale, which is the case if g satisfies the following system of equations:

$$g(x,0) = p_0(0)g(x,0) + p_+(0)g(x,1) - \mathbb{I}\{x = 0\},$$
$$g(x,y) = p_-(y)g(x,y-1) + p_0(y)g(x,y) + p_+(y)g(x,y+1) - \mathbb{I}\{y = x\},$$

for $y \geq 1$. Take $g(x,0) = g(x,1) = \cdots = g(x,x) = 0$. Then for $y = x$ we get

$$g(x,x+1) = g(x,x+1) - g(x,x) = \frac{1}{p_+(x)},$$

and, for $y \geq x+1$,

$$\begin{aligned} g(x,y+1) - g(x,y) &= \frac{p_-(y)}{p_+(y)}(g(x,y) - g(x,y-1)) \\ &= \prod_{z=x+1}^{y} \frac{p_-(z)}{p_+(z)}(g(x,x+1) - g(x,x)) \\ &= \frac{1}{p_+(x)} \prod_{z=x+1}^{y} \frac{p_-(z)}{p_+(z)}. \end{aligned}$$

Therefore, for $y \geq x+1$,

$$\begin{aligned} g(x,y) &= \sum_{u=x}^{y-1}(g(x,u+1) - g(x,u)) = \frac{1}{p_+(x)} \sum_{u=x}^{y-1} \prod_{z=x+1}^{u} \frac{p_-(z)}{p_+(z)} \\ &= \frac{1}{p_-(x)} \sum_{u=x}^{y-1} \prod_{z=x}^{u} \frac{p_-(z)}{p_+(z)}, \end{aligned}$$

which is increasing in y. This sequence is bounded under the condition (1.13). Then

$$g(x,\infty) := \lim_{y\to\infty} g(x,y) = \frac{1}{p_+(x)} \sum_{u=x}^{\infty} \prod_{z=x+1}^{u} \frac{p_-(z)}{p_+(z)} < \infty.$$

The sequence (1.14) is a martingale, so for all n, x, and x_0,

$$g(x,x_0) = \mathbb{E}_{x_0} Z_0 = \mathbb{E}_{x_0} Z_n = \mathbb{E}_{x_0} g(x, X_n) - \mathbb{E}_{x_0} \sum_{k=0}^{n-1} \mathbb{I}\{X_k = x\}$$

and hence

$$\sum_{k=0}^{n-1} \mathbb{P}_{x_0}\{X_k = x\} = \mathbb{E}_{x_0} g(x, X_n) - g(x, x_0) \ < \ g(x, \infty) \ < \ \infty.$$

The finiteness of the Green function implies the transience of $\{X_n\}$; hence $X_n \to \infty$ a.s. as $n \to \infty$. Thus, we get the following explicit representation for the renewal measure:

$$\begin{aligned} h_{x_0}(x) \ &= \ g(x, \infty) - g(x, x_0) = \frac{1}{p_+(x)} \sum_{u=x \vee x_0}^{\infty} \prod_{z=x+1}^{u} \frac{p_-(z)}{p_+(z)} \\ &= \frac{1}{p_-(x)} \sum_{u=x \vee x_0}^{\infty} \prod_{z=x}^{u} \frac{p_-(z)}{p_+(z)}. \end{aligned}$$

□

Now let us derive some asymptotics for $h_{x_0}(x)$ as $x \to \infty$.

Proposition 1.34 *Assume that*

$$\frac{2m_1(x)}{m_2(x)} \ = \ \frac{2(\varepsilon_+(x) + \varepsilon_-(x))}{2p + \varepsilon_+(x) - \varepsilon_-(x)} \sim r(x) \quad \text{as } x \to \infty, \tag{1.15}$$

where $r(x)$ is a differentiable decreasing function such that $r'(x)/r^2(x)$ has a limit at infinity. Then

$$h_{x_0}(x) \sim \frac{1}{pr(x)} \frac{1}{1 + \lim_{y\to\infty} r'(y)/r^2(y)} \quad \text{as } x \to \infty.$$

Proof We have

$$\prod_{z=x}^{u} \frac{p_-(z)}{p_+(z)} = \exp\left\{ \sum_{z=x}^{u} \log \frac{1 - \varepsilon_-(z)/p}{1 + \varepsilon_+(z)/p} \right\}.$$

The asymptotic equivalence (1.15) is equivalent to

$$\log \frac{1 - \varepsilon_-(x)/p}{1 + \varepsilon_+(x)/p} \sim -r(x) \quad \text{as } x \to \infty.$$

Fix an $\varepsilon > 0$. Then, for all sufficiently large x, we can write

$$-(1+\varepsilon) r(x) \ \le \ \log \frac{1 - \varepsilon_-(x)/p}{1 + \varepsilon_+(x)/p} \ \le \ -(1-\varepsilon) r(x).$$

Therefore, for such x, we have the following upper bound:

$$\begin{aligned} h_{x_0}(x) &\le \frac{1}{p_-(x)} \sum_{u=x}^{\infty} \exp\left\{-(1-\varepsilon)\sum_{z=x}^{u} r(z)\right\} \\ &\le \frac{1}{p_-(x)} \sum_{u=x}^{\infty} \exp\left\{-(1-\varepsilon)\int_x^{u+1} r(z)dz\right\} \\ &\le \frac{1}{p_-(x)} \int_x^{\infty} \exp\left\{-(1-\varepsilon)\int_x^{u} r(z)dz\right\}du, \end{aligned}$$

as $r(z)$ is a decreasing function. Setting

$$U_\varepsilon(x) = \int_x^{\infty} \exp\left\{-(1-\varepsilon)\int_0^{u} r(z)dz\right\}du$$

we observe that

$$\int_x^{\infty} \exp\left\{-(1-\varepsilon)\int_x^{u} r(z)dz\right\}du = \frac{U_\varepsilon(x)}{-U'_\varepsilon(x)}.$$

By L'Hôpital's rule and the equality $U''_\varepsilon(x) = -(1-\varepsilon)r(x)U'_\varepsilon(x)$ we have

$$\begin{aligned} \lim_{x\to\infty} \frac{U_\varepsilon(x)}{-U'_\varepsilon(x)/r(x)} &= \lim_{x\to\infty} \frac{U'_\varepsilon(x)}{-U''_\varepsilon(x)/r(x) + U'_\varepsilon(x)r'(x)/r^2(x)} \\ &= \frac{1}{1-\varepsilon+\lim_{x\to\infty} r'(x)/r^2(x)}. \end{aligned}$$

Therefore,

$$\limsup_{x\to\infty} h_{x_0}(x)r(x) \le \frac{1}{p}\frac{1}{1-\varepsilon+\lim_{x\to\infty} r'(x)/r^2(x)}.$$

Similarly, starting from the inequalities

$$\begin{aligned} h_{x_0}(x) &\ge \frac{1}{p_+(x)} \sum_{u=x}^{\infty} \exp\left\{-(1+\varepsilon)\sum_{z=x+1}^{u} r(z)\right\} \\ &\ge \frac{1}{p_+(x)} \sum_{u=x}^{\infty} \exp\left\{-(1+\varepsilon)\int_x^{u} r(z)dz\right\} \\ &\ge \frac{1}{p_+(x)} \int_x^{\infty} \exp\left\{-(1+\varepsilon)\int_x^{u} r(z)dz\right\}du, \end{aligned}$$

we get the lower bound

$$\liminf_{x\to\infty} h_{x_0}(x)r(x) \ge \frac{1}{p}\frac{1}{1+\varepsilon+\lim_{x\to\infty} r'(x)/r^2(x)}.$$

Since $\varepsilon > 0$ is arbitrary we arrive at the claim of the theorem. □

Example 1.35 Assume that $\varepsilon_+(x) \sim \mu_+/x$ and $\varepsilon_-(x) \sim \mu_-/x$ as $x \to \infty$. If $\mu := \mu_+ + \mu_- > p$ then (1.15) is valid, with $r(x) = \mu/px$, $r'(x)/r^2(x) \to -p/\mu$, and we deduce that

$$h_{x_0}(x) \sim \frac{x}{\mu - p} \quad \text{as } x \to \infty.$$

Example 1.36 Assume that $\varepsilon_+(x) \sim \mu_+/x^\alpha$ and $\varepsilon_-(x) \sim \mu_-/x^\alpha$ as $x \to \infty$. If $\mu := \mu_+ + \mu_- > 0$ and $\alpha \in (0,1)$ then (1.15) is valid with $r(x) = \mu/px^\alpha$, $r'(x)/r^2(x) \to 0$, and we deduce a Weibullian asymptotics for the renewal measure at infinity,

$$h_{x_0}(x) \sim \frac{x^\alpha}{\mu} \sim \frac{1}{m_1(x)} \quad \text{as } x \to \infty.$$

These two examples demonstrate the kind of asymptotic behaviour of the renewal measure that we could expect for general Markov chains; see Chapters 4 and 6.

We conclude this section by showing that the condition (1.13) is also necessary for the transience of nearest neigbour Markov chains. The transience of $\{X_n\}$ implies that, for all x, the sequence $\sum_{k=0}^{n-1} \mathbb{I}\{X_k = x\}$ monotonically converges a.s. and in L_1 as $n \to \infty$. Therefore, the sequence (1.14) satisfies $\mathbb{E} \min_n Z_n > -\infty$. This allows us to apply the martingale convergence theorem: Z_n converges almost surely to an integrable random variable Z_∞. Combining this with the convergence of $\sum_{k=0}^{n-1} \mathbb{I}\{X_k = x\}$, we infer that $g(x, X_n)$ converges almost surely too. If we assume now that (1.13) is not valid then

$$g(x,y) \uparrow g(x,\infty) = \infty \quad \text{as } y \to \infty,$$

and the irreducibility of $\{X_n\}$ would imply that

$$\limsup_{n\to\infty} g(x, X_n) = \infty \quad \text{a.s.}$$

This would contradict the convergence of $g(x, X_n)$, hence (1.13) is necessary for the transience of $\{X_n\}$.

An alternative approach to the classification of nearest neighbour Markov chains may be found in Karlin and Taylor [87, Section 3.7].

1.4.3 Harmonic Functions and *h*-Transforms

Consider $\{X_n\}$ killed at hitting zero, by setting $p_-(1) = 0$. The corresponding transition kernel is *substochastic*, which means that each row sums to a value not greater than 1. Let us construct a harmonic function for this kernel, that is, a non-negative solution V to the system of linear equations

$$V(x) = p_+(x)V(x+1) + p_0(x)V(x) + p_-(x)V(x-1), \quad x \ge 1, \quad (1.16)$$

with initial condition $V(0) = 0$.

Lemma 1.37 *For all $x \ge 1$,*

$$V(x) = V(1)\sum_{y=0}^{x-1}\prod_{k=1}^{y}\frac{p_-(k)}{p_+(k)}. \quad (1.17)$$

Proof Let τ_y be the first hitting time of y, that is,

$$\tau_y := \inf\{n \ge 1 : X_n = y\}.$$

Then the equations (1.16) with initial condition $V(0) = 0$ are equivalent to

$$V(x) = \mathbb{E}_x\{V(X_1);\ \tau_0 > 1\}, \quad x \ge 1, \quad (1.18)$$

which defines a harmonic function for the chain $\{X_n\}$ killed at hitting zero.

It is clear that (1.16) can be rewritten in the form

$$p_+(x)[V(x+1) - V(x)] = p_-(x)[V(x) - V(x-1)].$$

Consequently,

$$V(x+1) - V(x) = [V(1) - V(0)]\prod_{k=1}^{x}\frac{p_-(k)}{p_+(k)}, \quad x \ge 1. \quad (1.19)$$

Recalling that $V(0) = 0$, we then obtain the harmonic function V for the chain $\{X_n\}$ killed at hitting zero in closed form:

$$V(x) = \sum_{y=0}^{x-1}[V(y+1) - V(y)] = V(1)\sum_{y=0}^{x-1}\prod_{k=1}^{y}\frac{p_-(k)}{p_+(k)}.$$

□

The existence of a positive harmonic function allows us to transform a strictly substochastic transition kernel for the chain $\{X_n\}$ killed at hitting zero into a stochastic transition kernel. For every $x \ge 1$, define

$$\widehat{p}_+(x) := \frac{V(x+1)}{V(x)}p_+(x), \quad \widehat{p}_0(x) = p_0(x), \text{ and } \widehat{p}_-(x) := \frac{V(x-1)}{V(x)}p_-(x).$$

The new transition kernel $\widehat{P}$ is stochastic because, as follows from (1.16),

$$\widehat{p}_-(x) + \widehat{p}_0(x) + \widehat{p}_+(x) = 1 \quad \text{for all } x \ge 1.$$

This transformation is called Doob's h-transform, for a Markov chain killed at hitting zero. Let $\{\widehat{X}_n\}$ be a Markov chain on $\{1, 2, \ldots\}$ with transition kernel $\widehat{P}$.

Lemma 1.38 *The chain $\{\widehat{X}_n\}$ is always transient.*

Proof As shown in the previous subsection, it suffices to show that (1.13) holds for the transition probabilities $\widehat{P}$. We first apply the definition of $\widehat{P}$:

$$\sum_{u=1}^{\infty}\prod_{z=2}^{u}\frac{\widehat{p}_-(z)}{\widehat{p}_+(z)}=\sum_{u=1}^{\infty}\prod_{z=2}^{u}\frac{V(z-1)}{V(z+1)}\frac{p_-(z)}{p_+(z)}$$
$$=\sum_{u=1}^{\infty}\frac{V(1)V(2)}{V(u)V(u+1)}\prod_{z=2}^{u}\frac{p_-(z)}{p_+(z)}.$$

It follows from (1.19) that

$$\frac{1}{V(u)}-\frac{1}{V(u+1)}=\frac{V(u+1)-V(u)}{V(u)V(u+1)}=\frac{V(1)}{V(u)V(u+1)}\prod_{z=1}^{u}\frac{p_-(z)}{p_+(z)}.$$

Therefore,

$$\sum_{u=1}^{\infty}\prod_{z=2}^{u}\frac{\widehat{p}_-(z)}{\widehat{p}_+(z)}=\frac{p_+(1)}{p_-(1)}V(2)\sum_{u=1}^{\infty}\left(\frac{1}{V(u)}-\frac{1}{V(u+1)}\right)$$
$$\le\frac{p_+(1)}{p_-(1)}\frac{V(2)}{V(1)}<\infty,$$

which is equivalent to the transience of the transformed chain $\{\widehat{X}_n\}$. □

A standard application of Doob's h-transform is a random walk conditioned to stay positive. Let $\{X_n\}$ be a simple symmetric random walk on $\mathbb{Z}$, that is, $p_-(x)=p_+(x)=1/2$ for all $x\in\mathbb{Z}$. Then it follows from (1.17) that $V(x)=xV(1)$. As a result the transformed chain $\{\widehat{X}_n\}$ has transition probabilities

$$\widehat{p}_-(x)=\frac{x-1}{2x}=\frac{1}{2}-\frac{1}{2x},\quad \widehat{p}_+(x)=\frac{x+1}{2x}=\frac{1}{2}+\frac{1}{2x},\quad x\ge 1.$$

It is immediate from these formulae that the transformed chain has an asymptotically zero drift and unit second moment of jumps.

If the original Markov chain $\{X_n\}$ is recurrent then one can use the h-transform to connect the stationary measure π of $\{X_n\}$ with the Green function of $\{\widehat{X}_n\}$. The following representation for the invariant measure π via a cycle structure (generated by the returning time to the state 0) of the Markov chain $\{X_n\}$ is well known – see e.g. [126, Theorem 10.4.9] – for $x\ge 1$:

$$\pi(x)=\pi(0)\sum_{n=1}^{\infty}\mathbb{P}_0\{X_n=x,\ \tau_0>n\}$$
$$=\pi(0)p_+(0)\sum_{n=0}^{\infty}\mathbb{P}_1\{X_n=x,\ \tau_0>n\}.$$

Noting that $\mathbb{P}_1\{X_n = x,\ \tau_0 > n\} = \frac{V(1)}{V(x)}\mathbb{P}_1\{\widehat{X}_n = x\}$ for all $x, n \geq 1$, we obtain

$$\pi(x) = \frac{\pi(0)p_+(0)V(1)}{V(x)}\widehat{h}_1(x), \tag{1.20}$$

where

$$\widehat{h}_1(x) := \sum_{n=0}^{\infty} \mathbb{P}_1\{\widehat{X}_n = x\}, \quad x \geq 1.$$

Let us consider a couple of examples; we first discuss a drift of order $-\mu/x$.

Example 1.39 Let $\varepsilon_+(x) \sim -\mu_+/x$ and $\varepsilon_-(x) \sim -\mu_-/x$ as $x \to \infty$ in such a way that

$$\sum_{x=0}^{\infty}\left|\varepsilon_+(x) + \varepsilon_-(x) + \frac{\mu_+ + \mu_-}{x}\right| < \infty.$$

Let $\mu := \mu_+ + \mu_- > p$, so the chain is positive recurrent. As follows from (1.19), for all $x \geq 1$,

$$\begin{aligned}
V(x+1) - V(x) &= [V(1) - V(0)]\prod_{k=1}^{x}\frac{p_-(k)}{p_+(k)} \\
&= [V(1) - V(0)]\exp\left(\sum_{k=1}^{x}(\log p_-(k) - \log p_+(k))\right) \\
&= [V(1) - V(0)]\exp\left(\sum_{k=1}^{x}(\log(1 - \varepsilon_-(k)/p) - \log(1 + \varepsilon_+(k)/p))\right).
\end{aligned}$$

As in (1.5), we conclude that there is an asymptotic relation, for some c_1,

$$\begin{aligned}
V(x+1) - V(x) &\sim [V(1) - V(0)]\exp\left(-\frac{1}{p}\sum_{k=1}^{x}(\varepsilon_-(k) + \varepsilon_+(k)) + c_1\right) \\
&\sim [V(1) - V(0)]\exp\left(\frac{\mu_- + \mu_+}{p}\log x + c_2\right) \\
&\sim c_3 x^{\mu/p} \quad \text{as } x \to \infty.
\end{aligned}$$

Therefore, as $x \to \infty$,

$$\frac{V(x+1)}{V(x)} = 1 + \frac{V(x+1) - V(x)}{V(x)} = 1 + \frac{\mu/p + 1}{x} + o(1/x),$$

and

$$\frac{V(x-1)}{V(x)} = 1 - \frac{V(x) - V(x-1)}{V(x)} = 1 - \frac{\mu/p + 1}{x} + o(1/x).$$

Hence, the transition probabilities of the transformed Markov chain satisfy the relations

$$\begin{aligned}\widehat{p}_+(x) &:= \frac{V(x+1)}{V(x)}p_+(x) &= p + \frac{\mu_- + p}{x} + o(1/x),\\ \widehat{p}_-(x) &:= \frac{V(x-1)}{V(x)}p_-(x) &= p - \frac{\mu_+ + p}{x} + o(1/x).\end{aligned}$$

It follows from Example 1.35 with $\widehat{\mu}_+ = \mu_- + p$ and $\widehat{\mu}_- = \mu_+ + p$ that

$$\widehat{h}_1(x) \sim \frac{x}{\widehat{\mu}_+ + \widehat{\mu}_- - p} = \frac{x}{\mu + p},$$

which, on being substituted into (1.20), implies, as $x \to \infty$,

$$\pi(x) = c_3\frac{\widehat{h}_1(x)}{V(x)} \sim \frac{c_4}{x^{\mu/p}},$$

which coincides with the expression (1.11).

This relation between the stationary measure of a nearest neighbour Markov chain and the Green function of the transformed chain may be extended to the general case. We follow this approach in Chapter 8 to derive the power asymptotics of invariant probabilities of this type for a broad class of Markov chains on $\mathbb{R}$, those with asymptotically zero drift of order $-\mu/x$.

The second example concerns a drift of order $-\mu/x^\alpha$, $\alpha \in (0,1)$.

Example 1.40 Let $\varepsilon_+(x) \sim -\mu_+/x^\alpha$ and $\varepsilon_-(x) \sim -\mu_-/x^\alpha$ as $x \to \infty$ for some $\mu_+, \mu_- > 0$ and $\alpha \in (1/2, 1)$, in such a way that

$$\sum_{x=0}^{\infty}\left|\varepsilon_+(x) + \varepsilon_-(x) + \frac{\mu_+ + \mu_-}{x^\alpha}\right| < \infty.$$

Similarly to the last example, for some c_5,

$$\begin{aligned}V(x+1) - V(x) &\sim [V(1) - V(0)]\exp\left(-\frac{1}{p}\sum_{k=1}^{x}(\varepsilon_-(k) + \varepsilon_+(k)) + c_5\right)\\ &\sim c_6\exp\left(\frac{\mu_- + \mu_+}{p(1-\alpha)}x^{1-\alpha}\right) \quad \text{as } x \to \infty.\end{aligned}$$

Therefore, as $x \to \infty$,

$$\frac{V(x+1)}{V(x)} = 1 + \frac{\mu_+ + \mu_-}{px^\alpha} + o(1/x)$$

and

$$\frac{V(x-1)}{V(x)} = 1 - \frac{\mu_+ + \mu_-}{px^\alpha} + o(1/x).$$

Hence, the transition probabilities of the transformed Markov chain satisfy the relations

$$\begin{aligned}\widehat{p}_+(x) &:= \frac{V(x+1)}{V(x)}p_+(x) &= p + \frac{\mu_-}{x^\alpha} + O(1/x^{2\alpha}),\\ \widehat{p}_-(x) &:= \frac{V(x-1)}{V(x)}p_-(x) &= p - \frac{\mu_+}{x^\alpha} + O(1/x^{2\alpha}).\end{aligned}$$

It follows from Example 1.35 with $\widehat{\mu}_+ = \mu_-$ and $\widehat{\mu}_- = \mu_+$ that

$$\widehat{h}_1(x) \sim \frac{x^\alpha}{\widehat{\mu}_+ + \widehat{\mu}_-},$$

which, on being substituted into (1.20), implies a Weibullian asymptotic behaviour of the invariant probabilities, as $x \to \infty$:

$$\pi(x) = c_7 \frac{\widehat{h}_1(x)}{V(x)} \sim c_8 \exp\left(-\frac{\mu_- + \mu_+}{p(1-\alpha)}x^{1-\alpha}\right),$$

which coincides with the expression (1.12).

General Markov chains on $\mathbb{R}$ with asymptotically zero drift of order $-\mu/x^\alpha$, $\alpha \in (0,1)$, are considered in Chapter 9 where we again follow the above approach to derive the Weibullian-type asymptotics of invariant probabilities.

1.4.4 Down-Crossing Probabilities for Transient Chain

Let $\{X_n\}$ be transient, that is, its probability of hitting the origin, $\mathbb{P}_x\{\tau_0 < \infty\}$, is less than 1 for all $x \geq 1$. The goal of the following calculations is to find this probability.

The function $V(x)$ computed in (1.17) is increasing and bounded provided that the condition (1.13) holds. As it has already been noticed in (1.18), the sequence $V(X_{n\wedge\tau_0})$ is a bounded non-negative martingale, so, by the optional stopping theorem (see e.g. [56, Section 4.7]),

$$\begin{aligned}V(x) &= \mathbb{E}_x V(X_0) = \mathbb{E}_x V(X_{\tau_0})\\ &= V(0)\mathbb{P}_x\{\tau_0 < \infty\} + V(\infty)\mathbb{P}_x\{\tau_0 = \infty\}\end{aligned}$$

and hence

$$\mathbb{P}_x\{\tau_0 < \infty\} = \frac{V(\infty) - V(x)}{V(\infty) - V(0)} = \frac{\sum_{y=x}^{\infty} \prod_{k=1}^{y} \frac{p_-(k)}{p_+(k)}}{\sum_{y=0}^{\infty} \prod_{k=1}^{y} \frac{p_-(k)}{p_+(k)}}.$$

Owing to the left-continuity of the Markov chain, similarly we get, for all $0 \le \widehat{x} < x$,

$$\mathbb{P}_x\{\tau_{\widehat{x}} < \infty\} = \frac{V(\infty) - V(x)}{V(\infty) - V(\widehat{x})} = \frac{\sum_{y=x}^{\infty} \prod_{k=1}^{y} \frac{p_-(k)}{p_+(k)}}{\sum_{y=\widehat{x}}^{\infty} \prod_{k=1}^{y} \frac{p_-(k)}{p_+(k)}}. \tag{1.21}$$

Example 1.41 In the case where $\varepsilon_+(x) \sim \mu_+/x$ and $\varepsilon_-(x) \sim \mu_-/x$ as $x \to \infty$, $\mu := \mu_+ + \mu_- > p$, and

$$\sum_{x=0}^{\infty} \left| \varepsilon_+(x) + \varepsilon_-(x) - \frac{\mu}{x} \right| < \infty,$$

then similarly to (1.11) we derive that

$$\prod_{k=1}^{y} \frac{p_-(k)}{p_+(k)} \sim c_5 y^{-\mu/p} \quad \text{as } y \to \infty,$$

where $c_5 > 0$. Therefore, (1.21) implies that there exists a function $c(\widehat{x}) \to 1$ as $\widehat{x} \to \infty$ such that

$$\mathbb{P}_x\{\tau_{\widehat{x}} < \infty\} \sim c(\widehat{x})(\widehat{x}/x)^{\mu/p-1} \quad \text{as } x \to \infty, \text{ uniformly for all } \widehat{x} < x.$$

In particular,

$$\mathbb{P}_x\{\tau_{\widehat{x}} < \infty\} \sim (\widehat{x}/x)^{\mu/p-1} \quad \text{as } \widehat{x},\ x \to \infty,\ x > \widehat{x}.$$

Compare this result with Theorem 3.2 and Corollary 3.3, where a general transient Markov chain with a drift of order μ/x is studied.

Example 1.42 Assume that $\varepsilon_+(x) \sim \mu_+/x^\alpha$ and $\varepsilon_-(x) \sim \mu_-/x^\alpha$ as $x \to \infty$. If $\mu := \mu_+ + \mu_- > 0$, $\alpha \in (1/2, 1)$, and

$$\sum_{x=0}^{\infty} \left| \varepsilon_+(x) + \varepsilon_-(x) - \frac{\mu}{x^\alpha} \right| < \infty$$

then the series $\sum \varepsilon^2(x)$ is convergent and we obtain

$$\prod_{k=1}^{y} \frac{p_-(k)}{p_+(k)} \sim c_6 e^{-\mu y^{1-\alpha}/p(1-\alpha)} \quad \text{as } y \to \infty,$$

where $c_6 > 0$. Therefore, (1.21) implies a Weibullian asymptotic behaviour of the down-crossing probability, that is, there exists a function $c(\widehat{x}) \to 1$ as $\widehat{x} \to \infty$ such that

$$\begin{aligned}\mathbb{P}_x\{\tau_{\widehat{x}}<\infty\} &\sim c(\widehat{x})\frac{\sum_{u=x}^{\infty}\exp\left(-\mu u^{1-\alpha}/p(1-\alpha)\right)}{\sum_{u=\widehat{x}}^{\infty}\exp\left(-\mu u^{1-\alpha}/p(1-\alpha)\right)}\\ &\sim c(\widehat{x})\left(\frac{x}{\widehat{x}}\right)^{\alpha}\exp\left(\mu(\widehat{x}^{1-\alpha}-x^{1-\alpha})/p(1-\alpha)\right)\end{aligned}$$

as $x\to\infty$ uniformly for all $\widehat{x}<x$. In particular,

$$\mathbb{P}_x\{\tau_{\widehat{x}}<\infty\}\sim\left(\frac{x}{\widehat{x}}\right)^{\alpha}\exp\left(\mu(\widehat{x}^{1-\alpha}-x^{1-\alpha})/p(1-\alpha)\right)\quad\text{as }\widehat{x},\ x\to\infty,\ x>\widehat{x}.$$

Compare this result with Theorem 3.7, where a general transient Markov chain with a drift of order μ/x^{α}, $\alpha\in(1/2,1)$, is studied.

1.5 Heuristics Coming from Diffusion Processes

1.5.1 Diffusion with Bounded Smooth Infinitesimal Parameters

Another example where various characteristics are available in closed form is provided by diffusion processes on $\mathbb{R}$ that are continuous-time Markov processes with continuous paths. If they are sampled at non-random equally spaced time epochs they give us examples of Markov chains for which some characteristics are explicitly calculable.

Let us start with a result that demonstrates that the existence of an invariant probability measure for a diffusion process is equivalent to its positive recurrence.

Lemma 1.43 *For a diffusion process $\{X(t)\}$ with diffusion coefficient everywhere positive the following are equivalent:*

(i) *there is a stationary version of the process $\{X(t)\}$;*
(ii) *the process $\{X(t)\}$ is positive recurrent, that is, $\mathbb{E}_x\tau_y<\infty$ for all states x and y, where $\tau_y:=\inf\{t:X(t)=y\}$.*

Proof Let $\{X(t)\}$ possess an invariant probability measure π. Then the same is true for the slotted Markov chain $X_n=X(n)$, $n\in\mathbb{Z}^+$. Since the diffusion coefficient is everywhere positive, the jumps of $\{X_n\}$ are absolutely continuous with positive density function, so the chain $\{X_n\}$ is ψ-irreducible; see [126, Proposition 4.2.2]. Therefore, the existence of an invariant probability measure for $\{X_n\}$ implies the positive recurrence of any compact set B of positive Lebesgue measure, in the sense that $\mathbb{E}_x\tau_B<\infty$ for all x. Hence, B is positive recurrent for $\{X(t)\}$ too, which implies the positive recurrence of the diffusion process due to the continuity of its paths.

Vice versa, let $\{X(t)\}$ be positive recurrent. Then, for any two fixed distinct states x and y, the stopping time

$$\tau := \min\{t : X(t) = x \text{ and } X(s) = y \text{ for some } s < t\}$$

is finite on average given $X(0) = x$, $\mathbb{E}_x\tau < \infty$. In addition, $\tau > 0$. For that reason a measure

$$\begin{aligned}\mu(B) &:= \mathbb{E}_x \int_0^\tau \mathbb{I}\{X(t) \in B\}\,dt \\ &= \int_0^\infty \mathbb{P}_x\{X(t) \in B,\ \tau > t\}\,dt\end{aligned}$$

is non-zero and finite, $\mu(\mathbb{R}) = \mathbb{E}_x\tau \in (0, \infty)$. Let us show that it is invariant for $\{X(t)\}$, that is, for any $s > 0$ and any bounded continuous function $\varphi : \mathbb{R} \to \mathbb{R}$, we have

$$\int_{\mathbb{R}} \varphi(z)\mu(dz) = \int_{\mathbb{R}} \mathbb{E}\{\varphi(X(s)) \mid X(0) = z\}\mu(dz).$$

Indeed, the difference between the right- and left-hand side integrals equals

$$\begin{aligned}&\int_{\mathbb{R}} \mathbb{E}\{\varphi(X(s)) - \varphi(z) \mid X(0) = z\}\mu(dz) \\ &\quad = \int_{\mathbb{R}} \mathbb{E}\{\varphi(X(t+s)) - \varphi(X(t)) \mid X(t) = z\} \int_0^\infty \mathbb{P}_x\{X(t) \in dz,\ \tau > t\}\,dt \\ &\quad = \int_0^\infty \mathbb{E}_x\{\varphi(X(t+s)) - \varphi(X(t)),\ \tau > t\}\,dt,\end{aligned}$$

because $\{\tau > t\} = \overline{\{\tau \le t\}} \in \sigma(X_u,\ u \le t)$. Since

$$\begin{aligned}\int_0^\infty \mathbb{E}_x\{\varphi(X(t+s)),\ \tau > t\}\,dt &= \mathbb{E}_x \int_0^\tau \varphi(X(t+s))dt \\ &= \mathbb{E}_x \int_s^{\tau+s} \varphi(X(t))dt,\end{aligned}$$

we get

$$\begin{aligned}&\int_0^\infty \mathbb{E}_x\{\varphi(X(t+s)) - \varphi(X(t)),\ \tau > t\}\,dt \\ &\quad = \mathbb{E}_x \int_s^{\tau+s} \varphi(X(t))dt - \mathbb{E}_x \int_0^\tau \varphi(X(t))dt \\ &\quad = \mathbb{E}_x \int_\tau^{\tau+s} \varphi(X(t))dt - \mathbb{E}_x \int_0^s \varphi(X(t))dt \\ &\quad = 0,\end{aligned}$$

by the Markov property, since $X(\tau) = x$. □

Consider a diffusion process $X = \{X(t)\}$ on $\mathbb{R}$ with smooth drift $\mu(x)$ and diffusion coefficient $\sigma^2(x) > 0$. In the case of a stationary diffusion process, the invariant density function $p(x)$ solves the stationary Kolmogorov forward equation

$$0 = -\frac{d}{dx}(\mu(x)p(x)) + \frac{1}{2}\frac{d^2}{dx^2}(\sigma^2(x)p(x)),$$

which has the following solution:

$$p(x) = \frac{c}{\sigma^2(x)} \exp\left(\int_0^x \frac{2\mu(y)}{\sigma^2(y)} dy\right), \quad c > 0. \tag{1.22}$$

It follows that a diffusion process possesses a probabilistic invariant distribution – is positive recurrent – if and only if

$$\text{the function } \frac{1}{\sigma^2(x)} \exp\left(\int_0^x \frac{2\mu(y)}{\sigma^2(y)} dy\right) \quad \text{is integrable at } \pm\infty. \tag{1.23}$$

It is also known that the half-line $(-\infty,0]$ is recurrent for a diffusion process, in the sense that $\mathbb{P}_x\{X(t) \leq 0 \text{ for some } t\} = 1$ for all $x > 0$ if

$$\text{the function } \exp\left(-\int_0^x \frac{2\mu(y)}{\sigma^2(y)} dy\right) \quad \text{is not integrable at } \infty; \tag{1.24}$$

see, e.g. [88, Chapter 15, Theorem 7.3] or [34, Section 4.1]; and, vice versa, it is transient in the sense that $\mathbb{P}_x\{X(t) > 0 \text{ for all } t > 0\} > 0$ for all $x > 0$ if

$$\text{the function } \exp\left(-\int_0^x \frac{2\mu(y)}{\sigma^2(y)} dy\right) \quad \text{is integrable at } \infty, \tag{1.25}$$

see e.g. [88, Chapter 15, Lemma 6.1].

As one can see, the classification of diffusion processes relies heavily on the asymptotic behaviour of the ratio $2\mu(x)/\sigma^2(x)$ at infinity. In particular, if

$$\mu(x) \sim -\mu/x \text{ and } \sigma^2(x) \to \sigma^2 > 0 \qquad \text{as } x \to \infty \tag{1.26}$$

for some $\mu \in \mathbb{R}$ and $\sigma^2 > 0$ then:

- integrability at infinity in (1.23) holds for $2\mu > \sigma^2$;
- non-integrability at infinity in (1.24) holds for $2\mu > -\sigma^2$;
- integrability at infinity in (1.25) holds for $2\mu < -\sigma^2$.

Knowledge of the invariant probability density function in closed form, (1.22), allows us to analyse its asymptotic behaviour under various regularity conditions of the drift and diffusion coefficients at infinity.

Example 1.44 Let $\{X(t)\}$ possess a probabilistic invariant measure and let (1.26) hold with $2\mu > \sigma^2$. If

$$\int_1^\infty \left| \frac{\mu(x)}{\sigma^2(x)} + \frac{\mu}{\sigma^2 x} \right| dx < \infty$$

then (1.22) yields the following asymptotic equivalence, for some $c_1 > 0$,

$$p(x) \sim \frac{c_1}{x^{2\mu/\sigma^2}} \quad \text{as } x \to \infty.$$

Example 1.45 Let $\{X(t)\}$ possess a probabilistic invariant measure. If $\mu(x) \sim -\mu/x^\alpha$ and $\sigma^2(x) \to \sigma^2 > 0$ as $x \to \infty$ for some $\mu > 0$ and $\alpha \in (0,1)$, in such a way that

$$\int_1^\infty \left| \frac{\mu(x)}{\sigma^2(x)} + \frac{\mu}{\sigma^2 x^\alpha} \right| dx < \infty,$$

then

$$p(x) \sim c_2 \exp\left(-2\mu x^{1-\alpha}/\sigma^2(1-\alpha)\right) \quad \text{as } x \to \infty.$$

Let $\{X(t)\}$ be a diffusion process satisfying the condition (1.25), so that the negative half-line $(-\infty, 0]$ is transient. A harmonic function $h(x)$ for such a diffusion process with transition kernel $P(t,x,dy)$, that is, a solution to the equation

$$\left(\frac{\sigma^2(x)}{2}\frac{d^2}{dx^2} + \mu(x)\frac{d}{dx}\right)h(x) = 0, \tag{1.27}$$

is computable in closed form as follows:

$$h(x) = \int_x^\infty \exp\left(-\int_0^z \frac{2\mu(y)}{\sigma^2(y)}dy\right)dz, \quad x \in \mathbb{R}. \tag{1.28}$$

This is a positive decreasing function. By Itô's formula, the process $\{h(X(t))\}$ is a martingale; hence we can apply Doob's h-transform, which returns a new stochastic transition kernel

$$\widehat{P}(t,x,dy) := \frac{h(y)}{h(x)}P(t,x,dy).$$

Let us consider a diffusion process $\widehat{X} = \{\widehat{X}(t)\}$ with this transition kernel. The drift coefficient of $\widehat{X}$ is

$$\begin{aligned}\widehat{\mu}(x) &= \lim_{t\to 0}\frac{1}{t}\int (y-x)\frac{h(y)}{h(x)}P(t,x,dy)\\ &= \lim_{t\to 0}\frac{1}{t}\int (y-x)\left(1 + \frac{h'(x)}{h(x)}(y-x) + O((y-x)^2)\right)P(t,x,dy)\\ &= \mu(x) + \frac{h'(x)}{h(x)}\sigma^2(x),\end{aligned} \tag{1.29}$$

and, since $h'(x) < 0$, $\widehat{\mu}(x) < \mu(x)$. The diffusion coefficient does not change: $\widehat{\sigma}^2(x) = \sigma^2(x)$.

If, for some $\widetilde{c} > 3$,

$$\frac{2\mu(x)}{\sigma^2(x)} \geq \frac{\widetilde{c}}{x} \quad \text{for all sufficiently large } x,$$

then, under some mild additional condition,

$$-h'(x) \geq c_2 h(x)/x \qquad \text{for some } c_2 > 0,$$

and the set $(-\infty, 0]$ is positive recurrent for the transformed chain $\{\widehat{X}(t)\}$. Indeed, in this case

$$h(x) \leq \int_x^{\infty} \exp\left(c_3 - \int_1^z \frac{\widetilde{c}}{y} dy\right) dz \; = \; c_4 x^{1-\widetilde{c}},$$

hence the function

$$\begin{aligned}\exp\left(\int_0^x \frac{2\widehat{\mu}(y)}{\widehat{\sigma}^2(y)} dy\right) &= \exp\left(\int_0^x \frac{2\mu(y)}{\sigma^2(y)} dy + \int_0^x 2\frac{h'(y)}{h(y)} dy\right) \\ &= \frac{h^2(x)}{h^2(0)} \exp\left(\int_0^x \frac{2\mu(y)}{\sigma^2(y)} dy\right) \; = \; -\frac{h^2(x)}{h'(x)} \frac{1}{h^2(0)} \\ &\leq \frac{x h(x)}{c_2} \; \leq \; c_4 x x^{1-\widetilde{c}}/c_2\end{aligned}$$

is integrable at infinity because $\widetilde{c} > 3$ and the condition (1.23) for positive recurrence is met.

If, for some $\widetilde{c} \in (1, 3]$ and a function $p(x)$ that is absolutely integrable at infinity, we have

$$\frac{2\mu(x)}{\sigma^2(x)} = \frac{\widetilde{c}}{x} + p(x),$$

then the diffusion process $\{X(t)\}$ is transient by the criterion (1.25) and the transformed process $\{\widehat{X}(t)\}$ is null recurrent, because in this case

$$h'(x) \; \sim \; -\exp\left(c_5 - \int_1^x \frac{\widetilde{c}}{y} dy\right) = -e^{c_5} x^{-\widetilde{c}} \text{and}$$

$$h(x) = -\int_x^{\infty} h'(z) dz \; \sim \; c_6 x^{1-\widetilde{c}}.$$

Thus the function

$$\exp\left(\int_0^x \frac{2\widehat{\mu}(y)}{\widehat{\sigma}^2(y)} dy\right) = -\frac{h^2(x)}{h'(x)} \; \sim \; c_7 x^{2-\widetilde{c}}$$

is not integrable at infinity because $\widetilde{c} \in (1,3]$ and hence $\{\widehat{X}(t)\}$ is not positive recurrent, by (1.23), but it is still recurrent, by (1.24), because the function

$$\exp\left(-\int_0^x \frac{2\widehat{\mu}(y)}{\widehat{\sigma}^2(y)} dy\right) = -\frac{h'(x)}{h^2(x)} \sim \frac{x^{\widetilde{c}-2}}{c_7}$$

is not integrable at infinity either.

Conversely, let us consider a recurrent diffusion process $\{X(t)\}$ when $\tau = \tau_{(-\infty,0]} = \min\{t \geq 0 : X(t) \leq 0\}$ is finite with probability 1. Consider the process $Y(t) := X(t \wedge \tau)$, which is the original process stopped at the time of leaving the positive half-line. Its harmonic function solves (1.27) with $h(0) = 1$,

$$h(x) = 1 + \int_0^x \exp\left(-\int_0^z \frac{2\mu(y)}{\sigma^2(y)} dy\right) dz, \quad x \geq 0. \tag{1.30}$$

It is an increasing function tending to infinity as $x \to \infty$, due to the recurrence condition (1.24). By Itô's formula, the process $\{h(Y(t))\}$ is a martingale; hence we can apply Doob's h-transform, which returns a new stochastic transition kernel

$$\widehat{P}_Y(t,x,dy) := \frac{h(y)}{h(x)} P_Y(t,x,dy).$$

Let us consider a diffusion process $\{\widehat{Y}(t)\}$ with this transition kernel. The drift coefficient of $\{\widehat{Y}(t)\}$ is given in (1.29). Since the function $h(x)$ is increasing, $\widehat{\mu}(x) > \mu(x)$. The increase in the drift is so strong that the process $\{\widehat{Y}(t)\}$ is transient. Indeed, the function

$$\begin{aligned}
\exp\left(-\int_0^x \frac{2\widehat{\mu}(y)}{\widehat{\sigma}^2(y)} dy\right) &= \exp\left(-\int_0^x \frac{2\mu(y)}{\sigma^2(y)} dy - \int_0^x 2\frac{h'(y)}{h(y)} dy\right) \\
&= \frac{1}{h^2(x)} \exp\left(-\int_0^x \frac{2\mu(y)}{\sigma^2(y)} dy\right) \\
&= \frac{h'(x)}{h^2(x)} = \left(\frac{-1}{h(x)}\right)'
\end{aligned}$$

is integrable at infinity because $h(x) \to \infty$ and, therefore, the condition (1.25) for transience is met:

$$\int_z^\infty \exp\left(-\int_0^x \frac{2\widehat{\mu}(y)}{\widehat{\sigma}^2(y)} dy\right) dx = \frac{1}{h(z)} < \infty.$$

We follow up the idea of these calculations related to harmonic functions and changes of measure for diffusion processes in our tail analysis of invariant measures of Markov chains in Chapters 8 and 9.

1.5.2 Green Function for Transient Diffusion

Let $\{X(t)\}$ be a transient diffusion on $\mathbb{R}$ (or $\mathbb{R}^+$) with the following generator:

$$A = \mu(x)\frac{d}{dx} + \frac{\sigma^2(x)}{2}\frac{d^2}{dx^2}.$$

We consider a regular diffusion, in the sense of properties (i)–(iii) of [135, Chapter VII.3]. For its transience it is sufficient to assume that the following function,

$$U(x) := \int_x^\infty \exp\left\{-\int_0^v \frac{2\mu(y)}{\sigma^2(y)}dy\right\}dv, \tag{1.31}$$

is finite for all x, see (1.25); this function solves the homogeneous equation

$$AU = 0. \tag{1.32}$$

In this case $X(t) \to \infty$ a.s. and we are interested in the continuous-time analogue of the renewal (Green) function,

$$H_y(x, x+h] := \int_0^\infty \mathbb{P}_y\{X(t) \in (x, x+h]\}\,dt, \quad h > 0.$$

By Proposition 1.6 in Revuz and Yor [135, Chapter VII.1], the process

$$f(X(t)) - f(X(0)) - \int_0^t Af(X(s))ds$$

is a local martingale for a wide class of functions f. This suggests the following method for computing the renewal measure of $X(t)$. Fix x and h. Suppose we can find a bounded function $f(z) = f_{h,x}(z)$ such that $f(z) \to 0$ as $z \to \infty$ and

$$Af(z) = -\mathbb{I}\{z \in (x, x+h]\}. \tag{1.33}$$

Then the optional stopping theorem, see e.g. [56, Section 4.7], and a.s. convergence of $X(t) \to \infty$ as $t \to \infty$ give us the equality

$$f(y) = \mathbb{E}_y f(X(0)) = \mathbb{E}_y\left[\int_0^\infty \mathbb{I}\{X(t) \in (x, x+h]\}\,dt\right] = H_y(x, x+h],$$

which allows us to analyse H_y.

So, we need to solve the ordinary differential equation (1.33). To this end, consider

$$m(x) := \int_0^x \frac{2dv}{-U'(v)\sigma^2(v)} = \int_0^x \frac{2}{\sigma^2(v)}\exp\left\{\int_0^v \frac{2\mu(y)}{\sigma^2(y)}dy\right\}dv$$

and also

$$G_x(z) := \begin{cases} U(z)m(z) + \int_z^x U(v)m(dv), & z \le x, \\ U(z)m(x), & z > x. \end{cases}$$

We have

$$\frac{d}{dz}G_x(z) = \begin{cases} U'(z)m(z), & z \le x, \\ U'(z)m(x), & z > x \end{cases}$$

and

$$\frac{d^2}{dz^2}G_x(z) = \begin{cases} U''(z)m(z) - 2/\sigma^2(z), & z \le x, \\ U''(z)m(x), & z > x, \end{cases}$$

where we consider the left second derivative at $z = x$, which, together with (1.32), implies that

$$AG_x(z) = \begin{cases} -1, & z \le x, \\ 0, & z > x, \end{cases}$$

and hence the function

$$f(z) = G_{h,x}(z) := G_{x+h}(z) - G_x(z) \tag{1.34}$$

solves (1.33).

Alternatively, one can notice that $U(x)$ is the scale function, that $m(x)$ corresponds to the speed measure, and that (see [135, Chapter VII, Theorem 3.12])

$$AG_x(z) = \frac{d}{dm(z)}\left(\frac{dG_x(z)}{-dU(z)}\right).$$

Thus, it follows from (1.34) that, for $y < x$,

$$H_y(x, x+h] = f(y) = \int_x^{x+h} U(v)m(dv) = \int_x^{x+h} \frac{2U(v)dv}{-U'(v)\sigma^2(v)}.$$

More formally one can obtain the last equality from Corollary 3.8 and Exercise 3.20 in [135, Chapter VII.3].

If the function $W(v) := U(v)/U'(v)\sigma^2(v)$ is long-tailed at infinity, see Definition 1.26, then we get the following local renewal theorem for a process $X(t)$ starting at y:

$$H_y(x, x+h] \sim \frac{2U(x)}{-U'(x)\sigma^2(x)}h \quad \text{as } x \to \infty.$$

Assume that

$$\frac{2\mu(x)}{\sigma^2(x)} \sim r(x) \quad \text{as } x \to \infty, \tag{1.35}$$

for some differentiable function $r(x)$ such that the quotient $r'(x)/r^2(x)$ has a limit at infinity. Hence, we can apply L'Hôpital's rule and the equality $U'' = -rU'$ to obtain

$$\lim_{x\to\infty} \frac{U(x)}{-U'(x)/r(x)} = \lim_{x\to\infty} \frac{U'(x)}{-U''(x)/r(x) + U'(x)r'(x)/r^2(x)}$$
$$= \frac{1}{1+\lim_{x\to\infty} r'(x)/r^2(x)}.$$

Therefore, for any fixed $h > 0$,

$$H_y(x, x+h] \sim \frac{2}{\sigma^2(x)r(x)} \frac{1}{1+\lim_{y\to\infty} r'(y)/r^2(y)} h \quad \text{as } x \to \infty.$$

Example 1.46 If $\mu(x) \sim \mu/x$ and $\sigma^2(x) \to \sigma^2 > 0$ as $x \to \infty$ with $2\mu > \sigma^2$ then (1.35) is satisfied by $r(x) = 2\mu/\sigma^2 x$, $r'(x)/r^2(x) \to -\sigma^2/2\mu$, and we obtain

$$H_y(x, x+h] \sim \frac{2h}{2\mu - \sigma^2} x \quad \text{as } x \to \infty.$$

Example 1.47 If $\mu(x) \sim \mu/x^\alpha$, $\mu > 0$, $\alpha \in (0,1)$, and $\sigma^2(x) \to \sigma^2 > 0$ as $x \to \infty$ then (1.35) is satisfied by $r(x) = 2\mu/\sigma^2 x^\alpha$, $r'(x)/r^2(x) \to 0$, and we find that

$$H_y(x, x+h] \sim \frac{h}{\mu} x^\alpha \sim \frac{h}{\mu(x)} \quad \text{as } x \to \infty.$$

Note that this asymptotic behaviour of the renewal function does not depend on the diffusion coefficient, as for a process with constant positive drift.

1.5.3 Bessel Processes

A Bessel process is an important example of a diffusion process with asymptotically zero drift whose various probabilistic characteristics can be calculated in closed form; this provides some intuition for what can be expected for Markov chains. The simplest version of a Bessel process is defined as the Euclidean norm $\|B^{(d)}(t)\|$ of a d-dimensional Brownian motion $B^{(d)}(t)$, and it solves the stochastic differential equation

$$dX(t) = dY(t) + \frac{d-1}{2}\frac{dt}{X(t)} = dY(t) + \frac{2\nu+1}{2}\frac{dt}{X(t)}, \tag{1.36}$$

where the process $Y(t)$ is a one-dimensional Brownian motion. The parameter $\nu = (d-2)/2$ is called the *index* of X. By the same stochastic differential equation we define a Bessel process with an arbitrary index $\nu \in \mathbb{R}$. A Bessel process with a non-integer dimension naturally appears as the norm of a multi-dimensional Brownian motion in a cone, and the dimension is determined by the cone's geometry; see Corollary 3 in [54] and its proof.

In other words, X is a diffusion process with drift $(2\nu+1)/2x$ and diffusion coefficient 1. The intrinsic property of a Bessel process is that its drift is singular at the origin, which makes it impossible to apply the results of the last subsection.

The drift of the squared Bessel process $X^2(t)$ at any state equals $2\nu+2$, and this gives rise to the following classification; see e.g. [21, Appendix 1.21].

- If $\nu > 0$ then the process $\{X(t)\}$ is transient and there is a unique strong solution to equation (1.36). The case where the index $\nu = 0$ corresponds to the process $\sqrt{B_1^2 + B_2^2}$, which is null recurrent but the origin is never visited; hence there is again a unique strong solution to the equation (1.36).
- If $-1 \le \nu < 0$ then the hitting time of the origin from any state $x > 0$ is finite with probability 1 and has infinite mean. In the case $-1 < \nu < 0$, the origin is a repelling (instantaneously reflecting) state for X, so there is a weak solution to (1.36) which is not unique. In the case where the index is -1, the origin is an absorbing state.
- If $\nu < -1$ then the hitting time of the origin from any state $x > 0$ has finite mean $x^2/|2\nu + 2|$ and the origin is an absorbing state for $\{X(t)\}$, so there is no weak solution to (1.36).

In the first case, where $\nu \ge 0$, the transition density of $\{X(t)\}$ is well known, see e.g. [21, Appendix 1.21], and is given by the equalities

$$
\begin{aligned}
p_t(x,y) &= \frac{1}{t}\frac{y^{\nu+1}}{x^{\nu}} e^{-(x^2+y^2)/2t} I_\nu(xy/t),\\
p_t(0,y) &= \frac{y^{2\nu+1}}{2^\nu t^{\nu+1}\Gamma(\nu+1)} e^{-y^2/2t},
\end{aligned} \tag{1.37}
$$

where $I_\nu(z)$ is a modified Bessel function. The same formula is still valid for $\nu \in (-1,0)$ if we reflect the process $\{X(t)\}$ each time it reaches the origin.

In the positive recurrent case $\nu < -1$ or in the null recurrent case $\nu \in (-1,0)$, if we kill the process at 0 then the transition probability density function of $\{X(t)\}$ is

$$
p_t(x,y) = \frac{1}{t}\frac{y^{\nu+1}}{x^{\nu}} e^{-(x^2+y^2)/2t} I_{|\nu|}(xy/t).
$$

If $\nu \ge 0$ or $\nu \in (-1,0)$ and the process $\{X(t)\}$ is reflected each time it reaches the origin, the probability density function of $X(t)$ given $X(0) = 0$ is given by

$$p_t(x) \;=\; p_t(0,x) = \frac{1}{2^\nu \Gamma(\nu+1)} \frac{x^{2\nu+1}}{t^{\nu+1}} e^{-x^2/2t}. \tag{1.38}$$

In both cases the probability density function of $X^2(t)/t$ is

$$\frac{1}{2^{\nu+1}\Gamma(\nu+1)} x^\nu e^{-x/2},$$

which is a Gamma density function with mean $2(\nu+1)$ and variance $4(\nu+1)$.

In the transient case $\nu > 0$ we can write down the Green function h_0 of $\{X(t)\}$ in closed form by integrating (1.38):

$$h_0(y) = \int_0^\infty p_t(0,y)dt = \frac{y^{2\nu+1}}{2^\nu \Gamma(\nu+1)} \int_0^\infty \frac{1}{t^{\nu+1}} e^{-y^2/2t} dt \;=\; \frac{y}{\nu},$$

which indicates the asymptotic behaviour of the renewal measure that we can expect for transient Markov chains with drift of order c/x at infinity; see Section 4.8 for results in this area.

It follows from the representation of the α-potential density G_α of X in [21, Appendix 1.21] that, for all $x \geq 0$,

$$h_x(y) = \int_0^\infty p_t(x,y)dt = \frac{1}{\nu} \frac{y^{2\nu+1}}{\max(x,y)^{2\nu}},$$

which implies that the first hitting time $\tau_{[0,y]}$ of the compact set $[0,y]$ is finite with probability

$$\begin{aligned} \mathbb{P}_x\{\tau_{[0,y]} < \infty\} &= \mathbb{P}_x\{X(t) = y \text{ for some } t\} \\ &= \frac{h_x(y)}{h_y(y)} \;=\; \Big(\frac{y}{x}\Big)^{2\nu} \quad \text{for } x > y; \end{aligned} \tag{1.39}$$

this kind of result for transient Markov chains is discussed in Chapter 3.

For any ν, the function $h(x) = x^{-2\nu}$ is harmonic for $\{X(t)\}$ as it solves the equation

$$\Big(\frac{1}{2}\frac{d^2}{dx^2} + \frac{2\nu+1}{2x}\frac{d}{dx}\Big)h(x) = 0.$$

By Itô's formula, the process $\{h(X(t))\}$ is a local martingale. Let $y > 0$. If $\nu > 0$ then $h(x)$ is bounded on $[y,\infty)$ and if $\nu < 0$ then it is bounded on $[0,y]$. So, in either case we can apply the optional stopping time theorem (see e.g. [56, Section 4.7]) for martingales and conclude that, for $\nu > 0$ and $x > y$,

$$\begin{aligned} h(x) &= h(y)\mathbb{P}_x\{X(t) = y \text{ for some } t\} + h(\infty)\mathbb{P}_x\{X(t) \neq y \text{ for all } t\} \\ &= h(y)\mathbb{P}_x\{\tau_{[0,y]} < \infty\}, \end{aligned}$$

which agrees with (1.39).

If $\nu \le -1$, which corresponds to the origin being an absorbing state, then, for $x < y$,

$$\begin{aligned} h(x) &= h(y)\mathbb{P}_x\{X(t) = y \text{ for some } t\} + h(0)\mathbb{P}_x\{X(t) \ne y \text{ for all } t\} \\ &= h(y)\mathbb{P}_x\Big\{\sup_{t\ge 0} X(t) \ge y\Big\}, \end{aligned}$$

which implies that

$$\mathbb{P}_x\Big\{\sup_{t\ge 0} X(t) \ge y\Big\} = \frac{h(x)}{h(y)} = \Big(\frac{x}{y}\Big)^{2|\nu|}.$$

For recurrent Markov chains, the tail distribution of the trajectory supremum until the time of the first entry to a neighborhood of the origin is described in Theorem 8.26.

In conclusion, let us establish a link to Markov chains by sampling the process $\{X(t)\}$ at integer time epochs and getting a Markov chain $X_n := X(n)$ in this way; in the null recurrent case we assume a reflecting boundary condition. This Markov chain is of Lamperti type with mean drift $m_1(x)$ and second moment of jumps $m_2(x)$ satisfying the relations

$$m_1(x) \sim \frac{\nu + 1/2}{x} =: \frac{c}{x} \quad \text{and} \quad m_2(x) \to 1 \quad \text{as } x \to \infty. \tag{1.40}$$

Indeed, it follows from (1.37) that

$$\begin{aligned} \mathbb{E}_x X(1) &= \int_0^\infty \frac{y^{\nu+2}}{x^\nu} e^{-(x^2+y^2)/2} I_\nu(xy)dy \\ &= \frac{e^{-x^2/2}}{x^\nu} \int_0^\infty y^{\nu+2} e^{-y^2/2} I_\nu(xy)dy \\ &= \frac{e^{-x^2/2}}{x^\nu} \frac{\Gamma(\nu+3/2)}{\frac{x}{2}\Gamma(\nu+1)} e^{x^2/4} 2^{\nu/2} M_{-\nu/2-1,\nu/2}(x^2/2), \end{aligned}$$

where $M_\cdot(\cdot)$ is the Whittaker function; see [74, Formula 6.643(2)]. As $x \to \infty$,

$$M_{-\nu/2-1,\nu/2}(x^2/2) = \frac{\Gamma(\nu+1)}{\Gamma(\nu+3/2)} e^{x^2/4} (x^2/2)^{\nu/2+1} \Big(1 + \frac{2\nu+1}{2x^2} + O(1/x^4)\Big),$$

which gives

$$\mathbb{E}_x X(1) = x\Big(1 + \frac{2\nu+1}{2x^2} + O(1/x^4)\Big) \quad \text{as } x \to \infty;$$

this in turn yields the first relation in (1.40). In a similar way we conclude that the asymptotic behaviour of the higher moments of jumps, for any fixed $j \ge 1$, is given by

$$\mathbb{E}_x X^{2j}(1) = x^{2j} + 2j(\nu+j)x^{2j-2} + O(x^{2j-4}) \quad \text{as } x \to \infty. \tag{1.41}$$

Choosing $j = 1$ and using the formula for the first moment of $X(1)$, one obtains the second convergence in (1.40).

If the Bessel process $\{X(t)\}$ is transient or null recurrent, that is, if $\nu > -1$, then it follows from the distribution property of the Bessel process $\{X(t)\}$ discussed above that, for all n, X_n^2/n has a Γ-distribution with mean $2(\nu+1)$ and variance $4(\nu+1)$. In Sections 4.5 and 4.6 we discuss the convergence of X_n^2/n to a Γ-distribution for a general transient or null recurrent Markov chain with asymptotic drift of order c/x.

1.6 General Approach and Plan of the Book

One of the most popular examples of Markov chains with asymptotically zero drift is a driftless random walk conditioned to stay positive. This process is an h-transform of a random walk killed at leaving $\mathbb{R}^+$. If the second moment of the original random walk is finite then the transformed process has a drift of order $1/x$, that is, $xm_1(x) \to c_1 > 0$. But the second moment of the transformed process is finite if and only if the third moment of the original walk is finite; see the calculations in Section 11.1. Therefore, Lamperti's criterion for transience is not always applicable to this type of chain.

This observation motivates us to look for appropriate conditions for transience, null recurrence, and positive recurrence in terms of the truncated moments and tail probabilities of the jumps $\xi(x)$. For any $s > 0$ we denote the s-truncation of the kth moment of the jump at state x by

$$m_k^{[s]}(x) := \mathbb{E}\{\xi^k(x);\ |\xi(x)| \le s\}.$$

Another reason for considering truncated moments comes from the case where the drift function decays more slowly than $1/x$, say as $1/x^\beta$ with β between 0 and 1. In that case it is not practical to assume the boundedness or even the existence of a full second moment of jumps, whereas an appropriate restriction on the growth of a truncated second moment is a rational approach; see e.g. Section 5.1.

In Chapter 2 we introduce a classification of Markov chains with asymptotically zero drift, which relies on the relation between $m_1^{[s(x)]}$ and $m_2^{[s(x)]}$. Additional assumptions are expressed in terms of truncated moments of higher order and the tail probabilities of jumps. Another, more important, contrast with previous results on recurrence and transience is the fact that we do not use concrete Lyapunov test functions (such as x^2, $\log^a x$ or $x^2 \log x \log\log x$). Instead, we construct an abstract Lyapunov function that is motivated by the harmonic function of a diffusion process with drift $m_1(x)$ and diffusion coefficient $m_2(x)$; see Section 1.5 above.

The asymptotic behaviour of transient Markov chains and the tail analysis of recurrent Markov chains is discussed in Chapters 3–6 and 8–9 respectively. In Chapter 7, motivated by the exponential change-of-measure approach suggested by Cramér in the 1920s for the study of large deviations of sums of independent random variables in the context of risk processes, we suggest the following general strategy for the study of positive recurrent Markov chains with asymptotically zero drift.

- First, apply an appropriate Doob's h-transform to a process $\{X_n\}$ killed at its time of entry to the half-line $(-\infty, \widehat{x}]$ for some $\widehat{x} \in \mathbb{R}$, in order to change the sign of the drift from negative to positive so that we get a transition kernel that generates a transient embedded Markov chain. With necessity an appropriate change of measure is generated by a subexponential function, either regularly varying or Weibullian-type at infinity.
- Second, apply the limit results to the transient Markov chain that is obtained.
- Third, apply an inverse change of measure that makes it possible to identify the tail and local asymptotics of both the stationary and pre-stationary distributions of the original positive recurrent Markov chain.

In Chapter 10 we show that our approach also works for Markov chains with asymptotically negative drift bounded away from zero. We consider Markov chains that are asymptotically homogeneous in space, that is, Markov chains with jumps satisfying $\xi(x) \Rightarrow \xi$ as $x \to \infty$. This means that far from the origin one can approximate $\{X_n\}$ by a random walk, which makes it natural to apply an exponential change of measure in a similar way to how this is done for sums of independent random variables. We study the tail asymptotic behaviour of the stationary and pre-stationary distributions of $\{X_n\}$ in the case where the limiting random variable ξ has negative mean and satisfies the Cramér condition. It turns out that the tail behaviour of these distributions depends on the rate of convergence of $\xi(x)$ to ξ.

In Chapter 11 we consider some important applications of our results. Processes with asymptotically zero drift appear naturally in various stochastic models such as random billiards, see Menshikov et al. [125], and random polymers, see Alexander [5], Alexander and Zygouras [6], and De Coninck et al. [41]).

Such chains appear when we study critical and near-critical branching processes. In critical branching processes one typically observes a linearly growing second moment of jumps, but, on considering the square root of the process, one obtains bounded second moments and a drift that decreases to zero. Thus we can apply our theorems to this transformation. As a result we get limit theorems for population-size-dependent processes with migration of particles. To the best of our knowledge, there is no paper in the literature in which a combination of size dependence and migration has been considered.

We have also found that processes with asymptotically zero drift can be used in the study of risk processes with a reserve-dependent premium rate. More precisely, we have derived upper and lower bounds for ruin probabilities in the case when the premium rate approaches from above – as the risk reserve grows – the critical value for a model with constant rate.

Besides these two main examples we consider also random walks conditioned to stay positive and reflected random walks.

2

Lyapunov Functions and Classification of Markov Chains

As one can see from the results for diffusion processes in Section 1.5, their classification relies heavily on the asymptotic behaviour of the ratio $2m_1(x)/m_2(x)$ at infinity. Roughly speaking:

- if, for some $\varepsilon > 0$, $2m_1(x)/m_2(x) \le -(1+\varepsilon)/x$ for all sufficiently large x then some neighborhood of zero is positive recurrent;
- if, for some $\varepsilon > 0$, $2m_1(x)/m_2(x) \le (1-\varepsilon)/x$ for all sufficiently large x then some neighborhood of zero is recurrent;
- if, for some $\varepsilon > 0$, $2m_1(x)/m_2(x) \ge (1+\varepsilon)/x$ for all sufficiently large x then any compact set is transient.

For diffusion processes, the necessary and sufficient conditions for positive recurrence or recurrence or transience involving the ratio $2m_1(x)/m_2(x)$ are available; see (1.23)–(1.25). For Markov chains, similar necessary and sufficient conditions in terms of the ratio $2m_1(x)/m_2(x)$ are not as readily available as they are for diffusion processes.

In this chapter we introduce criteria for the transience, recurrence, and positive recurrence of discrete-time Markov chains. We do this by constructing Lyapunov functions that depend on the ratio of the truncated moments of the chain, which are motivated by functions (1.23)–(1.25). Let us recall the standard sufficient conditions for positive recurrence, recurrence, and transience in terms of test functions.

Theorem 2.1 ([126, Theorem 11.0.1]) *Let $L(x)$ be a non-negative test function such that, for some x_* and $\varepsilon > 0$,*

$$\mathbb{E}\{L(X_1) - L(x) \mid X_0 = x\} \le -\varepsilon \quad \textit{for all } x > x_*, \tag{2.1}$$

and let

$$\mathbb{E}\{L(X_1) \mid X_0 = x\} < \infty \quad \textit{for all } x \le x_*. \tag{2.2}$$

Then the set $(-\infty, x_*]$ *is positive recurrent.*

Theorem 2.2 ([126, Theorem 8.0.2]) *Let* $L(x)$ *be a non-negative test function that is unbounded at infinity and is such that, for some* x_*,

$$\mathbb{E}\{L(X_1) - L(x) \mid X_0 = x\} \le 0 \quad \textit{for all } x > x_*. \tag{2.3}$$

Then the set $(-\infty, x_*]$ *is recurrent.*

Theorem 2.3 ([126, Theorem 8.0.2]) *Let* $L(x)$ *be a non-negative bounded decreasing test function such that, for some* x_*, $L(x_*) > L(x)$ *for some* $x > x_*$ *and*

$$\mathbb{E}\{L(X_1) - L(x) \mid X_0 = x\} \le 0 \quad \textit{for all } x > x_*. \tag{2.4}$$

Then the set $(-\infty, x_*]$ *is transient.*

Here we refer to the book by Meyn and Tweedie [126]; for some bibliographical notes see the comments for this chapter.

2.1 Reference Drift Function

In this chapter, $r(x) > 0$ is a reference drift function. It is always assumed to be a decreasing continuous function that is non-integrable at infinity, that is, for $x \ge 0$,

$$R(x) := \int_0^x r(y)dy \to \infty \quad \text{as } x \to \infty; \tag{2.5}$$

in what follow we define $R(x) = 0$ for $x < 0$. The function $R(x)$ is concave on the positive half-line because $r(x)$ is assumed to be decreasing. Therefore, for all $h > -xr(x)$,

$$\begin{aligned} R(x + h/r(x)) &\le R(x) + R'(x)h/r(x) \\ &= R(x) + h. \end{aligned} \tag{2.6}$$

If, in addition, $r(x)$ is differentiable and, for some $c > 0$,

$$0 \ge r'(x) \ge -cr^2(x) \quad \text{for all } x \ge 0, \tag{2.7}$$

then, for all $x \ge 0$ and $h > 0$,

$$\begin{aligned}\frac{1}{r(x)} - \frac{1}{r(x+h/r(x))} &= \int_x^{x+h/r(x)} \frac{r'(y)}{r^2(y)}dy \\ &\ge -c\int_x^{x+h/r(x)} dy = -c\frac{h}{r(x)}.\end{aligned}$$

Therefore

$$r(x+h/r(x)) \ge \frac{r(x)}{1+ch}, \quad h > 0. \tag{2.8}$$

Similarly, for $h \in (0, 1/c)$ and x such that $x - h/r(x) \ge 0$,

$$r(x-h/r(x)) \le \frac{r(x)}{1-ch}. \tag{2.9}$$

The lower bound (2.8) implies that, for all $h > 0$,

$$\begin{aligned}R(x+h/r(x)) &= R(x) + \int_x^{x+h/r(x)} r(y)dy \\ &\ge R(x) + \frac{h}{r(x)}r(x+h/r(x)) \\ &\ge R(x) + \frac{h}{1+ch}.\end{aligned} \tag{2.10}$$

Together with the upper bound (2.6) this gives the two-sided bound

$$R(x) + \frac{h}{1+ch} \le R(x+h/r(x)) \le R(x)+h \quad \text{for all } h > 0. \tag{2.11}$$

Similarly,

$$R(x) - \frac{h}{1-ch} \le R(x-h/r(x)) \le R(x)-h, \tag{2.12}$$

where the first inequality is valid for $h \in (0, 1/c)$ and the second is valid for $h \in (0, xr(x))$.

So, $1/r(x)$ is a natural x-step responsible for a constant increase in the function $R(x)$. Moreover, (2.11) and (2.12) imply that, for any increasing function $s(x)$ of order $o(1/r(x))$,

$$R(x \pm s(x)) = R(x) + o(1) \quad \text{as } x \to \infty. \tag{2.13}$$

Notice also that (2.8) and (2.9) yield a similar relation for $r(x)$,

$$r(x \pm s(x)) \sim r(x) \quad \text{as } x \to \infty. \tag{2.14}$$

2.2 Positive Recurrence

2.2.1 Positive Recurrence Criteria Motivated by Diffusion Processes

In this section we will be interested in the sufficient conditions under which the set $(-\infty, x_*]$ is positive recurrent for some x_*, that is, $\mathbb{E}_x \tau_{(-\infty,x_*]} < \infty$ for all $x \le x_*$.

The conditions below are formulated in terms of truncated moments of jumps,

$$m_k^{[s]}(x) := \mathbb{E}\{\xi^k(x);\ |\xi(x)| \le s\}.$$

Let x_0 be such that

$$\frac{2m_1^{[x]}(x)}{m_2^{[x]}(x)} \le -r(x) \quad \text{for all } x > x_0. \tag{2.15}$$

For $r(x)$ decreasing not too fast – roughly speaking, if $r(x) > 1/x$ – this means that the drift towards the origin dominates the diffusion, and the corresponding Markov chain $\{X_n\}$ is positive recurrent.

In the theorem below it is shown that – similarly to diffusion processes – the chain $\{X_n\}$ is positive recurrent provided that:

$$\text{The function} \quad \frac{1}{b(x)}e^{-R(x)} = \frac{1}{b(x)}e^{-\int_0^x r(y)dy} \quad \text{is integrable,} \tag{2.16}$$

where $b(x) > 0$ is a differentiable function such that

$$\liminf_{x\to\infty} \frac{m_2^{[x]}(x)}{b(x)} > 0. \tag{2.17}$$

For Markov chains, we also need to impose some technical conditions on $r(x)$ and on the function

$$W(x) := e^{R(x)} \int_x^\infty \frac{1}{b(y)} e^{-R(y)} dy,$$

which is a well-defined function owing to (2.16).

In the next theorem sufficient conditions are given that guarantee that the test function

$$L(x) := \int_0^x W(y)dy, \quad x > 0, \tag{2.18}$$

with $L(x) = 0$ on $\mathbb{R}^-$, is appropriate for the application of Theorem 2.1. In particular, it agrees with the case $r(x) \equiv \varepsilon > 0$, where the most natural choice of the test function is a linear one, and with the case $r(x) = c/x$, where the most effective test function is x^2.

Theorem 2.4 *Let the drift condition (2.15) hold with some decreasing function $r(x) > 0$ such that the conditions (2.16) and (2.17) are satisfied and*

$$\mathbb{E}\{\xi(x)W(\xi(x));\ \xi(x) > 0\} < \infty \quad \textit{for all } x. \tag{2.19}$$

Let the following integrability conditions on positive jumps hold:

$$\mathbb{E}\{\xi^3(x);\ \xi(x) \in (0,x]\} = o(x^2/W(x)), \tag{2.20}$$

$$\mathbb{E}\{\xi(x)W(\xi(x));\ \xi(x) > x\} \to 0 \quad \textit{as } x \to \infty. \tag{2.21}$$

Assume that the function $W(x)$ is increasing and convex and satisfies the following conditions, for some constants c_1, c_2,

$$W(2x) \le c_1 W(x) \quad \textit{for all } x > 0, \tag{2.22}$$

$$|W'(x+y) - W'(x)| \le c_2 \frac{W(x)}{x^2}|y| \quad \textit{for all } x > 0,\ y \in [-x/2, x]. \tag{2.23}$$

Then there exists an x_* such that the set $(-\infty, x_*]$ is positive recurrent.

The conditions (2.20) and (2.21) are fulfilled if, for example, the function $x^2/W(x)$ increases and

$$\text{the family } \left\{\xi^+(x)W(\xi^+(x)),\ x \ge 0\right\} \text{ is uniformly integrable.} \tag{2.24}$$

The justification follows from Lemmas 2.24 and 2.26 below.

Corollary 2.5 *For some $\varepsilon > 0$ and $x_0 > 0$, let*

$$\frac{2m_1^{[x]}(x)}{m_2^{[x]}(x)} \le -\frac{1+\varepsilon}{x} \quad \textit{for all } x > x_0,$$

and let $\mathbb{E}\{\xi^2(x); \xi(x) > 0\} < \infty$ for all x. Let the truncated second moments of jumps $\mathbb{E}\{\xi^2(x); |\xi(x)| \le x\}$ be bounded away from zero,

$$\mathbb{E}\{\xi^3(x),\ \xi(x) \in [0,x]\} = o(x), \tag{2.25}$$

$$\mathbb{E}\{\xi^2(x),\ \xi(x) > x\} \to 0 \quad \textit{as } x \to \infty. \tag{2.26}$$

Then there exists an x_* such that the set $(-\infty, x_*]$ is positive recurrent.

Notice that both (2.25) and (2.26) hold provided that the family of random variables $\{(\xi^+(x))^2,\ x > 0\}$ is uniformly integrable.

Proof of Corollary 2.5. It follows from Theorem 2.4 if we take $r(x) = \frac{1+\varepsilon}{1+x}$ for $x > 0$ and $b(x) = 1$ then

$$R(x) = (1+\varepsilon)\log(1+x),$$

$$e^{-R(x)} = 1/(1+x)^{1+\varepsilon},$$

$$W(x) = (1+x)/\varepsilon.$$

This leads to the test function $L(x) = ((1+x)^2 - 1)/2\varepsilon$ for $x > 0$. Note in passing that $L(x) = x^2\mathbb{I}\{x > 0\}$ is also an appropriate test function. □

The last corollary relates to a quadratic Lyapunov function; its assumptions on jumps are too restrictive compared with the classical Lamperti criterion, which guarantees positive recurrence of the set $(-\infty, x_0]$ under the condition $2xm_1(x) + m_2(x) \le -\varepsilon$ for $x > x_0$ only. On the other hand, Corollary 2.5 imposes no conditions on the left tail distribution of $\xi(x)$ below the level $-x$.

Let $\log_{(m)} x$ denote the mth iteration of the logarithm of x:

$$\log_{(m)} x = \log\log_{(m-1)} x.$$

Corollary 2.6 *For some $m \in \mathbb{N}$ and $\varepsilon > 0$, let*

$$\frac{2m_1^{[x]}(x)}{m_2^{[x]}(x)} \le -\frac{1}{x} - \frac{1}{x\log x} - \cdots - \frac{1}{x\log x\cdots\log_{(m-1)} x} - \frac{1+\varepsilon}{x\log x\cdots\log_{(m)} x}$$

for all sufficiently large x; the denominators consist of products of x and iterated logarithms. Let

$$\mathbb{E}\{\xi^2(x)\log\xi(x)\cdots\log_{(m)}\xi(x);\ \log_{(m)}\xi(x) > 0\} < \infty \quad \textit{for all } x.$$

Let the truncated second moment $\mathbb{E}\{\xi^2(x); |\xi(x)| \le x\}$ be bounded away from zero, let

$$\mathbb{E}\{\xi(x)^3;\ \xi(x) \in [0,x]\} = o\Big(\frac{x}{\log x\cdots\log_{(m)} x}\Big), \tag{2.27}$$

and let

$$\mathbb{E}\{\xi^2(x)\log\xi(x)\cdots\log_{(m)}\xi(x);\ \xi(x) > x\} \to 0. \tag{2.28}$$

Then there exists an x_* such that the set $(-\infty, x_*]$ is positive recurrent.

Note that both (2.27) and (2.28) hold provided that the family

$$\{\xi^2(x)\log\xi(x)\cdots\log_{(m)}\xi(x)\mathbb{I}\{\log_{(m)}\xi(x) > 0\},\ x > 0\}$$

is uniformly integrable.

Proof of Corollary 2.6. Let $x = e^{(m)}$ be a solution to the equation $\log_{(m)} x = 1$. Consider (see (2.15), (2.16))

$$r(x) := \Big(\frac{1}{y} + \frac{1}{y\log y} + \cdots + \frac{1}{y\log y\cdots\log_{(m-1)} y} + \frac{1+\varepsilon}{y\log y\cdots\log_{(m)} y}\Big)\Big|_{y=e^{(m)}+x}$$

and $b(x) = 1$; then

$$R(x) = \Big(\log y + \log\log y + \cdots + \log_{(m)} y + (1+\varepsilon)\log_{(m+1)} y\Big)\Big|_{y=e^{(m)}+x} - \big(e^{(m-1)} + e^{(m-2)} + \cdots + 1\big),$$

$$e^{-R(x)} = \frac{e^{(m)}e^{(m-1)}\cdots 1}{y\,\log y \cdots \log_{(m-1)} y\, \log_{(m)}^{1+\varepsilon} y}\Big|_{y=e^{(m)}+x},$$

$$W(x) = \frac{1}{\varepsilon} y \log y \cdots \log_{(m-1)} y\, \log_{(m)} y\Big|_{y=e^{(m)}+x},$$

$$L(x) \sim \frac{1}{2\varepsilon} x^2 \log x \cdots \log_{(m-1)} x\, \log_{(m)} x.$$

□

The next corollary deals with the case when the second moment of jumps vanishes at infinity.

Corollary 2.7 *For some $\alpha > 0$, $c_1, c_2 > 0$, and $x_0 > 0$, let*

$$m_1^{[x]}(x) \le -c_1/x^{1+\alpha} \quad \text{for all } x > x_0,$$
$$m_2^{[x]}(x) \sim c_2/x^{\alpha} \quad \text{as } x \to \infty,$$
$$\mathbb{E}\{\xi^{2+\alpha}(x);\ \xi(x) > 0\} < \infty \quad \text{for all } x.$$

Let

$$\mathbb{E}\{\xi^3(x);\ \xi(x) \in [0,x]\} = o(x^{1-\alpha}), \tag{2.29}$$
$$\mathbb{E}\{\xi^{2+\alpha}(x);\ \xi(x) > x\} \to 0 \quad \text{as } x \to \infty. \tag{2.30}$$

If $2c_1/c_2 > 1+\alpha$ then there exists an x_* such that the set $(-\infty, x_*]$ is positive recurrent.

In the case $\alpha \in (0,1)$, both (2.29) and (2.30) hold provided that the family of random variables $\{(\xi^+(x))^{2+\alpha},\ x > 0\}$ is uniformly integrable.

Proof of Corollary 2.7. The corollary follows if we take $c \in (1+\alpha, 2c_1/c_2)$, $r(x) = \frac{c}{1+x}$ for $x > 0$, and $b(x) = 1/(1+x)^\alpha$; then $R(x) = c\log(1+x)$, $e^{-R(x)} = 1/(1+x)^c$,

$$W(x) = (1+x)^c \int_x^\infty \frac{(1+y)^\alpha}{(1+y)^c} dy = \frac{(1+x)^{\alpha+1}}{\alpha - c + 1},$$

and

$$L(x) = \frac{(1+x)^{2+\alpha} - 1}{(\alpha - c + 1)(2+\alpha)}.$$

□

The advantage of Theorem 2.4 is that it covers all functions considered in the corollaries above in a unified way; the main condition (2.16) is motivated by the existence condition (1.23) for the stationary density of a diffusion process. But, at the same time, this link to diffusion processes results in the need for finite second moments, which is natural in Corollaries 2.5 and 2.6, while there are other examples where to require the existence of the second moments of jumps is clearly excessive. In the next subsection we discuss amended moment conditions for drifts such as $-1/x^{\alpha}$, $0 < \alpha < 1$, which may be characterised by the convergence $xm_1(x) \to -\infty$ as $x \to \infty$.

Proof of Theorem 2.4. We consider the test function (2.18), for which we need to show (2.1) and (2.2). Since $W(x)$ is assumed to be increasing,

$$L(x) \le xW(x) \quad \text{for all } x > 0; \tag{2.31}$$

hence (2.2) follows from the condition (2.19), and it remains to show (2.1). By construction, $L'(x) = W(x)$ and

$$L''(x) = W'(x) \ = \ r(x)W(x) - 1/b(x). \tag{2.32}$$

Let us prove that the mean drift of $L(x)$ is negative and bounded away from zero for all sufficiently large x. First we analyse the Taylor expansion for the function L, with the Lagrange form of the remainder, here $x, x+y > 0$:

$$\begin{aligned} L(x+y) - L(x) &= L'(x)y + L''(x+\theta y)\frac{y^2}{2} \\ &= W(x)y + W'(x+\theta y)\frac{y^2}{2}, \end{aligned} \tag{2.33}$$

where $0 \le \theta = \theta(x,y) \le 1$. Since $W(x)$ is assumed convex, W' is increasing; hence, for all $y \in [-x, 0]$,

$$\begin{aligned} L(x+y) - L(x) &\le W(x)y + W'(x)\frac{y^2}{2} \\ &= W(x)y + r(x)W(x)\frac{y^2}{2} - \frac{y^2}{2b(x)}, \end{aligned} \tag{2.34}$$

as follows from (2.32). Next, by condition (2.23), for $y \in [0, x]$, we have

$$W'(x+\theta y) \le W'(x) + c_2\frac{W(x)}{x^2}y. \tag{2.35}$$

Substituting this into (2.33) we obtain, for all $y \in [0, x]$,

$$L(x+y) - L(x) \le W(x)y + r(x)W(x)\frac{y^2}{2} - \frac{y^2}{2b(x)} + c_3\frac{W(x)}{x^2}y^3. \tag{2.36}$$

Using the fact that L is increasing and the inequalities (2.31) and (2.22), we deduce that

$$L(x+y) \le L(2y) \le 2yW(2y) \le 2c_1 yW(y) \quad \text{for all } y > x. \tag{2.37}$$

Now we are ready to bound the mean drift of $\{L(X_n)\}$. We start with the following upper bound:

$$\begin{aligned}\mathbb{E}L(x+\xi(x)) - L(x) &\le \mathbb{E}\{L(x+\xi(x)) - L(x);\ \xi(x) \ge -x\}\\ &\le \mathbb{E}\{L(x+\xi(x)) - L(x);\ \xi(x) \in [-x,0]\}\\ &\quad + \mathbb{E}\{L(x+\xi(x)) - L(x);\ \xi(x) \in [0,x]\}\\ &\quad + \mathbb{E}\{L(x+\xi(x));\ \xi(x) > x\}.\end{aligned} \tag{2.38}$$

It follows from (2.34) that

$$\begin{aligned}&\mathbb{E}\{L(x+\xi(x)) - L(x);\ \xi(x) \in [-x,0]\}\\ &\quad \le W(x)\mathbb{E}\{\xi(x);\ \xi(x) \in [-x,0]\} + \frac{1}{2}r(x)W(x)\mathbb{E}\{\xi^2(x);\ \xi(x) \in [-x,0]\}\\ &\quad - \frac{1}{2b(x)}\mathbb{E}\{\xi^2(x);\ \xi(x) \in [-x,0]\}.\end{aligned} \tag{2.39}$$

It follows from (2.36) that

$$\begin{aligned}&\mathbb{E}\{L(x+\xi(x)) - L(x);\ \xi(x) \in [0,x]\}\\ &\quad \le W(x)\mathbb{E}\{\xi(x);\ \xi(x) \in [0,x]\} + \frac{1}{2}r(x)W(x)\mathbb{E}\{\xi^2(x);\ \xi(x) \in [0,x]\}\\ &\qquad - \frac{1}{2b(x)}\mathbb{E}\{\xi^2(x);\ \xi(x) \in [0,x]\} + c_3\frac{W(x)}{x^2}\mathbb{E}\{\xi^3(x);\ \xi(x) \in [0,x]\}\\ &\quad \le W(x)\mathbb{E}\{\xi(x);\ \xi(x) \in [0,x]\} + \frac{1}{2}r(x)W(x)\mathbb{E}\{\xi^2(x);\ \xi(x) \in [0,x]\}\\ &\qquad - \frac{1}{2b(x)}\mathbb{E}\{\xi^2(x);\ \xi(x) \in [0,x]\} + o(1) \quad \text{as } x \to \infty,\end{aligned} \tag{2.40}$$

owing to the condition (2.20). Finally, it follows from (2.37) by the condition (2.21) that

$$\begin{aligned}\mathbb{E}\{L(x+\xi(x));\ \xi(x) > x\} &\le 2c_1\mathbb{E}\{\xi(x)W(\xi(x));\ \xi(x) > x\}\\ &\to 0 \quad \text{as } x \to \infty.\end{aligned} \tag{2.41}$$

Substituting the upper bounds (2.39)–(2.41) into (2.38) we deduce that

$$\begin{aligned}
&\mathbb{E}\{L(x+\xi(x))-L(x)\}\\
&\quad\le W(x)\mathbb{E}\{\xi(x);\ |\xi(x)|\le x\}+\frac{1}{2}r(x)W(x)\mathbb{E}\{\xi^2(x);\ |\xi(x)|\le x\}\\
&\qquad-\frac{1}{2b(x)}\mathbb{E}\{\xi^2(x);\ |\xi(x)|\le x\}+o(1)\\
&\quad=W(x)\frac{m_2^{[x]}(x)}{2}\Big(\frac{2m_1^{[x]}(x)}{m_2^{[x]}(x)}+r(x)\Big)-\frac{1}{2b(x)}m_2^{[x]}(x)+o(1)\\
&\quad\le-\frac{1}{2b(x)}m_2^{[x]}(x)+o(1)\quad\text{as } x\to\infty,
\end{aligned}$$

owing to (2.15). Then (2.17) implies (2.1) for all sufficiently large x and the proof is complete. □

2.2.2 Non-Diffusive Positive Recurrence Criteria in the Case $xm_1(x)\to-\infty$

If the drift approaches the value of zero at a rate slower than $1/x$, say $1/x^\alpha$ with $\alpha\in(0,1)$, then it is possible to relax the positive recurrence conditions motivated by diffusion processes.

For some decreasing function $r(x)$ and $x_0>0$, let

$$m_1^{[x/2]}(x)\le -r(x)\quad\text{for all } x\ge x_0. \tag{2.42}$$

Define

$$W(x):=\int_0^x \min\Big(1,\ \frac{1}{yr(y)}\Big)dy,\quad x>0.$$

Let $xr(x)$ be increasing to infinity; then W is concave and

$$W(x)\ge\frac{1}{r(x)}\quad\text{for all sufficiently large } x. \tag{2.43}$$

Consider a test function L defined as $L(x)=0$ for all $x\le 0$ and

$$L(x):=\int_0^x W(y)dy\quad\text{for } x>0.$$

Since the second moment of jumps is not assumed finite, there is no diffusion-motivated intuition behind this test function.

Theorem 2.8 *Let the drift condition* (2.42) *hold for some decreasing function* $r(x)>0$ *such that* $xr(x)$ *is increasing to infinity. Assume that the jumps satisfy the following integrability conditions:*

$$\mathbb{E}\{\xi(x)W(\xi(x));\ \xi(x) > x/2\} \to 0, \tag{2.44}$$

$$\mathbb{E}\{\xi^2(x);\ |\xi(x)| \le x/2\} = o(xr(x)) \quad \text{as } x \to \infty. \tag{2.45}$$

Let

$$\mathbb{E}\{\xi(x)W(\xi(x));\ \xi(x) > 0\} < \infty \quad \text{for all } x. \tag{2.46}$$

Then there exists an x_* such that the set $(-\infty, x_*]$ is positive recurrent.

Owing to (2.43), conditions (2.44) and (2.45) are fulfilled if, for example:

$$\text{The family } \big\{|\xi(x)|W(|\xi(x)|),\ x \ge 0\big\} \text{ is uniformly integrable.} \tag{2.47}$$

A justification follows from Lemmas 2.24 and 2.26.

Corollary 2.9 *For some $\alpha \in (0,1)$, $\varepsilon > 0$ and $x_0 > 0$, let*

$$\mathbb{E}\{\xi(x);\ |\xi(x)| \le x/2\} \le -\varepsilon/x^\alpha \quad \text{for all } x > x_0.$$

Also, as $x \to \infty$, let

$$\mathbb{E}\{\xi^{1+\alpha}(x);\ \xi(x) > x/2\} \to 0, \tag{2.48}$$

$$\mathbb{E}\{\xi^2(x);\ |\xi(x)| \le x/2\} = o(x^{1-\alpha}), \tag{2.49}$$

and

$$\mathbb{E}\{\xi^{1+\alpha}(x);\ \xi(x) > 0\} < \infty \quad \text{for all } x. \tag{2.50}$$

Then there exists an x_* such that the set $(-\infty, x_*]$ is positive recurrent.

Both (2.49) and (2.48) hold provided that the family of random variables $\{|\xi(x)|^{1+\alpha},\ x > 0\}$ is uniformly integrable.

Proof of Corollary 2.9. The corollary follows if we take $r(x) = \varepsilon/(1+x)^\alpha$ for $x > 0$; then $W(x) \sim c_1 x^\alpha$ and $L(x) \sim c_2 x^{1+\alpha}$. □

Proof of Theorem 2.8. By construction, $L'(x) = W(x)$ and

$$L''(x) = W'(x) = \min\Big(1, \frac{1}{xr(x)}\Big) > 0 \quad \text{is decreasing;} \tag{2.51}$$

in particular, $W(x)$ is a concave function.

Since W is increasing, $L(x) \le xW(x)$ for $x > 0$; hence (2.2) follows from the concavity of W and the condition (2.46), and it remains to show that the mean drift of $L(x)$ is negative and bounded away from zero for all sufficiently large x. We start with the following upper bound:

$$\begin{aligned}
\mathbb{E}L(x+\xi(x))-L(x) &\le \mathbb{E}\{L(x+\xi(x))-L(x);\ \xi(x)\ge -x/2\}\\
&\le \mathbb{E}\{L(x+\xi(x))-L(x);\ |\xi(x)|\le x/2\}\\
&\quad + \mathbb{E}\{L(x+\xi(x));\ \xi(x)>x/2\}\\
&=: E_1(x)+E_2(x). \qquad (2.52)
\end{aligned}$$

Let us estimate the first term on the right-hand side via a Taylor expansion:

$$\begin{aligned}
&E_1(x)\\
&\quad = L'(x)\mathbb{E}\{\xi(x);\ |\xi(x)|\le x/2\}+\frac{1}{2}\mathbb{E}\{L''(x+\theta\xi(x))\xi^2(x);\ |\xi(x)|\le x/2\}\\
&\quad = W(x)\mathbb{E}\{\xi(x);\ |\xi(x)|\le x/2\}+\frac{1}{2}\mathbb{E}\{W'(x+\theta\xi(x))\xi^2(x);\ |\xi(x)|\le x/2\},
\end{aligned}$$

where $0\le\theta=\theta(x,\xi(x))\le 1$. Since W' is decreasing and

$$W'(x/2)=\frac{2}{xr(x/2)}\le\frac{2}{xr(x)},$$

we deduce that

$$\begin{aligned}
E_1(x) &\le W(x)\mathbb{E}\{\xi(x);\ |\xi(x)|\le x/2\}+\frac{1}{2}W'(x/2)\mathbb{E}\{\xi^2(x);\ |\xi(x)|\le x/2\}\\
&\le W(x)\mathbb{E}\{\xi(x);\ |\xi(x)|\le x/2\}+\frac{1}{xr(x)}\mathbb{E}\{\xi^2(x);\ |\xi(x)|\le x/2\}.
\end{aligned}$$

The condition (2.45) allows us to conclude that

$$E_1(x)\le W(x)\mathbb{E}\{\xi(x);\ |\xi(x)|\le x/2\}+o(1)\quad \text{as } x\to\infty. \qquad (2.53)$$

In order to estimate the second expectation on the right-hand side of (2.52), we first note that, since the function $\min(1,1/yr(y))$ is decreasing, we get

$$W(3x)\le 3W(x),$$

and therefore

$$L(3x)\le 9xW(x),$$

because $L(x)\le xW(x)$. Hence,

$$\begin{aligned}
E_2(x) &\le \mathbb{E}\{L(3\xi(x));\ \xi(x)>x/2\}\\
&\le 9\mathbb{E}\{\xi(x)W(\xi(x));\ \xi(x)>x/2\}\ \to\ 0\quad \text{as } x\to\infty, \qquad (2.54)
\end{aligned}$$

owing to the condition (2.44). Substituting (2.53) and (2.54) into (2.52) we get

$$\begin{aligned}
\mathbb{E}L(x+\xi(x))-L(x) &\le W(x)\mathbb{E}\{\xi(x);\ |\xi(x)|\le x/2\}+o(1)\\
&\le -W(x)r(x)+o(1)\quad \text{as } x\to\infty,
\end{aligned}$$

by (2.42). The inequality (2.43) implies that the drift of $\{L(X_n)\}$ is negative and bounded away from zero for all sufficiently large x. □

2.3 Non-Positivity

In this section we will be interested in conditions that provide a kind of *non-positivity* of a Markov chain $\{X_n\}$, that is, conditions for the existence of an x_* such that $\mathbb{E}_x \tau_{(-\infty,x_*]} = \infty$ for some $x \le x_*$. Below we show the even stronger result, that $\mathbb{E}_y \tau_{(-\infty,x_*]} = \infty$ for all $y > x_*$.

As follows from the condition (1.23) for the positive recurrence of a diffusion process, the condition for the non-positivity of a diffusion process just negates (1.23), so it occurs when:

$$\text{The function} \quad \frac{1}{m_2(x)} \exp\left(\int_0^x \frac{2m_1(y)}{m_2(y)} dy\right) \quad \text{is not integrable at infinity.} \tag{2.55}$$

One could expect that, in terms of test functions, the existence of a non-negative function L such that, for some x_* and $\varepsilon > 0$, $\mathbb{E}\{L(X_1) - L(X_0) \mid X_0 = x\} \ge \varepsilon$ for all $x > x_*$ would imply the non-positivity of $\{X_n\}$; however, the negation of (2.1) does not by itself imply this, as follows from the following counterexample. Let $\{X_n\}$ be a Markov chain on $\mathbb{Z}^+$ with transition probabilities

$$p(x,y) := \begin{cases} 1/2 & \text{if } y = 2(x+1), \\ 1/2 & \text{if } y = 0. \end{cases}$$

Then $m_1(x) = 1$ for any x, while this chain is geometrically ergodic since the returning time to zero is geometrically distributed with success probability $1/2$. This simple counterexample of Doeblin type shows that to conclude non positivity we need to ensure some compactness conditions on the jumps; see below.

Fix an increasing function $s(x) \le x/2$. Let

$$\frac{2m_1^{[s(x)]}(x)}{m_2^{[s(x)]}(x)} \ge -r(x) \quad \text{for all } x > x_0, \tag{2.56}$$

for a decreasing function $r(x) > 0$. In the next theorem we show that the chain $\{X_n\}$ is not positive recurrent provided that

$$\text{the function} \quad e^{-R(x)} = e^{-\int_0^x r(y)dy} \quad \text{is not integrable at infinity.} \tag{2.57}$$

This is motivated by the condition (2.55) for the non-positivity of diffusion processes. It turns out to be very close to guaranteeing the non-positivity of $\{X_n\}$, but we still need some additional technical conditions on $r(x)$ and on the function

$$W(x) := e^{R(x)} \int_0^x e^{-R(y)} dy,$$

which grows at least as x. Proving non-positivity is perhaps the hardest problem we consider in this chapter.

Theorem 2.10 *Let the drift condition (2.56) hold for some differentiable decreasing function $r(x) = O(1/x)$ such that the condition (2.57) is satisfied. Assume that the twice differentiable function $W(x)$ is convex and satisfies the conditions (2.22) and (2.23). Let negative jumps satisfy the following integrability conditions:*

$$\mathbb{E}\{|\xi(x)|^3,\ \xi(x) \in [-s(x),0]\} = o(x^2/W(x)), \tag{2.58}$$

$$\mathbb{P}\{\xi(x) \le -s(x)\} = o(1/xW(x)) \quad \text{as } x \to \infty, \tag{2.59}$$

and, additionally,

$$m_1(x) \ge -c_3/x, \quad c_3 \in (0,\infty), \quad \text{for all } x > x_0, \tag{2.60}$$

$$c_4 := \sup_{x>0} m_2(x) < \infty, \tag{2.61}$$

$$\liminf_{x\to\infty} m_2^{[s(x)]}(x) > 0. \tag{2.62}$$

Then there is an x_* such that $\mathbb{E}_x \tau_{(-\infty,y]} = \infty$ for all $x > y > x_*$.

We require the bounds (2.60) and (2.61) on the full moments of jumps in order to derive a square integrable martingale from $\{X_n\}$, which is needed for our proof.

The conditions (2.58) and (2.59) are fulfilled for some $s(x) = o(x)$ if, for example, the function $W(x)$ is regularly varying at infinity and the family of random variables $\{\xi^-(x)W(\xi^-(x)),\ x \ge 0\}$ is uniformly integrable; sufficiency then follows from Lemmas 2.24 and 2.26.

Corollary 2.11 *For some $\varepsilon > 0$ and $x_0 > 0$, let*

$$\frac{2m_1^{[s(x)]}(x)}{m_2^{[s(x)]}(x)} \ge -\frac{1-\varepsilon}{x} \quad \text{for all } x > x_0.$$

Let the conditions (2.60)–(2.62) hold, that is,

$$\mathbb{E}\{\xi^3(x),\ \xi(x) \in [-s(x),0]\} = o(x), \tag{2.63}$$

$$\mathbb{P}\{\xi(x) \le -s(x)\} = o(1/x^2) \quad \text{as } x \to \infty, \tag{2.64}$$

Then there is an x_* such that $\mathbb{E}_x \tau_{(-\infty,y]} = \infty$ for all $x > y > x_*$.

Both (2.63) and (2.64) hold for some $s(x) = o(x)$ provided that the family of random variables $\{(\xi^-(x))^2,\ x > 0\}$ is uniformly integrable.

Proof of Corollary 2.11. It follows from Theorem 2.10 that if we take $r(x) = \frac{1-\varepsilon}{1+x}$ for $x > 0$ then

$$\begin{aligned} R(x) &= (1-\varepsilon)\log(1+x), \\ e^{-R(x)} &= 1/(1+x)^{1-\varepsilon}, \\ W(x) &= (1+x)/\varepsilon, \end{aligned}$$

which implies the test function $L(x) = ((1+x)^2 - 1)/2\varepsilon$; see the proof of Theorem 2.10 below. □

Corollary 2.12 *For some $m \in \mathbb{N}$ and $\varepsilon > 0$, let*

$$\frac{2m_1^{[s(x)]}(x)}{m_2^{[s(x)]}(x)} \geq -\frac{1}{x} - \frac{1}{x\log x} - \cdots - \frac{1}{x\log x \cdots \log_{(m-1)} x} - \frac{1-\varepsilon}{x\log x \cdots \log_{(m)} x}$$

for all sufficiently large x. Let the conditions (2.60)–(2.62) *hold, let*

$$\mathbb{E}\{\xi(x)^3;\ \xi(x) \in [-s(x), 0]\} = o(x/\log x \cdots \log_{(m)} x), \tag{2.65}$$

and let

$$\mathbb{P}\{\xi(x) \leq -s(x)\} = o(1/x^2 \log x \cdots \log_{(m)} x). \tag{2.66}$$

Then there is an x_* such that $\mathbb{E}_x \tau_{(-\infty, y]} = \infty$ for all $x > y > x_*$.

Both (2.65) and (2.66) hold provided that the family of random variables

$$\{(\xi^-(x))^2 \log \xi^-(x) \cdots \log_{(m)} \xi^-(x)\, \mathbb{I}\{\log_{(m)} \xi^-(x) > 0\},\ x > 0\}$$

is uniformly integrable.

Proof of Corollary 2.12. Consider

$$\begin{aligned} r(x) := \Big(\frac{1}{y} + \frac{1}{y\log y} + \cdots \\ + \frac{1}{y\log y \cdots \log_{(m-1)} y} + \frac{1-\varepsilon}{y\log y \cdots \log_{(m)} y}\Big)\Big|_{y=e^{(m)}+x}, \end{aligned}$$

where $\log_{(m)} e^{(m)} = 1$. Then

$$\begin{aligned} R(x) &= \big(\log y + \log\log y + \cdots + \log_{(m)} y + (1-\varepsilon)\log_{(m+1)} y\big)\Big|_{y=e^{(m)}+x} \\ &\quad - e^{(m-1)} - e^{(m-2)} - \cdots - 1, \\ e^{-R(x)} &= \frac{e^{(m)} e^{(m-1)} \cdots 1}{y\ \log y \cdots \log_{(m-1)} y\ \log_{(m)}^{1-\varepsilon} y}\Big|_{y=e^{(m)}+x}, \end{aligned}$$

$$W(x) = \frac{1}{\varepsilon} y \log y \cdots \log_{(m-1)} y \ \log_{(m)} y \Big|_{y=e^{(m)}+x},$$

$$L(x) \sim \frac{1}{2\varepsilon} x^2 \log x \cdots \log_{(m-1)} x \ \log_{(m)} x.$$

□

Proof of Theorem 2.10. Consider a non-negative test function $L(x)$ defined to be zero on the negative half-line and as

$$L(x) := \int_0^x W(y) dy \quad \text{for all } x \geq 0.$$

First let us prove that the mean drift of $L(x)$ is positive and bounded away from zero for all sufficiently large x; more precisely, that, for some x_* and $\varepsilon > 0$,

$$\mathbb{E}\{L(x + \xi(x)) - L(x);\ \xi(x) \leq s(x)\} \geq \varepsilon \quad \text{for all } x > x_*. \tag{2.67}$$

Having this in mind, we analyse the Taylor expansion for the function L with Lagrange form of remainder, here $x, x + y > 0$:

$$\begin{aligned} L(x + y) - L(x) &= L'(x)y + L''(x + \theta y)\frac{y^2}{2} \\ &= W(x)y + W'(x + \theta y)\frac{y^2}{2}, \end{aligned} \tag{2.68}$$

where $0 \leq \theta = \theta(x, y) \leq 1$. Since $W(x)$ is assumed to be convex, W' is increasing, hence

$$\begin{aligned} L(x + y) - L(x) &\geq W(x)y + W'(x)\frac{y^2}{2} \\ &= W(x)y + r(x)W(x)\frac{y^2}{2} + \frac{y^2}{2} \quad \text{for all } y > 0. \end{aligned} \tag{2.69}$$

We deduce from (2.23) that

$$W'(x + \theta y) \ \geq \ W'(x) - c_2 W(x)|y|/x^2 \quad \text{for } y \in [-x/2, 0];$$

hence it follows from (2.68) that

$$\begin{aligned} &L(x + y) - L(x) \\ &\geq W(x)y + r(x)W(x)\frac{y^2}{2} + \frac{y^2}{2} - c_2\frac{W(x)}{x^2}|y|^3 \ \text{ for all } y \in [-x/2, 0]. \end{aligned} \tag{2.70}$$

Now we are ready to estimate the mean drift of $L(X)$. Since L is non-negative and non-decreasing, the following lower bound holds:

$$\begin{aligned}\mathbb{E}L(x+\xi(x)) - L(x) \ge -L(x)\mathbb{P}\{\xi(x) \le -s(x)\} \\ + \mathbb{E}\{L(x+\xi(x)) - L(x);\ \xi(x) \in [-s(x),0]\} \\ + \mathbb{E}\{L(x+\xi(x)) - L(x);\ \xi(x) \in [0,s(x)]\}. \quad (2.71)\end{aligned}$$

It follows from (2.69) that

$$\begin{aligned}&\mathbb{E}\{L(x+\xi(x)) - L(x);\ \xi(x) \in [0,s(x)]\} \\ &\quad \ge W(x)\mathbb{E}\{\xi(x);\ \xi(x) \in [0,s(x)]\} + \frac{1}{2}r(x)W(x)\mathbb{E}\{\xi^2(x);\ \xi(x) \in [0,s(x)]\} \\ &\quad + \frac{1}{2}\mathbb{E}\{\xi^2(x);\ \xi(x) \in [0,s(x)]\}. \qquad (2.72)\end{aligned}$$

It follows from (2.70) that

$$\begin{aligned}&\mathbb{E}\{L(x+\xi(x)) - L(x);\ \xi(x) \in [-s(x),0]\} \\ &\quad \ge W(x)\mathbb{E}\{\xi(x);\ \xi(x) \in [-s(x),0]\} + \frac{1}{2}r(x)W(x)\mathbb{E}\{\xi^2(x);\ \xi(x) \in [-s(x),0]\} \\ &\quad + \frac{1}{2}\mathbb{E}\{\xi^2(x);\ \xi(x) \in [-s(x),0]\} - c_2\frac{W(x)}{x^2}\mathbb{E}\{|\xi(x)|^3;\ \xi(x) \in [-s(x),0]\} \\ &\quad \ge W(x)\mathbb{E}\{\xi(x);\ \xi(x) \in [-s(x),0]\} + \frac{1}{2}r(x)W(x)\mathbb{E}\{\xi^2(x);\ \xi(x) \in [-s(x),0]\} \\ &\quad + \frac{1}{2}\mathbb{E}\{\xi^2(x);\ \xi(x) \in [-s(x),0]\} + o(1) \quad \text{as } x \to \infty, \qquad (2.73)\end{aligned}$$

owing to the condition (2.58). Finally, it follows from (2.59) and the inequality $L(x) \le xW(x)$ that the first term on the right-hand side of (2.71) tends to zero as $x \to \infty$. Together with the lower bounds (2.72) and (2.73), this implies that

$$\begin{aligned}&\mathbb{E}\{L(x+\xi(x)) - L(x);\ \xi(x) \le s(x)\} \\ &\quad \ge W(x)m_1^{[s(x)]}(x) + \frac{1}{2}r(x)W(x)m_2^{[s(x)]}(x) + \frac{1}{2}m_2^{[s(x)]}(x) + o(1) \\ &\quad \ge m_2^{[s(x)]}(x) + o(1) \quad \text{as } x \to \infty,\end{aligned}$$

owing to (2.56). Then (2.62) implies (2.67) for all sufficiently large x, say, for $x > x_*$.

Let $x_0 > x_*$ and let $x_1 > x_0 + s(x_0)$. Consider an auxiliary Markov chain $\{Y_n\}$ living on $(-\infty, x_1 + s(x_1)]$ whose jumps $\eta(x)$ satisfy

$$x + \eta(x) = \min\{x + \xi(x), x_1 + s(x_1)\},$$

so that the trajectories of $\{X_n\}$ and $\{Y_n\}$ coincide until the first time that $\{X_n\}$ leaves the set $(-\infty, x_1]$. By the construction of $\{Y_n\}$ and because $s(x)$ increases, we also have

$$\mathbb{E}\{L(x+\eta(x)) - L(x);\ \eta(x) \le s(x)\} \ge \varepsilon \qquad \text{for all } x \in (x_*, x_1]. \quad (2.74)$$

Consider the following stopping time:

$$\begin{aligned}\theta &:= \min\{n \ge 1 :\ Y_n \le x_* \text{ or } Y_n > x_1\}\\ &= \min\{n \ge 1 :\ X_n \le x_* \text{ or } X_n > x_1\},\end{aligned}$$

and define another auxiliary Markov chain Z_n that equals Y_n for all $n \le \theta$ and $Z_n = Y_\theta$ for all $n > \theta$; it follows from (2.74) that the process $L(Z_n) - \varepsilon(\theta \wedge n)$ is a submartingale. From the optional stopping time theorem (see e.g. [56, Section 4.7]) we have

$$\mathbb{E}\{\theta \mid Y_0 = x_0\} \le \frac{L(x_1 + s(x_1)) - L(x_0)}{\varepsilon} < \infty.$$

Then, since the submartingale $\{L(Z_n)\}$ is bounded,

$$\mathbb{E}\{L(Z_\theta) \mid Y_0 = x_0\} \ge \mathbb{E}\{L(Z_0) \mid Y_0 = x_0\} = L(x_0).$$

On the other hand,

$$\begin{aligned}&\mathbb{E}\{L(Z_\theta) \mid Y_0 = x_0\}\\ &\quad \le L(x_*)\mathbb{P}\{Z_\theta \le x_* \mid Y_0 = x_0\} + L(x_1 + s(x_1))\mathbb{P}\{Z_\theta > x_1 \mid Y_0 = x_0\}\\ &\quad \le L(x_*) + L(x_1 + s(x_1))\mathbb{P}\{Z_\theta > x_1 \mid Y_0 = x_0\}.\end{aligned}$$

Therefore

$$\mathbb{P}\{Z_\theta > x_1 \mid Y_0 = x_0\} \ge \frac{L(x_0) - L(x_*)}{L(x_1 + s(x_1))}.$$

The condition (2.22) implies that

$$\begin{aligned}L(2x) &= \int_0^{2x} W(y)dy = 2\int_0^x W(2y)dy\\ &\le 2c_1 \int_0^x W(y)dy = 2c_1 L(x) \quad \text{for all } x > 0; \end{aligned} \quad (2.75)$$

hence

$$\mathbb{P}\{Z_\theta > x_1 \mid Y_0 = x_0\} \ge \frac{L(x_0) - L(x_*)}{2c_1 L(x_1)}.$$

So, for all $x_1 > x_0 + s(x_0)$,

$$\mathbb{P}\{X_\theta > x_1 \mid X_0 = x_0\} \ge \frac{L(x_0) - L(x_*)}{2c_1 L(x_1)}; \quad (2.76)$$

in words, starting at point x_0, the chain $\{X_n\}$ exceeds the level x_1 before touching the set $(-\infty, x_*]$ with probability not less than the ratio on the right-hand side of (2.76).

Consider now a starting state $x_1 > 2x_*$, a stopping time

$$\tau = \tau_{(-\infty, x_1/2]} = \min\{n : X_n \le x_1/2\},$$

and a stopped Markov chain $\widehat{X}_n = X_{n\wedge\tau}$ with initial state $\widehat{X}_0 = x_1$ and with jumps $\widehat{\xi}(x)$ defined as $\widehat{\xi}(x) = \xi(x)$ for all $x > x_1/2$ and $\widehat{\xi}(x) = 0$ for all $x \le x_1/2$. Denote $\widehat{m}_1(x) := \mathbb{E}\widehat{\xi}(x)$; by the condition (2.60) we have

$$\widehat{m}_1(x) \ge -2c_3/x_1 \quad \text{for all } x \in \mathbb{R}. \tag{2.77}$$

Given $\widehat{X}_0 = x_1$, the process

$$M_n := \widehat{X}_n - x_1 - \sum_{k=0}^{n-1} \widehat{m}_1(\widehat{X}_k) = \sum_{k=0}^{n-1} (\xi(\widehat{X}_k) - \widehat{m}_1(\widehat{X}_k))$$

is a square integrable – by (2.61) – martingale, $M_0 = 0$. Then, by (2.77),

$$\widehat{X}_n = x_1 + M_n + \sum_{k=0}^{n-1} \widehat{m}_1(\widehat{X}_k) \ge x_1 + M_n - 2c_3 n/x_1,$$

which implies, for $n \le x_1^2/8c_3$, that

$$\begin{aligned}
\mathbb{P}\{\widehat{X}_n \le x_1/2 \mid \widehat{X}_0 = x_1\} &= \mathbb{P}\{M_n \le -x_1/2 + 2c_3 n/x_1\} \\
&\le \mathbb{P}\{M_n \le -x_1/4\} \\
&\le 16\frac{\mathbb{E}M_n^2}{x_1^2} \le 16c_4\frac{n}{x_1^2},
\end{aligned}$$

owing to Chebyshev's inequality and the upper bound for the second moment of a square integrable martingale, $\mathbb{E}M_n^2 \le c_4 n$, which follows from (2.61). Hence, for $n \le x_1^2/32c_4$,

$$\mathbb{P}\{\widehat{X}_n > x_1/2 \mid \widehat{X}_0 = x_1\} \ge 1/2.$$

Since $\{\widehat{X}_n\}$ is the process $\{X_n\}$ stopped when it enters $(-\infty, x_1/2]$, the event $\widehat{X}_n > x_1/2$ yields $\tau \ge n$, so

$$\mathbb{P}\{\tau_{(-\infty, x_1/2]} \ge x_1^2/32c_4 \mid X_0 = x_1\} \ge 1/2.$$

So, starting at the point x_0, with the probability estimated from below given in (2.76), $\{X_n\}$ reaches the level x_1 before it enters $(-\infty, x_*]$, and it then does not drop below the level $x_1/2$ within a time interval of length $x_1^2/32c_4$ with probability at least $1/2$. Therefore

$$\mathbb{P}\{\tau_{(-\infty, x_*]} \ge x_1^2/32c_4 \mid X_0 = x_0\} \ge \frac{L(x_0) - L(x_*)}{4c_1 L(x_1)}.$$

Thus, owing to (2.75),

$$\begin{aligned}\mathbb{P}\{\tau_{(-\infty,x_*]} \ge j \mid X_0 = x_0\} &\ge \frac{L(x_0) - L(x_*)}{4c_1 L(\sqrt{32c_4 j})}\\ &\ge c_5 \frac{L(x_0) - L(x_*)}{L(\sqrt{j})}, \quad c_6 < \infty.\end{aligned}$$

It remains to prove that the function $1/L(\sqrt{x})$ is not integrable. Indeed, since $L(y) \le yW(y)$,

$$\int_1^\infty \frac{1}{L(\sqrt{x})} dx = 2\int_1^\infty \frac{y}{L(y)} dy \ \ge\ 2\int_1^\infty \frac{1}{W(y)} dy.$$

Taking into account that

$$\frac{1}{W(y)} = \frac{e^{-R(y)}}{\int_0^y e^{-R(z)} dz} = \frac{d}{dy} \log \int_0^y e^{-R(z)} dz,$$

we conclude the non-integrability of $1/L(\sqrt{x})$, from (2.57). Therefore

$$\sum_{j=1}^\infty \mathbb{P}\{\tau_{(-\infty,x_*]} \ge j \mid X_0 = x_0\} = \infty;$$

hence $\mathbb{E}_{x_0} \tau_{(-\infty,x_*]}$ cannot be finite. □

2.4 Recurrence and Null Recurrence

2.4.1 Recurrence

Assume that, for some decreasing function $r(x) \downarrow 0$,

$$\frac{2m_1^{[x]}(x)}{m_2^{[x]}(x)} \le r(x) \quad \text{for all } x > x_0. \tag{2.78}$$

The main condition for recurrence is that the function

$$e^{-R(x)} = e^{-\int_0^x r(y)dy} \quad \text{is non-integrable at infinity.} \tag{2.79}$$

This condition is motivated by the recurrence condition (1.24) for diffusion processes and turns out to be very close to guaranteeing the recurrence of $\{X_n\}$. As for positive recurrence, proving the recurrence of a Markov chain is more difficult than for a diffusion process and it requires some additional regularity conditions on $r(x)$ and moment-like conditions on jumps.

In the next theorem we formulate the conditions for recurrence in terms of a decreasing function $\widetilde{r}(x)$ dominating $r(x)$, $\widetilde{r}(x) > r(x)$, such that the function $e^{-\widetilde{R}(x)}$ is also non-integrable where

$$\widetilde{R}(x) := \int_0^x \widetilde{r}(y)dy. \tag{2.80}$$

Consider a function $\widetilde{L}(x)$ that is zero for negative x and is defined as

$$\widetilde{L}(x) := \int_0^x e^{-\widetilde{R}(y)}dy \quad \text{for all } x \geq 0;$$

it is an unboundedly increasing function because $e^{-\widetilde{R}(x)}$ is assumed non-integrable at infinity. When we apply the next general theorem to a particular regular function r in Corollaries 2.14 and 2.15 below, we need to choose $\widetilde{r}$ sufficiently greater than r that the difference $\widetilde{r} - r$ decreases sufficiently slowly to satisfy the conditions (2.82) and (2.83); on the other hand, a larger function $\widetilde{r}(x)$ produces smaller values of $e^{-\widetilde{R}(x)}$, so the choice of a suitable $\widetilde{r}$ is a rather delicate task in each particular case.

Theorem 2.13 *Let the drift condition* (2.78) *hold. Let*

$$\widetilde{r}'(x) = O(1/x^2) \quad \text{as } x \to \infty. \tag{2.81}$$

Let positive jumps satisfy the following integrability conditions: as $x \to \infty$,

$$\mathbb{E}\{\xi^3(x);\ \xi(x) \in (0, x]\} = o\big(x^2(\widetilde{r}(x) - r(x))m_2^{[x]}(x)\big), \tag{2.82}$$

$$\mathbb{E}\{\widetilde{L}(\xi(x));\ \xi(x) \geq x\} = o\Big((\widetilde{r}(x) - r(x))e^{-\widetilde{R}(x)}m_2^{[x]}(x)\Big). \tag{2.83}$$

If the function $\widetilde{L}(x) \to \infty$ as $x \to \infty$ then there exists an x_* such that the set $(-\infty, x_*]$ is recurrent.

Corollary 2.14 *For some $\varepsilon > 0$ and $x_0 > 0$, let*

$$\frac{2m_1^{[x]}(x)}{m_2^{[x]}(x)} \leq \frac{1-\varepsilon}{x} \quad \text{for all } x > x_0.$$

As $x \to \infty$, let

$$\mathbb{E}\{\xi^3(x);\ \xi(x) \in [0, x]\} = o(x m_2^{[x]}(x)), \tag{2.84}$$

$$\mathbb{E}\{\xi^{\varepsilon/2}(x);\ \xi(x) \geq x\} = o(m_2^{[x]}(x)/x^{2-\varepsilon/2}). \tag{2.85}$$

Then there exists an x_* such that the set $(-\infty, x_*]$ is recurrent.

It follows from Lemmas 2.24 and 2.26 that both (2.84) and (2.85) hold provided that the family of random variables $\{(\xi^+(x))^2,\ x > 0\}$ is uniformly integrable. As far as applications are concerned, we will apply this result to show the recurrence of state-dependent near-critical branching processes with migration in Theorem 11.4.

Proof of Corollary 2.14. The corollary follows if we take $\widetilde{r}(x) = \frac{1-\varepsilon/2}{1+x}$ for $x > 0$, which dominates $r(x) = (1-\varepsilon)/x$. Then

$$\widetilde{R}(x) = (1-\varepsilon/2)\log(1+x),$$
$$e^{-\widetilde{R}(x)} = 1/(1+x)^{1-\varepsilon/2},$$

which implies the test function $\widetilde{L}(x) = 2((1+x)^{\varepsilon/2} - 1)/\varepsilon$. □

Corollary 2.15 *For some $m \in \mathbb{N}$ and $\varepsilon > 0$, let*

$$\frac{2m_1^{[x]}(x)}{m_2^{[x]}(x)} \le \frac{1}{x} + \frac{1}{x\log x} + \cdots + \frac{1}{x\log x\cdots\log_{(m-1)} x} + \frac{1-\varepsilon}{x\log x\cdots\log_{(m)} x}$$

for all sufficiently large x. As $x \to \infty$, let

$$\mathbb{E}\{\xi(x)^3;\ \xi(x) \in [0,x]\} = o\Big(\frac{x m_2^{[x]}(x)}{\log x\cdots\log_{(m)} x}\Big), \tag{2.86}$$

and

$$\mathbb{E}\{\log_{(m)}^{\varepsilon/2} \xi(x);\ \xi(x) > x\Big\} = o\Big(\frac{m_2^{[x]}(x)}{x^2\log x\cdots\log_{(m-1)} x\ \log_{(m)}^{1-\varepsilon/2} x}\Big). \tag{2.87}$$

Then there exists an x_* such that the set $(-\infty, x_*]$ is recurrent.

Both (2.86) and (2.87) hold provided that the family of random variables

$$\{\xi^2(x)\log\xi(x)\cdots\log_{(m)}\xi(x)\mathbb{I}\{\log_{(m)}\xi(x) > 0\},\ x > 0\}$$

is uniformly integrable; see Lemmas 2.24 and 2.26 below for a justification.

Proof of Corollary 2.15. Consider

$$\widetilde{r}(x) := \Big(\frac{1}{y} + \frac{1}{y\log y} + \cdots + \frac{1}{y\log y\ldots\log_{(m-1)} y} + \frac{1-\varepsilon/2}{y\log y\ldots\log_{(m)} y}\Big)\Big|_{y=e^{(m)}+x},$$

where $\log_{(m)} e^{(m)} = 1$. Then

$$\widetilde{R}(x) = \Big(\log y + \log\log y + \cdots + \log_{(m)} y + (1-\varepsilon/2)\log_{(m+1)} y\Big)\Big|_{y=e^{(m)}+x} - e^{(m-1)} - e^{(m-2)} - \cdots - 1,$$

$$r(x) - \widetilde{r}(x) = O\Big(\frac{1}{x\log x\cdots\log_{(m)} x}\Big),$$

$$e^{-\widetilde{R}(x)} = \frac{e^{(m)}\,e^{(m-1)}\cdots 1}{y\,\log y \cdots \log_{(m-1)} y\,\log_{(m)}^{1-\varepsilon/2} y}\bigg|_{y=e^{(m)}+x},$$

$$\widetilde{L}(x) = \frac{2}{\varepsilon}\Big(\log_{(m)}^{\varepsilon/2}(e^{(m)}+x) - 1\Big).$$

□

Proof of Theorem 2.13. Following Theorem 2.2, we construct a non-negative increasing unbounded test function whose mean drift is non-positive outside the set $(-\infty, x_*]$, for some x_*.

Let us prove that the increasing Lyapunov function $\widetilde{L}(x)$ constructed above is an appropriate test function. Since $\widetilde{L}(x)$ is increasing, for $x > 0$ we have

$$\begin{aligned}
&\mathbb{E}\widetilde{L}(x+\xi(x)) - \widetilde{L}(x)\\
&\quad\le \mathbb{E}\{\widetilde{L}(x+\xi(x)) - \widetilde{L}(x);\ \xi(x) \ge -x\}\\
&\quad\le \mathbb{E}\{\widetilde{L}(x+\xi(x)) - \widetilde{L}(x);\ |\xi(x)| \le x\} + \mathbb{E}\{\widetilde{L}(x+\xi(x));\ \xi(x) > x\}\\
&\quad\le \widetilde{L}'(x)m_1^{[x]}(x) + \frac{1}{2}\widetilde{L}''(x)m_2^{[x]}(x) + \frac{1}{6}\mathbb{E}\{\xi^3(x)\widetilde{L}'''(x+\theta\xi(x)); |\xi(x)| \le x\}\\
&\qquad + \mathbb{E}\{\widetilde{L}(2\xi(x));\ \xi(x) > x\},
\end{aligned} \tag{2.88}$$

where $0 \le \theta = \theta(x,\xi(x)) \le 1$, by a Taylor expansion, with the remainder in Lagrange form.

The derivative $\widetilde{L}'(x) = e^{-\widetilde{R}(x)}$ is decreasing, so $\widetilde{L}(x)$ is concave on $\mathbb{R}^+$. Thus $\widetilde{L}(2x) \le 2\widetilde{L}(x)$ and hence the fourth term on the right-hand side of (2.88) may be bounded above as follows:

$$\mathbb{E}\{\widetilde{L}(2\xi(x));\ \xi(x) > x\} = o\Big((r(x) - \widetilde{r}(x))e^{-\widetilde{R}(x)}m_2^{[x]}(x)\Big), \tag{2.89}$$

owing to the condition (2.83).

By construction, $\widetilde{L}'(x) = e^{-\widetilde{R}(x)}$ and $\widetilde{L}''(x) = -\widetilde{r}(x)e^{-\widetilde{R}(x)}$, so the sum of the first and second terms on the right-hand side of (2.88) is given by

$$\frac{1}{2}e^{-\widetilde{R}(x)}m_2^{[x]}(x)\Big(\frac{2m_1^{[x]}(x)}{m_2^{[x]}(x)} - \widetilde{r}(x)\Big) \le -\frac{1}{2}e^{-\widetilde{R}(x)}\big(\widetilde{r}(x) - r(x)\big)m_2^{[x]}(x), \tag{2.90}$$

owing to (2.78). Again by the construction of $\widetilde{L}$,

$$\widetilde{L}'''(x) = (-\widetilde{r}'(x) + \widetilde{r}^2(x))e^{-\widetilde{R}(x)},$$

hence $\widetilde{L}'''(x) \ge 0$ for all x since $\widetilde{r}' \le 0$ and, for all x and $y > 0$,

$$\begin{aligned}
\widetilde{L}'''(x+y) &\le (-\widetilde{r}'(x+y) + \widetilde{r}^2(x))e^{-\widetilde{R}(x)}\\
&\le (c_1/x^2 + \widetilde{r}^2(x))e^{-\widetilde{R}(x)}\\
&\le c_2 e^{-\widetilde{R}(x)}/x^2,
\end{aligned}$$

owing to (2.81), which implies in particular that $\widetilde{r}(x) = O(1/x)$. Hence,

$$\begin{aligned}\mathbb{E}\{\widetilde{L}'''(x+\theta\xi(x))\xi^3(x);\ |\xi(x)|\le x\} &\le \mathbb{E}\{\widetilde{L}'''(x+\theta\xi(x))\xi^3(x);\ \xi(x)\in[0,x]\}\\ &\le c_2\frac{e^{-\widetilde{R}(x)}}{x^2}\mathbb{E}\{\xi(x)^3;\ \xi(x)\in[0,x]\}\\ &= o\big(e^{-\widetilde{R}(x)}(\widetilde{r}(x)-r(x))m_2^{[x]}(x)\big),\end{aligned} \tag{2.91}$$

by the condition (2.82). Substituting (2.89)–(2.91) into (2.88) we finally obtain

$$\mathbb{E}\widetilde{L}(x+\xi(x)) - \widetilde{L}(x) \le -\frac{1+o(1)}{2}e^{-\widetilde{R}(x)}\big(\widetilde{r}(x)-r(x)\big)m_2^{[x]}(x) \quad \text{as } x\to\infty,$$

where the right-hand side is negative for all sufficiently large x, say for $x > x_*$. Hence Theorem 2.3 applies, as required. □

2.4.2 Null Recurrence

Combining Corollaries 2.14 and 2.11 we get the following conditions for null recurrence.

Corollary 2.16 *For some $\varepsilon > 0$ and $x_0 > 0$, let*

$$\left|\frac{2m_1^{[s(x)]}(x)}{m_2^{[s(x)]}(x)}\right| \le \frac{1-\varepsilon}{x} \quad \textit{for all } x > x_0.$$

Let the conditions (2.60) *and* (2.62) *hold, and let the family of random variables $\{(\xi^2(x)),\ x > 0\}$ be uniformly integrable. Then there is an x_* such that $\mathbb{P}_x\{\tau_{(-\infty,x_*]} < \infty\}$ but $\mathbb{E}_x\tau_{(-\infty,x_*]} = \infty$ for all initial states $x > x_*$.*

Combining Corollaries 2.15 and 2.12 we get another set of conditions for null recurrence. As far as applications are concerned, we will apply this result to show the null recurrence of state-dependent near-critical branching processes with migration in Theorem 11.5, the null recurrence of stochastic difference equations in Theorem 11.17, and the null recurrence of ALOHA networks in Theorem 11.18.

Corollary 2.17 *For some $m \in \mathbb{N}$ and $\varepsilon > 0$, let*

$$\left|\frac{2m_1^{[x]}(x)}{m_2^{[x]}(x)}\right| \le \frac{1}{x} + \frac{1}{x\log x} + \cdots + \frac{1}{x\log x\cdots\log_{(m-1)}x} + \frac{1-\varepsilon}{x\log x\cdots\log_{(m)}x}$$

for all sufficiently large x. Let the conditions (2.60) *and* (2.62) *hold, and let the family*

$$\{\xi^2(x)\log\xi(x)\cdots\log_{(m)}\xi(x),\ x > 0\} \quad \textit{be uniformly integrable.}$$

Then there is an x_* such that $\mathbb{P}_x\{\tau_{(-\infty,x_*]} < \infty\}$ is finite a.s. but $\mathbb{E}_x\tau_{(-\infty,x_*]} = \infty$ for all initial states $x > x_*$.

2.5 Transience

2.5.1 Condition Motivated by Diffusion

Fix an increasing function $s(x) \to \infty$ as $x \to \infty$ such that $s(x) = o(x)$. Assume that, for some decreasing function $r(x) > 0$,

$$\frac{2m_1^{[s(x)]}(x)}{m_2^{[s(x)]}(x)} \geq r(x) \quad \text{for } x > x_0; \tag{2.92}$$

in general, this means that the drift to the right dominates the diffusion and thus the Markov chain $\{X_n\}$ is transient provided that $r(x)$ decreases sufficiently slowly – roughly speaking, if $r(x) > 1/x$.

The main condition in the next theorem is that the function

$$e^{-R(x)} = e^{-\int_0^x r(y)dy} \quad \text{is integrable;} \tag{2.93}$$

the theorem is motivated by the transience condition (1.25) for a diffusion process and turns out to be very close to guaranteeing the transience of $\{X_n\}$. As for positive recurrence, proving the transience of a Markov chain is more complicated than for a diffusion process and it requires some additional regularity conditions on $r(x)$ together with moment-like conditions on jumps.

Theorem 2.18 *Let the drift condition* (2.92) *hold with a decreasing function $r(x) > 0$, $r(x) = O(1/x)$, such that the condition* (2.93) *is satisfied. Let a decreasing differentiable function $\widetilde{r}(x) \leq r(x)$ be such that*

$$\widetilde{r}\,'(x) = O(1/x^2), \tag{2.94}$$

$$\widetilde{R}(x) := \int_0^x \widetilde{r}(y)dy \to \infty \quad \textit{as } x \to \infty, \tag{2.95}$$

$$e^{-\widetilde{R}(x-s(x))} = O\big(e^{-\widetilde{R}(x)}\big) \quad \textit{as } x \to \infty, \tag{2.96}$$

and let the function $e^{-\widetilde{R}(x)}$ be integrable. Let the negative jumps satisfy the following conditions: as $x \to \infty$,

$$\mathbb{E}\{|\xi(x)|^3;\ \xi(x) \in [-s(x),0]\} = o\big(x^2(r(x) - \widetilde{r}(x))m_2^{[s(x)]}(x)\big), \tag{2.97}$$

$$\mathbb{P}\{\xi(x) \leq -s(x)\} = o\Big((r(x) - \widetilde{r}(x))e^{-\widetilde{R}(x)}m_2^{[s(x)]}(x)\Big). \tag{2.98}$$

Then, for all $x \in \mathbb{R}$,

$$\mathbb{P}_y\{X_n > x \textit{ for all } n \geq 0\} \to 1 \quad \textit{as } y \to \infty. \tag{2.99}$$

If, in addition, for some $x_0 \in \mathbb{R}$,

$$\mathbb{P}_{x_0}\left\{\limsup_{n\to\infty} X_n = \infty\right\} = 1 \tag{2.100}$$

then

$$\mathbb{P}_{x_0}\left\{\lim_{n\to\infty} X_n = \infty\right\} = 1. \tag{2.101}$$

The condition (2.100) (which was first proposed in this framework by Lamperti [111]) can be restated as follows: for any N the exit time from the set $(-\infty, N]$ is finite with probability 1. In this way it is clear that, for a countable Markov chain, irreducibility implies (2.100). For a Markov chain on a general state space, the related topic is ψ-irreducibility; see [126, Sections 4 and 8].

If, for instance, $r(x) = 1/x^\alpha$ for some $\alpha \in (0,1)$ then $\exp(-R(x)) = \exp\left(-x^{1-\alpha}/(1-\alpha)\right)$ and the condition (2.96) fails for an $s(x)$ growing faster than x^α. Hence (2.96) allows us to consider an arbitrary $s(x)$ of order $o(x)$ only in the case where the drift is of order $O(1/x)$; see the corollaries below. In the next subsection we present conditions that are appropriate for a drift characterised by the convergence $xm_1(x) \to \infty$ as $x \to \infty$.

Corollary 2.19 *For some* $\varepsilon > 0$, *let*

$$\frac{2m_1^{[s(x)]}(x)}{m_2^{[s(x)]}(x)} \ge \frac{1+\varepsilon}{x}$$

for all sufficiently large x. *Let the truncated second moments* $m_2^{[s(x)]}(x)$ *be bounded away from zero and infinity, let*

$$\mathbb{E}\{|\xi(x)|^3;\ \xi(x) \in [-s(x), 0]\} = o(x) \quad \text{as } x \to \infty, \tag{2.102}$$

and let

$$\mathbb{P}\{\xi(x) \le -s(x)\} = o(1/x^2 \log^{1+\varepsilon} x) \qquad \text{as } x \to \infty. \tag{2.103}$$

Then (2.99) *holds and the condition* (2.100) *implies* (2.101).

As follows from Lemma 2.24, both (2.102) and (2.103) hold for some $s(x) = o(x)$ provided that

$$\sup_{x>0} \mathbb{E}\{\xi^2(x) \log^{1+2\varepsilon} |\xi(x)|;\ \xi(x) < -1\} < \infty.$$

We will apply this result to show the transience of ALOHA networks in Theorem 11.18.

Proof of Corollary 2.19. The corollary follows if we take

$$r(x) := \frac{1+\varepsilon}{1+x} \quad \text{and} \quad \widetilde{r}(x) := \frac{1}{1+x} + \frac{1+\varepsilon}{(1+x)\log(1+x)};$$

then

$$\widetilde{R}(x) = \log(1+x) + (1+\varepsilon)\log\log(1+x),$$
$$e^{-\widetilde{R}(x)} = 1/((1+x)\log^{1+\varepsilon}(1+x)),$$

while $r(x) - \widetilde{r}(x) = O(1/x)$. □

Corollary 2.20 *For some $m \in \mathbb{N}$ and $\varepsilon > 0$, let*

$$\frac{2m_1^{[s(x)]}(x)}{m_2^{[s(x)]}(x)} \geq \frac{1}{x} + \frac{1}{x\log x} + \cdots + \frac{1}{x\log x \cdots \log_{(m-1)} x} + \frac{1+\varepsilon}{x\log x \cdots \log_{(m)} x}$$

for all sufficiently large x. Let the truncated second moments $m_2^{[s(x)]}(x)$ be bounded away from zero and infinity; let, as $x \to \infty$,

$$\mathbb{E}\{|\xi(x)|^3;\ \xi(x) \in [-s(x), 0]\} = o\Big(\frac{x}{\log x \cdots \log_{(m)} x}\Big) \tag{2.104}$$

and let

$$\mathbb{P}\{\xi(x) \leq -s(x)\} = o\Big(\frac{1}{x^2\log^2 x \cdots \log^2_{(m)} x\ \log^{1+\varepsilon}_{(m+1)} x}\Big). \tag{2.105}$$

Then (2.99) *holds and the condition* (2.100) *implies* (2.101).

As follows from Lemma 2.24, both the conditions (2.104) and (2.105) hold for some $s(x) = o(x)$ if

$$\sup_{x>0} \mathbb{E}\xi^2(x)\log^2|\xi(x)| \cdots \log^2_{(m)}|\xi(x)|\log^{1+2\varepsilon}_{(m+1)}|\xi(x)|\mathbb{I}\{\log_{(m+1)}(-\xi(x)) > 0\} < \infty.$$

Proof of Corollary 2.20. Consider

$$r(x) := \Big(\frac{1}{y} + \frac{1}{y\log y} + \cdots + \frac{1+\varepsilon}{y\log y \cdots \log_{(m)} y}\Big)\Big|_{y=e^{(m)}+x}$$

and

$$\widetilde{r}(x) := \Big(\frac{1}{y} + \frac{1}{y\log y} + \cdots + \frac{1}{y\log y \cdots \log_{(m)} y} + \frac{1+\varepsilon}{y\log y \cdots \log_{(m+1)} y}\Big)\Big|_{y=e^{(m)}+x},$$

where $\log_{(m)} e^{(m)} = 1$. Then

$$r(x) - \widetilde{r}(x) = O\Big(\frac{1}{x \log x \cdots \log_{(m)} x}\Big),$$

$$\widetilde{R}(x) = \Big(\log y + \log\log y + \cdots + \log_{(m+1)} y + (1+\varepsilon)\log_{(m+2)} y\Big)\Big|_{y=e^{(m)}+x} - e^{(m-1)} - e^{(m-2)} - \cdots - 1,$$

and

$$e^{-\widetilde{R}(x)} = \frac{e^{(m)}\, e^{(m-1)} \cdots 1}{y \log y \cdots \log_{(m)} y \,\log_{(m+1)}^{1+\varepsilon} y}\Big|_{y=e^{(m)}+x}.$$

□

Proof of Theorem 2.18. We follow Theorem 2.3 in order to prove transience; so, we construct a non-negative bounded test function $L_*(x) \downarrow 0$ such that $\{L_*(X_n)\}$ is a supermartingale.

Consider a decreasing function

$$\widetilde{L}(x) := \int_x^\infty e^{-\widetilde{R}(y)} dy \quad \text{for all } x \geq 0,$$
$$\widetilde{L}(x) := \widetilde{L}(0) \quad \text{for all } x < 0,$$

which is well defined owing to the assumption that $e^{-\widetilde{R}(x)}$ is integrable; this function is bounded, $\widetilde{L}(x) \leq \widetilde{L}(0) < \infty$.

Let us prove that the mean drift of $\widetilde{L}(x)$ is negative for all sufficiently large x. Since $\widetilde{L}(x)$ is decreasing, we have

$$\begin{aligned}
&\mathbb{E}\widetilde{L}(x+\xi(x)) - \widetilde{L}(x)\\
&\quad \leq \mathbb{E}\{\widetilde{L}(x+\xi(x)) - \widetilde{L}(x);\ \xi(x) \leq s(x)\}\\
&\quad \leq \widetilde{L}(0)\mathbb{P}\{\xi(x) < -s(x)\} + \mathbb{E}\{\widetilde{L}(x+\xi(x)) - \widetilde{L}(x);\ |\xi(x)| \leq s(x)\}\\
&\quad = \widetilde{L}(0)\mathbb{P}\{\xi(x) < -s(x)\} + \widetilde{L}'(x) m_1^{[s(x)]}(x) + \frac{1}{2}\widetilde{L}''(x) m_2^{[s(x)]}(x)\\
&\qquad + \frac{1}{6}\mathbb{E}\{\widetilde{L}'''(x+\theta\xi(x))\xi^3(x);\ |\xi(x)| \leq s(x)\},
\end{aligned}$$

where $0 \leq \theta = \theta(x,\xi(x)) \leq 1$, by a Taylor expansion with the remainder in Lagrange form. By construction, $\widetilde{L}'(x) = -e^{-\widetilde{R}(x)} < 0$, $\widetilde{L}''(x) = \widetilde{r}(x)e^{-\widetilde{R}(x)} > 0$, and

$$\widetilde{L}'''(x+y) = (\widetilde{r}\,'(x+y) - \widetilde{r}^{\,2}(x+y))e^{-\widetilde{R}(x+y)} < 0, \qquad (2.106)$$

since $r' \leq 0$, and

$$\widetilde{L}'''(x+y) = O\big(e^{-\widetilde{R}(x)}/x^2\big) \qquad (2.107)$$

as $x \to \infty$ uniformly for all $|y| \le s(x) = o(x)$, owing to (2.94), the fact that $\widetilde{r}(x) \le r(x) = O(1/x)$, and (2.96). Hence,

$$\begin{aligned}
&\mathbb{E}\{\widetilde{L}'''(x+\theta\xi(x))\xi^3(x); |\xi(x)| \le s(x)\} \\
&\quad \le \mathbb{E}\{\widetilde{L}'''(x+\theta\xi(x))\xi^3(x); \xi(x) \in [-s(x), 0]\} \\
&\quad \le c_1 \frac{e^{-\widetilde{R}(x)}}{x^2} \mathbb{E}\{|\xi(x)|^3; \xi(x) \in [-s(x), 0]\} \\
&\quad = o\big(e^{-\widetilde{R}(x)}(r(x) - \widetilde{r}(x)) m_2^{[s(x)]}(x)\big),
\end{aligned}$$

by the condition (2.97), and, therefore,

$$\begin{aligned}
&\mathbb{E}\widetilde{L}(x+\xi(x)) - \widetilde{L}(x) \\
&\quad \le \widetilde{L}(0)\mathbb{P}\{\xi(x) \le -s(x)\} - e^{-\widetilde{R}(x)}\Big(m_1^{[s(x)]}(x) - \frac{1}{2}\widetilde{r}(x) m_2^{[s(x)]}(x)\Big) \\
&\qquad + o\big(e^{-\widetilde{R}(x)}(r(x) - \widetilde{r}(x))\big) m_2^{[s(x)]}(x) \\
&\quad \le \widetilde{L}(0)\mathbb{P}\{\xi(x) \le -s(x)\} - e^{-\widetilde{R}(x)} \frac{m_2^{[s(x)]}(x)}{2}(1+o(1))\big(r(x) - \widetilde{r}(x)\big),
\end{aligned}$$

by (2.92) and the fact that $r(x) - \widetilde{r}(x) \ge 0$. Now applying the condition (2.98) we conclude that the right-hand side is negative for all sufficiently large x, so there exists a sufficiently large x_* such that

$$\mathbb{E}\widetilde{L}(x+\xi(x)) - \widetilde{L}(x) \le 0 \quad \text{for all } x \ge x_*.$$

Now take $L_*(x) := \min(\widetilde{L}(x), \widetilde{L}(x_*))$. Then

$$\mathbb{E}L_*(x+\xi(x)) - L_*(x) \le \mathbb{E}\widetilde{L}(x+\xi(x)) - \widetilde{L}(x) \le 0$$

for all $x \ge x_*$ and

$$\mathbb{E}L_*(x+\xi(x)) - L_*(x) = \mathbb{E}\{\widetilde{L}(x+\xi(x)) - \widetilde{L}(x_*); x+\xi(x) \ge x_*\} \le 0$$

for all $x < x_*$. Therefore, $\{L_*(X_n)\}$ constitutes a positive bounded supermartingale. Thus Doob's inequality for non-negative supermartingales (see, e.g. [63, Chapter VII.9]) implies (2.99).

For (2.101), we apply Doob's convergence theorem, by which $L_*(X_n)$ has an a.s. limit as $n \to \infty$. Owing to the condition (2.100), this limit equals $L_*(\infty) = 0$, and the proof is complete. □

2.5.2 An Alternative Approach to Transience

Again let us fix some increasing function $s(x) = o(x)$.

Theorem 2.21 *For some $\varepsilon > 0$ and $x_0 > 0$, let the drift satisfy*

$$\frac{2m_1^{[s(x)]}(x)}{m_2^{[s(x)]}(x)} \geq \frac{1+\varepsilon}{x} \quad \text{for all } x > x_0, \tag{2.108}$$

and let the negative jumps be such that

$$\mathbb{P}\{\xi(x) < -s(x)\} \leq p(x) m_1^{[s(x)]}(x), \tag{2.109}$$

where the decreasing function $p(x) > 0$ is integrable. Then (2.99) *follows. If, in addition, the irreducibily condition* (2.100) *holds then* (2.101) *is valid.*

Clearly the condition (2.109) is weaker than (2.103). Regarding applications, we apply this result to show the transience of a random walk conditioned to stay positive in Section 11.1, the transience of state-dependent near-critical branching processes with migration in Theorem 11.3, and the transience of level-dependent collective risk processes in Theorem 11.10.

Corollary 2.22 *For some $\alpha \in (0,1)$, $\varepsilon > 0$ and $x_0 > 0$, let*

$$\mathbb{E}\{\xi(x); |\xi(x)| \leq s(x)\} \geq \frac{\varepsilon}{x^\alpha} \quad \text{for all } x > x_0.$$

Also, as $x \to \infty$, let

$$\mathbb{P}\{\xi(x) \leq -s(x)\} = o(p(x)/x^\alpha), \tag{2.110}$$

$$\mathbb{E}\{\xi^2(x),\ |\xi(x)| \leq s(x)\} = o(x^{1-\alpha}), \tag{2.111}$$

where the decreasing function $p(x) > 0$ is integrable. Then (2.99) *follows. If, in addition, the irreducibily condition* (2.100) *holds, then* (2.101) *is valid.*

Notice that both (2.110) and (2.111) hold for some $s(x) = o(x)$ provided that the family of random variables $\{|\xi(x)|^{1+\alpha},\ x > 0\}$ possesses an integrable majorant; see Lemmas 2.33 and 2.26.

Proof of Theorem 2.21. By Lemma 2.28, there exists a more slowly decreasing function $p_1(x)$ that is still integrable and for which $p_1(x)/p(x) \to \infty$, so we can strengthen the condition (2.109) to the following one:

$$\mathbb{P}\{\xi(x) < -s(x)\} = o\big(p(x) m_1^{[s(x)]}(x)\big) \quad \text{as } x \to \infty. \tag{2.112}$$

Since $p(x)$ is decreasing and integrable at infinity, by Lemma 2.29 below there exists a continuous decreasing integrable function $V_1(x)$, regularly varying at infinity with index -1 and such that $p(x) \leq V_1(x)$. Take

$$V(x) := \int_x^\infty V_2(y)dy, \quad \text{where} \quad V_2(x) := \int_x^\infty \frac{V_1(y)}{y}dy.$$

By Theorem 1(a) from [63, Chapter VIII, Section 9] we know that V_2 is regularly varying at infinity with index -1 and $V_2(x) \sim V_1(x)$ as $x \to \infty$. Since V_1 is integrable, the non-negative decreasing function $V(x)$ is bounded, $V(0) < \infty$, and $V(x)$ is slowly varying, by the same reference.

Let us prove that the mean drift of $V(x)$ is negative for all sufficiently large x. Since $V(x)$ is decreasing, we have

$$\begin{aligned}\mathbb{E}V(x+\xi(x)) - V(x) &\le \mathbb{E}\{V(x+\xi(x)) - V(x); \xi(x) \le s(x)\}\\ &\le V(0)\mathbb{P}\{\xi(x) < -s(x)\} + \mathbb{E}\{V(x+\xi(x)) - V(x); |\xi(x)| \le s(x)\}\\ &= V(0)\mathbb{P}\{\xi(x) < -s(x)\} + V'(x)\mathbb{E}\{\xi(x); |\xi(x)| \le s(x)\}\\ &+ \frac{1}{2}\mathbb{E}\{V''(x+\theta\xi(x))\xi^2(x); |\xi(x)| \le s(x)\},\end{aligned}$$

where $0 \le \theta = \theta(x,\xi(x)) \le 1$, by a Taylor expansion with the remainder in Lagrange form. By the construction, $V'(x) = -V_2(x)$ and

$$V''(x+y) = \frac{V_1(x+y)}{x+y} = (1+o(1))\frac{V_1(x)}{x}$$

as $x \to \infty$ uniformly for $|y| \le s(x)$. Hence,

$$\begin{aligned}&\mathbb{E}V(x+\xi(x)) - V(x)\\ &\le V(0)\mathbb{P}\{\xi(x) \le -s(x)\} - V_2(x)m_1^{[s(x)]}(x) + (1+o(1))\frac{V_1(x)}{2x}m_2^{[s(x)]}(x).\end{aligned}$$

The first term on the right-hand side is of order $o(V_1(x)m_1^{[s(x)]}(x))$ by (2.112) and the inequality $p(x) \le V_1(x)$. The third term is not greater than

$$(1+o(1))V_1(x)\frac{m_1^{[s(x)]}(x)}{1+\varepsilon}$$

because of the condition (2.108). Then

$$\begin{aligned}\mathbb{E}V(x+\xi(x)) - V(x) \le -V_1(x)m_1^{[s(x)]}(x) + V_1(x)\frac{m_1^{[s(x)]}(x)}{1+\varepsilon}\\ + o(V_1(x)m_1^{[s(x)]}(x)).\end{aligned}$$

Thus there exists an x_* sufficiently large that

$$\mathbb{E}V(x+\xi(x)) - V(x) \le -\frac{\varepsilon}{1+2\varepsilon}m_1^{[s(x)]}(x)V_1(x) \quad \text{for all } x \ge x_*.$$

Then the rest of the proof is the same as the proof of Theorem 2.18. □

2.6 Auxiliary Lemmas on Dominating Functions and Random Variables

We will have a repeated need to construct majorants for functions or random variables that satisfy certain properties. In this section we collect all results of this type that are required in our calculations.

Definition 2.23 A family $\{\xi_\theta,\ \theta \in \Theta\}$ of positive random variables on $(\Omega, \mathcal{F}, \mathbb{P})$ is called *uniformly integrable* if

$$\sup_{\theta\in\Theta} \mathbb{E}\{\xi_\theta;\ \xi_\theta > A\} \ \to\ 0 \quad \text{as } A \to \infty.$$

Equivalently, $\{\xi_\theta,\ \theta \in \Theta\}$ is called uniformly integrable if

$$\sup_{\theta\in\Theta} \mathbb{E}\xi_\theta \ < \infty$$

and, for any $\varepsilon > 0$ there exists a $\delta > 0$ such that

$$\sup_{\theta\in\Theta} \mathbb{E}\{\xi_\theta;\ B\} \ \le\ \varepsilon \quad \text{whenever } \mathbb{P}\{B\} \le \delta.$$

Lemma 2.24 *Let $\xi_\theta \ge 0$ be a family of positive random variables indexed by $\theta \in \Theta$. Then the following statements are equivalent:*

(i) *the family $\{\xi_\theta,\ \theta \in \Theta\}$ is uniformly integrable;*
(ii) *there exists an increasing non-negative function $g(x) \to \infty$ such that*

$$\sup_{\theta\in\Theta} \mathbb{E}\xi_\theta g(\xi_\theta) \ <\ \infty.$$

Proof (i) $\Rightarrow$ (ii). Uniform integrability implies the existence of an increasing sequence $n_k \to \infty$, $k \ge 0$, such that $n_0 = 0$ and

$$\mathbb{E}\{\xi_\theta;\ \xi_\theta > n_k\} \ \le\ 1/k^2 \quad \text{for all } \theta \in \Theta \text{ and } k \ge 1.$$

Define an increasing unbounded function $g(x)$ by $g(0) = 0$ and

$$g(x) := \sum_{k=0}^{\infty} (k+1)\mathbb{I}\{x \in (n_k, n_{k+1}]\}, \quad x > 0. \tag{2.113}$$

The expectation of $\xi_\theta g(\xi_\theta)$ may be bounded as follows:

$$\begin{aligned}
\mathbb{E}\xi_\theta g(\xi_\theta) &= \sum_{k=0}^{\infty} (k+1)\mathbb{E}\{\xi_\theta;\ \xi_\theta \in (n_k, n_{k+1}]\} \\
&= \sum_{k=0}^{\infty} \mathbb{E}\{\xi_\theta;\ \xi_\theta > n_k\} \ \le\ \mathbb{E}\xi_\theta + \sum_{k=1}^{\infty} 1/k^2,
\end{aligned}$$

where the right-hand side is uniformly bounded for all $\theta \in \Theta$; this completes the proof of the direct implication.

The implication (ii) $\Rightarrow$ (i) is immediate. □

Lemma 2.25 *Let $\mathcal{F}_{\theta,n}$ be a filtration indexed by $\theta \in \Theta$. Let $Y_{\theta,n}$, $n \geq 0$, be a family of increasing processes, $Y_{\theta,n+1} \geq Y_{\theta,n}$ for all n and θ, while $Y_{\theta,0} = 0$. Let the family of conditional distributions of $Y_{\theta,n+1} - Y_{\theta,n}$ given $\mathcal{F}_{\theta,n}$ be uniformly integrable a.s. for all $n \geq 0$, $\theta \in \Theta$. Let τ_θ be a family of stopping times with respect to $\mathcal{F}_{\theta,n}$. Then the following holds true:*

(i) *If*

$$\text{the family } \{\tau_\theta,\ \theta \in \Theta\} \text{ is uniformly integrable,} \tag{2.114}$$

then the family of random variables Y_{θ,τ_θ}, $\theta \in \Theta$, is uniformly integrable too.

(ii) *If, for some E_θ,*

$$\text{the family } \{\tau_\theta / E_\theta,\ \theta \in \Theta\} \text{ is uniformly integrable,} \tag{2.115}$$

then the family of random variables $Y_{\theta,\tau_\theta}/E_\theta$, $\theta \in \Theta$, is uniformly integrable too.

Proof First, let us show that

$$\mathbb{E}Y_{\theta,\tau_\theta} \leq C\mathbb{E}\tau_\theta, \tag{2.116}$$

where

$$C := \sup_{n,\theta,\omega} \mathbb{E}\{Y_{\theta,n+1} - Y_{\theta,n} \mid \mathcal{F}_{\theta,n}\} < \infty.$$

Indeed,

$$\begin{aligned}
\mathbb{E}Y_{\theta,\tau_\theta} &= \mathbb{E}\sum_{k=0}^{\infty}(Y_{\theta,k+1} - Y_{\theta,k})\mathbb{I}\{k < \tau_\theta\} \\
&= \mathbb{E}\sum_{k=0}^{\infty}\mathbb{E}\{(Y_{\theta,k+1} - Y_{\theta,k})\mathbb{I}\{k < \tau_\theta\} \mid \mathcal{F}_{\theta,k}\} \\
&= \mathbb{E}\sum_{k=0}^{\infty}\mathbb{I}\{k < \tau_\theta\}\mathbb{E}\{Y_{\theta,k+1} - Y_{\theta,k} \mid \mathcal{F}_{\theta,k}\},
\end{aligned}$$

because $\{k < \tau_\theta\} = \overline{\{k \geq \tau_\theta\}} \in \mathcal{F}_{\theta,k}$. Hence,

$$\mathbb{E}Y_{\theta,\tau_\theta} \leq C\mathbb{E}\sum_{k=0}^{\infty}\mathbb{I}\{k < \tau_\theta\} = C\mathbb{E}\tau_\theta,$$

and (2.116) follows. Similarly, for any natural number N,

$$\mathbb{E}\{Y_{\theta,\tau_\theta} - Y_{\theta,N};\ \tau_\theta > N\} \le C\mathbb{E}\{\tau_\theta - N;\ \tau_\theta > N\}, \qquad (2.117)$$

because

$$\begin{aligned}
\mathbb{E}\{Y_{\theta,\tau_\theta} - Y_{\theta,N};\ \tau_\theta > N\} &= \mathbb{E}\sum_{k=N}^{\infty}(Y_{\theta,k+1} - Y_{\theta,k})\mathbb{I}\{k < \tau_\theta\} \\
&= \mathbb{E}\sum_{k=N}^{\infty}\mathbb{E}\{(Y_{\theta,k+1} - Y_{\theta,k})\mathbb{I}\{k < \tau_\theta\} \mid \mathcal{F}_{\theta,k}\} \\
&= \mathbb{E}\sum_{k=N}^{\infty}\mathbb{I}\{k < \tau_\theta\}\mathbb{E}\{Y_{\theta,k+1} - Y_{\theta,k} \mid \mathcal{F}_{\theta,k}\} \\
&\le C\sum_{k=N}^{\infty}\mathbb{P}\{k < \tau_\theta\} \;=\; C\mathbb{E}\{\tau_\theta - N;\ \tau_\theta > N\}.
\end{aligned}$$

Under the uniform integrability condition (2.114), it follows from (2.116) that $\mathbb{E}Y_{\theta,\tau_\theta}$ is bounded. Further, for any natural N and any event $B \in \mathcal{F}$,

$$\begin{aligned}
\mathbb{E}\{Y_{\theta,\tau_\theta};\ B\} &= \mathbb{E}\{Y_{\theta,\tau_\theta};\ \tau_\theta \le N,\ B\} + \mathbb{E}\{Y_{\theta,N};\ \tau_\theta > N,\ B\} \\
&\quad + \mathbb{E}\{Y_{\theta,\tau_\theta} - Y_{\theta,N};\ \tau_\theta > N,\ B\} \\
&\le 2\mathbb{E}\{Y_{\theta,N};\ B\} + \mathbb{E}\{Y_{\theta,\tau_\theta} - Y_{\theta,N};\ \tau_\theta > N\}, \qquad (2.118)
\end{aligned}$$

since the process Y_θ is increasing. For any fixed N, the first expectation on the right-hand side tends to zero as $\mathbb{P}\{B\} \to 0$ owing to the uniform integrability of the jumps of Y_θ; this follows because

$$\sup_\theta \mathbb{E}Y_{\theta,N} \le CN \;<\; \infty,$$

because of (2.116) with $\tau = N$, and

$$\mathbb{E}\{Y_{\theta,N};\ B\} = \sum_{k=0}^{N-1}\mathbb{E}\{Y_{\theta,k+1} - Y_{\theta,k};\ B\}.$$

The second expectation on the right-hand side of (2.118) tends to zero as $N \to \infty$ uniformly for all θ owing to (2.117) and the uniform integrability of $\{\tau_\theta\}$.

Under the condition (2.115), (2.116) implies that $\mathbb{E}Y_{\theta,\tau_\theta}/E_\theta$ is bounded. Further, for any natural number N and event B,

$$\mathbb{E}\left\{\frac{Y_{\theta,\tau_\theta}}{E_\theta};\ B\right\} \le 2\mathbb{E}\left\{\frac{Y_{\theta,NE_\theta}}{E_\theta};\ B\right\} + \mathbb{E}\left\{\frac{Y_{\theta,\tau_\theta} - Y_{\theta,NE_\theta}}{E_\theta};\ \tau_\theta > NE_\theta\right\}, \qquad (2.119)$$

since the process Y_θ is increasing. For any fixed N, the first expectation on the right-hand side tends to zero as $\mathbb{P}\{B\} \to 0$ due to the uniform integrability of the jumps of Y_θ, because

$$\sup_\theta \mathbb{E}Y_{\theta,NE_\theta}/E_\theta \le CN \ < \infty$$

by (2.116) with $\tau = NE_\theta$, and because

$$\mathbb{E}\left\{\frac{Y_{\theta,NE_\theta}}{E_\theta};\ B\right\} = \sum_{k=0}^{NE_\theta-1} \mathbb{E}\left\{\frac{Y_{\theta,k+1} - Y_{\theta,k}}{E_\theta};\ B\right\}.$$

The second expectation on the right-hand side of (2.119) tends to zero as $N \to \infty$ uniformly for all θ due to (2.117) and the uniform integrability of $\{\tau_\theta/E_\theta\}$. □

Lemma 2.26 *Let $p > 0$ and $V(x) \le x^p$ be a function such that both $V(x)$ and $x^p/V(x)$ are increasing and unbounded. If the family of random variables $\{V(|\xi_\theta|),\ \theta \in \Theta\}$ is uniformly integrable then*

$$\sup_{\theta\in\Theta} \mathbb{E}\{|\xi_\theta|^p;\ |\xi_\theta| \le x\} = o\left(\frac{x^p}{V(x)}\right) \quad \text{as } x \to \infty.$$

Proof Fix an $A < x$. Then, for all $\theta \in \Theta$,

$$\begin{aligned}\mathbb{E}\{|\xi_\theta|^p;\ |\xi_\theta| \le x\} &\le A^p + \mathbb{E}\{|\xi_\theta|^p;\ A < |\xi_\theta| \le x\}\\ &= A^p + \mathbb{E}\left\{\frac{|\xi_\theta|^p}{V(|\xi_\theta|)} V(|\xi_\theta|);\ A < |\xi_\theta| \le x\right\}\\ &\le A^p + \frac{x^p}{V(x)}\mathbb{E}\{V(|\xi_\theta|);\ |\xi_\theta| > A\},\end{aligned}$$

as the function $y^p/V(y)$ is increasing. Since $x^p/V(x) \to \infty$, for any fixed A we have

$$\limsup_{x\to\infty} \frac{V(x)}{x^p} \sup_{\theta\in\Theta} \mathbb{E}\{|\xi_\theta|^p;\ |\xi_\theta| \le x\} \le \sup_{\theta\in\Theta} \mathbb{E}\{V(|\xi_\theta|);\ |\xi_\theta| > A\},$$

and the conclusion follows by letting $A \to \infty$, owing to the uniform integrability of the family $\{V(|\xi_\theta|),\ \theta \in \Theta\}$ and the convergence $V(y) \uparrow \infty$. □

Lemma 2.27 *Let $\alpha \in (0,1]$ and $\gamma \ge \alpha$. Let a family of positive random variables $\{\xi_\theta,\ \theta \in \Theta\}$ possess a majorant Ξ with $(\gamma+1-\alpha)$th moment finite, that is, $\mathbb{E}\Xi^{\gamma+1-\alpha} < \infty$ and*

$$\xi_\theta \le_{st} \Xi \quad \text{for all } \theta \in \Theta.$$

Then there exists a decreasing function $p(x)$ that is integrable at infinity and such that

$$\sup_{\theta\in\Theta} \mathbb{E}\{\xi_\theta^{\gamma+1};\ \xi_\theta \le x\} = o(x^{1+\alpha} p(x)) \quad \text{as } x \to \infty.$$

Proof Fubini's theorem yields that

$$\begin{aligned}
\mathbb{E}\{\xi_\theta^{\gamma+1};\ \xi_\theta \le x\} &= \int_0^x \mathbb{P}\{\xi_\theta \in du\} \int_0^u (\gamma+1) y^\gamma dy \\
&= (\gamma+1) \int_0^x y^\gamma \mathbb{P}\{\xi_\theta \in (y,x]\} dy \\
&\le (\gamma+1) \int_0^x y^\gamma \mathbb{P}\{\Xi > y\} dy,
\end{aligned}$$

by the majorisation condition. Therefore, by the Markov inequality,

$$\begin{aligned}
\mathbb{E}\{\xi_\theta^{\gamma+1};\ \xi_\theta \le x\} &\le (\gamma+1) \int_0^x y^\alpha \mathbb{E}\{\Xi^{\gamma-\alpha};\ \Xi > y\} dy \\
&= (\gamma+1) x^{1+\alpha} p(x),
\end{aligned}$$

where

$$p(x) := \frac{1}{x^{1+\alpha}} \int_0^x y^\alpha \mathbb{E}\{\Xi^{\gamma-\alpha};\ \Xi > y\} dy.$$

The finiteness of $\mathbb{E}\Xi^{\gamma+1-\alpha}$ implies the integrability at infinity of $p(x)$. Indeed,

$$\begin{aligned}
\int_0^\infty p(x)dx &= \int_0^\infty \frac{dx}{x^{1+\alpha}} \int_0^x y^\alpha \mathbb{E}\{\Xi^{\gamma-\alpha};\ \Xi > y\} dy \\
&= \int_0^\infty y^\alpha \mathbb{E}\{\Xi^{\gamma-\alpha};\ \Xi > y\} dy \int_y^\infty \frac{dx}{x^{1+\alpha}} \\
&= \frac{1}{\alpha} \int_0^\infty \mathbb{E}\{\Xi^{\gamma-\alpha};\ \Xi > y\} dy \\
&= \frac{\mathbb{E}\Xi^{\gamma+1-\alpha}}{\alpha} < \infty,
\end{aligned}$$

by the moment condition on Ξ. In addition, the function $p(x)$ is decreasing because

$$\begin{aligned}
&\frac{d}{dx} \frac{1}{x^{1+\alpha}} \int_0^x y^\alpha \mathbb{E}\{\Xi^{\gamma-\alpha};\ \Xi > y\} dy \\
&\quad = -\frac{1+\alpha}{x^{2+\alpha}} \int_0^x y^\alpha \mathbb{E}\{\Xi^{\gamma-\alpha};\ \Xi > y\} dy + \frac{1}{x} \mathbb{E}\{\Xi^{\gamma-\alpha};\ \Xi > x\} \\
&\quad \le -\frac{1+\alpha}{x^{2+\alpha}} \mathbb{E}\{\Xi^{\gamma-\alpha};\ \Xi > x\} \int_0^x y^\alpha dy + \frac{1}{x} \mathbb{E}\{\Xi^{\gamma-\alpha};\ \Xi > x\} \\
&\quad = 0.
\end{aligned}$$

The proof is complete owing to Lemma 2.28. □

Lemma 2.28 *Let $p(x) > 0$ be a decreasing function that is integrable at infinity. Then there exists a decreasing function $p_1(x) > 0$ that is integrable at infinity and such that $p_1(x)/p(x) \to \infty$ as $x \to \infty$.*

Proof Without loss of generality we can assume that p is a left-continuous function. Since $p(x)$ is integrable at infinity, there exists an increasing sequence $x_k \to \infty$, $k \geq 0$, such that $x_0 = 0$ and

$$\int_{x_k}^{\infty} p(y)dy \leq 1/k^3 \quad \text{for all } k \geq 1.$$

Since $p(x)$ is decreasing, a sequence x_k may be chosen in such a way that

$$(k+2)p(x_{k+1}) < (k+1)p(x_k) \quad \text{for all } k \geq 1,$$

Owing to this condition the sequence y_k such that $x_k < y_k < x_{k+1}$ for all k is well defined:

$$y_k := \sup\{x \geq x_k : (k+1)p(x) \geq (k+2)p(x_{k+1})\}.$$

Define a function $p_1(x)$ as follows:

$$p_1(x) := \begin{cases} (k+1)p(x) & \text{for } x \in [x_k, y_k], \\ (k+2)p(x_{k+2}) & \text{for } x \in (y_k, y_{k+1}], \end{cases}$$

which is decreasing by construction. Since

$$p_1(x) \geq (k+1)p(x) \quad \text{for all } x \in [x_k, x_{k+1}],$$

the function $p_1(x)$ satisfies the condition $p_1(x)/p(x) \to \infty$ as $x \to \infty$. Lastly, its integral may be bounded as follows:

$$\begin{aligned} \int_{x_1}^{\infty} p_1(x)dx &= \sum_{k=1}^{\infty} \int_{x_k}^{x_{k+1}} p_1(x)dx \\ &= \sum_{k=1}^{\infty}(k+1) \int_{x_k}^{y_k} p(x)dx + \sum_{k=1}^{\infty}(k+2)p(x_{k+1})(x_{k+1} - y_k) \\ &\leq \sum_{k=1}^{\infty}(k+2) \int_{x_k}^{x_{k+1}} p(x)dx \\ &\leq \sum_{k=1}^{\infty} \frac{k+2}{k^3} < \infty, \end{aligned}$$

where the last bound is due to the choice of x_k, and this completes the proof. □

Lemma 2.29 (Denisov [46]) *Let $p(x) > 0$ be a decreasing function which is integrable at infinity. Then there exists a decreasing function $p_1(x) > 0$ that is integrable at infinity, dominates $p(x)$, and is regularly varying at infinity with index -1.*

Lemma 2.30 *Let $p(x) > 0$ be a decreasing function that is integrable at infinity. Then, for any $k \geq 1$, there exists a decreasing function $p_k(x) \geq p(x)$ that is integrable at infinity, is k times differentiable, and, for all $j \leq k$, satisfies*

$$\frac{d^j}{dx^j} p_k(x) = O(1/x^{1+j}) \quad \text{as } x \to \infty.$$

Proof Consider a decreasing function $p_k(x)$ defined by the equality

$$p_k(x) := 2^k \int_{x/2}^{\infty} dy_k \int_{y_k/2}^{\infty} dy_{k-1} \cdots \int_{y_3/2}^{\infty} dy_2 \int_{y_2/2}^{\infty} \frac{p(y_1)}{y_1^k} dy_1.$$

First, since the function $p(x)/x^k$ is decreasing,

$$\int_{y_2/2}^{\infty} \frac{p(y_1)}{y_1^k} dy_1 \geq \int_{y_2/2}^{y_2} \frac{p(y_1)}{y_1^k} dy_1 \geq \frac{y_2}{2} \frac{p(y_2)}{y_2^k} = \frac{1}{2} \frac{p(y_2)}{y_2^{k-1}},$$

so the repetition of this lower bound eventually leads to the inequalities

$$p_k(x) \geq 2^k \int_{x/2}^{x} \frac{1}{2^{k-1}} \frac{p(y_k)}{y_k} dy_k \geq 2^k \frac{x}{2} \frac{1}{2^{k-1}} \frac{p(x)}{x} = p(x).$$

Second, $p_k(x)$ is integrable at infinity because

$$\int_{y_2/2}^{\infty} \frac{p(y_1)}{y_1^k} dy_1 \leq p(y_2/2) \int_{y_2/2}^{\infty} \frac{1}{y_1^k} dy_1 = O\Big(\frac{p(y_2/2)}{y_2^{k-1}}\Big),$$

and hence after $k-1$ steps we arrive at the upper bound

$$p_k(x) \leq c \int_{x/2}^{\infty} \frac{p(y_k/2^{k-1})}{y_k} dy_k, \quad c < \infty,$$

where the integral on the right-hand side is integrable with respect to x, since

$$\begin{aligned} \int_0^{\infty} dx \int_{x/2}^{\infty} \frac{p(y/2^{k-1})}{y} dy &= \int_0^{\infty} \frac{p(y/2^{k-1})}{y} dy \int_0^{2y} dx \\ &= 2 \int_0^{\infty} p(y/2^{k-1}) dy < \infty. \end{aligned}$$

Third,

$$\frac{d^k}{dx^k} p_k(x) = -\frac{2^k}{2} \frac{d^{k-1}}{dx^{k-1}} \int_{x/4}^{\infty} dy_{k-1} \cdots \int_{y_3/2}^{\infty} dy_2 \int_{y_2/2}^{\infty} \frac{p(y_1)}{y_1^k} dy_1$$

$$\vdots$$

$$= (-1)^k \frac{2^k}{2 \times 4 \times \cdots \times 2^k} \frac{p(x/2^k)}{(x/2^k)^k} = O(p(x/2^k)/x^k) \qquad \text{as } x \to \infty.$$

Since $p(x)$ is decreasing and integrable at infinity, $p(x) = O(1/x)$ as $x \to \infty$, so $p_k^{(k)}(x) = O(1/x^{1+k})$. Integrating the kth derivative $k - j$ times we find that the jth derivative of $p_k(x)$ is not greater than the $(k - j)$th integral of c/x^{1+k}, which is of order $O(1/x^{1+j})$. This completes the proof. □

Lemma 2.31 *Let $r(x) > 0$ be a decreasing function such that $xr(x) \to \infty$ as $x \to \infty$. Then there exists a decreasing function $\widetilde{r}(x) \leq r(x)$ such that $x\widetilde{r}(x) \to \infty$ and*

$$\frac{d}{dx}\widetilde{r}(x) \geq -c\widetilde{r}(x)/x \quad \textit{for some } c < \infty. \tag{2.120}$$

Proof Since $xr(x) \to \infty$, there exists a function $a(x) \to \infty$ as $x \to \infty$ such that $a(x)$ increases, while $a(x)/x$ decreases, and

$$xr(x) \geq a(x) \quad \text{for all } x > 0. \tag{2.121}$$

Consider the function

$$\begin{aligned} \widetilde{r}(x) &:= x \int_x^{\infty} \frac{a(y)}{y^3} dy \\ &\leq x \int_x^{\infty} \frac{r(y)}{y^2} dy \\ &\leq xr(x) \int_x^{\infty} \frac{1}{y^2} dy = r(x), \end{aligned}$$

which is decreasing because

$$\begin{aligned} \frac{d}{dx}\widetilde{r}(x) &= \int_x^{\infty} \frac{a(y)}{y^3} dy - \frac{a(x)}{x^2} \\ &\leq \frac{a(x)}{x} \int_x^{\infty} \frac{1}{y^2} dy - \frac{a(x)}{x^2} \leq 0, \end{aligned}$$

since $a(y)/y$ is decreasing. On the other hand, since $a(y)$ is increasing,

$$\widetilde{r}(x) \geq xa(x) \int_x^{\infty} \frac{1}{y^3} dy = \frac{a(x)}{2x},$$

so that

$$\frac{d}{dx}\widetilde{r}(x) \ge -\frac{a(x)}{x^2} \ge -2\frac{\widetilde{r}(x)}{x};$$

hence (2.120) follows. □

Lemma 2.32 *Let $\xi \ge 0$ be a random variable and let $V(x) \ge 0$ be an increasing function such that $\mathbb{E}V(\xi) < \infty$. Let $U(x) \ge 0$ be a function such that the function $f(x) := V(x)/xU(x)$ is increasing and satisfies the condition*

$$\sup_{x>1} \frac{f(2x)}{f(x)} < \infty. \tag{2.122}$$

Then there exists an increasing function $s(x) \to \infty$ of order $o(x)$ such that

$$\mathbb{E}\{U(\xi);\ \xi > s(x)\} = o(p(x)xU(x)/V(x)) \quad \text{as } x \to \infty,$$

where $p(x)$ is a decreasing function that is integrable at infinity and is determined only by ξ and $V(x)$.

Proof Since $\mathbb{E}V(\xi) < \infty$, the decreasing function

$$p_1(x) := \mathbb{E}\{V(\xi)/\xi;\ \xi > x\}$$

is integrable at infinity. Then, by Lemmas 2.28 and 2.29,

$$\mathbb{E}\{V(\xi)/\xi;\ \xi > x\} = o(p(x)) \quad \text{as } x \to \infty,$$

where the decreasing function $p(x)$ is integrable and regularly varying at infinity with index -1. Hence, since $V(x)/xU(x)$ is increasing, we have,

$$\begin{aligned}\mathbb{E}\{U(\xi);\ \xi > x\} &= \mathbb{E}\Big\{\frac{U(\xi)\xi}{V(\xi)} V(\xi)/\xi;\ \xi > x\Big\}\\ &\le \frac{\mathbb{E}\{V(\xi)/\xi;\ \xi > x\}}{V(x)/xU(x)}\\ &= o(p(x)xU(x)/V(x)) \quad \text{as } x \to \infty.\end{aligned}$$

Therefore, for any $n \in \mathbb{N}$,

$$\mathbb{E}\{U(\xi);\ \xi > x/n\} = o(p(x)xU(x)/V(x)) \quad \text{as } x \to \infty$$

because the function $p(x)$ is regularly varying at infinity and because (2.122) holds. Hence, there exists an increasing sequence $x_n \to \infty$ such that

$$\mathbb{E}\{U(\xi);\ \xi > x/n\} \le p(x)xU(x)/nV(x) \quad \text{for all } x \ge x_n.$$

Then the level function $s(x) = \frac{x}{n}\mathbb{I}\{x \in (x_n, x_{n+1}]\}$ is of order $o(x)$ and delivers the stated result. □

Lemma 2.33 *Let $\xi \ge 0$ be a random variable with finite γth moment for some $\gamma \in [1, \infty)$. Let $\alpha \in [1/\gamma, 1]$. Then, for all $\beta \in [0, \gamma - 1/\alpha]$, there exists an increasing function $s(x) \to \infty$ of order $o(x^\alpha)$ such that*

$$\mathbb{E}\{\xi^\beta;\ \xi > s(x)\} = o(p(x)/x^{\alpha(\gamma-\beta)-1}) \quad \textit{as } x \to \infty,$$

where $p(x)$ is a decreasing function that is integrable at infinity and is determined only by ξ, γ, and α.

Proof Put $\eta = \xi^{1/\alpha}$ and $V(x) = x^{\alpha\gamma}$. As follows from Lemma 2.32 with $U(x) = x^{\alpha\beta}$, since $\mathbb{E}\xi^\gamma = \mathbb{E}V(\eta) < \infty$ there exists a function $p(x)$ that is integrable at infinity and regularly varying at infinity with index -1 and a function $s(x) = o(x)$ such that

$$\begin{aligned}\mathbb{E}\{\eta^{\alpha\beta};\ \eta > s(x)\} &= o(p(x)xU(x)/V(x)) \\ &= o(p(x)/x^{\alpha(\gamma-\beta)-1}) \quad \text{as } x \to \infty,\end{aligned}$$

which can be rewritten as

$$\mathbb{E}\{\xi^\beta;\ \xi > s^\alpha(x)\} = o(p(x)/x^{\alpha(\gamma-\beta)-1}) \quad \text{as } x \to \infty,$$

and the proof is complete. □

We also need a generalisation of the last result to truncation levels $s(x)$ of a more general form. To this end we prove the following result.

Lemma 2.34 *Let $\xi \ge 0$ be a random variable and let $V(x) \ge 0$, $V(x) \to \infty$, be a strictly increasing function such that $\mathbb{E}V(\xi) < \infty$ and*

$$c_V := \sup_{x>1} V(2x)/V(x) < \infty. \tag{2.123}$$

Let $g(x) \ge 0$, $g(x) \to \infty$, be an increasing function such that

$$\sup_{x>1} g(2x)/g(x) < \infty. \tag{2.124}$$

Then there exists an increasing function $s(x) \to \infty$ of order $o(V^{-1}(xg(x)))$ such that

$$\mathbb{P}\{\xi > s(x)\} = o(p(x)/g(x)) \quad \textit{as } x \to \infty,$$

where $p(x)$ is a decreasing function that is integrable at infinity.

Proof Since V is strictly increasing and g is increasing, the function $f(x) := V^{-1}(xg(x))$ is strictly increasing too and, owing to the condition (2.123),

$$\frac{f(x/c_V)}{f(x)} = \frac{V^{-1}(xg(x/c_V)/c_V)}{V^{-1}(xg(x))} \le \frac{V^{-1}(xg(x)/c_V)}{V^{-1}(xg(x))} \le \frac{1}{2}. \tag{2.125}$$

Let us define a random variable η such that $f(\eta) = \xi$. Then the probability in question may be represented as

$$\mathbb{P}\{\xi > f(x)\} = \mathbb{P}\{f(\eta) > f(x)\} = \mathbb{P}\{\eta > x\}.$$

Since $V(\xi) = V(f(\eta)) = \eta g(\eta)$ and $\mathbb{E}V(\xi) < \infty$, it follows that $\mathbb{E}\eta g(\eta) < \infty$ too. Hence,

$$p_1(x) := \mathbb{E}\{g(\eta);\ \eta > x\}$$

is integrable at infinity. Then, by Lemmas 2.28 and 2.29,

$$\mathbb{E}\{g(\eta);\ \eta > x\} = o(p(x)) \quad \text{as } x \to \infty,$$

where the decreasing function $p(x)$ is integrable and regularly varying at infinity with index -1. Therefore,

$$\mathbb{P}\{\eta > x\} \le \frac{\mathbb{E}\{g(\eta);\ \eta > x\}}{g(x)} = o(p(x)/g(x)) \quad \text{as } x \to \infty.$$

This implies that, for any $n \in \mathbb{N}$,

$$\mathbb{P}\{\eta > x/n\} = o(p(x/n)/g(x/n)) = o(p(x)/g(x)) \quad \text{as } x \to \infty$$

because the function $p(x)$ is regularly varying at infinity and because the condition (2.124) holds. Equivalently, for any $n \in \mathbb{N}$,

$$\mathbb{P}\{\xi > f(x/n)\} = o(p(x)/g(x)) \quad \text{as } x \to \infty.$$

Together with (2.125) this implies the existence of a truncation level $s(x) = o(f(x))$, which completes the proof. □

Taking $V(x) = x^2$ we get the following corollary.

Corollary 2.35 *Let $\xi \ge 0$ be a random variable with finite second moment. Let $g(x) \ge 0$, $g(x) \to \infty$, be an increasing function satisfying the condition (2.124). Then there exists an increasing function $s(x) \to \infty$ of order $o(\sqrt{xg(x)})$ such that*

$$\mathbb{P}\{\xi > s(x)\} = o(p(x)/g(x)) \quad \text{as } x \to \infty,$$

where $p(x)$ is a decreasing function that is integrable at infinity.

Lemma 2.36 *Let $\xi \ge 0$ be a random variable and let $V(x)$ be a non-negative function such that $\mathbb{E}V(\xi)\log(1+\xi) < \infty$. Then there exists an increasing function $s(x) \to \infty$ of order $o(x)$ such that*

$$\mathbb{E}\{V(\xi);\ \xi > s(x)\} = o(p(x)x) \quad \text{as } x \to \infty,$$

where $p(x)$ is a decreasing function that is integrable at infinity.

Proof The claim follows almost immediately because

$$\begin{aligned}\int_1^\infty \frac{\mathbb{E}\{V(\xi);\ \xi > x\}}{x}dx &= \int_1^\infty \frac{dx}{x}\int_x^\infty V(y)\mathbb{P}\{\xi \in dy\}\\ &= \int_1^\infty V(y)\mathbb{P}\{\xi \in dy\}\int_1^y \frac{dx}{x}\\ &= \int_1^\infty V(y)(\log y)\mathbb{P}\{\xi \in dy\} < \infty.\end{aligned}$$

Hence, by Lemmas 2.28 and 2.29,

$$\mathbb{E}\{V(\xi);\ \xi > x\} = o(p(x)x) \quad \text{as } x \to \infty,$$

where the decreasing function $p(x)$ is integrable and regularly varying at infinity with index -1. Then using the arguments as in Lemma 2.32 completes the proof. □

Lemma 2.37 *Let $\xi_1, \ldots, \xi_n$ be independent random variables with zero mean and finite variance. Denote $S_n := \xi_1 + \cdots + \xi_n$. Then, for all x, $y > 0$,*

$$\mathbb{P}\{S_n > x\} \le e^{x/y}\Big(\frac{\mathbb{V}\mathrm{ar}\, S_n}{xy}\Big)^{x/y} + \sum_{i=1}^n \mathbb{P}\{\xi_i > y\}, \tag{2.126}$$

and, for all $x > \max(y, 2\sqrt{\mathbb{V}\mathrm{ar}\, S_n})$,

$$\begin{aligned}&\mathbb{E}\{S_n^2;\ S_n > x\}\\ &\le e^{x/y}\Big(\frac{\mathbb{V}\mathrm{ar}\, S_n}{xy}\Big)^{x/y}x^2 + \sum_{i=1}^n \mathbb{E}\{\xi_i^2;\ \xi_i > y\} + \mathbb{V}\mathrm{ar}\, S_n \sum_{i=1}^n \mathbb{P}\{\xi_i > y\}.\end{aligned} \tag{2.127}$$

Proof The inequality (2.126) is due to Fuk and Nagaev; see e.g. Corollary 1.11 in [129] or Theorem 4 in [70]. This inequality allows us to obtain a bound similar to (2.127) as follows. For any $x > y$, the function $z^{1-2x/y}$ is integrable at infinity with respect to z, so

$$\begin{aligned}\mathbb{E}\{S_n^2;\ S_n > x\} &= x^2\mathbb{P}\{S_n > x\} + 2\int_x^\infty z\mathbb{P}\{S_n > z\}\,dz\\ &\le e^{x/y}(\mathbb{V}\mathrm{ar}\, S_n)^{x/y}\Big[\Big(\frac{1}{xy}\Big)^{x/y} + 2\int_x^\infty z\Big(\frac{1}{z^2y/x}\Big)^{x/y}dz\Big]\\ &\quad + \sum_{i=1}^n\Big[x^2\mathbb{P}\{\xi_i > y\} + 2\int_x^\infty z\mathbb{P}\Big\{\xi_i > z\frac{y}{x}\Big\}\,dz\Big]\\ &= e^{x/y}\Big(\frac{\mathbb{V}\mathrm{ar}\, S_n}{xy}\Big)^{x/y}\Big(\frac{x^2}{x/y-1}+1\Big) + (x/y)^2\sum_{i=1}^n \mathbb{E}\{\xi_i^2;\ \xi_i > y\}.\end{aligned}$$

Let us now prove (2.127) following the method of the proof of (2.126) from [70, Theorem 4]. We start with the following upper bounds:

$$\begin{aligned}
&\mathbb{E}\{S_n^2;\ S_n > x\} \\
&\quad \le \mathbb{E}\{S_n^2;\ S_n > x,\ \xi_i \le y \text{ for all } i \le n\} + \sum_{i=1}^{n} \mathbb{E}\{S_n^2;\ S_n > x,\ \xi_i > y\} \\
&\quad \le \mathbb{E}\{T_n^2;\ T_n > x\} + \sum_{i=1}^{n} \mathbb{E}\{S_n^2;\ S_n > x,\ \xi_i > y\}, \qquad (2.128)
\end{aligned}$$

where $T_n = \eta_1 + \cdots + \eta_n$, and $\eta_i = \xi_i \mathbb{I}\{\xi_i \le y\}$, so $\mathbb{E}\eta_i \le 0$. Since T_n is bounded by ny, all its positive exponential moments are finite; hence, for all $\lambda > 0$,

$$\begin{aligned}
\mathbb{E}\{T_n^2;\ T_n > x\} &= \mathbb{E}\left\{\frac{e^{\lambda T_n}}{e^{\lambda T_n}/T_n^2};\ T_n > x\right\} \\
&\le \frac{\mathbb{E}e^{\lambda T_n}}{e^{\lambda x}/x^2} \quad \text{for all } x \ge 2/\lambda,
\end{aligned}$$

because the function $e^{\lambda x}/x^2$ is increasing in the range $x \ge 2/\lambda$. Further,

$$\begin{aligned}
\mathbb{E}e^{\lambda \eta_i} &= 1 + \lambda \mathbb{E}\eta_i + \mathbb{E}(e^{\lambda \eta_i} - 1 - \lambda \eta_i) \\
&\le 1 + \lambda \mathbb{E}\eta_i + \frac{e^{\lambda y} - 1 - \lambda y}{y^2} \mathbb{E}\eta_i^2,
\end{aligned}$$

since $\eta_i \le y$ and the function $(e^z - 1 - z)/z^2$ is increasing in $z \in \mathbb{R}$. Thus,

$$\begin{aligned}
\mathbb{E}e^{\lambda \eta_i} &\le 1 + \frac{e^{\lambda y} - 1 - \lambda y}{y^2} \mathbb{V}\mathrm{ar}\, \xi_i \\
&\le \exp\left(\frac{e^{\lambda y} - 1 - \lambda y}{y^2} \mathbb{V}\mathrm{ar}\, \xi_i\right) \le \exp\left(\frac{e^{\lambda y} - 1}{y^2} \mathbb{V}\mathrm{ar}\, \xi_i\right),
\end{aligned}$$

and furthermore

$$\mathbb{E}e^{\lambda T_n} \le \exp\left(\frac{e^{\lambda y} - 1}{y^2} \mathbb{V}\mathrm{ar}\, S_n\right).$$

Take

$$\lambda = \frac{1}{y} \log\left(\frac{xy}{\mathbb{V}\mathrm{ar}\, S_n} + 1\right),$$

so that $x > 2/\lambda$, because the above relation is equivalent to

$$\frac{xy}{\mathbb{V}\mathrm{ar}\, S_n} + 1 > e^{2y/x},$$

which is satisfied due to $x > \max(y, 2\sqrt{\mathbb{V}\mathrm{ar}\, S_n})$. Then $\mathbb{E}e^{\lambda T_n} \le e^{x/y}$, so

$$\begin{aligned}\frac{\mathbb{E}e^{\lambda T_n}}{e^{\lambda x}} &\le e^{x/y} e^{-\frac{x}{y}\log(xy/(\mathbb{V}\mathrm{ar}\, S_n)+1)} \\ &\le e^{x/y}\Big(\frac{\mathbb{V}\mathrm{ar}\, S_n}{xy}\Big)^{x/y}. \qquad (2.129)\end{aligned}$$

By the independence of the ξ_i,

$$\begin{aligned}\mathbb{E}\{S_n^2;\ S_n > x,\ \xi_n > y\} &\le \mathbb{E}\{(S_{n-1}+X_n)^2;\ \xi_n > y\} \\ &= \mathbb{E}\{\mathbb{E}\{(S_{n-1}+\xi_n)^2 \mid \xi_n\};\ \xi_n > y\} \\ &= \mathbb{E}\{\mathbb{V}\mathrm{ar}\, S_{n-1} + \xi_n^2;\ \xi_n > y\}.\end{aligned}$$

Therefore

$$\mathbb{E}\{S_n^2;\ S_n > x,\ \xi_n > y\} \le \mathbb{V}\mathrm{ar}\, S_{n-1}\mathbb{P}\{\xi_n > y\} + \mathbb{E}\{\xi_n^2;\ \xi_n > y\},$$

which implies that

$$\sum_{i=1}^n \mathbb{E}\{S_n^2;\ S_n > x,\ \xi_i > y\} \le \mathbb{V}\mathrm{ar}\, S_n \sum_{i=1}^n \mathbb{P}\{\xi_i > y\} + \sum_{i=1}^n \mathbb{E}\{\xi_i^2;\ \xi_i > y\}. \qquad (2.130)$$

Substituting (2.129) and (2.130) into (2.128) we conclude the proof of the upper bound for the tail second moment of S_n. □

Lemma 2.38 *Let $\xi_1, \ldots, \xi_n$ be independent random variables with zero mean and finite absolute moments of order $p \ge 2$. Denote $S_n := \xi_1 + \cdots + \xi_n$. Then, for some C_p which depends only on p,*

$$\mathbb{E}|S_n|^p \le C_p n^{p/2-1} \sum_{i-1}^n \mathbb{E}|\xi_i|^p. \qquad (2.131)$$

If $\mathbb{V}\mathrm{ar}\,\xi_i < \infty$ for all i then, for all $p \le 2$,

$$\mathbb{E}|S_n|^p \le \Big(\sum_{i=1}^n \mathbb{V}\mathrm{ar}\,\xi_i\Big)^{p/2}. \qquad (2.132)$$

In particular, if the ξ_i are independent and identically distributed random variables with finite moment of order $p \vee 2$ then

$$\mathbb{E}|S_n|^p \le Cn^{p/2} \quad \text{for all } n \ge 1 \text{ and } p > 0, \qquad (2.133)$$

where

$$C = C(p, \xi_1) = \begin{cases} C_p\mathbb{E}\xi_1^p & \text{if } p > 2, \\ (\mathbb{V}\mathrm{ar}\,\xi_1)^{p/2} & \text{if } p \le 2. \end{cases}$$

Proof For $p \geq 2$, the result goes back to Dharmadhikari and Jogdeo [48, Theorem 2].

For $p \leq 2$ the function $x^{p/2}$ is concave, so

$$\mathbb{E}|S_n|^p \leq (\mathbb{E}S_n^2)^{p/2} = \Big(\sum_{i=1}^n \mathbb{V}\mathrm{ar}\,\xi_i\Big)^{p/2},$$

by the independence of the ξ_i. □

2.7 Comments for Chapter 2

The Lyapunov function approach for proving the positive recurrence, recurrence, or transience of countable Markov chains goes back to Foster [68], and was rediscovered by Moustafa in [127]. A state-dependent variation was given by Malyshev and Menshikov in [119].

The first classification of nearest neighbour Markov chains with drift of order c/x was made by Harris [77] and by Hodges and Rosenblatt [79].

A regular study of processes with asymptotically zero drift on $\mathbb{R}^+$ was initiated by Lamperti in a series of papers [111, 112, 113]. In [111, Theorem 2.2] he showed that if $\limsup X_n = \infty$ and $\mathbb{E}|\xi(x)|^{2+\delta}$ are bounded for some positive δ then

- $2xm_1(x) \leq m_2(x) + O(x^{-\delta})$ yields the recurrence of X_n,
- $2xm_1(x) \geq (1+\varepsilon)m_2(x)$ yields the transience of X_n.

In [113, Theorem 2.1] he proved that $2xm_1(x) + m_2(x) \leq -\varepsilon$ is sufficient for the positive recurrence of $\{X_n\}$. It was shown in [111, Theorem 3.1] that $2xm_1(x) + m_2(x) \geq \varepsilon$ implies that $\{X_n\}$ is non-positive (either null-recurrent or transient) provided $xm_1(x)$ and $m_2(x)$ are bounded and $m_4(x) = o(x^2)$.

These criteria were improved later by Menshikov, Asymont, and Yasnogorodskii [124]. Instead of assuming the existence of the $(2+\delta)$th bounded moment they assumed that $\mathbb{E}\xi^2(x)\log^{2+\delta}(1+|\xi(x)|)$ is bounded. Moreover, they established more precise classifications for positive recurrence, null recurrence, and transience on the basis of iterated logarithms; these are improved further in Corollaries 2.6, 2.12, 2.15, and 2.20.

Corollary 2.7 on positive recurrence in the absence of second moments goes back to Korshunov [101, Theorem 5]. Corollary 2.19 on transience in the absence of second moments is due to Menshikov and Wade [123, Theorem 2.1]; we prove it under minimal moment conditions. Sandrić [139, Theorem 1.3] suggested some sufficient conditions for the recurrence of a chain with drift of order c/x^β where the jumps have momented of order $1+\beta$ that are infinite, so results like Corollary 2.9 do not work; these conditions apply only under the assumption that the tails of jumps are regularly varying.

We are aware of two different approaches to proving non-positivity; one is due to Lamperti [113] and the other goes back to Menshikov, Asymont, and Yasnogorodskii [124]. In Theorem 2.10 we followed the first approach, significantly improving the non-positivity results from both [113] and [124].

3

Down-Crossing Probabilities for Transient Markov Chain

In this chapter we consider a (right) transient Markov chain $\{X_n\}$ taking values in $\mathbb{R}$, that is, for any fixed $\widehat{x} \in \mathbb{R}$,

$$\mathbb{P}_x\{\tau_B < \infty\} \to 0 \quad \text{as } x \to \infty,$$

where $\tau_B := \min\{n \geq 1 : X_n \in B\}$, $B := (-\infty, \widehat{x}\,]$. We are interested in the rate of convergence to zero of this probability as $x \to \infty$. It clearly depends on the asymptotic properties of the drift of $\{X_n\}$ at infinity.

The standard approach to the understanding of the rate of decay of down-crossing probabilities for transient chains is via the construction of a bounded decreasing function U such that $\{U(X_n)\}$ is a supermartingale outside B. Then there is an upper bound on the down-crossing probability:

$$\mathbb{P}_x\{\tau_B < \infty\} \leq \frac{U(x)}{U(\widehat{x})}, \quad x > \widehat{x}.$$

Therefore, the results available in the literature (see e.g. [61] or [121]) in which one finds a Lyapunov function that proves transience automatically provide some upper bounds on the down-crossing probabilities, while they are not explicitly stated in these works. However, the upper bounds obtained in this way are only logarithmically sharp.

The down-crossing probability, as a function of x, is a harmonic function for the chain $\{X_n\}$ killed at hitting the set B. So, the aim of this chapter is to find functions $U_\pm$ such that $\{U_+(X_n)\}$ is a bounded submartingale while $\{U_-(X_n)\}$ is also a bounded supermartingale and such that both U_+ and U_- are as close to the function $\mathbb{P}_x\{\tau_B < \infty\}$ as possible; hence both are asymptotically harmonic. That allows us to derive upper and lower bounds for down-crossing probabilities which are precise up to a constant factor; see below Theorem 3.2 for the case $m_1(x) \sim c/x$ and Theorems 3.7 and 3.10 below for the case $m_1(x)x \to \infty$.

3.1 Markov Chains with Asymptotically Zero Drift: Slow Decay of Down-Crossing Probability

We start with the following result, which states that, for almost any Markov chain with asymptotically zero drift, the down-crossing probability decays more slowly than any exponential function.

Theorem 3.1 *Let a Markov chain $\{X_n\}$ on $\mathbb{R}$ be such that*

$$\limsup_{x\to\infty} \mathbb{E}\{\xi(x);\ \xi(x) > -x\} \le 0 \tag{3.1}$$

and, in addition,

$$\liminf_{x\to\infty} \mathbb{E}\{\xi^2(x);\ \xi(x) \in (-x,0)\} > 0. \tag{3.2}$$

Then there exists an $\widehat{x}$ such that, for $B := (-\infty,\widehat{x}\,]$ and all $\lambda > 0$,

$$e^{\lambda x}\mathbb{P}_x\{\tau_B < \infty\} \to \infty \quad \text{as } x \to \infty.$$

Proof Let $\lambda > 0$. Consider a bounded decreasing function

$$U_\lambda(x) := \min(e^{-\lambda x}, 1).$$

For all $x > 0$,

$$\begin{aligned}\mathbb{E}(U_\lambda(x+\xi(x)) - U_\lambda(x)) &\ge \mathbb{E}\{e^{-\lambda(x+\xi(x))} - e^{-\lambda x};\ x+\xi(x) > 0\}\\ &= e^{-\lambda x}\mathbb{E}\{e^{-\lambda\xi(x)} - 1;\ x+\xi(x) > 0\}.\end{aligned}$$

Since $e^{-y} \ge 1 - y$ for all y and $e^{-y} \ge 1 - y + y^2/2$ for all $y < 0$, we have

$$\begin{aligned}&\mathbb{E}\{e^{-\lambda\xi(x)} - 1;\ x+\xi(x) > 0\}\\ &\quad\ge -\lambda\mathbb{E}\{\xi(x);\ x+\xi(x) > 0\} + \frac{\lambda^2}{2}\mathbb{E}\{\xi^2(x);\ \xi(x) \in (-x,0)\}.\end{aligned}$$

Then, owing to the conditions (3.1) and (3.2), there exists a sufficiently large $\widehat{x}_\lambda > 0$ such that

$$\mathbb{E}(U_\lambda(x+\xi(x)) - U_\lambda(x)) \ge 0 \quad \text{for all } x > \widehat{x}_\lambda.$$

Therefore, the process $\{U_\lambda(X_{n\wedge\tau_{B_\lambda}})\}$ is a bounded submartingale, where $B_\lambda := (-\infty,\widehat{x}_\lambda]$. Hence, by the optional stopping theorem (see e.g. [56, Section 4.7]), for $z > \widehat{x}_\lambda$ and $x \in (\widehat{x}_\lambda, z)$,

$$\mathbb{E}_x U_\lambda(X_{\tau_{B_\lambda}\wedge\tau_{(z,\infty)}}) \ge \mathbb{E}_x U_\lambda(X_0) = U_\lambda(x).$$

Letting $z \to \infty$ we conclude that

$$\begin{aligned}
&\mathbb{E}_x\{U_\lambda(X_{\tau_{B_\lambda}});\ \tau_{B_\lambda} < \infty\}\\
&\quad = \lim_{z\to\infty} \mathbb{E}_x\{U_\lambda(X_{\tau_{B_\lambda}});\ \tau_{B_\lambda} < \tau_{(z,\infty)}\}\\
&\quad = \lim_{z\to\infty} \mathbb{E}_x U_\lambda(X_{\tau_{B_\lambda}\wedge\tau_{(z,\infty)}}) - \lim_{z\to\infty} \mathbb{E}_x\{U_\lambda(X_{\tau_{(z,\infty)}});\ \tau_{B_\lambda} > \tau_{(z,\infty)}\}\\
&\quad \geq U_\lambda(x) - 0 = U_\lambda(x).
\end{aligned}$$

On the other hand, since U_λ is bounded by 1,

$$\mathbb{E}_x\{U_\lambda(X_{\tau_{B_\lambda}});\ \tau_{B_\lambda} < \infty\} \leq \mathbb{P}_x\{\tau_{B_\lambda} < \infty\}.$$

This allows us to deduce the lower bound

$$\mathbb{P}_x\{\tau_{B_\lambda} < \infty\} \geq U_\lambda(x) = e^{-\lambda x} \quad \text{for all } x > \widehat{x}_\lambda,$$

and hence the theorem's conclusion follows with $\widehat{x} = \widehat{x}_1$ and $B := (-\infty, \widehat{x}_1]$, because, by the Markov property, for all $\lambda < 1$ and $x > \widehat{x}_1$,

$$\mathbb{P}_x\{\tau_{B_1} < \infty\} \geq \mathbb{P}_x\{\tau_{B_\lambda} < \infty\} \inf_{y\in(\widehat{x}_1,\widehat{x}_\lambda]} \mathbb{P}_y\{\tau_{B_1} < \infty\}$$

and

$$\begin{aligned}
\mathbb{P}_y\{\tau_{B_1} < \infty\} &\geq U_1(y) = e^{-y}\\
&\geq e^{-\widehat{x}_\lambda} > 0 \quad \text{for all } y \in (\widehat{x}_1, \widehat{x}_\lambda].
\end{aligned}$$ □

Let us show by an example that the condition (3.2), which imposes a kind of non-degeneracy of jumps, is essential for the conclusion to hold. Consider the skip-free Markov chain $\{X_n\}$ on $\mathbb{Z}^+$ described in Section 1.4, for which $\xi(x)$ takes the values -1, 1, or 0 only, with probabilities $p_-(x)$, $p_+(x)$, and $p_0(x)$ respectively; $p_-(0) = 0$. The probability of hitting zero is computed in (1.21),

$$\mathbb{P}_x\{\tau_0 < \infty\} = \frac{\sum_{y=x}^{\infty} \prod_{k=1}^{y} \frac{p_-(k)}{p_+(k)}}{\sum_{y=0}^{\infty} \prod_{k=1}^{y} \frac{p_-(k)}{p_+(k)}} \quad \text{for all } x > 0.$$

Consider the case where $p_+(x) := 1/(x+1)$ and $p_-(x) := 1/(2(x+1))$. In this case the drift is asymptotically zero while the probability of hitting zero is exponentially decreasing, $1/2^x$. Clearly, the condition (3.2) fails here.

3.2 Drift of Order $1/x$

In this section $r(x) > 0$ is a bounded decreasing differentiable function satisfying (2.7) with $c = 1$, that is,

$$0 \geq r'(x) \geq -r^2(x) \quad \text{for all } x \geq 0, \tag{3.3}$$

which yields

$$r(x) \geq \frac{1}{c_1 + x} \quad \text{for all } x \geq 0,$$

where $c_1 = 1/r(0)$. Then, in particular,

$$R(x) := \int_0^x r(y)dy \to \infty \quad \text{as } x \to \infty; \tag{3.4}$$

in what follows we define $R(x) = 0$ for $x < 0$. The increasing function $R(x)$ is concave on the positive half-line because $r(x)$ is decreasing. As shown in (2.11) and (2.12),

$$R(x) + \frac{h}{1+h} \leq R(x + h/r(x)) \leq R(x) + h, \tag{3.5}$$

$$R(x) - \frac{h}{1-h} \leq R(x - h/r(x)) \leq R(x) - h. \tag{3.6}$$

Then, as already discussed in Section 2.1, $1/r(x)$ is a natural x-step responsible for the constant increase of the function $R(x)$ and, for any increasing function $s(x)$ of order $o(1/r(x))$,

$$R(x \pm s(x)) = R(x) + o(1), \tag{3.7}$$

$$r(x \pm s(x)) \sim r(x) \quad \text{as } x \to \infty. \tag{3.8}$$

Fix an increasing function $s(x) \to \infty$ as $x \to \infty$ such that $x - s(x)$ increases and $s(x) = o(x)$.

Specifically, in this section we consider a transient Markov chain $\{X_n\}$ whose jumps are such that

$$m_2^{[s(x)]}(x) \to b > 0 \quad \text{and} \quad m_1^{[s(x)]}(x) \sim \mu/x \quad \text{as } x \to \infty, \tag{3.9}$$

where $\mu \geq b/2$. If $\mu > b/2$ then $\{X_n\}$ is transient, under some minor additional conditions; see Theorem 2.21. If $\mu = b/2$ then $\{X_n\}$ can still be transient, provided that there exists an appropriate logarithmic expansion of the first two truncated moments of jumps; see Corollary 2.20 for details. In addition, we assume that

$$\frac{2m_1^{[s(x)]}(x)}{m_2^{[s(x)]}(x)} = r(x) + o(p(x)) \quad \text{as } x \to \infty \tag{3.10}$$

for some decreasing positive function $r(x) \to 0$ satisfying $r(x)x \to 2\mu/b \geq 1$ as $x \to \infty$ and some decreasing integrable function $p(x) \geq 0$. Since $p(x)$ is decreasing and integrable, $p(x)x \to 0$ as $x \to \infty$. We also assume that

$$r'(x) \sim -r(x)/x \sim -(b/2\mu)r^2(x) \quad \text{and} \quad p'(x) = O(r^2(x)). \tag{3.11}$$

It follows from Lemma 2.30 that the condition on $p'(x)$ is always satisfied for a properly chosen function p. Since $xr(x) \sim 2\mu/b \geq 1$,

$$R(x) = \int_0^x r(y)dy \sim \frac{2\mu}{b}\log x \quad \text{as } x \to \infty.$$

Assume that the function $e^{-R(x)}$ is integrable at infinity, which automatically holds if $2\mu/b > 1$. This allows us to define the following bounded decreasing function, which plays the most important rôle in our analysis of the down-crossing probability for a transient Markov chain:

$$U(x) := \int_x^\infty e^{-R(y)}dy \quad \text{for } x \geq 0; \tag{3.12}$$

$U(x) = U(0)$ for $x \leq 0$. As follows from the sequel, $U(X_n)$ is almost a martingale; see Corollary 3.6 below.

We have $U(x) \to 0$ as $x \to \infty$. According to our assumptions,

$$r(x) = \frac{2\mu}{b}\frac{1}{x} + \frac{\varepsilon(x)}{x},$$

where $\varepsilon(x) \to 0$ as $x \to \infty$. In view of the representation theorem for slowly varying functions, there exists a function $\ell(x)$, slowly varying at infinity, such that $e^{-R(x)} = x^{-\rho-1}\ell(x)$ and $U(x) \sim x^{-\rho}\ell(x)/\rho$, where $\rho := 2\mu/b - 1 \geq 0$.

The main result in this subsection is the following theorem, which provides lower and upper bounds for the down-crossing probability of transient Markov chains with asymptotically zero drift described above.

Theorem 3.2 *Let the drift conditions* (3.9) *and* (3.10) *be valid with* $\mu \geq b/2$ *and* $r(x)$ *satisfying the regularity condition* (3.11). *Let the function* $e^{-R(x)}$ *be integrable at infinity and* $\{X_n\}$ *be a transient Markov chain. For some increasing* $s(x) = o(x)$, *let the following integrability condition hold:*

$$\mathbb{E}\{|\xi(x)|^3;\ |\xi(x)| \leq s(x)\} = o(p(x)/r^2(x)) \quad \text{as } x \to \infty. \tag{3.13}$$

If the right jump tails satisfy an upper bound

$$\mathbb{P}\{\xi(x) > s(x)\} = o(p(x)e^{-R(x)}/U(x)) \quad \text{as } x \to \infty, \tag{3.14}$$

then there exist a constant $c_1 > 0$ *and a level* $\widehat{x}$ *such that*

$$\mathbb{P}_x\{X_n \leq x_0 \text{ for some } n\} \geq c_1\frac{U(x)}{U(x_0)} \quad \text{for all } x > x_0 \geq \widehat{x}$$

and, uniformly for all $x > x_0$,

$$\mathbb{P}_x\{X_n \le x_0 \text{ for some } n\} \ \ge \ (1+o(1))\frac{U(x)}{U(x_0)} \quad \text{as } x_0 \to \infty.$$

If the negative jumps satisfy the condition

$$\mathbb{E}\big\{U(x+\xi(x));\ \xi(x) < -s(x)\big\} = o(p(x)e^{-R(x)}) \quad \text{as } x \to \infty, \tag{3.15}$$

then there exist a constant $c_2 < \infty$ and a level $\widehat{x}$ such that

$$\mathbb{P}_x\{X_n \le x_0 \text{ for some } n\} \ \le \ c_2\frac{U(x)}{U(x_0)} \quad \text{for all } x > x_0 \ge \widehat{x}$$

and, uniformly for all $x > x_0$,

$$\mathbb{P}_x\{X_n \le x_0 \text{ for some } n\} \ \le \ (1+o(1))\frac{U(x)}{U(x_0)} \quad \text{as } x_0 \to \infty.$$

We may compare this with the results on down-crossing probabilities for Bessel processes, see (1.39), or for nearest neighbour Markov chains, see Section 1.4.4.

In Theorem 11.10, we will apply the last result to derive bounds for the ruin probability of level-dependent collective risk processes.

In the case $\rho = 2\mu/b - 1 > 0$ the last asymptotic results may be specified as follows.

Corollary 3.3 *Let $\{X_n\}$ be a transient Markov chain. Let the drift conditions* (3.9) *and* (3.10) *be valid with $\mu > b/2$ and $r(x)$ satisfying the regularity condition* (3.11). *For some increasing $s(x) = o(x)$, let the following integrability condition hold:*

$$\mathbb{E}\big\{|\xi(x)|^3;\ |\xi(x)| \le s(x)\big\} = o(p(x)x^2) \quad \text{as } x \to \infty.$$

If the right jump tails satisfy the upper bound

$$\mathbb{P}\{\xi(x) > s(x)\} = o(p(x)/x) \quad \text{as } x \to \infty$$

and the negative jumps satisfy the condition

$$\mathbb{E}\big\{U(x+\xi(x));\ \xi(x) < -s(x)\big\} = o(p(x)e^{-R(x)}) \quad \text{as } x \to \infty$$

then, for any $\varepsilon > 0$,

$$\mathbb{P}_x\{X_n \le \gamma x \text{ for some } n\} \ \to \ \gamma^\rho \quad \text{as } x \to \infty$$

uniformly for all $\gamma \in (\varepsilon, 1)$.

To specify the asymptotics in the case $\rho = 2\mu/b - 1 = 0$, we need to consider the logarithmic expansions of the first two truncated moments of jumps. We assume that, for some $m \in \mathbb{N}$ and $\varepsilon > 0$,

$$r(x) = \left(\frac{1}{y} + \frac{1}{y\log y} + \cdots + \frac{1}{y\log y \cdots \log_{(m-1)} y} + \frac{1+\varepsilon}{y\log y \cdots \log_{(m)} y}\right)\bigg|_{y=x+e^{(m)}}. \tag{3.16}$$

Then

$$R(x) = \big(\log y + \log\log y + \cdots + \log_{(m)} y + (1+\varepsilon)\log_{(m+1)} y\big)\big|_{y=x+e^{(m)}} - \big(e^{(m-1)} + e^{(m-2)} + \cdots + 1\big)$$

and

$$U(x) = \frac{e^{(m)}e^{(m-1)}\cdots 1}{\varepsilon \log^{\varepsilon}_{(m)}(x+e^{(m)})}.$$

Corollary 3.4 *Let the drift conditions* (3.9) *and* (3.10) *be valid with* $\mu = b/2$ *and* $r(x)$ *satisfying* (3.16) *and the regularity condition* (3.11). *Let* $\{X_n\}$ *be a transient Markov chain. For some increasing* $s(x) = o(x)$, *let the following integrability condition hold:*

$$\mathbb{E}\big\{|\xi(x)|^3;\ |\xi(x)| \le s(x)\big\} = o(p(x)x^2) \quad \text{as } x \to \infty.$$

If the right jump tails satisfy the upper bound

$$\mathbb{P}\{\xi(x) > s(x)\} = o(p(x)/x\log x \cdots \log_{(m)} x) \quad \text{as } x \to \infty$$

and the negative jumps satisfy the condition

$$\mathbb{E}\big\{1/\log^{\varepsilon}_{(m)}(x+\xi(x));\ \xi(x) < -s(x)\big\} = o(p(x)/x\log x \cdots \log^{1+\varepsilon}_{(m)} x)$$

as $x \to \infty$, *then, uniformly for all* $x > x_0$,

$$\mathbb{P}_x\{X_n \le x_0 \text{ for some } n\} \sim \left(\frac{\log_{(m)} x_0}{\log_{(m)} x}\right)^{\varepsilon} \quad \text{as } x_0 \to \infty.$$

To prove Theorem 3.2, first let us prove some auxiliary results. We start by defining the decreasing Lyapunov functions needed. Without loss of generality we assume that $p(x) \le r(x)$ for all x. Consider the functions $r_+(x) := r(x) + p(x)$ and $r_-(x) := r(x) - p(x)$ and let

$$\begin{aligned} R_\pm(x) &:= \int_0^x r_\pm(y)dy, \\ U_\pm(x) &:= \int_x^\infty e^{-R_\pm(y)}dy, \quad x \ge 0, \end{aligned} \tag{3.17}$$

with $U_\pm(x) = U_\pm(0)$ for $x \le 0$. We have $0 \le r_-(x) \le r(x) \le r_+(x)$, $0 \le R_-(x) \le R(x) \le R_+(x)$, and $U_-(x) \ge U(x) \ge U_+(x) > 0$. Since

$$C_p := \int_0^\infty p(y)dy$$

is finite, we have

$$R_\pm(x) = R(x) \pm C_p + o(1) \quad \text{as } x \to \infty. \tag{3.18}$$

Therefore,

$$U_\pm(x) \sim e^{\mp C_p} U(x) \to 0 \quad \text{as } x \to \infty. \tag{3.19}$$

Lemma 3.5 *If the integrability conditions* (3.13) *and* (3.14) *hold, then, as* $x \to \infty$,

$$\mathbb{E}\{U_+(x+\xi(x)) - U_+(x);\ \xi(x) \ge -s(x)\} \ge p(x)(1+o(1))e^{-R_+(x)}. \tag{3.20}$$

If the integrability conditions (3.13) *and* (3.15) *hold, then, as* $x \to \infty$,

$$\mathbb{E}U_-(x+\xi(x)) - U_-(x) \le -p(x)(1+o(1))e^{-R_-(x)}. \tag{3.21}$$

Since the function U_+ is decreasing, the lower bound (3.20) yields that

$$\mathbb{E}U_+(x+\xi(x)) - U_+(x) \ge p(x)(1+o(1))e^{-R_+(x)},$$

which is symmetric to (3.21). However, it is stated as in (3.20) because we apply it to truncated Markov chains; see the part of the proof of Theorem 3.2 concerning the lower bound.

Proof of Lemma 3.5. We start with the following decomposition:

$$\begin{aligned}\mathbb{E}U_\pm(x+\xi(x)) - U_\pm(x) &= \mathbb{E}\{U_\pm(x+\xi(x)) - U_\pm(x);\ \xi(x) < -s(x)\}\\ &\quad + \mathbb{E}\{U_\pm(x+\xi(x)) - U_\pm(x);\ |\xi(x)| \le s(x)\}\\ &\quad + \mathbb{E}\{U_\pm(x+\xi(x)) - U_\pm(x);\ \xi(x) > s(x)\}.\end{aligned} \tag{3.22}$$

Here the third term on the right-hand side is negative because $U_\pm$ is decreasing and it may be bounded below as follows:

$$\begin{aligned}\mathbb{E}\{U_\pm(x+\xi(x)) - U_\pm(x);\ \xi(x) > s(x)\} &\ge -U_\pm(x)\mathbb{P}\{\xi(x) > s(x)\}\\ &= o\big(p(x)e^{-R_\pm(x)}\big),\end{aligned} \tag{3.23}$$

provided the condition (3.14) holds and owing to the relations (3.18) and (3.19). Furthermore, the first term on the right-hand side of (3.22) is positive and possesses the following upper bound:

$$\begin{aligned}
&\mathbb{E}\{U_\pm(x+\xi(x)) - U_\pm(x);\ \xi(x) < -s(x)\} \\
&\quad \le \mathbb{E}\{U_\pm(x+\xi(x));\ \xi(x) < -s(x)\} \\
&\quad = o\big(p(x)e^{-R_\pm(x)}\big),
\end{aligned} \tag{3.24}$$

provided the condition (3.15) holds and owing to the relations (3.18) and (3.19). To estimate the second term on the right-hand side of (3.22), we make use of a Taylor expansion:

$$\begin{aligned}
&\mathbb{E}\{U_\pm(x+\xi(x)) - U_\pm(x);\ |\xi(x)| \le s(x)\} \\
&\quad = U'_\pm(x)\mathbb{E}\{\xi(x); |\xi(x)| \le s(x)\} + \frac{1}{2}U''_\pm(x)\mathbb{E}\{\xi^2(x); |\xi(x)| \le s(x)\} \\
&\qquad + \frac{1}{6}\mathbb{E}\big\{U'''_\pm(x+\theta\xi(x))\xi^3(x); |\xi(x)| \le s(x)\big\},
\end{aligned} \tag{3.25}$$

where $0 \le \theta = \theta(x,\xi(x)) \le 1$. By the construction of $U_\pm$,

$$U'_\pm(x) = -e^{-R_\pm(x)}, \quad U''_\pm(x) = r_\pm(x)e^{-R_\pm(x)} = (r(x) \pm p(x))e^{-R_\pm(x)}. \tag{3.26}$$

Thus it follows that

$$\begin{aligned}
&U'_\pm(x)m_1^{[s(x)]}(x) + \frac{1}{2}U''_\pm(x)m_2^{[s(x)]}(x) \\
&\quad = e^{-R_\pm(x)}\Big(-m_1^{[s(x)]}(x) + (r(x) \pm p(x))\frac{m_2^{[s(x)]}(x)}{2}\Big) \\
&\quad = \frac{m_2^{[s(x)]}(x)}{2}e^{-R_\pm(x)}\Big(-\frac{2m_1^{[s(x)]}(x)}{m_2^{[s(x)]}(x)} + r(x) \pm p(x)\Big) \\
&\quad = \pm\frac{m_2^{[s(x)]}(x)}{2}e^{-R_\pm(x)}p(x)(1+o(1)),
\end{aligned} \tag{3.27}$$

by (3.10). Finally, let us estimate the last term in (3.25). Notice that, by the condition (3.11) on the derivative of $r(x)$ and $p(x)$,

$$\begin{aligned}
U'''_\pm(x) &= \big(r'(x) \pm p'(x) - (r(x) \pm p(x))^2\big)e^{-R_\pm(x)} \\
&= O(r^2(x))e^{-R_\pm(x)},
\end{aligned}$$

hence, owing to (3.7) and (3.8),

$$U'''_\pm(x+y) = O(r^2(x))e^{-R_\pm(x)}$$

as $x \to \infty$ uniformly for $|y| \le s(x)$, which implies that

$$\begin{aligned}
&\big|\mathbb{E}\big\{U'''_\pm(x+\theta\xi(x))\xi^3(x); |\xi(x)| \le s(x)\big\}\big| \\
&\quad \le c_1 r^2(x)\mathbb{E}\big\{|\xi^3(x)|;\ |\xi(x)| \le s(x)\big\}e^{-R_\pm(x)}.
\end{aligned}$$

Thus, in view of (3.13),

$$\left|\mathbb{E}\left\{U_{\pm}'''(x+\theta\xi(x))\xi^3(x);\ |\xi(x)|\le s(x)\right\}\right| = o\left(p(x)e^{-R_{\pm}(x)}\right). \tag{3.28}$$

Substituting (3.27) and (3.28) into (3.25), we obtain that

$$\begin{aligned}\mathbb{E}\{U_{\pm}(x+\xi(x)) - U_{\pm}(x);\ |\xi(x)|\le s(x)\}&\\ = \pm m_2^{[s(x)]}(x)p(x)(1+o(1))e^{-R_{\pm}(x)}.&\end{aligned} \tag{3.29}$$

Substituting (3.23) – or (3.24) – and (3.29) into (3.22), we finally come to the desired conclusion. □

Lemma 3.5 implies the following result.

Corollary 3.6 *Under the conditions of Lemma 3.5, there exists an $\widehat{x}$ such that, for all $x > \widehat{x}$,*

$$\begin{aligned}\mathbb{E}U_-(x+\xi(x)) - U_-(x) &\le 0,\\ \mathbb{E}\{U_+(x+\xi(x)) - U_+(x);\ \xi(x)\ge -s(x)\} &\ge 0.\end{aligned}$$

Proof of Theorem 3.2. The process $U_-(X_n)$ is bounded above by $U_-(0)$. Let $\widehat{x}$ be any level guaranteed by the last corollary and let $x_0 \ge \widehat{x}$, $B = (-\infty, x_0]$, and $\tau_B = \min\{n\ge 1 : X_n \in B\}$.

By Corollary 3.6, $U_-(X_{n\wedge\tau_B})$ is a bounded supermartingale. Hence, by the optional stopping theorem (see e.g. [56, Section 4.7]), for $z > \widehat{x}$ and $x \in (\widehat{x}, z)$,

$$\mathbb{E}_x U_-(X_{\tau_B\wedge\tau_{(z,\infty)}}) \le \mathbb{E}_x U_-(X_0) = U_-(x).$$

Letting $z\to\infty$ we conclude that

$$\begin{aligned}&\mathbb{E}_x\{U_-(X_{\tau_B});\ \tau_B<\infty\}\\ &\quad - \lim_{z\to\infty}\mathbb{E}_x\{U_-(X_{\tau_B});\ \tau_B < \tau_{(z,\infty)}\}\\ &= \lim_{z\to\infty}\mathbb{E}_x U_-(X_{\tau_B\wedge\tau_{(z,\infty)}}) - \lim_{z\to\infty}\mathbb{E}_x\{U_-(X_{\tau_{(z,\infty)}});\ \tau_B > \tau_{(z,\infty)}\}\\ &\le U_-(x) - 0 = U_-(x).\end{aligned}$$

On the other hand, since U_- is decreasing,

$$\mathbb{E}_x\{U_-(X_{\tau_B});\ \tau_B<\infty\} \ge U_-(x_0)\mathbb{P}_x\{\tau_B<\infty\}.$$

Therefore

$$\mathbb{P}_x\{\tau_B<\infty\} \le \frac{U_-(x)}{U_-(x_0)}, \tag{3.30}$$

which implies both upper bounds of the theorem, by (3.19).

On the other hand, let

$$U_{+0}(x) := \begin{cases} U_+(x_0 - s(x_0)) & \text{if } x \le x_0 - s(x_0), \\ U_+(x) & \text{if } x > x_0 - s(x_0). \end{cases}$$

As $x - s(x)$ is increasing,

$$\mathbb{E}\{U_{+0}(x + \xi(x));\ \xi(x) \ge -s(x)\} = \mathbb{E}\{U_+(x + \xi(x));\ \xi(x) \ge -s(x)\}$$

for all $x > x_0$. Therefore the process $\{U_{+0}(X_{n\wedge\tau_B})\}$ is a bounded submartingale owing to the lower bound provided by Corollary 3.6. Hence, again by the optional stopping theorem, for $x > x_0$,

$$\mathbb{E}_x\{U_{+0}(X_{\tau_B});\ \tau_B < \infty\} \ge \mathbb{E}_x U_{+0}(X_0) = U_+(x).$$

On the other hand, since U_{+0} is bounded by $U_+(x_0 - s(x_0))$,

$$\mathbb{E}_x\{U_{+0}(X_{\tau_B});\ \tau_B < \infty\} \le U_+(x_0 - s(x_0))\mathbb{P}_x\{\tau_B < \infty\}.$$

This allows us to deduce a lower bound:

$$\mathbb{P}_x\{\tau_B < \infty\} \ge \frac{U_+(x)}{U_+(x_0 - s(x_0))},$$

which completes the proof of both lower bounds, owing to (3.7) and (3.19). □

3.3 The Case where $xm_1(x) \to \infty$ but $m_1(x) = o(1/\sqrt{x})$

In this section we consider a transient Markov chain $\{X_n\}$ whose jumps are such that

$$m_2^{[s(x)]}(x) \to b > 0 \text{ and } xm_1^{[s(x)]}(x) \to \infty \qquad \text{as } x \to \infty, \tag{3.31}$$

for some increasing function $s(x) = o(x)$, which implies transience subject to some minor additional conditions; see Theorem 2.21. In addition, we assume that

$$\frac{2m_1^{[s(x)]}(x)}{m_2^{[s(x)]}(x)} = r(x) + o(p(x)) \quad \text{as } x \to \infty \tag{3.32}$$

for a decreasing positive differentiable function $r(x) \to 0$ satisfying $r(x)x \to \infty$ as $x \to \infty$ and some decreasing differentiable function $p(x) \ge 0$ which is assumed to be integrable,

$$C_p := \int_0^\infty p(x)dx < \infty. \tag{3.33}$$

Since $p(x)$ is decreasing and integrable, $p(x)x \to 0$ as $x \to \infty$.

In this subsection we consider the case where $r(x)=o(1/\sqrt{x})$; more precisely,

$$r^2(x) = o(p(x)) \quad \text{as } x \to \infty. \tag{3.34}$$

We also assume that

$$p'(x) \;=\; o(r^2(x)) \text{ and } r'(x) = o(r^2(x)) \quad \text{as } x \to \infty. \tag{3.35}$$

In view of (3.31), the condition (3.32) is equivalent to

$$-m_1^{[s(x)]}(x) + \frac{m_2^{[s(x)]}(x)}{2} r(x) = o(p(x)) \quad \text{as } x \to \infty. \tag{3.36}$$

Define the increasing function $R(x)$ as in (3.4). Since $xr(x) \to \infty$, the function $e^{-R(x)}$ is integrable at infinity. This allows us to define the decreasing function $U(x)$ as in (3.12), and this plays a key rôle in the next result.

Theorem 3.7 *Let $\{X_n\}$ be a transient Markov chain whose first two moments of jumps truncated at some level $s(x) = o(1/r(x))$ satisfy* (3.31) *and* (3.32) *while $r(x)$ satisfies* (3.34). *Assume the regularity condition* (3.35). *Let the following integrability condition on jumps hold:*

$$\mathbb{E}\{|\xi(x)|^3;\ |\xi(x)| \le s(x)\} = o(p(x)/r^2(x)) \quad \textit{as } x \to \infty. \tag{3.37}$$

If the right jump tails satisfy the upper bound

$$\mathbb{P}\{\xi(x) > s(x)\} = o(p(x)r(x)) \quad \textit{as } x \to \infty \tag{3.38}$$

then there exist a constant $c_1 > 0$ and a level $\widehat{x}$ such that

$$\mathbb{P}_x\{X_n \le x_0 \textit{ for some } n\} \;\ge\; c_1 \frac{U(x)}{U(x_0)} \quad \textit{for all } x > x_0 \ge \widehat{x}$$

and, uniformly for all $x > x_0$,

$$\mathbb{P}_x\{X_n \le x_0 \textit{ for some } n\} \;\ge\; (1+o(1)) \frac{U(x)}{U(x_0)} \quad \textit{as } x_0 \to \infty.$$

If the negative jumps satisfy the condition

$$\mathbb{E}\{U(x+\xi(x));\ \xi(x) < -s(x)\} = o\big(p(x)e^{-R(x)}\big) \quad \textit{as } x \to \infty, \tag{3.39}$$

then there exist a constant $c_2 < \infty$ and a level $\widehat{x}$ such that

$$\mathbb{P}_x\{X_n \le x_0 \textit{ for some } n\} \;\le\; c_2 \frac{U(x)}{U(x_0)} \quad \textit{for all } x > x_0 \ge \widehat{x}$$

and, uniformly for all $x > x_0$,

$$\mathbb{P}_x\{X_n \le x_0 \textit{ for some } n\} \;\le\; (1+o(1)) \frac{U(x)}{U(x_0)} \quad \textit{as } x_0 \to \infty.$$

Notice that the right-hand side of (3.37) may be bounded away from 0 only in the case where $p(x)/r^2(x) \to \infty$, which is equivalent to the condition (3.34).

To prove Theorem 3.7, we consider the same functions $r_\pm(x)$, $R_\pm(x)$, and $U_\pm(x)$ as in the previous subsection. The only difference is that, owing to (3.35), we have

$$\frac{U'(x)}{((1/r(x))e^{-R(x)})'} = \frac{-e^{-R(x)}}{(-r'(x)/r^2(x) - 1)e^{-R(x)}} \to 1 \quad \text{as } x \to \infty,$$

so L'Hôpital's rule yields

$$U(x) \sim \frac{1}{r(x)} e^{-R(x)} \quad \text{as } x \to \infty. \tag{3.40}$$

Then, similarly to Lemma 3.5 the following result holds.

Lemma 3.8 *If the integrability conditions* (3.37) *and* (3.38) *hold, then, as* $x \to \infty$,

$$\mathbb{E}\{U_+(x + \xi(x)) - U_+(x);\ \xi(x) \ge -s(x)\} \ge \frac{b + o(1)}{2} p(x) e^{-R_+(x)}. \tag{3.41}$$

If the integrability conditions (3.37) *and* (3.39) *hold then*

$$\mathbb{E}U_-(x + \xi(x)) - U_-(x) \le -\frac{b + o(1)}{2} p(x) e^{-R_-(x)} \quad \textit{as } x \to \infty. \tag{3.42}$$

Proof The calculations are the same as in Lemma 3.5 apart from the estimation of the third derivative of $U_\pm$. By the condition (3.35) on the derivatives of $r(x)$ and $p(x)$,

$$\begin{aligned} U_\pm'''(x) &= \big(r'(x) \pm p'(x) + (r(x) \pm p(x))^2\big) e^{-R_\pm(x)} \\ &= O\big(r^2(x) e^{-R_\pm(x)}\big). \end{aligned}$$

As is shown in (3.7), $R(x + s(x)) = R(x) + o(1)$ for any $s(x) = o(1/r(x))$. Therefore

$$\begin{aligned} &\big|\mathbb{E}\{U_\pm'''(x + \theta\xi(x))\xi^3(x); |\xi(x)| \le s(x)\}\big| \\ &\le c_1 r^2(x) \mathbb{E}\{|\xi^3(x)|;\ |\xi(x)| \le s(x)\} e^{-R_\pm(x)} \\ &= o\big(p(x) e^{-R_\pm(x)}\big), \end{aligned} \tag{3.43}$$

owing to the condition (3.37) on the third absolute moment.

This upper bound makes it possible to conclude that the desired results hold, in the same way as in Lemma 3.5. □

Lemma 3.8 implies the following result.

Corollary 3.9 *There exists an $\widehat{x}$ such that, for all $x > \widehat{x}$,*

$$\mathbb{E}U_-(x + \xi(x)) - U_-(x) \le 0,$$
$$\mathbb{E}\{U_+(x + \xi(x)) - U_+(x);\ \xi(x) \ge -s(x)\} \ge 0.$$

This corollary allows us to conclude the proof of Theorem 3.7 in the same way as for Theorem 3.2.

3.4 General Case where $xm_1(x) \to \infty$

If $r(x)$ decreases more slowly than $1/\sqrt{x}$ then the function $r^2(x)$ is not integrable and, since $U'''_\pm(x)$ is of order $r^2(x)e^{-R_\pm(x)}$, it does not possess a bound like $o\big(p(x)e^{-R_\pm(x)}\big)$. So, the last term in the Taylor expansion (3.25) is not negligible and instead makes a significant contribution to the drift of $U_\pm$. If $r(x)$ is sandwiched between $1/\sqrt{x}$ and $1/\sqrt[3]{x}$ then we need to consider a Taylor expansion that includes the fourth derivative of $U_\pm$ and, consequently, the fourth moment of the jumps. The more slowly $r(x)$ decreases, the higher the moments of jumps required.

So, in this subsection we use the same setting as in the previous subsection, but now we consider the general case and do not assume that $r(x) = o(1/\sqrt{x})$. Instead, we assume that, for some $\gamma \in \{2, 3, 4, \ldots\}$,

$$r^\gamma(x) = o(p(x)) \quad \text{as } x \to \infty \tag{3.44}$$

and

$$-m_1^{[s(x)]}(x) + \sum_{j=2}^{\gamma} (-1)^j \frac{m_j^{[s(x)]}(x)}{j!} r^{j-1}(x) = o(p(x)) \quad \text{as } x \to \infty, \tag{3.45}$$

where $p(x)$ is a decreasing integrable function. We further assume that the function $r(x)$ is γ times differentiable and, for all $1 \le k \le \gamma - 1$,

$$r^{(k)}(x) = o(r^\gamma(x)), p^{(k)}(x) = o(r^\gamma(x)) \quad \text{as } x \to \infty. \tag{3.46}$$

If $r(x) \sim c/x^\alpha$, where $\gamma\alpha < 2$, then it follows from Lemma 2.30 that the condition on the derivatives of $p(x)$ is always satisfied for a properly chosen function p, so the condition (3.46) on the derivatives of p does not restrict generality under this specific choice of $r(x)$.

In the next result, we consider the same functions $R(x)$ and $U(x)$ as in the previous subsection.

Theorem 3.10 *Let $\{X_n\}$ be a transient Markov chain whose first γ moments of jumps truncated at some level $s(x) = o(1/r(x))$ satisfy the conditions* (3.31) *and* (3.45)*, where* γ *is defined in* (3.44)*. Assume the regularity condition* (3.46) *and the integrability condition*

$$\mathbb{E}\big\{|\xi(x)|^{\gamma+1};\ |\xi(x)| \le s(x)\big\} = o(p(x)/r^\gamma(x)) \quad \text{as } x \to \infty. \tag{3.47}$$

If the right jump tails satisfy the upper bound

$$\mathbb{P}\{\xi(x) > s(x)\} = o(p(x)r(x)) \quad \text{as } x \to \infty, \tag{3.48}$$

then there exist a constant $c_1 > 0$ *and a level* $\widehat{x}$ *such that*

$$\mathbb{P}_x\{X_n \le x_0 \text{ for some } n\} \ \ge\ c_1 \frac{U(x)}{U(x_0)} \quad \text{for all } x > x_0 \ge \widehat{x}$$

and, uniformly for all $x > x_0$,

$$\mathbb{P}_x\{X_n \le x_0 \text{ for some } n\} \ \ge\ (1+o(1))\frac{U(x)}{U(x_0)} \quad \text{as } x_0 \to \infty.$$

If the negative jumps satisfy the condition

$$\mathbb{E}\big\{U(x+\xi(x));\ \xi(x) < -s(x)\big\} = o\big(p(x)e^{-R(x)}\big) \quad \text{as } x \to \infty \tag{3.49}$$

then there exist a constant $c_2 < \infty$ *and a level* $\widehat{x}$ *such that*

$$\mathbb{P}_x\{X_n \le x_0 \text{ for some } n\} \ \le\ c_2 \frac{U(x)}{U(x_0)} \quad \text{for all } x > x_0 \ge \widehat{x}$$

and, uniformly for all $x > x_0$,

$$\mathbb{P}_x\{X_n \le x_0 \text{ for some } n\} \ \le\ (1+o(1))\frac{U(x)}{U(x_0)} \quad \text{as } x_0 \to \infty.$$

Notice that the right-hand side of (3.47) may be bounded away from 0 only in the case where $p(x)/r^\gamma(x) \to \infty$, which is equivalent to the condition (3.44).

In Theorem 11.12, we will apply the last result to derive bounds for the ruin probability in level-dependent collective risk processes. We consider there the case where $r(x) \sim \theta/x^\alpha$ for some $\alpha \in (0,1)$ and assume asymptotic expansions for the first γ moments of jumps with respect to the powers of $1/x^\alpha$ and show that $U(x)$ is then a product of Weibull-type functions.

We consider the same functions $r_\pm(x)$, $R_\pm(x)$, and $U_\pm(x)$ as in the previous subsection and get the following result, which is similar to Lemma 3.8.

Lemma 3.11 *If the integrability conditions* (3.47) *and* (3.48) *hold then, as* $x \to \infty$,

$$\mathbb{E}\{U_+(x+\xi(x)) - U_+(x);\ \xi(x) \ge -s(x)\} \ge \frac{b+o(1)}{2} p(x)e^{-R_+(x)}. \tag{3.50}$$

If the integrability conditions (3.47) *and* (3.49) *hold then*

$$\mathbb{E}U_-(x+\xi(x)) - U_-(x) \le -\frac{b+o(1)}{2} p(x) e^{-R_-(x)} \quad \text{as } x \to \infty. \tag{3.51}$$

Proof We start with the decomposition (3.22), where the first and third terms on the right-hand side possess the same bounds as in the proof of Lemma 3.8.

To estimate the second term on the right-hand side of (3.22), we make use of a Taylor expansion with $\gamma + 1$ terms:

$$\begin{aligned} &\mathbb{E}\{U_\pm(x+\xi(x)) - U_\pm(x);\ |\xi(x)| \le s(x)\} \\ &= \sum_{k=1}^{\gamma} \frac{U_\pm^{(k)}(x)}{k!} m_k^{[s(x)]}(x) + \mathbb{E}\left\{\frac{U_\pm^{(\gamma+1)}(x+\theta\xi(x))}{(\gamma+1)!} \xi^{\gamma+1}(x); |\xi(x)| \le s(x)\right\}, \end{aligned} \tag{3.52}$$

where $0 \le \theta = \theta(x, \xi(x)) \le 1$. By the construction of $U_\pm$,

$$U'_\pm(x) = -e^{-R_\pm(x)}, \quad U''_\pm(x) = r_\pm(x) e^{-R_\pm(x)} = (r(x) \pm p(x)) e^{-R_\pm(x)}, \tag{3.53}$$

and, for $k = 3, \ldots, \gamma + 1$,

$$\begin{aligned} U_\pm^{(k)}(x) &= -(e^{-R_\pm(x)})^{(k-1)} \\ &= (-1)^k \big(r_\pm^{k-1}(x) + o(p(x))\big) e^{-R_\pm(x)} \quad \text{as } x \to \infty, \end{aligned}$$

where we know that the remainder term in the parentheses on the right is of order $o(p(x))$ by the conditions (3.46) and (3.44). By the definition of $r_\pm(x)$,

$$r_\pm^{k-1}(x) - (r(x) \pm p(x))^{k-1} = r^{k-1}(x) + o(p(x)) \quad \text{for all } k \ge 3,$$

which implies the relation

$$U_\pm^{(k)}(x) = (-1)^k \big(r^{k-1}(x) + o(p(x))\big) e^{-R_\pm(x)} \quad \text{as } x \to \infty. \tag{3.54}$$

It follows from the equalities (3.53) and (3.54) that

$$\begin{aligned} &\sum_{k=1}^{\gamma} \frac{U_\pm^{(k)}(x)}{k!} m_k^{[s(x)]}(x) \\ &= e^{-R_\pm(x)} \left(\sum_{k=1}^{\gamma} (-1)^k \frac{r^{k-1}(x)}{k!} m_k^{[s(x)]}(x) + o(p(x)) \pm p(x) \frac{m_2^{[s(x)]}(x)}{2} \right) \\ &= e^{-R_\pm(x)} \left(o(p(x)) \pm p(x) \frac{m_2^{[s(x)]}(x)}{2} \right), \end{aligned} \tag{3.55}$$

by the condition (3.45). Owing to the condition (3.46) on the derivatives of $r(x)$ and to (3.44), we have

$$U_\pm^{(\gamma+1)}(x) = (-1)^{\gamma+1}(r^\gamma(x) + o(r^\gamma(x)))e^{-R_\pm(x)}.$$

Then, similarly to (3.43), the last term in (3.52) possesses the following bound:

$$\begin{aligned}
&\Big|\mathbb{E}\Big\{\frac{U_\pm^{(\gamma+1)}(x+\theta\xi(x))}{(\gamma+1)!}\xi^{\gamma+1}(x);\ |\xi(x)|\le s(x)\Big\}\Big| \\
&\quad \le O\big(r^\gamma(x)e^{-R_\pm(x)}\big)\mathbb{E}\big\{|\xi(x)|^{\gamma+1};\ |\xi(x)|\le s(x)\big\} \\
&\quad = o\big(p(x)e^{-R_\pm(x)}\big),
\end{aligned}$$

by the condition (3.47). Therefore, it follows from (3.52) and (3.55) that

$$\begin{aligned}
&\mathbb{E}\{U_\pm(x+\xi(x)) - U_\pm(x);\ |\xi(x)|\le s(x)\} \\
&\quad = \pm p(x)\frac{m_2^{[s(x)]}(x)}{2}e^{-R_\pm(x)} + o\big(p(x)e^{-R_\pm(x)}\big).
\end{aligned}$$

Together with (3.23), (3.24), and (3.22) this completes the proof. □

Lemma 3.11 implies an analogue of Corollary 3.9 that allows us to conclude the proof of Theorem 3.10 in the same way as that of Theorem 3.2.

3.5 Upper Bound for Down-Crossing Probability

Now we produce some upper bounds for the down-crossing probability for a transient Markov chain that are rough versions of the more precise bounds derived in the previous sections. The main goal is to obtain upper bounds under weaker moment conditions than above.

Assume that there exists an $\widehat{x}$ such that

$$\frac{2m_1^{[s(x)]}(x)}{m_2^{[s(x)]}(x)} \ge r(x) > \frac{1}{x} \quad \text{for all } x > \widehat{x}, \tag{3.56}$$

where a decreasing differentiable function $r(x)$ satisfies the condition

$$r'(x) \ge -(1-\varepsilon)r^2(x), \quad \varepsilon > 0, \quad \text{for all } x > \widehat{x}. \tag{3.57}$$

Then the drift to the right dominates the diffusion and the corresponding Markov chain X is typically transient; see Theorem 2.21.

Theorem 3.12 *Assume that the drift of* $\{X_n\}$ *possesses the lower bound* (3.56), *with some* $r(x)$ *satisfying the condition* (3.57), *and* $s(x) = o(1/r(x))$. *For some* $\delta < \varepsilon$, *let*

$$\mathbb{E}\{e^{-\delta R(x+\xi(x))};\ \xi(x) < -s(x)\} = o\big(r^2(x)e^{-\delta R(x)}m_2^{[s(x)]}(x)\big) \quad \text{as } x \to \infty. \tag{3.58}$$

Then there exists an x_* such that, for all $y > x \ge x_*$,

$$\mathbb{P}_y\{X_n \le x \text{ for some } n \ge 1\} \le e^{\delta(R(x)-R(y))}.$$

In particular, owing to (3.6), *for any fixed $h > 0$ we have*

$$\mathbb{P}_x\{X_n \le x - h/r(x) \text{ for some } n \ge 1\} \le e^{-\delta h/2} \quad \text{for sufficiently large } x.$$

The condition (3.57) is satisfied for $r(x) = (1+2\varepsilon)/(1+x)$, hence the following corollary.

Corollary 3.13 *Assume that the drift of $\{X_n\}$ possesses the lower bound* (3.56) *with $r(x) = (1+2\varepsilon)/x$ for some $\varepsilon \in (0,1/2]$ and $s(x) = o(x)$. For some $\delta \in (0,\varepsilon)$ let*

$$\mathbb{E}\{(x+\xi(x))^{-\delta};\ \xi(x) < -s(x)\} = o\big(m_2^{[s(x)]}(x)/x^{2+\delta}\big) \quad \text{as } x \to \infty. \tag{3.59}$$

Then there exists an x_* such that, for all $y > x \ge x_*$,

$$\mathbb{P}_y\{X_n \le x \text{ for some } n \ge 1\} \le \left(\frac{1+x}{1+y}\right)^{\delta}.$$

The condition (3.57) is also satisfied for $r(x) = c/(1+x)^\beta$, $c > 0$, $\beta \in (0,1)$, for any $\varepsilon \in (0,1)$. Thus the following corollary holds true.

Corollary 3.14 *Assume that the drift of $\{X_n\}$ possesses the lower bound* (3.56) *with $r(x) = c/(1+x)^\beta$ for some $c > 0$, $\beta \in (0,1)$, and $s(x) = o(x^\beta)$. For some $\delta > 0$ let*

$$\mathbb{E}\{e^{-\delta(x+\xi(x))^{1-\beta}};\ \xi(x) < -s(x)\} = o\big(m_2^{[s(x)]}(x)e^{-\delta x^{1-\beta}}\big) \quad \text{as } x \to \infty. \tag{3.60}$$

Then there exists an x_* such that, for all $y > x \ge x_*$,

$$\mathbb{P}_y\{X_n \le x \text{ for some } n \ge 1\} \le e^{\delta(x^{1-\beta}-y^{1-\beta})}.$$

Proof of Theorem 3.12. Consider a decreasing test function $W(x) := e^{-\delta R(x)}$ which is bounded by 1. Let us prove that the mean drift of $W(x)$ is negative for all sufficiently large x. Indeed, since the function $W(x)$ is decreasing,

$$\begin{aligned}\mathbb{E}W(x+\xi(x)) - W(x) &\le \mathbb{E}\{W(x+\xi(x)) - W(x);\ \xi(x) \le s(x)\}\\ &\le \mathbb{E}\{W(x+\xi(x));\ \xi(x) < -s(x)\}\\ &\quad + W'(x)\mathbb{E}\{\xi(x);\ |\xi(x)| \le s(x)\}\\ &\quad + \frac{1}{2}W''(x+\theta\xi(x))\mathbb{E}\{\xi^2(x);\ |\xi(x)| \le s(x)\}\\ &=: E_1 + E_2 + E_3,\end{aligned} \tag{3.61}$$

where $0 \le \theta = \theta(x, \xi(x)) \le 1$, by a Taylor expansion. By the condition (3.58), the first term on the right-hand side is of order

$$E_1 = o\big(r^2(x)W(x)m_2^{[s(x)]}(x)\big) \quad \text{as } x \to \infty. \tag{3.62}$$

The second term on the right-hand side of (3.61) is

$$\begin{aligned}E_2 &= -\delta r(x)W(x)m_1^{[s(x)]}(x)\\ &\le -\frac{\delta}{2}r^2(x)W(x)m_2^{[s(x)]}(x) \quad \text{for } x \ge \widehat{x},\end{aligned} \tag{3.63}$$

owing to (3.56). In order to bound the third term on the right-hand side of (3.61), we first notice that, owing to (3.57),

$$\begin{aligned}W''(x) &= \delta\big(\delta r^2(x) - r'(x)\big)W(x)\\ &\le \delta(\delta+1-\varepsilon)r^2(x)W(x) \quad \text{for } x \ge 0.\end{aligned}$$

By (3.8) and (3.7),

$$W''(x+y) \le \delta(\delta+1-\varepsilon)(1+o(1))r^2(x)W(x)$$

as $x \to \infty$ uniformly for all $|y| \le s(x) = o(1/r(x))$. Thus

$$E_3 \le \frac{\delta}{2}(\delta+1-\varepsilon)(1+o(1))r^2(x)W(x)m_2^{[s(x)]}(x) \quad \text{as } x \to \infty. \tag{3.64}$$

Substituting (3.62)–(3.64) into (3.61) we deduce that

$$\mathbb{E}W(x+\xi(x)) - W(x) \le \frac{\delta}{2}\big(\delta-\varepsilon+o(1)\big)r^2(x)W(x)m_2^{[s(x)]}(x) \quad \text{as } x \to \infty.$$

Then there exists an x_* sufficiently large that

$$\mathbb{E}W(x+\xi(x)) - W(x) < 0 \quad \text{for all } x \ge x_*.$$

Now take $W_*(x) := \min(W(x), W(x_*))$ so that $\{W_*(X_n)\}$ constitutes a positive bounded supermartingale with respect to the filtration $\{\mathcal{F}_n\} = \{\sigma(X_k, k \le n)\}$. Hence we may apply Doob's inequality for non-negative supermartingales and deduce that, for all $y \ge x \ge 0$ (so that $W_*(y) \le W_*(x)$),

$$\mathbb{P}\Big\{\sup_{n\ge1} W_*(X_n) \ge W_*(x) \Big| X_0 = y\Big\} \le \frac{\mathbb{E}_y W_*(X_0)}{W_*(x)} = e^{\delta(R_*(x)-R_*(y))},$$

which is equivalent to the first claim of the theorem. □

Notice that the condition (3.57) fails for functions $r(x)$ that are asymptotically equivalent to $1/x$, which arise when we consider the case of iterated logarithms; see e.g. Corollary 2.20. To cope with such functions, we introduce a decreasing twice-differentiable function $\widetilde{r}(x) > 0$ such that $\widetilde{r} \le r$, and, for some $\varepsilon > 0$,

$$\widetilde{r}\,'(x) \ge -\widetilde{r}^2(x)\Big(\frac{r(x)}{\widetilde{r}(x)} - \varepsilon\Big) \quad \text{for } x \ge \widehat{x}, \tag{3.65}$$

which, in particular, implies $\widetilde{r}\,'(x) \ge -\widetilde{r}(x)r(x)$. Notice that, for $\widetilde{r}(x) = r(x)$, the condition (3.65) reduces to (3.57). We also assume that

$$\widetilde{r}\,''(x) \;=\; O(\widetilde{r}(x)r^2(x)) \quad \text{as } x \to \infty. \tag{3.66}$$

Denote

$$\widetilde{R}(x) := \int_0^x \widetilde{r}(y)dy \quad \text{for all } x > 0,$$

and $\widetilde{R}(x) := 0$ for all $x \le 0$.

Theorem 3.15 *Assume that the drift of $\{X_n\}$ possesses the lower bound (3.56) with the function $r(x)$ satisfying (3.3), $\widetilde{r}(x)$ satisfying (3.65)–(3.66), and $s(x) = o(\widetilde{r}(x)/r^2(x))$. For some $\delta < \varepsilon$ let*

$$\mathbb{E}\{e^{-\delta\widetilde{R}(x+\xi(x))};\ \xi(x) < -s(x)\} = o\big(\widetilde{r}^{\,2}(x)e^{-\delta\widetilde{R}(x)}m_2^{[s(x)]}(x)\big) \text{ as } x \to \infty. \tag{3.67}$$

Then there exists an x_* such that, for all $y > x \ge x_*$,

$$\mathbb{P}_y\{X_n \le x \text{ for some } n \ge 1\} \le e^{\delta(\widetilde{R}(x)-\widetilde{R}(y))}.$$

The condition (3.65) is satisfied for

$$r(x) = \Big(\frac{1}{y} + \cdots + \frac{1}{y\log y\cdots\log_{(m-1)} y} + \frac{1+\varepsilon}{y\log y\cdots\log_{(m)} y}\Big)\Big|_{y=e^{(m)}+x},$$

for some $\varepsilon > 0$, $m \ge 1$, and

$$\widetilde{r}(x) = \frac{1}{y\log y\cdots\log_{(m)} y}\Big|_{y=e^{(m)}+x}.$$

In this case

$$\widetilde{R}(x) = \log_{(m+1)}(e^{(m)} + x),$$

and hence the following corollary holds true.

Corollary 3.16 *Assume that the drift of $\{X_n\}$ possesses the lower bound (3.56) with $r(x)$ as defined above and $s(x) = o(x/\log x \cdots \log_{(m)} x)$. For some $\delta \in (0, \varepsilon)$, as $x \to \infty$, let*

$$\mathbb{E}\{\log_{(m)}^{-\delta}(x + \xi(x));\ \xi(x) < -s(x)\} = o\big(m_2^{[s(x)]}(x)/x^2 \log^2 x \cdots \log_{(m)}^{2+\delta} x\big). \tag{3.68}$$

Then there exists an x_* such that, for all $y > x \geq x_*$,

$$\mathbb{P}_y\{X_n \leq x \text{ for some } n \geq 1\} \leq \left(\frac{\log_{(m)}(e^{(m)} + x)}{\log_{(m)}(e^{(m)} + y)}\right)^{\delta}.$$

Proof of Theorem 3.15. Let us consider a decreasing test function $\widetilde{W}(x) := e^{-\delta \widetilde{R}(x)}$ which is bounded by 1 and let us prove that the mean drift of $\widetilde{W}(x)$ is negative for all sufficiently large x. Indeed, since the function $\widetilde{W}(x)$ is decreasing,

$$\begin{aligned}
\mathbb{E}\widetilde{W}(x + \xi(x)) - \widetilde{W}(x) &\leq \mathbb{E}\{\widetilde{W}(x + \xi(x)) - \widetilde{W}(x);\ \xi(x) \leq s(x)\} \\
&\leq \mathbb{E}\{\widetilde{W}(x + \xi(x));\ \xi(x) < -s(x)\} \\
&\quad + \widetilde{W}'(x)\mathbb{E}\{\xi(x);\ |\xi(x)| \leq s(x)\} \\
&\quad + \frac{1}{2}\widetilde{W}''(x)\mathbb{E}\{\xi^2(x);\ |\xi(x)| \leq s(x)\} \\
&\quad + \frac{1}{6}\mathbb{E}\{\widetilde{W}'''(x + \theta\xi(x))\xi^3(x);\ |\xi(x)| \leq s(x)\} \\
&=: E_1 + E_2 + E_3 + E_4,
\end{aligned} \tag{3.69}$$

where $0 \leq \theta = \theta(x, \xi(x)) \leq 1$, using a Taylor expansion. By the same arguments as in the last proof, as $x \to \infty$ we have

$$E_1 = o\big(\widetilde{r}^2(x)\widetilde{W}(x)m_2^{[s(x)]}(x)\big), \tag{3.70}$$

$$E_2 \leq -\frac{\delta}{2}\widetilde{r}(x)r(x)W(x)m_2^{[s(x)]}(x), \tag{3.71}$$

$$E_3 = \frac{\delta}{2}(\delta\widetilde{r}^2(x) - \widetilde{r}'(x))\widetilde{W}(x)m_2^{[s(x)]}(x). \tag{3.72}$$

Next, owing to (3.65), (3.66), and the inequality $\widetilde{r} \leq r$,

$$\begin{aligned}
|\widetilde{W}'''(x)| &= \delta\big|-\delta^2\widetilde{r}^3(x) + 3\delta\widetilde{r}(x)\widetilde{r}'(x) - \widetilde{r}''(x)\big|\widetilde{W}(x) \\
&\leq c\widetilde{r}(x)r^2(x)\widetilde{W}(x) \quad \text{for some } c < \infty.
\end{aligned}$$

By (3.3), $r(x+y) \sim r(x)$ and, by (3.66), $\widetilde{r}(x+y) \sim \widetilde{r}(x)$, $\widetilde{R}(x+y) \sim \widetilde{R}(x)$, and $\widetilde{W}(x+y) \sim \widetilde{W}(x)$ as $x \to \infty$ uniformly for all $|y| \leq s(x) = o(\widetilde{r}(x)/r^2(x))$, which implies

$$|\widetilde{W}'''(x + y)| \leq c_1\widetilde{r}(x)r^2(x)\widetilde{W}(x)$$

as $x \to \infty$ uniformly for all $|y| \le s(x)$. Then

$$\begin{aligned}|E_4| &\le c_1\widetilde{r}(x)r^2(x)\widetilde{W}(x)\mathbb{E}\{|\xi^3(x)|;\ |\xi(x)| \le s(x)\}\\ &\le c_1 s(x)\widetilde{r}(x)r^2(x)\widetilde{W}(x)m_2^{[s(x)]}(x)\\ &= o(\widetilde{r}^2(x))\widetilde{W}(x)m_2^{[s(x)]}(x) \quad \text{as } x \to \infty, \end{aligned} \tag{3.73}$$

since $s(x) = o(\widetilde{r}(x)/r^2(x))$. Substituting (3.70)–(3.73) into (3.69) we deduce that

$$\begin{aligned}&\mathbb{E}\widetilde{W}(x+\xi(x)) - \widetilde{W}(x)\\ &\quad\le \frac{\delta}{2}\big(-r(x)\widetilde{r}(x) + \delta\widetilde{r}^2(x) - \widetilde{r}'(x) + o(\widetilde{r}^2(x))\big)\widetilde{W}(x)m_2^{[s(x)]}(x)\\ &\quad\le \frac{\delta}{2}\big((\delta-\varepsilon)\widetilde{r}^2(x) + o(\widetilde{r}^2(x))\big)\widetilde{W}(x)m_2^{[s(x)]}(x) \quad \text{as } x \to \infty,\end{aligned}$$

owing to the condition (3.65). Then there exists an x_* sufficiently large that

$$\mathbb{E}\widetilde{W}(x+\xi(x)) - \widetilde{W}(x) < 0 \quad \text{for all } x \ge x_*,$$

which concludes the proof in the same way as in Theorem 3.12. □

3.6 Comments for Chapter 3

The only asymptotic result on down-crossing probabilities for transient Markov chains with asymptotically zero drift of which we are aware of was obtained by Vatutin [144] in the context of critical branching processes with immigration. He derived asymptotics for the probability of hitting zero for such processes, which agree with our lower and upper bounds presented in Theorem 3.2 for general Markov chains. A reduction of a critical branching process with immigration to a Markov chain with drift of order c/x and bounded second moment of jumps via an $\sqrt{x}$-transform is discussed in Section 11.3.

4

Limit Theorems for Transient and Null Recurrent Markov Chains with Drift Proportional to $1/x$

Assume that the first two moments of the jumps of a Markov chain $\{X_n\}$ demonstrate regular behaviour at infinity, namely

$$m_1(x) \sim \mu/x, \quad m_2(x) \to b > 0 \quad \text{as } x \to \infty.$$

Then, as follows from Corollaries 2.16 and 2.19 under additional technical conditions,

- if $\mu \in (-b/2, b/2)$ then $\{X_n\}$ is null recurrent and $X_n \to \infty$ in probability as $n \to \infty$, say if $\{X_n\}$ is countable;
- if $\mu > b/2$ then $\{X_n\}$ is transient and $X_n \to \infty$ with probability 1 as $n \to \infty$.

It turns out that in both cases X_n increases at a rate $\sqrt{n}$. More precisely, the following weak convergence is observed:

$$\frac{X_n^2}{bn} \Rightarrow \Gamma_{1/2+\mu/b,2} \quad \text{as } n \to \infty.$$

This is the main topic that we discuss in this chapter, and we include results concerning the renewal function, which is well defined in the transient case.

4.1 Truncation of Jumps

In what follows, we repeatedly make use of a truncation technique for proving various limit theorems. The idea behind this is that if we truncate jumps at a sufficiently high level then we get a new Markov chain whose trajectories agree with that of the original chain with high probability.

Let $\{B(x) \subseteq \mathbb{R},\ x \in \mathbb{R}\}$ be a collection of Borel sets. Given a Markov chain $\{X_n\}$ with jumps $\xi(x)$, consider a modified Markov chain $\{\widetilde{X}_n\}$ whose jumps $\widetilde{\xi}(x)$ are defined by

$$\widetilde{\xi}(x) = \begin{cases} \xi(x) & \text{if } \xi(x) \in B(x), \\ \text{any value} & \text{if } \xi(x) \notin B(x). \end{cases}$$

Here our standard choice is either $B(x) = [-s(x), s(x)]$ or $[-s(x), \infty)$ and 'any value' is taken to be 0, which corresponds to the truncation of the original jumps $\xi(x)$ at levels $-s(x)$ or $s(x)$.

In this section, we prove a coupling that allows us to compare two Markov chains which have asymptotically equal jumps. This result is used each time we want to simplify our calculations relating to the characteristics of $\{X_n\}$. We formulate the result in a more general setting as follows.

Let $Y = \{Y_n\}$ and $Z = \{Z_n\}$ be two Markov chains with jumps $\eta(x)$ and $\zeta(x)$ respectively. Denote by H_y^Z the renewal measure generated by the chain Z with initial state $Z_0 = y$, that is,

$$H_y^Z(A) := \sum_{n=0}^{\infty} \mathbb{P}_y\{Z_n \in A\}, \quad A \in \mathcal{B}(\mathbb{R}).$$

Lemma 4.1 *Assume that the random variables $\eta(x)$ and $\zeta(x)$ can be constructed on the same probability space in such a way that*

$$\mathbb{P}\{\eta(x) \neq \zeta(x)\} \le p(x)v(x) \quad \textit{for all } x, \tag{4.1}$$

where $v(x) > 0$ and $p(x) > 0$ are decreasing functions and $p(x)$ is integrable at infinity. Also, for all $z \in \mathbb{R}$, let

$$\mathbb{P}\{Z_n > z \textit{ for all } n \ge 0 \mid Z_0 = y\} \to 1 \quad \textit{as } y \to \infty, \tag{4.2}$$

and, for some $c < \infty$ and an increasing function $l(x) > 0$ satisfying $l(x + l(x)) \le c_1 l(x)$ for all x,

$$H_y^Z(x, x + l(x)] \le c\frac{l(x)}{v(x)} \quad \textit{for all } y \textit{ and } x. \tag{4.3}$$

Then, for any $\varepsilon > 0$ there exists an x_ε such that the chains $\{Y_n\}$ and $\{Z_n\}$ can be constructed on the same probability space in such a way that

$$\mathbb{P}\{Y_n = Z_n \textit{ for all } n \ge 0\} \ge 1 - \varepsilon \quad \textit{provided } Y_0 = Z_0 \ge x_\varepsilon. \tag{4.4}$$

Proof Let us construct a probability space and sequences of independent random fields $\{\eta_n(x), x \in \mathbb{R}\}_{n\ge 0}$ and $\{\zeta_n(x), x \in \mathbb{R}\}_{n \ge 0}$ on this space such that

$$\mathbb{P}\{\eta_n(x) \neq \zeta_n(x)\} \le p(x)v(x) \quad \text{for all } x \in \mathbb{R} \text{ and } n \ge 0, \tag{4.5}$$

which is possible owing to (4.1). Then let us define Markov chains $\{Y_n\}$ and $\{Z_n\}$ as follows: $Y_0 = Z_0$,

$$Y_{n+1} = Y_n + \eta_{n+1}(Y_n), \quad Z_{n+1} = Z_n + \zeta_{n+1}(Z_n), \quad n \ge 0.$$

Fix an $\varepsilon > 0$. For any z,

$$\begin{aligned}
&\mathbb{P}\{Y_n \neq Z_n \text{ for some } n \mid Z_0 = y\} \\
&\quad \leq \mathbb{P}\{Z_n \leq z + l(z) \text{ for some } n \mid Z_0 = y\} \\
&\qquad + \mathbb{P}\{Y_n \neq Z_n \text{ for some } n,\ Z_n > z + l(z) \text{ for all } n \mid Z_0 = y\}.
\end{aligned}$$

Owing to (4.2), there exists an $y_1(z)$ such that

$$\mathbb{P}\{Z_n \leq z + l(z) \text{ for some } n \mid Z_0 = y\} \leq \varepsilon/2 \quad \text{for all } y > y_1(z).$$

Given $Y_0 = Z_0 > z + l(z)$,

$$\begin{aligned}
&\mathbb{P}\{Y_n \neq Z_n \text{ for some } n,\ Z_n > z + l(z) \text{ for all } n \mid Z_0 = y\} \\
&\quad \leq \mathbb{P}\{\eta_{n+1}(Y_n) \neq \zeta_{n+1}(Z_n),\ Y_n = Z_n \text{ for some } n, \\
&\qquad Z_n > z + l(z) \text{ for all } n \mid Z_0 = y\}.
\end{aligned}$$

The probability on the right-hand side does not exceed the following sum:

$$\begin{aligned}
&\sum_{n=0}^{\infty} \mathbb{P}\{\eta_{n+1}(Y_n) \neq \zeta_{n+1}(Z_n),\ Y_n = Z_n > z + l(z) \mid Z_0 = y\} \\
&\quad = \int_{z+l(z)}^{\infty} \mathbb{P}\{\eta(x) \neq \zeta(x)\} H_y^Z(dx) \\
&\quad \leq \int_{z+l(z)}^{\infty} p(x)v(x) H_y^Z(dx),
\end{aligned}$$

by the condition (4.1). The last integral tends to 0 as $z \to \infty$. Indeed, both functions $p(z)$ and $v(x)$ are decreasing, hence

$$\int_{z+l(z)}^{\infty} p(x)v(x) H_y^Z(dx) \leq \sum_{i=1}^{\infty} p(x_i)v(x_i) H_y^Z(x_i, x_{i+1}],$$

where $x_0 := z$ and $x_{i+1} := x_i + l(x_i)$ for $i \geq 0$. Then, by the condition (4.3) on H_y^Z and the property $l(x + l(x)) \leq c_1 l(x)$,

$$\begin{aligned}
\int_{z+l(z)}^{\infty} p(x)v(x) H_y^Z(dx) &\leq c \sum_{i=1}^{\infty} p(x_i) l(x_i) \\
&= c \sum_{i=1}^{\infty} p(x_i) l(x_{i-1} + l(x_{i-1})) \\
&\leq c c_1 \sum_{i=1}^{\infty} p(x_i) l(x_{i-1}) \\
&= c c_1 \sum_{i=1}^{\infty} p(x_i)(x_i - x_{i-1}).
\end{aligned}$$

The function $p(x)$ is decreasing; therefore

$$\sum_{i=1}^{\infty} p(x_i)(x_i - x_{i-1}) \leq \int_z^{\infty} p(u)du \ \to\ 0 \quad \text{as } z \to \infty$$

because $p(x)$ is integrable. Hence,

$$\int_{z+l(z)}^{\infty} p(x)v(x)H_y^Z(dx) \to 0 \quad \text{as } z \to \infty \text{ uniformly for all } y, \tag{4.6}$$

which implies convergence to 0 of the integral from z to ∞. Then the integral from z to ∞ is less than $\varepsilon/2$ for a sufficiently large $z = z(\varepsilon)$, and this concludes the proof with $x_\varepsilon = y_1(z(\varepsilon))$. □

Assume that

$$\mathbb{P}\Big\{\limsup_{n\to\infty} Y_n = \infty\Big\} = 1 \tag{4.7}$$

and, for any distribution of Z_0,

$$Z_n \stackrel{\text{a.s.}}{\to} \infty \quad \text{as } n \to \infty. \tag{4.8}$$

Then, under the conditions of Lemma 4.1,

$$Y_n \stackrel{\text{a.s.}}{\to} \infty \quad \text{as } n \to \infty. \tag{4.9}$$

Indeed, given any $\varepsilon \in (0,1)$, by Lemma 4.1 there exists a level x_ε such that (4.4) holds. By the condition (4.7), the stopping time

$$\tau_\varepsilon := \min\{n \geq 0 : Y_n \geq x_\varepsilon\} \tag{4.10}$$

is finite with probability 1. Set $Z_0 = Y_{\tau_\varepsilon}$. Since then $Z_0 \geq x_\varepsilon$, it follows from (4.4) that, for all A,

$$\mathbb{P}\Big\{\liminf_{n\to\infty} Y_{\tau_\varepsilon+n} > A\Big\} \geq \mathbb{P}\Big\{\liminf_{n\to\infty} Z_n > A\Big\} - \varepsilon,$$

which, owing to (4.8), implies that, for all A,

$$\mathbb{P}\Big\{\liminf_{n\to\infty} Y_{\tau_\varepsilon+n} > A\Big\} \geq 1 - \varepsilon.$$

Therefore, owing to the finiteness of τ_ε,

$$\mathbb{P}\Big\{\liminf_{n\to\infty} Y_n > A\Big\} \geq 1 - \varepsilon,$$

for all $\varepsilon > 0$ and $A < \infty$. Owing to the arbitrary choice of $\varepsilon > 0$, we have

$$\mathbb{P}\Big\{\liminf_{n\to\infty} Y_n > A\Big\} = 1 \quad \text{for all } A < \infty;$$

hence (4.9) follows.

If, instead of (4.8), for any distribution of Z_0, it holds that

$$Z_n \xrightarrow{p} \infty \quad \text{as } n \to \infty \tag{4.11}$$

then

$$Y_n \xrightarrow{p} \infty \quad \text{as } n \to \infty. \tag{4.12}$$

To show this convergence, we again consider the stopping time (4.10) and define the same $Z_0 = Y_{\tau_\varepsilon}$. Since τ_ε is finite, there exists an N such that

$$\mathbb{P}\{\tau_\varepsilon > N\} \le \varepsilon.$$

Then, for $n > N$,

$$\begin{aligned}\mathbb{P}\{Y_n > A\} &\ge 1 - \mathbb{P}\{\tau_\varepsilon > N\} - \mathbb{P}\{\tau_\varepsilon \le N,\ Y_n \le A\}\\ &\ge 1 - \varepsilon - \sum_{k=0}^{N} \mathbb{P}\{\tau_\varepsilon = k,\ Z_{n-k} \le A\} - \varepsilon,\end{aligned}$$

owing to (4.4). Therefore,

$$\mathbb{P}\{Y_n > A\} \ge 1 - 2\varepsilon - \sum_{k=0}^{N} \mathbb{P}\{Z_{n-k} \le A\},$$

where each of the probabilities $\mathbb{P}\{Z_{n-k} \le A\}$ tends to zero as $n \to \infty$ uniformly for all $k \le N$. Thus,

$$\liminf_{n\to\infty} \mathbb{P}\{Y_n > A\} \ge 1 - 2\varepsilon$$

and (4.12) follows because of the arbitrary choice of $\varepsilon > 0$.

In particular, if, for some function $V(x)$, a centring sequence a_n, and a normalising sequence c_n we have that

$$\frac{V(Z_n) - a_n}{c_n} \quad \text{converges as } n \to \infty \text{ a.s. or weakly,}$$

then

$$\frac{V(Y_n) - a_n}{c_n} \quad \text{converges as } n \to \infty \text{ a.s. or weakly.}$$

4.2 Upper Bound for Average Up-Crossing Time for Transient Chain

Let us define

$$L(x,n) := \sum_{k=0}^{n-1} \mathbb{I}\{X_k \ge x\}. \tag{4.13}$$

The next theorem is devoted to the properties of $L(x, T(t))$, where $T(t)$ is the first up-crossing time:

$$T(t) := \min\{n \geq 1 : X_n > t\}.$$

Let $v(z) \downarrow 0$ be a decreasing function. Denote

$$V(u) := \int_0^u \frac{1}{v(z)} dz \quad \text{for } u \geq 0, \tag{4.14}$$

and $V(u) = 0$ for $u < 0$. Since the function $1/v(z)$ is increasing, V is convex.

Theorem 4.2 *For some increasing function $s(x) > 0$ and for some $\widehat{x} \geq 0$, let*

$$\mathbb{E}\{\xi(x);\ \xi(x) \leq s(x)\} \geq v(x) \quad \textit{for all } x \geq \widehat{x}. \tag{4.15}$$

Then, for all $t \geq y \geq \widehat{x}$,

$$\mathbb{E}_y L(\widehat{x}, T(t)) \leq V(t + s(t)) - V(y) \ = \ \int_y^{t+s(t)} \frac{1}{v(z)} dz. \tag{4.16}$$

Further, the family of random variables

$$\frac{1}{V(t + s(t)) - V(x)} L(x, T(t)), \quad t \geq y \geq x \geq \widehat{x},\ X_0 = y, \tag{4.17}$$

is uniformly integrable.

Proof Let us consider the following continuous test function:

$$\widehat{V}(u) \ := \ V(\widehat{x} \vee u) \ = \ \begin{cases} V(\widehat{x}) & \text{if } u < \widehat{x}, \\ V(u) & \text{if } u \geq \widehat{x}. \end{cases}$$

This function is convex, as is V, so

$$\mathbb{E}_u\{\widehat{V}(X_1) - \widehat{V}(u);\ X_1 - u \leq s(u)\} \geq \widehat{V}'(u)\mathbb{E}\{\xi(u);\ \xi(u) \leq s(u)\},$$

where the right derivative of $\widehat{V}$ equals

$$\widehat{V}'(u) = \begin{cases} 0 & \text{if } u < \widehat{x}, \\ 1/v(u) & \text{if } u \geq \widehat{x}. \end{cases}$$

Therefore

$$\mathbb{E}_u\{\widehat{V}(X_1) - \widehat{V}(u);\ X_1 - u \leq s(u)\} \geq \begin{cases} 1 & \text{if } u \geq \widehat{x}, \\ 0 & \text{if } u < \widehat{x}, \end{cases}$$

by the condition (4.15). Since the function $u + s(u)$ is increasing,

$$\mathbb{E}_u\{\widehat{V}(X_1) - \widehat{V}(u);\ X_1 \leq t + s(t)\} \geq \begin{cases} 1 & \text{if } u \in [\widehat{x}, t], \\ 0 & \text{if } u < \widehat{x}, \end{cases} \tag{4.18}$$

Thus the process $Y_n := \widehat{V}(X_n \wedge (t + s(t)))$ satisfies the inequality

$$\mathbb{E}_y Y_{T(t)} \ge V(y) + \mathbb{E}_y \sum_{k=0}^{T(t)-1} \mathbb{I}\{X_k \ge \widehat{x}\}, \quad y \in [\widehat{x}, t], \tag{4.19}$$

due to the following adapted version of the proof of Dynkin's formula (see e.g. [126, Theorem 11.3.1] and note that $\mathcal{F}_n = \sigma(Y_0, \ldots, Y_n)$):

$$\begin{aligned}
\mathbb{E}_y Y_{T(t)} &= \mathbb{E}_y Y_0 + \mathbb{E}_y \sum_{n=1}^{\infty} \mathbb{I}\{n \le T(t)\}(Y_n - Y_{n-1}) \\
&= V(y) + \mathbb{E}_y \sum_{n=1}^{\infty} \mathbb{E}\{\mathbb{I}\{n \le T(t)\}(Y_n - Y_{n-1}) \mid \mathcal{F}_{n-1}\} \\
&= V(y) + \mathbb{E}_y \sum_{n=1}^{\infty} \mathbb{I}\{T(t) \ge n\} \mathbb{E}\{Y_n - Y_{n-1} \mid \mathcal{F}_{n-1}\},
\end{aligned}$$

because $\{n \le T(t)\} = \overline{\{T(t) \le n-1\}} \in \mathcal{F}_{n-1}$. Hence, (4.18) implies that

$$\begin{aligned}
\mathbb{E}_y Y_{T(t)} &\ge V(y) + \mathbb{E}_y \sum_{n=1}^{\infty} \mathbb{I}\{T(t) \ge n, X_{n-1} \ge \widehat{x}\} \\
&= V(y) + \mathbb{E}_y \sum_{n=1}^{T(t)} \mathbb{I}\{X_{n-1} \ge \widehat{x}\},
\end{aligned}$$

and the inequality (4.19) follows.

On the other hand, $Y_{T(t)} \le V(t + s(t))$ by the construction of $\{Y_n\}$. Hence

$$\mathbb{E}_y Y_{T(t)} \le \widehat{V}(t + s(t)) = V(t + s(t)), \tag{4.20}$$

because $t > \widehat{x}$, which together with (4.19) yields

$$\mathbb{E}_y L(\widehat{x}, T(t)) \le V(t + s(t)) - V(y),$$

and the upper bound (4.16) follows.

Now let us proceed with the proof of the uniform integrability in (4.17); the latter is equivalent to the following convergence:

$$\sup_{\widehat{x} \le x \le y \le t} \mathbb{E}_y \left\{ \frac{L(x, T(t))}{V(t + s(t)) - V(x)}; \; \frac{L(x, T(t))}{V(t + s(t)) - V(x)} > A \right\} \to 0 \tag{4.21}$$

as $A \to \infty$. For $N \in \mathbb{N}$, define θ_N to be the following stopping time:

$$\theta_N = \theta_N(x) := \inf\left\{ n : L(x, n) = \sum_{k=0}^{n-1} \mathbb{I}\{X_k \ge x\} = N \right\} - 1.$$

Similarly to (4.19),

$$\begin{aligned}\mathbb{E}_y Y_{T(t)} &\geq \mathbb{E}_y Y_{\theta_N \wedge T(t)} + \mathbb{E}_y \sum_{n=\theta_N+1}^{T(t)-1} \mathbb{I}\{X_n \geq x\} \\ &= \mathbb{E}_y Y_{\theta_N \wedge T(t)} + \mathbb{E}_y\{L(x,T(t)) - N;\ L(x,T(t)) > N\}. \quad (4.22)\end{aligned}$$

Therefore

$$\begin{aligned}&\mathbb{E}_y\{L(x,T(t)) - N;\ L(x,T(t)) > N\} \\ &\quad \leq \mathbb{E}_y(Y_{T(t)} - Y_{\theta_N \wedge T(t)}) \\ &\quad = \mathbb{E}_y\{Y_{T(t)} - Y_{\theta_N};\ T(t) > \theta_N\} \\ &\quad \leq \mathbb{E}_y\{\widehat{V}(X_{T(t)}) - \widehat{V}(X_{\theta_N}));\ T(t) > \theta_N\},\end{aligned}$$

by the definition of $\{Y_n\}$. Taking into account that $X_{T(t)} \leq t + s(t)$ and $X_{\theta_N} \geq x \geq \widehat{x}$, we deduce that

$$\begin{aligned}&\mathbb{E}_y\{L(x,T(t));\ L(x,T(t)) > N\} \\ &\quad \leq (V(t+s(t)) - V(x))\mathbb{P}_y\{L(x,T(t)) > N\}. \quad (4.23)\end{aligned}$$

Taking

$$N := [A(V(t+s(t)) - V(x))],$$

we obtain from (4.23) that the mean in (4.21) is not greater than

$$\mathbb{P}_y\{L(x,T(t)) > N\},$$

which in its turn is not greater than

$$\frac{\mathbb{E}_y L(x,T(t))}{N+1},$$

by the Markov inequality. Owing to the already proven upper bound (4.16), for $y \geq x$ we have

$$\frac{\mathbb{E}_y L(x,T(t))}{N+1} \leq \frac{V(t+s(t)) - V(y)}{A(V(t+s(t)) - V(x))} \leq \frac{1}{A},$$

and the proof of the uniform integrability (4.17) is complete. □

4.3 Transient Chain: Integro-Local Upper Bound for Renewal Function

A transient Markov chain $\{X_n\}$ visits any bounded set finitely many times only. Thus, as noted in Section 1.4.2, for countable Markov chains the renewal functions

$$H_y(x,x+h] := \mathbb{E}_y \sum_{n=0}^{\infty} \mathbb{I}\{x < X_n \le x+h\} = \sum_{n=0}^{\infty} \mathbb{P}_y\{x < X_n \le x+h\},$$

$$H(x,x+h] := \sum_{n=0}^{\infty} \mathbb{P}\{x < X_n \le x+h\} = \int_0^{\infty} H_y(x,x+h]\mathbb{P}\{X_0 \in dy\},$$

are well defined for all $x \in \mathbb{R}$ and $h > 0$. For general Markov chains, they are also well defined under some minor technical conditions. In the next result we derive upper bounds for these renewal functions. As shown in what follows, under some regularity conditions the upper bounds derived are asymptotically correct up to a constant multiplier.

Theorem 4.3 *Let the drift of* $\{X_n\}$ *possess the lower bound* (3.56) *with some* $r(x)$ *satisfying* (3.57) *and increasing function* $s(x) = o(1/r(x))$. *Assume* (4.15) *holds for some decreasing* $v(x)$ *satisfying*

$$c_v := \sup_{x>0} \frac{v(x)}{v(x+1/r(x))} < \infty. \tag{4.24}$$

Assume also an upper bound for the left tail

$$\mathbb{P}\{\xi(x) \le -s(x)\} \le p(x)v(x) \quad \text{for all } x \ge \widehat{x}, \tag{4.25}$$

where the decreasing function $p(x) > 0$ *is integrable at infinity. Then the family of random variables*

$$v(x)r(x)\sum_{n=0}^{\infty} \mathbb{I}\{x < X_n \le x+1/r(x)\}, \quad x \ge \widehat{x},\ X_0 = y,$$

is uniformly integrable.

In particular, there exists a $c_1 < \infty$ *such that*

$$H_y(x,x+1/r(x)] \le \frac{c_1}{v(x)r(x)},$$

for all $x \ge \widehat{x}$ *and* y, *and, further,*

$$H_y(\widehat{x},x] \le c_1 \int_{\widehat{x}}^{x+1/r(x)} \frac{dz}{v(z)}.$$

These upper bounds are rather accurate for $y \le x$. In the opposite case, $y > x$, sharper bounds can be obtained by combining the upper bounds for the renewal function in Theorem 4.3 with estimates for the down-crossing probabilities, that is, either with Theorem 3.12 or with the exact asymptotic results in Chapter 3.

Proof Considering the first entry of $\{X_n\}$ into the segment $(x, x+1/r(x)]$, we see that the first claim is equivalent to the uniform integrability of the family

$$v(x)r(x)\sum_{n=0}^{\infty}\mathbb{I}\{x < X_n \le x+1/r(x)\},\ x \ge \widehat{x},\ X_0 = y,\ y \in (x, x+1/r(x)]. \tag{4.26}$$

First let us consider a Markov chain $\{Y_n\}$ with jumps

$$\eta(x) := \max(\xi(x),\ -s(x)).$$

This Markov chain satisfies the conditions (3.58), because $\eta(x) \ge -s(x)$, and (3.56). So, Theorem 3.12 applies to the chain $\{Y_n\}$ with $\delta < \varepsilon$, where ε was defined in (3.57); hence

$$\mathbb{P}\{Y_n \le x \text{ for some } n \ge 1 \mid Y_0 = y\} \le e^{\delta(R(x)-R(y))} \text{ for all } y > x \ge x_*, \tag{4.27}$$

where x_* is given by Theorem 3.12. Without loss of generality we can assume that $x_* > \widehat{x}$. Consider the stopping time

$$T^Y(t) = \min\{n \ge 1 : Y_n > t\},$$

where

$$t := \begin{cases} x + 2/r(x) & \text{for } x \ge x_*, \\ x_* + 2/r(x_*) & \text{for } x \in [\widehat{x}, x_*]. \end{cases}$$

For any $Y_0 = y \subset (x, x+1/r(x)]$, we have

$$v(x)r(x)\sum_{n=0}^{T^Y(t)-1}\mathbb{I}\{x < Y_n \le x+1/r(x)\} \le v(x)r(x)\sum_{n=0}^{T^Y(t)-1}\mathbb{I}\{Y_n > x\}. \tag{4.28}$$

It follows from the convexity of the function $V(x)$ defined in (4.14) that

$$\begin{aligned} V(t+s(t)) - V(x) &\le V'(t+s(t))(t+s(t)-x) \\ &= \frac{t+s(t)-x}{v(t+s(t))}. \end{aligned}$$

Thus

$$\frac{1}{V(t+s(t)) - V(x)} \ge \frac{v(t+s(t))}{t+s(t)-x}.$$

For all sufficiently large x, $s(x) \le 1/r(x)$ and hence $s(t) \le 1/r(t)$. In addition, $t \le x + 2/r(x)$ for $x \ge x_*$. Therefore, for all sufficiently large x,

$$\begin{aligned}\frac{v(x)}{v(t+s(t))} &= \frac{v(x)}{v(x+1/r(x))}\frac{v(x+1/r(x))}{v(t)}\frac{v(t)}{v(t+s(t))}\\ &\le \frac{v(x)}{v(x+1/r(x))}\frac{v(x+1/r(x))}{v(x+2/r(x))}\frac{v(t)}{v(t+1/r(t))}\\ &\le c_v^3, \end{aligned} \tag{4.29}$$

by (4.24). Further,

$$r(x)(t+s(t)-x) = r(x)(2/r(x)+s(t)) \to 2 \quad \text{as } x \to \infty. \tag{4.30}$$

Therefore, there exists a $\gamma > 0$ such that

$$\frac{1}{V(t+s(t))-V(x)} \ge \gamma v(x)r(x) \quad \text{for all } x \ge \widehat{x},$$

which, on being applied to (4.28), yields

$$\begin{aligned} &v(x)r(x)\sum_{n=0}^{T^Y(t)-1}\mathbb{I}\{x < Y_n \le x+1/r(x)\}\\ &\quad \le \frac{1}{\gamma}\frac{1}{V(t+s(t))-V(x)}\sum_{n=0}^{T^Y(t)-1}\mathbb{I}\{Y_n > x\}. \end{aligned}$$

Finally, on the right-hand side the family of random variables with respect to $x \ge \widehat{x}$, $Y_0 = y$, $y \in (x, x+1/r(x)]$ is uniformly integrable, owing to Theorem 4.2 applied to the chain $\{Y_n\}$. So, the family of random variables

$$v(x)r(x)\sum_{n=0}^{T^Y(t)-1}\mathbb{I}\{x < Y_n \le x+1/r(x)\}, \quad x \ge \widehat{x},$$

is uniformly integrable too.

Further, after the stopping time $T^Y(t)$, the chain $\{Y_n\}$ falls below the level

$$t_1 := \begin{cases} x+1/r(x) & \text{for } x \ge x_*,\\ x_*+1/r(x_*) & \text{for } x \in [\widehat{x}, x_*], \end{cases}$$

with probability $e^{\delta(R(t_1)-R(t))}$ at the most; see (4.27), which is applicable because $t_1 > x_*$. Since the function $R(x)$ is concave,

$$\begin{aligned}\exp\left(\delta(R(x+1/r(x))-R(x+2/r(x)))\right) &\le \exp\left(-\delta R'(x+2/r(x))/r(x)\right)\\ &= \exp\left(-\delta r(x+2/r(x))/r(x)\right). \end{aligned}$$

As can be seen in (2.8), $r(x+2/r(x))/r(x) \ge 1/(1+2c)$ for all $x \ge 0$; hence we conclude that

$$\sup_{x\ge 0}\exp\left(\delta(R(x+1/r(x))-R(x+2/r(x)))\right) \le \exp\left(-\delta/(1+2c)\right) < 1.$$

Therefore, for all $y \ge t$,

$$\mathbb{P}\{Y_n \le t_1 \text{ for some } n \ge 1 \mid Y_0 = y\} \le \exp(-\delta/(1+2c)) < 1.$$

Hence, we obtain by the Markov property that the family

$$v(x)r(x)\sum_{n=0}^{\infty}\mathbb{I}\{x < Y_n \le x + 1/r(x)\}$$

is dominated by at most a geometric number at the most of summands taken from a uniformly integrable family of random variables. This yields the first claim of theorem for the chain $\{Y_n\}$, by Lemma 2.25(i) with the σ-algebra $\mathcal{F}_n$ generated by the history of the chain up to the nth time that it falls below the level t_1. In particular, for some $c_3 < \infty$,

$$H_y^Y(x, x+1/r(x)] \le \frac{c_3}{v(x)r(x)} \quad \text{for all } x \ge \widehat{x} \text{ and } y. \tag{4.31}$$

Further, in order to pass from $\{Y_n\}$ to $\{X_n\}$, we first notice that these two chains may be constructed on the same probability space as that described at the beginning of the proof of Lemma 4.1. This makes possible the following calculations. For all $x < y$,

$$\begin{aligned}
&\mathbb{P}\{X_n \le x \text{ for some } n \ge 1 \mid X_0 = Y_0 = y\}\\
&\quad \le \mathbb{P}\{Y_n < x \text{ for some } n \ge 1 \mid Y_0 - y\}\\
&\qquad + \mathbb{P}\{X_n \ne Y_n \text{ for some } n \ge 1,\ Y_k > x \text{ for all } k \ge 1 \mid X_0 = Y_0 = y\}\\
&\quad \le \mathbb{P}\{Y_n \le x \text{ for some } n \ge 1 \mid Y_0 = y\}\\
&\qquad + \mathbb{P}\{X_n \ne Y_n \text{ for some } n \ge 1 \mid X_0 = Y_0 = y\}.
\end{aligned}$$

The second probability on the right-hand side tends to 0 as $y \to \infty$ by Lemma 4.1. Lemma 4.1 is applicable by (4.25) and because the upper bound (4.27) implies (4.2) and the upper bound (4.31) implies (4.3) with $l(x) = 1/r(x)$, owing to (2.8). Together with (4.27) this yields that

$$\mathbb{P}\{X_n \le x \text{ for some } n \ge 1 \mid X_0 = y\} \le e^{\delta(R(x)-R(y))} + o(1) \tag{4.32}$$

as $x \to \infty$ uniformly for all $y > x$. In particular, there exists a sufficiently large $x_0 \ge x_*$ such that, for some $q < 1$,

$$\mathbb{P}\{X_n \le x + 1/r(x) \text{ for some } n \ge 1 \mid X_0 = y\} \le q \tag{4.33}$$

for all $x \ge x_0$ and $y \ge x + 2/r(x)$.

In the same way as for $\{Y_n\}$, we now consider a stopping time

$$T(t) = \min\{n \ge 1 : X_n > t\},$$

where

$$t := \begin{cases} x + 2/r(x) & \text{for } x \ge x_0, \\ x_0 + 2/r(x_0) & \text{for } x \in [\widehat{x}, x_0]. \end{cases}$$

Similarly to the chain $\{Y_n\}$, the family with respect to $x \ge \widehat{x}$, $X_0 = y$, $y \in (x, x + 1/r(x)]$ of random variables

$$v(x)r(x) \sum_{n=0}^{T(t)-1} \mathbb{I}\{x < X_n \le x + 1/r(x)\}$$

is uniformly integrable too, owing to the application of Theorem 4.2 to $\{X_n\}$.

Further, after the stopping time $T(t)$ the chain $\{X_n\}$ falls below the level

$$t_1 := \begin{cases} x + 1/r(x) & \text{for } x \ge x_0, \\ x_0 + 1/r(x_0) & \text{for } x \in [\widehat{x}, x_0], \end{cases}$$

with probability $q < 1$ at the most, see (4.33) which is applicable because $t_1 \ge x_0$. For the same reasons as for the Markov chain $\{Y_n\}$,

$$v(x)r(x) \sum_{n=0}^{\infty} \mathbb{I}\{x < X_n \le x + 1/r(x)\}$$

is majorised by at most a geometric number of summands taken from a uniformly integrable family of random variables. This yields the first claim of the theorem for the chain $\{X_n\}$, by Lemma 2.25(i) with the σ-algebra $\mathcal{F}_n$ generated by the history of the chain up to the nth time it falls below the level t_1.

The second claim of the theorem follows if we consider the points $x_0 := \widehat{x}$, $x_{n+1} := x_n + 1/r(x_n)$ and then, by the first result,

$$H_y(\widehat{x}, x] \le \sum_{n=0}^{N-1} H_y(x_n, x_{n+1}] \le c_1 \sum_{n=0}^{N-1} \frac{1}{v(x_n)r(x_n)},$$

where $N := \min\{n \ge 1 : x_n > x\}$, so $x_N \le x + 1/r(x)$. Since $1/v(z)$ is increasing, we finally get

$$\sum_{n=0}^{N-1} \frac{1}{v(x_n)r(x_n)} \le \sum_{n=0}^{N-1} \int_{x_n}^{x_n+1/r(x_n)} \frac{dz}{v(z)}$$
$$= \int_{\widehat{x}}^{x_N} \frac{dz}{v(z)} \le \int_{\widehat{x}}^{x+1/r(x)} \frac{dz}{v(z)}. \qquad \square$$

Now we consider the case where iterated logarithms play a rôle. Assume that there exist $\varepsilon > 0$, $m \ge 1$, and $\widehat{x}$ such that, for all $x > \widehat{x}$,

$$\begin{aligned}\frac{2m_1^{[s(x)]}(x)}{m_2^{[s(x)]}(x)} &\ge r(x)\\ &= \Big(\frac{1}{y} + \cdots + \frac{1}{y\log y \cdots \log_{(m-1)} y} + \frac{1+\varepsilon}{y\log y \cdots \log_{(m)} y}\Big)\Big|_{y=e^{(m)}+x}. \end{aligned} \tag{4.34}$$

Theorem 4.4 *Let the drift of $\{X_n\}$ possess the lower bound* (4.34) *for some increasing function $s(x) = o(x/\log x \cdots \log_{(m)} x)$. Assume* (4.15) *holds for $v(x) = \gamma/x$, $\gamma > 0$. Assume also that the upper bound for the left tail, for some $\delta < \varepsilon$, is given by*

$$\mathbb{P}\{\xi(x) \le -s(x)\} = o(m_2^{[s(x)]}/x^2\log^2 x \cdots \log^2_{(m-1)} x \log^{2+\delta}_{(m)} x) \text{ for all } x \ge \widehat{x}. \tag{4.35}$$

Then the family of random variables

$$\frac{1}{x^2\log x \cdots \log_{(m)} x}\sum_{n=0}^{\infty} \mathbb{I}\{\widehat{x} < X_n \le x\}, \quad x > \widehat{x},\ X_0 = y,$$

is uniformly integrable. In particular, there exists a $c < \infty$ such that

$$H_y(\widehat{x}, x] \le c_1 x^2 \log x \cdots \log_{(m)} x \quad \textit{for all } x > \widehat{x} \textit{ and } y.$$

Proof By the same arguments as in the proof of Theorem 4.3, we see that the first conclusion is equivalent to the uniform integrability of the family

$$\frac{1}{x^2\log x \cdots \log_{(m)} x}\sum_{n=0}^{\infty} \mathbb{I}\{x < X_n \le 2x\},\ x > \widehat{x},\ X_0 = y,\ y \in (x, 2x]. \tag{4.36}$$

The Markov chain $\{X_n\}$ satisfies the conditions (3.68) owing to (4.35). So, Corollary 3.16 applies; hence

$$\mathbb{P}\{X_n \le x \text{ for some } n \ge 1 \mid X_0 = y\} \le \left(\frac{\log_{(m)}(e^{(m)}+x)}{\log_{(m)}(e^{(m)}+y)}\right)^{\delta} \tag{4.37}$$

for all $y > x \ge x_*$, where x_* is given by Corollary 3.16. Without loss of generality we assume that $x_* > \widehat{x}$. In a similar way to its introduction for the Markov chain $\{Y_n\}$ in the proof of Theorem 4.3, let us consider the stopping time

$$T^X(t) = \min\{n \ge 1 : X_n > t\},$$

where

$$t := \begin{cases} 3x & \text{for } x \geq x_*, \\ 3x_* & \text{for } x \in [\widehat{x}, x_*]. \end{cases}$$

As was concluded previously for $\{Y_n\}$, the family of random variables

$$\frac{1}{x^2} \sum_{n=0}^{T^X(t)-1} \mathbb{I}\{x < X_n \leq 2x\}, \quad x > \widehat{x},$$

is uniformly integrable.

Further, after the stopping time $T^X(t)$ the chain $\{X_n\}$ falls below the level

$$t_1 := \begin{cases} 2x & \text{for } x \geq x_*, \\ 2x_* & \text{for } x \in [\widehat{x}, x_*] \end{cases}$$

with probability (4.27) at most; this is applicable because $t_1 > x_*$. Observe that, for $x > x_*$,

$$\left(\frac{\log_{(m)}(e^{(m)} + 2x)}{\log_{(m)}(e^{(m)} + 3x)} \right)^{\delta} \leq 1 - \frac{c_2}{\log x \cdots \log_{(m)} x} \quad \text{for some } c_2 > 0.$$

Therefore, for all $y \geq t$,

$$\mathbb{P}\{X_n \leq t_1 \text{ for some } n \geq 1 \mid X_0 = y\} \leq 1 - \frac{c_2}{\log x \cdots \log_{(m)} x}.$$

Hence, we obtain by the Markov property that the family

$$\frac{1}{x^2} \sum_{n=0}^{\infty} \mathbb{I}\{x < X_n \leq 2x\}$$

is dominated by at most a geometric number – with success probability $c_2/\log x \cdots \log_{(m)} x$ – of summands taken from a uniformly integrable family of random variables. This yields the first claim of theorem, by Lemma 2.25(ii) with $E_x = \log x \cdots \log_{(m)} x$. In particular, for some $c_3 < \infty$,

$$H_y^X(x, 2x] \leq c_3 x^2 \log x \cdots \log_{(m)} x \quad \text{for all } x > \widehat{x} \text{ and } y. \qquad \square$$

4.4 Factorisation Result for Renewal Function with Weights

In this section, either $n(x) \equiv \infty$ or $n(x) \to \infty$ as $x \to \infty$. Let $A(x) \subset \mathbb{R}$ be a family of Borel sets.

For a function $q(z) \geq 0$ on $\mathbb{R}$, we will look at the impact of $q(z)$ on the asymptotic behaviour of the partial renewal measure with weights

$$\sum_{n=0}^{n(x)} \mathbb{E}\{e^{-\sum_{k=0}^{n-1} q(X_k)};\ X_n \in A(x)\}, \tag{4.38}$$

in comparison with the behaviour of

$$\sum_{n=0}^{n(x)} \mathbb{P}\{X_n \in A(x)\}.$$

Lemma 4.5 *Let $a(x) > 0$ be a function on $\mathbb{R}^+$. Let the family of random variables*

$$a(x) \sum_{n=0}^{n(x)} \mathbb{I}\{X_n \in A(x)\}, \quad x > 0,\ X_0 = z, \tag{4.39}$$

be uniformly integrable and let there exist a $c > 0$ such that, for all $N \in \mathbb{Z}^+$ and $z \in \mathbb{R}$,

$$a(x) \sum_{n=N}^{n(x)} \mathbb{P}_z\{X_n \in A(x)\} \to c \quad \textit{as } x \to \infty. \tag{4.40}$$

If $q(z) \geq 0$ then

$$a(x) \sum_{n=0}^{n(x)} \mathbb{E}\{e^{-\sum_{k=0}^{n-1} q(X_k)};\ X_n \in A(x)\} \to c\mathbb{E}e^{-\sum_{k=0}^{\infty} q(X_k)} \quad \textit{as } x \to \infty.$$

Proof The conditions (4.39) and (4.40) imply that

$$a(x) \sum_{n=N}^{n(x)} \mathbb{P}\{X_n \in A(x)\} \to c \quad \text{as } x \to \infty$$

for any distribution of X_0 and for all N. Therefore, for any fixed $N \in \mathbb{N}$,

$$a(x) \sum_{n=0}^{N-1} \mathbb{P}\{X_n \in A(x)\} \to 0 \quad \text{as } x \to \infty.$$

Then

$$\begin{aligned}
&a(x) \sum_{n=0}^{n(x)} \mathbb{E}\{e^{-\sum_{k=0}^{n-1} q(X_k)};\ X_n \in A(x)\} - c\mathbb{E}e^{-\sum_{k=0}^{\infty} q(X_k)} \\
&= a(x)\Bigg(\sum_{n=0}^{n(x)} \mathbb{E}\{e^{-\sum_{k=0}^{n-1} q(X_k)};\ X_n \in A(x)\} \\
&\qquad - \mathbb{E}e^{-\sum_{k=0}^{\infty} q(X_k)} \sum_{n=0}^{n(x)} \mathbb{P}\{X_n \in A(x)\}\Bigg) + o(1)
\end{aligned}$$

$$= a(x)\Big(\mathbb{E}\sum_{n=N}^{n(x)}\Big(e^{-\sum_{k=0}^{n-1}q(X_k)} - \mathbb{E}e^{-\sum_{k=0}^{\infty}q(X_k)}\Big)\mathbb{I}\{X_n \in A(x)\}\Big) + o(1).$$

In turn, the mean on the right-hand side equals the following sum of the mean values of random variables:

$$\sum_{n=N}^{n(x)} = \zeta_1(x,N) + \zeta_2(x,N) + \zeta_3(x,N),$$

where

$$\zeta_1(x,N) := \sum_{n=N}^{n(x)}\Big(e^{-\sum_{k=0}^{N-1}q(X_k)} - \mathbb{E}e^{-\sum_{k=0}^{N-1}q(X_k)}\Big)\mathbb{I}\{X_n \in A(x)\},$$

$$\zeta_2(x,N) := \sum_{n=N}^{n(x)}\Big(e^{-\sum_{k=0}^{n-1}q(X_k)} - e^{-\sum_{k=0}^{N-1}q(X_k)}\Big)\mathbb{I}\{X_n \in A(x)\},$$

$$\zeta_3(x,N) := \sum_{n=N}^{n(x)}\Big(\mathbb{E}e^{-\sum_{k=0}^{N-1}q(X_k)} - \mathbb{E}e^{-\sum_{k=0}^{\infty}q(X_k)}\Big)\mathbb{I}\{X_n \in A(x)\}.$$

By the condition (4.39), both families of random variables $\{a(x)\zeta_2(x,N), x > 0, N \ge 1\}$ and $\{a(x)\zeta_3(x,N), x > 0, N \ge 1\}$ are uniformly integrable. Then, taking into account that $q(z) \ge 0$ implies the convergence

$$e^{-\sum_{k=0}^{N-1}q(X_k)} \overset{\text{a.s.}}{\to} e^{-\sum_{k=0}^{\infty}q(X_k)} \quad \text{as } N \to \infty, \tag{4.41}$$

we conclude that both $\sup_x a(x)|\mathbb{E}\zeta_2(x,N)|$ and $\sup_x a(x)|\mathbb{E}\zeta_3(x,N)|$ go to 0 as $N \to \infty$. This would prove the required result if we can show in addition that, for any fixed N,

$$a(x)\mathbb{E}\zeta_1(x,N) \to 0 \quad \text{as } x \to \infty. \tag{4.42}$$

Indeed, conditioning on $X_0, \ldots, X_{N-1}$ leads to the equality

$$\begin{aligned}
&\mathbb{E}\zeta_1(x,N)\\
&= \mathbb{E}\Big\{\Big(e^{-\sum_{k=0}^{N-1}q(X_k)} - \mathbb{E}e^{-\sum_{k=0}^{N-1}q(X_k)}\Big)\mathbb{E}\Big\{\sum_{n=N}^{n(x)}\mathbb{I}\{X_n \in A(x)\}\Big| X_0, \ldots, X_{N-1}\Big\}\Big\}\\
&= \mathbb{E}\Big\{\Big(e^{-\sum_{k=0}^{N-1}q(X_k)} - \mathbb{E}e^{-\sum_{k=0}^{N-1}q(X_k)}\Big)\mathbb{E}_{X_{N-1}}\sum_{n=N}^{n(x)}\mathbb{I}\{X_n \in A(x)\}\Big\},
\end{aligned}$$

by the Markov property. By the uniform integrability (4.39), the family of random variables

$$a(x)\Big(e^{-\sum_{k=0}^{N-1} q(X_k)} - \mathbb{E}e^{-\sum_{k=0}^{N-1} q(X_k)}\Big)\mathbb{E}_{X_{N-1}} \sum_{n=N}^{n(x)} \mathbb{I}\{X_n \in A(x)\}, \quad x > 0,$$

is uniformly integrable too. By the condition (4.40) we have

$$a(x)\mathbb{E}_{X_{N-1}} \sum_{n=N}^{n(x)} \mathbb{I}\{X_n \in A(x)\} \stackrel{\text{a.s.}}{\to} c \quad \text{as } x \to \infty.$$

This allows us to conclude that

$$a(x)\mathbb{E}\zeta_1(x,N) \to c\mathbb{E}\Big(e^{-\sum_{k=0}^{N-1} q(X_k)} - \mathbb{E}e^{-\sum_{k=0}^{N-1} q(X_k)}\Big) = 0 \quad \text{as } x \to \infty,$$

and (4.42) follows, which completes the proof. □

Lemma 4.6 *Let $g_n : \mathbb{R}^{n+1} \to \mathbb{R}$ be a sequence of uniformly bounded functions and let $E \in \mathbb{R}$ be a number such that, for all $N \in \mathbb{N}$ and $z_0, \ldots, z_N$,*

$$\mathbb{E}\{g_n(X_0, \ldots, X_n) \mid X_0 = z_0, \ldots, X_N = z_N\} \to E \quad \text{as } n \to \infty. \tag{4.43}$$

If $q(z) \geq 0$ then

$$\mathbb{E}e^{-\sum_{k=0}^{n-1} q(X_k)} g_n(X_0, \ldots, X_n) \to E\,\mathbb{E}e^{-\sum_{k=0}^{\infty} q(X_k)} \quad \text{as } n \to \infty.$$

Proof Fix any $N \in \mathbb{N}$. Then

$$\begin{aligned}
&\Big|\mathbb{E}e^{-\sum_{k=0}^{n-1} q(X_k)} g_n(X_0, \ldots, X_n) - \mathbb{E}g_n(X_0, \ldots, X_n)\mathbb{E}e^{-\sum_{k=0}^{\infty} q(X_k)}\Big| \\
&\quad \leq \Big|\mathbb{E}\Big(e^{-\sum_{k=0}^{N-1} q(X_k)} - \mathbb{E}e^{-\sum_{k=0}^{N-1} q(X_k)}\Big) g_n(X_0, \ldots, X_n)\Big| \\
&\quad\quad + \|g_n\|_\infty \mathbb{E}\Big|e^{-\sum_{k=0}^{n-1} q(X_k)} - e^{-\sum_{k=0}^{N-1} q(X_k)}\Big| \\
&\quad\quad + \|g_n\|_\infty \Big|\mathbb{E}e^{-\sum_{k=0}^{N-1} q(X_k)} - \mathbb{E}e^{-\sum_{k=0}^{\infty} q(X_k)}\Big| \\
&\quad =: |E_1(N,n)| + E_2(N,n) + E_3(N).
\end{aligned}$$

We have $E_2(N,n) \to 0$ and $E_3(N) \to 0$ as n, $N \to \infty$ by the dominated convergence in (4.41) because $q(z) \geq 0$. Further, conditioning on X_0, ..., X_{N-1} leads to the following equality and convergence:

$$\begin{aligned}
&E_1(N,n) \\
&= \mathbb{E}\Big\{\Big(e^{-\sum_{k=0}^{N-1} q(X_k)} - \mathbb{E}e^{-\sum_{k=0}^{N-1} q(X_k)}\Big)\mathbb{E}\{g_n(X_0, \ldots, X_n) \mid X_0, \ldots, X_{N-1}\}\Big\} \\
&\to E\,\mathbb{E}\Big(e^{-\sum_{k=0}^{N-1} q(X_k)} - \mathbb{E}e^{-\sum_{k=0}^{N-1} q(X_k)}\Big) \quad \text{as } n \to \infty,
\end{aligned}$$

by the condition (4.43), which allows us to conclude that, for any fixed N,

$$E_1(N,n) \to 0 \quad \text{as } n \to \infty,$$

and the proof is complete. □

Lemma 4.7 *Let p be a number between 0 and 1 and let $A_n \subset \mathbb{R}$ be a sequence of Borel sets such that, for all z,*

$$\mathbb{P}_z\{X_n \in A_n\} \to p \quad \text{as } n \to \infty. \tag{4.44}$$

If $q(z) \geq 0$ then

$$\mathbb{E}e^{-\sum_{k=0}^{n-1} q(X_k)}\mathbb{I}\{X_n \in A_n\} \to p\,\mathbb{E}e^{-\sum_{k=0}^{\infty} q(X_k)} \quad \text{as } n \to \infty.$$

Proof Take $g_n(X_0,\dots,X_n) = \mathbb{I}\{X_n \in A_n\}$; this is a bounded function satisfying the condition (4.43) with $E = p$ because

$$\mathbb{P}\{X_n \in A \mid X_0,\dots,X_N\} = \mathbb{P}\{X_n \in A \mid X_N\},$$

by the Markov property, and because

$$\mathbb{P}\{X_n \in A \mid X_N\} \stackrel{\text{a.s.}}{\to} p \quad \text{as } n \to \infty,$$

by the condition (4.44). □

4.5 Convergence to Γ-Distribution for Transient Chain

In this section we will be interested in the growth rate of a Markov chain $\{X_n\}$ on $\mathbb{R}$ that tends to infinity with probability 1 as $n \to \infty$, which happens when the chain is transient.

Theorem 4.8 *Suppose that there exist $b > 0$ and $\mu > b/2$ such that, for some increasing function $s(x) = o(x)$,*

$$m_1^{[s(x)]}(x) \sim \mu/x \ \text{ and } \ m_2^{[s(x)]}(x) \to b \quad \text{as } x \to \infty \tag{4.45}$$

and that there exists an $\widehat{x}$ such that, for all $x > \widehat{x}$,

$$\mathbb{P}\{|\xi(x)| > s(x)\} \leq p(x)/x, \tag{4.46}$$

$$\mathbb{E}\{|\xi(x)|;\ \xi(x) \leq -s(x)\} \leq p(x), \tag{4.47}$$

where the decreasing function $p(x) > 0$ is integrable at infinity. If

$$\limsup_{n\to\infty} X_n = \infty \quad \text{with probability } 1 \tag{4.48}$$

then X_n^2/nb converges weakly to a $\Gamma_{1/2+\mu/b,2}$-distribution with mean $1+2\mu/b$ and variance $2(1+2\mu/b)$ whose probability density function is

$$\frac{1}{\Gamma(1/2+\mu/b)2^{1/2+\mu/b}}x^{\mu/b-1/2}e^{-x/2}, \quad x>0.$$

As far as applications are concerned, we will apply this result to show convergence to a Γ distribution for a random walk conditioned to stay positive in Section 11.1 and convergence to a Γ-distribution for state-dependent branching processes with migration in Theorem 11.6.

Let us give a sufficient condition for (4.46) and (4.47) to hold. If the family $\{|\xi(x)|,\ x\ge 0\}$ possesses a majorant Ξ, that is, $|\xi(x)|\le_{\rm st}\Xi$ for all x, and the majorant is square integrable, $\mathbb{E}\Xi^2<\infty$, then there exists an increasing function $s(x)=o(x)$ such that (4.46) and (4.47) hold; see Lemma 2.33 with $\gamma=2$, $\alpha=1$, and $\beta=0,1$. Hence the following result.

Corollary 4.9 *Assume that, for some $b>0$ and $\mu>-b/2$, $m_1(x)\sim\mu/x$ and $m_2(x)\to b$ as $x\to\infty$. Assume that the family $\{|\xi(x)|,\ x\in\mathbb{R}\}$ possesses a square integrable majorant Ξ, that is, $\mathbb{E}\Xi^2<\infty$ and $\xi^2(x)\le_{st}\Xi$ for all x. If the condition* (4.48) *holds then X_n^2/nb converges weakly to a Γ-distribution with mean $1+2\mu/b$ and variance $2(1+2\mu/b)$.*

Proof of Theorem 4.8. The proof is based on the method of moments; see e.g. Durrett [56, Theorem 3.3.26].

Consider a modified Markov chain $\{\widetilde{X}_n\}$ on the same probability space as $\{X_n\}$ with jumps $\widetilde{\xi}(x)=\xi(x)\mathbb{I}\{|\xi(x)|\le s(x)\}$. If $\{\widetilde{X}_n\}$ does not satisfy the weak irreducibility condition (4.48) then we can increase the value of $s(x)$ on some set bounded above in such a way that $\{\widetilde{X}_n\}$ does satisfy (4.48). Indeed, it follows from the condition (4.45) that there exist a sufficiently high level x_0 and an $\varepsilon>0$ such that $\mathbb{P}\{\xi(x)\ge\varepsilon\}>0$ for all $x\ge x_0$. Then it suffices to increase $s(x)$ on the set $(-\infty,x_0]$ to ensure the condition (4.48) for $\{\widetilde{X}_n\}$.

Since (4.45) holds with $\mu>b/2$, $\{\widetilde{X}_n\}$ satisfies the condition (2.108) for any $\varepsilon\in(0,2\mu/b-1)$. Moreover, (4.46) implies (2.109) with a possibly slower decreasing $p(x)$ which is still integrable. Therefore, Theorem 2.21 is applicable to $\{\widetilde{X}_n\}$, so we conclude the transience and the convergence, for all z:

$$\mathbb{P}\{\widetilde{X}_n>z \text{ for all } n\ge 0 \mid X_0=y\}\to 1 \quad \text{as } y\to\infty.$$

By Theorem 4.3, there exist c and x_* such that

$$H_y^{\widetilde{X}}(x,2x)\le cx^2 \quad \text{for all } x>x_*.$$

So, all the conditions of Lemma 4.1 are satisfied for the chains $Y = X$ and $Z = \widetilde{X}$. By Theorem 2.21, the chain $Z = \widetilde{X}$ tends to infinity as $n \to \infty$, so it suffices to prove weak convergence to the same Γ-distribution for the process $\{Z_n\}$ with jumps $\zeta(x) = \xi(x)\mathbb{I}\{|\xi(x)| \le s(x)\}$; see the discussion at the end of Section 4.1. That is, it is sufficient to show that

$$\frac{Z_n^2}{nb} \Rightarrow \Gamma_{(2\mu+b)/2b,2} \quad \text{as } n \to \infty. \tag{4.49}$$

For all x, it holds that

$$\mathbb{E}\zeta(x) = m_1^{[s(x)]}(x) \quad \text{and} \quad \mathbb{E}\zeta^2(x) = m_2^{[s(x)]}(x). \tag{4.50}$$

In addition, the inequality $|\zeta(x)| \le s(x) = o(x)$ implies that, for all $j \ge 3$,

$$|\mathbb{E}\zeta^j(x)| \le m_2^{[s(x)]}(x)s^{j-2}(x) = o(x^{j-2}) \quad \text{as } x \to \infty. \tag{4.51}$$

Let us compute the mean of the increment of Z_n^{2i}. For $i = 1$ we have

$$\begin{aligned}\mathbb{E}\{Z_{n+1}^2 - Z_n^2 \mid Z_n = x\} &= \mathbb{E}(2x\zeta(x) + \zeta^2(x))\\ &= 2\mu + b + o(1) \quad \text{as } x \to \infty,\end{aligned}$$

by (4.50) and (4.45). Applying now the convergence of Z_n to infinity we get

$$\mathbb{E}(Z_{n+1}^2 - Z_n^2) \to 2\mu + b \quad \text{as } n \to \infty.$$

Hence,

$$\mathbb{E}Z_n^2 \sim (2\mu + b)n \quad \text{as } n \to \infty. \tag{4.52}$$

For $i \ge 2$, we have

$$\begin{aligned}&\mathbb{E}\{Z_{n+1}^{2i} - Z_n^{2i} \mid Z_n = x\}\\ &\quad = \mathbb{E}\left(2ix^{2i-1}\zeta(x) + i(2i-1)x^{2i-2}\zeta^2(x) + \sum_{l=3}^{2i} x^{2i-l}\zeta^l(x)\binom{2i}{l}\right)\\ &\quad = i(2\mu + (2i-1)b + o(1))x^{2i-2} + \sum_{l=3}^{2i} x^{2i-l}\mathbb{E}\zeta^l(x)\binom{2i}{l}\end{aligned} \tag{4.53}$$

as $x \to \infty$, by (4.50). Owing to (4.51),

$$\sum_{l=3}^{2i} x^{2i-l}\mathbb{E}\zeta^l(x)\binom{2i}{l} = \sum_{l=3}^{2i} x^{2i-l}o(x^{l-2}) = o(x^{2i-2}) \quad \text{as } x \to \infty.$$

Substituting this into (4.53) with $x = Z_n$ and taking into account the convergence $Z_n \to \infty$, we deduce that

$$\mathbb{E}\{Z_{n+1}^{2i} - Z_n^{2i}\} = i(2\mu + (2i-1)b + o(1))\mathbb{E}Z_n^{2i-2} \quad \text{as } n \to \infty. \tag{4.54}$$

In particular, for $i = 2$ we get

$$\mathbb{E}\{Z_{n+1}^4 - Z_n^4\} = 2(2\mu + 3b + o(1))\mathbb{E}Z_n^2$$
$$\sim 2(2\mu + 3b)(2\mu + b)n \quad \text{as } n \to \infty,$$

owing to (4.52). This implies that

$$\mathbb{E}Z_n^4 \sim (2\mu + 3b)(2\mu + b)n^2 \quad \text{as } n \to \infty.$$

By induction, we deduce from (4.54) that, for all $i \geq 1$,

$$\mathbb{E}Z_n^{2i} \sim (nb)^i \prod_{k=1}^{i}(2\mu/b + 2k - 1) \quad \text{as } n \to \infty,$$

which yields the convergence of all moments of Z_n^2/nb to that of a Γ-distribution with mean $1 + 2\mu/b$ and variance $2(1 + 2\mu/b)$, as any Γ-distribution is uniquely determined by all its moments. Hence (4.49) is proven and the proof of Theorem 4.8 is complete. □

4.6 Convergence to Γ-Distribution for Non-Positive Chain

The next result, on the convergence to a Γ-distribution, covers both transient and null recurrent chains.

Theorem 4.10 *Assume that, for some $b > 0$ and $\mu > -b/2$,*

$$m_1(x) \sim \mu/x \text{ and } m_2(x) \to b \quad \text{as } x \to \infty \tag{4.55}$$

and that the family $\{\xi^2(x),\ x \in \mathbb{R}\}$ possesses an integrable majorant Ξ, that is, $\mathbb{E}\Xi < \infty$ and

$$\xi^2(x) \leq_{\text{st}} \Xi \quad \text{for all } x. \tag{4.56}$$

If $X_n \to \infty$ in probability as $n \to \infty$ then X_n^2/nb converges weakly to a Γ-distribution with mean $1 + 2\mu/b$ and variance $2(1 + 2\mu/b)$.

The main difference between this result and Theorem 4.8 is that here we impose conditions on the asymptotic behaviour of the first two *full* moments of jumps, $m_1(x)$ and $m_2(x)$. Further, as we commented after Theorem 4.8, (4.56) implies (4.46). The rationale behind these more restrictive assumptions is that the renewal function of any null recurrent chain is infinite, hence we cannot use time-homogeneous truncations as in the proof of Theorem 4.8. In order to prove Theorem 4.10 we introduce a truncation of jumps which depends not only on the spatial coordinate x but also on time n.

Concerning applications, we apply this result in Theorem 11.7 to show convergence to a Γ-distribution for state-dependent branching processes with migration in the null recurrent case, in Theorem 11.16 to show convergence to a Γ-distribution for null recurrent stochastic difference equations, and in Theorem 11.18 to show convergence to a Γ-distribution for the ALOHA network.

Proof For any $n \in \mathbb{N}$, consider a new Markov chain $Y_k(n)$, $k = 0, 1, 2, \dots$, with transition probabilities depending on the parameter n, and whose jump $\eta(n,x)$ is just the original jump $\xi(x)$ truncated at levels $\pm(x \vee \sqrt{n})$ depending on both state x and time n, that is,

$$\eta(n,x) = \begin{cases} \xi(x) & \text{if } |\xi(x)| \le x \vee \sqrt{n}, \\ 0 & \text{otherwise.} \end{cases}$$

Given $Y_0(n) = X_0$, the probability of discrepancy between the trajectories of $\{Y_k(n)\}$ and $\{X_k\}$ by time n is, at most,

$$\begin{aligned} \mathbb{P}\{Y_k(n) \ne X_k \text{ for some } k \le n\} &\le \sum_{k=0}^{n-1} \mathbb{P}\{|X_{k+1} - X_k| \ge \sqrt{n}\} \\ &\le n\mathbb{P}\{\Xi \ge n\} \\ &\le \mathbb{E}\{\Xi; \Xi \ge n\} \to 0 \text{ as } n \to \infty. \end{aligned} \tag{4.57}$$

Since $X_n \to \infty$ in probability, (4.57) implies that, for every c,

$$\inf_{n > n_0, k \in [n_0, n]} \mathbb{P}\{Y_k(n) > c\} \to 1 \quad \text{as } n_0 \to \infty. \tag{4.58}$$

By the choice of truncation level,

$$|\xi(x) - \eta(n,x)| \le |\xi(x)|\mathbb{I}\{|\xi(x)| > x\}.$$

Therefore, by the condition (4.56),

$$\mathbb{E}\eta(n,x) = \mathbb{E}\xi(x) + o(1/x) \quad \text{as } x \to \infty \text{ uniformly for all } n \tag{4.59}$$

and

$$\mathbb{E}\eta^2(n,x) = \mathbb{E}\xi^2(x) + o(1) \text{ as } x \to \infty \text{ uniformly for all } n. \tag{4.60}$$

In addition, the inequality $|\eta(n,x)| \le x \vee \sqrt{n}$ and the condition (4.56) imply that, for all $j \ge 3$,

$$\mathbb{E}\eta^j(n,x) = o(x^{j-2} + n^{(j-2)/2}) \quad \text{as } x \to \infty \text{ uniformly for all } n. \tag{4.61}$$

Let us evaluate the mean of the increment of $Y_k^j(n)$. For $j = 2$ we have

$$\begin{aligned} \mathbb{E}\{Y_{k+1}^2(n) - Y_k^2(n) | Y_k(n) = x\} &= \mathbb{E}(2x\eta(n,x) + \eta^2(n,x)) \\ &= 2\mu + b + o(1) \end{aligned}$$

as $x \to \infty$ uniformly for all n, by (4.59) and (4.60). Now applying (4.58) we get

$$\mathbb{E}(Y_{k+1}^2(n) - Y_k^2(n)) \to 2\mu + b \quad \text{as } k, n \to \infty,\ k \le n.$$

Hence,

$$\mathbb{E}Y_n^2(n) \sim (2\mu + b)n \quad \text{as } n \to \infty. \tag{4.62}$$

Now let $j = 2i$, $i \ge 2$. We have

$$\begin{aligned}
&\mathbb{E}\{Y_{k+1}^{2i}(n) - Y_k^{2i}(n) | Y_k(n) = x\} \\
&= \mathbb{E}\left(2ix^{2i-1}\eta(n,x) + i(2i-1)x^{2i-2}\eta^2(n,x) + \sum_{l=3}^{2i} x^{2i-l}\eta^l(n,x)\binom{2i}{l}\right) \\
&= i(2\mu + (2i-1)b + o(1))x^{2i-2} + \sum_{l=3}^{2i} x^{2i-l}\mathbb{E}\eta^l(n,x)\binom{2i}{l}
\end{aligned} \tag{4.63}$$

as $x \to \infty$ uniformly for all n, by (4.59) and (4.60). Owing to (4.61),

$$\begin{aligned}
\sum_{l=3}^{2i} x^{2i-l}\mathbb{E}\eta^l(n,x)\binom{2i}{l} &= \sum_{l=3}^{2i} x^{2i-l} o(x^{l-2} + n^{(l-2)/2}) \\
&= o(x^{2i-2}) + \sum_{l=3}^{2i} x^{2i-l} o(n^{(l-2)/2})
\end{aligned}$$

as $x \to \infty$ uniformly for all n. Substituting this into (4.63) with $x = Y_k(n)$ and taking into account (4.58), we deduce that

$$\begin{aligned}
\mathbb{E}\{Y_{k+1}^{2i}(n) - Y_k^{2i}(n)\} &= i(2\mu + (2i-1)b + o(1))\mathbb{E}Y_k^{2i-2}(n) \\
&\quad + \sum_{l=3}^{2i} \mathbb{E}Y_k^{2i-l}(n) o(n^{(l-2)/2}).
\end{aligned} \tag{4.64}$$

In particular, for $j = 2i = 4$ we get

$$\begin{aligned}
\mathbb{E}\{Y_{k+1}^4(n) - Y_k^4(n)\} &= 2(2\mu + 3b)\mathbb{E}Y_k^2(n) + \mathbb{E}Y_k(n) o(\sqrt{n}) + o(n) \\
&\sim 2(2\mu + 3b)(2\mu + b)n,
\end{aligned}$$

owing to (4.62). This implies that

$$\mathbb{E}Y_n^4(n) \sim (2\mu + 3b)(2\mu + b)n^2 \quad \text{as } n \to \infty.$$

By induction, we deduce from (4.64) that

$$\mathbb{E}Y_n^{2i}(n) \sim (nb)^i \prod_{k=1}^{i} (2\mu/b + 2k - 1) \quad \text{as } n \to \infty,$$

which yields – by the method of moments – that $Y_n^2(n)/nb$ converges weakly to a Γ-distribution with mean $1 + 2\mu/b$ and variance $2(1 + 2\mu/b)$. Together with (4.57) this completes the proof of the theorem. □

4.7 Functional Convergence to Bessel Process for Non-Positive Chain

Once the weak convergence of X_n^2/n to a Γ-distribution is proven, it is natural to guess whether a diffusion approximation to X_n^2/n by a Bessel process exists. This question was originally answered in the positive by Lamperti in [112]. In the next theorem the result of Lamperti is given under minimal moment conditions; our proof is based on the method of moments, as is the proof of weak convergence to a Γ-distribution.

Introduce a family of piecewise-constant processes

$$X^{(n)}(t) = \frac{X_{[tn]}}{\sqrt{bn}}, \quad t \in [0, 1],$$

so that $X^{(n)}(t) \in D[0, 1]$, where $D[0, 1]$ is the space of real-valued functions on $[0, 1]$ that are right continuous with left limits.

Theorem 4.11 *Suppose that either $\mu > b/2$ and the conditions of Theorem 4.8 hold or $\mu > -b/2$ and the conditions of Theorem 4.10 hold. Then the process $\{X^{(n)}(t)\}$ converges weakly in $D[0, 1]$ to a Bessel process $Bes(t)$ starting at zero, with reflecting boundary condition in the null recurrent case, with drift μ/bx and diffusion coefficient 1; that is, $f(X^{(n)}(\cdot)) \Rightarrow f(Bes(\cdot))$ as $n \to \infty$ for all bounded functionals $f : D[0, 1] \to \mathbb{R}$ that are continuous in the Skorokhod topology.*

Notice that, since the limiting process is continuous, the last result is equivalent to weak convergence in the space $C[0, 1]$ if we define $\{X^{(n)}(t)\}$ as a continuous piecewise-linear process whose trajectory connects the points $(k/n, X_k/\sqrt{bn})$ by segments; for a justification see e.g. Ethier and Kurtz [60, Proposition 10.4].

All the arguments in the proof below will still be valid if we consider a triangular-array setting where the initial distribution of the chain depends on n in such a way that, for some $x_0 \in \mathbb{R}^+$,

$$X_0^{(n)}/\sqrt{bn} \xrightarrow{p} x_0 \quad \text{as } n \to \infty.$$

Then the process $\{X^{(n)}(t)\}$ converges weakly in $D[0, 1]$ to a Bessel process $Bes(t)$ with starting point x_0, drift μ/bx, and diffusion coefficient 1. In turn this implies that, if

$$X_0^{(n)}/\sqrt{bn} \Rightarrow \nu \quad \text{as } n \to \infty$$

for some probability distribution ν on $\mathbb{R}^+$, then the process $\{X^{(n)}(t)\}$ converges weakly in $D[0,1]$ to a Bessel process $Bes(t)$ with initial distribution ν.

We apply this result in Section 11.1 to show convergence to a Bessel process for a random walk conditioned to stay positive, in Theorems 11.3 and 11.7 to show convergence to a Bessel process for state-dependent branching processes with migration, in Theorem 11.16 to show convergence to a Bessel process for null recurrent stochastic difference equations, and in Theorem 11.18 to show convergence to a Bessel process for the ALOHA network.

Proof Let the conditions of Theorem 4.8 hold; then, as in the proof of that theorem, it is sufficient to prove the weak convergence to a Bessel process of the sequence of $D[0,1]$-processes $\{Z^{(n)}(t)\}$ defined by

$$Z^{(n)}(t) = \frac{Z_{[tn]}}{\sqrt{bn}}, \quad t \in [0,1],$$

where the process $\{Z_k\}$ was defined in Section 4.1.

By Prokhorov's theorem, we need to prove the weak convergence of finite-dimensional distributions and tightness in $D[0,1]$. We start with finite-dimensional distributions. By the method of moments, it suffices to prove that, for any sequence of time epochs $t_1 < t_2 < \cdots < t_k$ and natural numbers i_1, $i_2, \ldots, i_k$, the mixed moment

$$\mathbb{E}Z^{(n)}(t_1)^{2i_1} \cdots Z^{(n)}(t_k)^{2i_k} \tag{4.65}$$

converges to that of the Bessel process Bes, that is, to

$$\mathbb{E}Bes^{2i_1}(t_1) \cdots Bes^{2i_k}(t_k). \tag{4.66}$$

Indeed, conditioning on $Z^{(n)}(t_1), \ldots, Z^{(n)}(t_{k-1})$ yields the equality

$$\begin{aligned}
&\mathbb{E}\{Z^{(n)}(t_1)^{2i_1} \cdots Z^{(n)}(t_{k-1})^{2i_{k-1}} \\
&\quad \times [Z^{(n)}(t_k)^{2i_k} - Z^{(n)}(t_{k-1})^{2i_k} + Z^{(n)}(t_{k-1})^{2i_k}] \mid Z^{(n)}(t_1), \ldots, Z^{(n)}(t_{k-1})\} \\
&\quad = Z^{(n)}(t_1)^{2i_1} \cdots Z^{(n)}(t_{k-1})^{2i_{k-1}+2i_k} \\
&\quad + Z^{(n)}(t_1)^{2i_1} \cdots Z^{(n)}(t_{k-1})^{2i_{k-1}} \mathbb{E}\left\{ \frac{Z_{[nt_k]}^{2i_k} - Z_{[nt_{k-1}]}^{2i_k}}{(nb)^{i_k}} \,\middle|\, Z_{[nt_{k-1}]} \right\}.
\end{aligned}$$

The conditional expectation in the second term on the right-hand side equals

$$\sum_{j=[nt_{k-1}]}^{[nt_k]-1} \mathbb{E}\left\{ \frac{Z_{j+1}^{2i_k} - Z_j^{2i_k}}{(nb)^{i_k}} \,\middle|\, Z_{[nt_{k-1}]}(n) \right\},$$

where, by (4.54), the jth term in the sum may be evaluated as follows:

$$\mathbb{E}\left\{\frac{Z_{j+1}^{2i_k}-Z_j^{2i_k}}{(nb)^{i_k}}\,\middle|\,Z_{[nt_{k-1}]}(n)\right\}=(c_{i_k}+o(1))\mathbb{E}\left\{\frac{Z_j^{2i_k-2}}{(nb)^{i_k}}\,\middle|\,Z_{[nt_{k-1}]}\right\},$$

with $c_i=i(2\mu+(2i-1)b)$. In the case $i_k=1$ we get

$$\mathbb{E}\left\{\frac{Z_{j+1}^2-Z_j^2}{nb}\,\middle|\,Z_{[nt_{k-1}]}\right\}=\frac{c_1+o(1)}{nb}\quad\text{as } n\to\infty \text{ uniformly for all } j,$$

so that

$$\mathbb{E}\left\{\frac{Z_{[nt_k]}^2-Z_{[nt_{k-1}]}^2}{nb}\,\middle|\,Z_{[nt_{k-1}]}\right\}\to\frac{c_1(t_k-t_{k-1})}{b}\qquad\text{as } n\to\infty,$$

and hence

$$\begin{aligned}&\mathbb{E}Z^{(n)}(t_1)^{2i_1}\cdots Z^{(n)}(t_{k-1})^{2i_{k-1}}Z^{(n)}(t_k)^2\\&\quad=\mathbb{E}Z^{(n)}(t_1)^{2i_1}\cdots Z^{(n)}(t_{k-1})^{2i_{k-1}+2}\\&\qquad+\frac{c_1(t_k-t_{k-1})}{b}\mathbb{E}Z^{(n)}(t_1)^{2i_1}\cdots Z^{(n)}(t_{k-1})^{2i_{k-1}}+o(1).\end{aligned}$$

In the case $i_k=2$ we find, as in the proof of Theorem 4.8,

$$\begin{aligned}&\mathbb{E}\left\{\frac{Z_{j+1}^4-Z_j^4}{(nb)^2}\,\middle|\,Z_{[nt_{k-1}]}\right\}\\&=(c_2+o(1))\mathbb{E}\left\{\frac{Z_j^2}{(nb)^2}\,\middle|\,Z_{[nt_{k-1}]}\right\}\\&=(c_2+o(1))\mathbb{E}\left\{\frac{Z_j^2-Z_{[nt_{k-1}]}^2}{(nb)^2}\,\middle|\,Z_{[nt_{k-1}]}\right\}+(c_2+o(1))\frac{Z_{[nt_{k-1}]}^2}{(nb)^2}\\&=\frac{c_2c_1+o(1)}{(nb)^2}(j-[nt_{k-1}])+(c_2+o(1))\frac{Z_{[nt_{k-1}]}^2}{(nb)^2},\end{aligned}$$

so that, as $n\to\infty$,

$$\begin{aligned}&\mathbb{E}\left\{\frac{Z_{[nt_k]}^4-Z_{[nt_{k-1}]}^4}{(nb)^2}\,\middle|\,Z_{[nt_{k-1}]}\right\}\\&=(c_2c_1+o(1))\frac{(t_k-t_{k-1})^2}{2b^2}+(c_2+o(1))\frac{t_k-t_{k-1}}{b}\frac{Z_{[nt_{k-1}]}^2}{nb},\end{aligned}\tag{4.67}$$

and hence

$$\begin{aligned}
&\lim_{n\to\infty} \mathbb{E}Z^{(n)}(t_1)^{2i_1}\cdots Z^{(n)}(t_{k-1})^{2i_{k-1}}Z^{(n)}(t_k)^4 \\
&\quad= \lim_{n\to\infty} \mathbb{E}Z^{(n)}(t_1)^{2i_1}\cdots Z^{(n)}(t_{k-1})^{2i_{k-1}+4} \\
&\quad+ c_2\frac{t_k-t_{k-1}}{b}\lim_{n\to\infty} \mathbb{E}Z^{(n)}(t_1)^{2i_1}\cdots Z^{(n)}(t_{k-1})^{2i_{k-1}+2} \\
&\quad+ c_2c_1\frac{(t_k-t_{k-1})^2}{2b^2}\lim_{n\to\infty} \mathbb{E}Z^{(n)}(t_1)^{2i_1}\cdots Z^{(n)}(t_{k-1})^{2i_{k-1}}.
\end{aligned}$$

Similar relations hold for all $i_k\in\mathbb{N}$, with a clear pattern; for instance, for $i_k=3$,

$$\begin{aligned}
&\lim_{n\to\infty} \mathbb{E}Z^{(n)}(t_1)^{2i_1}\cdots Z^{(n)}(t_{k-1})^{2i_{k-1}}Z^{(n)}(t_k)^6 \\
&\quad= \lim_{n\to\infty} \mathbb{E}Z^{(n)}(t_1)^{2i_1}\cdots Z^{(n)}(t_{k-1})^{2i_{k-1}+6} \\
&\quad+ c_3\frac{t_k-t_{k-1}}{b}\lim_{n\to\infty} \mathbb{E}Z^{(n)}(t_1)^{2i_1}\cdots Z^{(n)}(t_{k-1})^{2i_{k-1}+4} \\
&\quad+ c_3c_2\frac{(t_k-t_{k-1})^2}{2b^2}\lim_{n\to\infty} \mathbb{E}Z^{(n)}(t_1)^{2i_1}\cdots Z^{(n)}(t_{k-1})^{2i_{k-1}+2} \\
&\quad+ c_3c_2c_1\frac{(t_k-t_{k-1})^3}{3!\,b^3}\lim_{n\to\infty} \mathbb{E}Z^{(n)}(t_1)^{2i_1}\cdots Z^{(n)}(t_{k-1})^{2i_{k-1}}.
\end{aligned}$$

Now let us show how to approximate the mixed even moments (4.66) via time discretisation of the Bessel process $Bes(t)$, for any $\mu > -b/2$. Consider a Markov chain $Bes_j := Bes(j)$. On the one hand, by the self-similarity and continuity of a Bessel process, we have

$$\begin{aligned}
\frac{(Bes([nt_1]),\ldots,Bes([nt_k]))}{\sqrt{n}} &=_{\rm st} (Bes([nt_1]/n),\ldots,Bes([nt_k]/n)) \\
&\Rightarrow (Bes(t_1),\ldots,Bes(t_k)) \quad \text{as } n\to\infty,
\end{aligned}$$

which implies the convergence of the mixed even moments

$$\mathbb{E}\left(\frac{Bes_{[nt_1]}}{\sqrt{n}}\right)^{2i_1}\cdots\left(\frac{Bes_{[nt_k]}}{\sqrt{n}}\right)^{2i_k} \to \mathbb{E}Bes^{2i_1}(t_1)\cdots Bes^{2i_k}(t_k)$$

as $n\to\infty$. On the other hand, the mean drift of the chain Bes_n is of order μ/bx and the second moment of jumps converges to 1 as $x\to\infty$, see (1.40); in the null recurrent case (1.40) is applicable because we are assuming a reflecting boundary condition for $X(t)$. In addition, (1.41) holds. Therefore, a relation similar to (4.54) follows, for all $i\ge 1$:

$$\mathbb{E}\{Bes_{n+1}^{2i}-Bes_n^{2i}\} = i(2\mu+(2i-1)b+o(1))\mathbb{E}Bes_n^{2i-2}.$$

So, all the calculations carried out for the evaluation of the mixed even moments of $Z^{(n)}(t)$ are applicable to those for Bes_n. Therefore, the mixed even moments (4.65) of $\{Z^{(n)}(t)\}$ converge to the corresponding mixed even moments (4.66) of the Bessel process $Bes(t)$; hence the weak convergence of the finite-dimensional distributions of $\{Z^{(n)}(t)\}$ follows by the method of moments.

Now it only remains to prove tightness. For that it is enough to show that there exists a $c < \infty$ such that, for all $0 \le t_1 < t_2 < t_3 \le 1$,

$$\mathbb{E}(Z^{(n)}(t_2)^2 - Z^{(n)}(t_1)^2)^2(Z^{(n)}(t_3)^2 - Z^{(n)}(t_2)^2)^2 \le c(t_3 - t_1)^2; \tag{4.68}$$

see e.g. Billingsley [16, Theorem 15.6]. Let us bound this expectation. Since we can always modify the chain $\{Z_n\}$ below any specific level, there is no loss of generality if we assume that, for all x,

$$\mathbb{E}\{Z_1^2 - Z_0^2 \mid Z_0 = x\} > 0, \tag{4.69}$$

$$\mathbb{E}\{Z_1^4 - Z_0^4 \mid Z_0 = x\} > 0. \tag{4.70}$$

Conditioning on $Z^{(n)}(t_1)$ and $Z^{(n)}(t_2)$ yields the following expression for the left-hand side of (4.68):

$$\begin{aligned}
&\mathbb{E}(Z^{(n)}(t_2)^2 - Z^{(n)}(t_1)^2)^2\mathbb{E}\{(Z^{(n)}(t_3)^2 - Z^{(n)}(t_2)^2)^2 \mid Z^{(n)}(t_1), Z^{(n)}(t_2)\}\\
&\quad = \mathbb{E}(Z^{(n)}(t_2)^2 - Z^{(n)}(t_1)^2)^2\mathbb{E}\{(Z^{(n)}(t_3)^2 - Z^{(n)}(t_2)^2)^2 \mid Z^{(n)}(t_2)\}.
\end{aligned}$$

In turn, the conditional expectation may be bounded as follows:

$$\begin{aligned}
&\mathbb{E}\{(Z^{(n)}(t_3)^2 - Z^{(n)}(t_2)^2)^2 \mid Z^{(n)}(t_2)\}\\
&\quad = \mathbb{E}\{Z^{(n)}(t_3)^4 - Z^{(n)}(t_2)^4 \mid Z^{(n)}(t_2)\}\\
&\qquad - 2Z^{(n)}(t_2)^2\mathbb{E}\{Z^{(n)}(t_3)^2 - Z^{(n)}(t_2)^2 \mid Z^{(n)}(t_2)\}\\
&\quad \le \mathbb{E}\{Z^{(n)}(t_3)^4 - Z^{(n)}(t_2)^4 \mid Z^{(n)}(t_2)\},
\end{aligned}$$

owing to (4.69). The calculations leading to (4.67) also imply that, for some $c_1 < \infty$,

$$\mathbb{E}\{Z^{(n)}(t_3)^4 - Z^{(n)}(t_2)^4 \mid Z^{(n)}(t_2)\} \le c_1(t_3 - t_2)Z^{(n)}(t_2)^2.$$

Therefore

$$\mathbb{E}\{(Z^{(n)}(t_3)^2 - Z^{(n)}(t_2)^2)^2 \mid Z^{(n)}(t_2)\} \le c_1(t_3 - t_2)Z^{(n)}(t_2)^2. \tag{4.71}$$

Further,

$$\begin{aligned}
&\mathbb{E}(Z^{(n)}(t_2)^2 - Z^{(n)}(t_1)^2)^2Z^{(n)}(t_2)^2\\
&\quad = \mathbb{E}(Z^{(n)}(t_2)^6 - Z^{(n)}(t_1)^6) - \mathbb{E}(Z^{(n)}(t_2)^4 - Z^{(n)}(t_1)^4)Z^{(n)}(t_1)^2\\
&\qquad - \mathbb{E}((Z^{(n)}(t_2)^2 - Z^{(n)}(t_1)^2)^2Z^{(n)}(t_1)^2 - \mathbb{E}(Z^{(n)}(t_2)^2 - Z^{(n)}(t_1)^2)Z^{(n)}(t_1)^4\\
&\quad \le \mathbb{E}(Z^{(n)}(t_2)^6 - Z^{(n)}(t_1)^6) - \mathbb{E}(Z^{(n)}(t_2)^4 - Z^{(n)}(t_1)^4)Z^{(n)}(t_1)^2\\
&\qquad - \mathbb{E}(Z^{(n)}(t_2)^2 - Z^{(n)}(t_1)^2)Z^{(n)}(t_1)^4\\
&\quad \le \mathbb{E}(Z^{(n)}(t_2)^6 - Z^{(n)}(t_1)^6)
\end{aligned}$$

because the second and third terms on the right-hand side of the first inequality are negative owing to the assumptions (4.70) and (4.69). Hence

$$\mathbb{E}(Z^{(n)}(t_2)^2 - Z^{(n)}(t_1)^2)^2 Z^{(n)}(t_2)^2 \le c_2(t_2 - t_1),$$

which together with (4.71) implies (4.68). Thus the diffusion approximation follows under the conditions of Theorem 4.8.

Under the conditions of Theorem 4.10 the proof is the same but it then starts with the time-dependent truncation of jumps. □

4.8 Integral Renewal Theorem for Transient Chain with Gamma Limit

The next result determines the asymptotic behaviour of the renewal functions $H_y(x)$ and $H(x)$ in the case of convergence to a Γ-distribution in the transient case. The proof is based on the preliminary upper bound given in Theorem 4.3.

Theorem 4.12 *Under the conditions of Theorem 4.8, for any initial distribution of the chain $\{X_n\}$, we have*

$$\sum_{n=0}^{[Bx^2]} \mathbb{P}\{X_n \in (\widehat{x}, x]\} = (I(B) + o(1))x^2$$

as $x \to \infty$ uniformly for all $B \ge 0$, where

$$I(B) := \int_0^B \Gamma(1/z)dz \;=\; B\Gamma(1/B) + \int_{1/B}^{\infty} \frac{1}{z}\gamma(z)dz, \quad I(\infty) = \frac{1}{2\mu - b},$$

$\widehat{x}$ is defined in Theorem 4.8, and $\Gamma(t)$ and $\gamma(t)$ denote the cumulative distribution function and the probability density function respectively of the Γ-distribution with mean $2\mu + b$ and variance $(2\mu + b)2b$. In particular,

$$H(\widehat{x}, x] \sim \frac{1}{2\mu - b} x^2 \quad as\ x \to \infty. \tag{4.72}$$

We apply this result in Section 11.1 to derive the asymptotics of the renewal measure for a random walk conditioned to stay positive.

Proof By Theorem 4.8, for every fixed $B > 0$ we have

$$\sum_{n=0}^{[Bx^2]} \mathbb{P}\{X_n \in (\widehat{x}, x]\} = \sum_{n=0}^{[Bx^2]} (\Gamma(x^2/n) + o(1))$$
$$= \sum_{n=0}^{[Bx^2]} \Gamma(x^2/n) + o(x^2) \quad \text{as } x \to \infty.$$

Since

$$\sum_{n=0}^{[Bx^2]} \Gamma(x^2/n) \sim x^2 \int_0^B \Gamma(1/z)dz \qquad \text{as } x \to \infty,$$

we conclude that, for any fixed $B > 0$,

$$\sum_{n=0}^{[Bx^2]} \mathbb{P}\{X_n \in (\widehat{x}, x]\} \sim I(B)x^2 \qquad \text{as } x \to \infty. \tag{4.73}$$

Since the sum is increasing in B, it remains for us to prove that (4.72) holds. First, since

$$\int_0^B \Gamma(1/z)dz \to \frac{1}{2\mu - b} \qquad \text{as } B \to \infty,$$

we conclude that the lower bound is

$$\liminf_{x\to\infty} \frac{H(\widehat{x}, x]}{x^2} \geq \frac{1}{2\mu - b}. \tag{4.74}$$

Second, for an arbitrary y, let us now prove the matching upper bound,

$$\limsup_{x\to\infty} \frac{H_y(\widehat{x}, x]}{x^2} \leq \frac{1}{2\mu - b}. \tag{4.75}$$

For any $A > 1$, $T(Ax)$ is the first up-crossing time of the level Ax. By the Markov property,

$$\begin{aligned} H_y(\widehat{x}, x] &\leq \mathbb{E}_y \sum_{n=0}^{T(Ax)-1} \mathbb{I}\{X_n \in (\widehat{x}, x]\} \\ &\quad + \mathbb{P}\{X_n \leq x \text{ for some } n \mid X_0 > Ax\} \sup_{z\leq x} H_z(\widehat{x}, x] \\ &\leq \mathbb{E}_y \sum_{n=0}^{T(Ax)-1} \mathbb{I}\{X_n \in (\widehat{x}, x]\} \\ &\quad + \left(e^{\delta(R(x)-R(Ax))} + o(1)\right) \sup_{z\leq x} H_z(\widehat{x}, x] \end{aligned} \tag{4.76}$$

as $x \to \infty$ uniformly for all $A > 1$, owing to (4.32) with $R(x)$ determined by $r(x) = \gamma/x$ with $\gamma \in (0, 2\mu/b)$; hence

$$e^{\delta(R(x)-R(Ax))} = 1/A^{\delta\gamma}.$$

Thus, applying the upper bound proven in Theorem 4.3 to the right-hand side of (4.76), we deduce that, for some $c < \infty$,

$$H_y(\widehat{x}, x] \le \mathbb{E}_y \sum_{n=0}^{T(Ax)-1} \mathbb{I}\{X_n \in (\widehat{x}, x]\} + \big(c/A^{\delta\gamma} + o(1)\big)x^2 \tag{4.77}$$

as $x \to \infty$ uniformly for all $A > 1$. The expectation of the sum on the right-hand side of (4.77) may be bounded as follows: for $C > 1$,

$$\mathbb{E}_y \sum_{n=0}^{T(Ax)-1} \mathbb{I}\{X_n \in (\widehat{x}, x]\} \le \mathbb{E}_y \sum_{n=0}^{[CA^2x^2]} \mathbb{I}\{X_n \in (\widehat{x}, x]\}$$
$$+ \mathbb{E}_y\Big\{ \sum_{n=0}^{T(Ax)-1} \mathbb{I}\{X_n > \widehat{x}\}; T(Ax) > CA^2x^2\Big\}.$$

The second term on the right-hand side is not greater than

$$\mathbb{E}_y\Big\{L(\widehat{x}, T(Ax));\ X_n \le \widehat{x} \text{ for some } n \ge A^2x^2\Big\}$$
$$+ \mathbb{E}_y\Big\{L(\widehat{x}, T(Ax));\ X_n > \widehat{x} \text{ for all } n \in [A^2x^2, T(Ax)-1],\ T(Ax) > CA^2x^2\Big\}$$
$$\le \mathbb{E}_y\Big\{L(\widehat{x}, T(Ax));\ X_n \le \widehat{x} \text{ for some } n \ge A^2x^2\Big\}$$
$$+ \mathbb{E}_y\Big\{L(\widehat{x}, T(Ax));\ L(\widehat{x}, T(Ax)) > (C-1)A^2x^2\Big\}.$$

Since the conditions of Theorem 4.2 are met if $v(x) = \mu/2x$, the family of random variables

$$\frac{L(\widehat{x}, T(Ax))}{(Ax)^2}$$

is uniformly integrable, so, for any fixed $A > 1$,

$$\sup_{x>\widehat{x},\ y} \frac{1}{x^2}\mathbb{E}_y\Big\{L(\widehat{x}, T(Ax));\ L(\widehat{x}, T(Ax)) > (C-1)A^2x^2\Big\} \le \psi(C),$$

where $\psi(C) \to 0$ as $C \to \infty$. Since $X_n \to \infty$ with probability 1,

$$\mathbb{P}\{X_n \le \widehat{x} \text{ for some } n \ge A^2x^2\} \to 0 \quad \text{as } x \to \infty.$$

Therefore, again by the uniform integrability of the family $L(\widehat{x}, T(Ax))/(Ax)^2$,

$$\frac{1}{x^2}\mathbb{E}_y\Big\{L(\widehat{x}, T(Ax));\ X_n \le \widehat{x} \text{ for some } n \ge A^2x^2\Big\} \to 0 \quad \text{as } x \to \infty.$$

Taken altogether, this yields

$$\limsup_{x\to\infty}\sup_y \frac{1}{x^2}\mathbb{E}_y\Big\{\sum_{n=0}^{T(Ax)-1}\mathbb{I}\{X_n>\widehat{x}\}; T(Ax)>CA^2x^2\Big\}\le\psi(C);$$

hence, uniformly for all y,

$$\limsup_{x\to\infty}\frac{1}{x^2}\mathbb{E}_y\sum_{n=0}^{T(Ax)-1}\mathbb{I}\{X_n\in(\widehat{x},x]\}\le\mathbb{E}_y\sum_{n=0}^{[CA^2x^2]}\mathbb{I}\{X_n\in(\widehat{x},x]\}+\psi(C),$$

which, on being substituted into (4.77), gives

$$\limsup_{x\to\infty}\frac{H_y(\widehat{x},x]}{x^2}\le\limsup_{x\to\infty}\frac{1}{x^2}\mathbb{E}_y\sum_{n=0}^{[CA^2x^2]}\mathbb{I}\{X_n\in(\widehat{x},x]\}+\psi(C)+c/A^{\delta\gamma}.$$

As has already been shown,

$$\frac{1}{x^2}\sum_{n=0}^{[CA^2x^2]}\mathbb{P}_y\{X_n\in(\widehat{x},x]\}\to I(CA^2)\quad\text{as } x\to\infty,$$

which implies the following upper bound, for each fixed A, $C>1$:

$$\limsup_{x\to\infty}\frac{H_y(\widehat{x},x]}{x^2}\le I(CA^2)+\psi(C)+c/A^{\delta\gamma}.$$

Now letting first $C\to\infty$ and then $A\to\infty$, we get the required upper bound (4.75). The lower bound (4.74) and upper bound (4.75) yield the following equivalence, for every fixed y:

$$H_y(\widehat{x},x]\sim\frac{1}{2\mu-b}x^2\qquad\text{as } x\to\infty.$$

Together with the bound of Theorem 4.3, which is uniform in y, this completes the proof of (4.72) and hence the result follows. □

The next result will be used later to find the tail asymptotics for a stationary measure when $\{X_n\}$ is recurrent.

Theorem 4.13 *Let the conditions of Theorem 4.8 hold. Then, for $q(z)\ge 0$ and any distribution of X_0,*

$$\sum_{n=0}^{[Bx^2]}\mathbb{E}\big\{e^{-\sum_{k=0}^{n-1}q(X_k)};\ X_n\in(\widehat{x},x]\big\}=(I(B)+o(1))x^2\,\mathbb{E}e^{-\sum_{k=0}^{\infty}q(X_k)}$$

as $x\to\infty$ uniformly for all $B\in[0,\infty]$, where $I(B)$ was defined in Theorem 4.12.

Proof We may apply Lemma 4.5 because the condition (4.39) is guaranteed by Theorem 4.3 and the condition (4.40) by Theorem 4.12. □

4.9 Local Renewal Theorem for Transient Chain on $\mathbb{Z}$ with Gamma Limit

In this section we discuss a local version of the renewal theorem in the case of convergence to a Γ-distribution. We will do this for a lattice Markov chain. Without loss of generality, let the minimal lattice on which $\{X_n\}$ is living be $\mathbb{Z}$. It is unclear whether the local renewal theorem would be valid if we assumed only a regular asymptotic behaviour of the moments of jumps. It is very likely that it can be proven only for a Markov chain that is asymptotically homogeneous in space, as defined in Definition 1.23, that is, if we assume weak convergence of the jumps $\xi(x)$ to some random variable ξ on $\mathbb{Z}$:

$$\xi(x) \Rightarrow \xi \quad \text{as } x \to \infty. \tag{4.78}$$

Theorem 4.14 *Let there exist $b > 0$ and $\mu > b/2$ such that*

$$m_1(x) \sim \mu/x \text{ and } m_2(x) \to b \quad \text{as } x \to \infty \tag{4.79}$$

and

$$\limsup_{n\to\infty} X_n = \infty \quad \text{with probability } 1.$$

Furthermore, we assume the convergence (4.78). Let $\mathbb{Z}$ be the minimal lattice for ξ, and let the limit ξ satisfy

$$\mathbb{E}\xi = 0, \quad \mathbb{E}\xi^2 = b. \tag{4.80}$$

In addition, let the jumps $\xi(x)$ be bounded below and above by J uniformly for all $x \in \mathbb{Z}^+$, that is,

$$|\xi(x)| \le J \qquad \text{for all } x \in \mathbb{Z}^+. \tag{4.81}$$

Then

$$h(x) := H\{x\} \sim \frac{2}{2\mu - b} x \quad \text{as } x \to \infty. \tag{4.82}$$

Moreover,

$$\mathbb{P}\Big\{\sum_{n=0}^{\infty} \mathbb{I}\{X_n = x\} > N\Big\} = c_1(x)\Big(1 - \frac{c_2(x)}{x}\Big)^N, \tag{4.83}$$

where $c_1(x) \to 1$ *and* $c_2(x) \to \mu - b/2 > 0$ *as* $x \to \infty$, *so the family of random variables*

$$\frac{1}{x}\sum_{n=0}^{\infty}\mathbb{I}\{X_n = x\}, \quad x \in \{1,2,3,\ldots\}, \tag{4.84}$$

is uniformly integrable.

More general results are derived in Chapter 6, via a different technique based on the martingale approach.

Proof Consider a stopping time

$$\tau(x) := \inf\{n \ge 1 : X_n \le x\}.$$

Since $\{X_n\}$ is transient, $\mathbb{P}_{x+k}\{\tau(x) = \infty\} > 0$ for all $k \ge 1$. First, let us understand the asymptotic behaviour of this probability as x grows. To this end, let us fix a $\delta \in (0, 2\mu/b - 1)$ and define two decreasing functions

$$U_\pm(x) := \frac{1}{(2\mu/b - 1 \pm \delta)x^{2\mu/b-1\pm\delta}}, \quad x \ge 1.$$

By the mean value theorem, for all x and $j \in \mathbb{Z}$ there is a $\theta \in (0,1)$ such that

$$U_\pm(x+j) - U_\pm(x) = -\frac{j}{(x+\theta j)^{2\mu/b\pm\delta}} \sim -\frac{j}{x^{2\mu/b\pm\delta}} \quad \text{as } x \to \infty,$$

which implies that

$$U_\pm(x+j) - U_\pm(x) \sim -j(2\mu/b - 1 \pm \delta)\frac{U_\pm(x)}{x}.$$

Then, since the $\xi(x)$ are bounded below, we obtain for all fixed $k \ge 1$ that

$$\begin{aligned}&\mathbb{E}_{x+k}\left\{U_\pm(X_{\tau(x)}) - U_\pm(x+k);\ \tau(x) < \infty\right\}\\ &\quad \sim (2\mu/b - 1 \pm \delta)\frac{U_\pm(x+k)}{x}\mathbb{E}_{x+k}\{x + k - X_{\tau(x)};\ \tau(x) < \infty\}.\end{aligned} \tag{4.85}$$

Let us compute the drift of $U_\pm(X_n)$. Since the jumps are bounded, by a Taylor expansion we have

$$\begin{aligned}&\mathbb{E}(U_\pm(x+\xi(x)) - U_\pm(x))\\ &= U'_\pm(x)m_1(x) + \frac{1}{2}m_2(x)U''_\pm(x)m_2(x) + O(U'''_\pm(x))\\ &= -x^{-2\mu/b\mp\delta}m_1(x) + (\mu/b \pm \delta/2)x^{-2\mu/b-1\mp\delta}m_2(x) + O(x^{-2\mu/b-2\mp\delta})\\ &= \pm(\delta b/2 + o(1))x^{-2\mu/b-1\mp\delta} \quad \text{as } x \to \infty.\end{aligned}$$

Therefore, the sequence $U_-(X_{n\wedge\tau(x)})$ is a supermartingale for all sufficiently large x. So, by the optional stopping theorem (see e.g. [56, Section 4.7]),

$$\mathbb{E}_{x+k}\{U_-(X_{\tau(x)});\ \tau(x)<\infty\} \le U_-(x+k).$$

This is equivalent to

$$\mathbb{E}_{x+k}\left\{U_-(X_{\tau(x)}) - U_-(x+k);\ \tau(x)<\infty\right\} \le U_-(x+k)\mathbb{P}_{x+k}\{\tau(x)=\infty\}.$$

Now using (4.85), we obtain, for all sufficiently large x,

$$\mathbb{P}_{x+k}\{\tau(x)=\infty\} \ \ge\ \frac{2\mu/b-1-2\delta}{x}\mathbb{E}_{x+k}\{x+k-X_{\tau(x)};\ \tau(x)<\infty\}. \tag{4.86}$$

Since $\{U_+(X_{n\wedge\tau(x)})\}$ is a submartingale for all sufficiently large x,

$$\mathbb{E}_{x+k}\{U_+(X_{\tau(x)});\ \tau(x)<\infty\} \ge U_+(x+k).$$

This implies that, for all sufficiently large x,

$$\mathbb{P}_{x+k}\{\tau(x)=\infty\} \ \le\ \frac{2\mu/b-1+2\delta}{x}\mathbb{E}_{x+k}\{x+k-X_{\tau(x)};\ \tau(x)<\infty\}.$$

Combining this lower bound with (4.86) and given the arbitrary choice of $\delta>0$, we conclude that, as $x\to\infty$,

$$\mathbb{P}_{x+k}\{\tau(x)=\infty\} \ =\ \frac{2\mu/b-1+o(1)}{x}\mathbb{E}_{x+k}\{x+k-X_{\tau(x)};\ \tau(x)<\infty\}. \tag{4.87}$$

Now let us determine the limit of $\mathbb{E}_{x+k}\{x+k-X_{\tau(x)};\ \tau(x)<\infty\}$. Let the ξ_k, $k\ge 1$, be independent copies of the random variable ξ. Define $S_0=0$, $S_k := S_{k-1}+\xi_k$ for $k\ge 1$, and

$$\theta_j := \min\{k\ge 1 : S_k < -j\},\quad \psi_j = -S_{\theta_j}.$$

Assumption (4.78) implies that, for every $n\ge 1$, $(X_1-X_0, X_2-X_0,\dots, X_n-X_0)$ converges weakly, as $X_0\to\infty$, to $(S_1,S_2,\dots,S_n)$. In particular,

$$\mathbb{E}_{x+k}\{x+k-X_{\tau(x)};\ \tau(x)\le n\} \ \to\ \mathbb{E}\{\psi_{k-1};\ \theta_k\le n\},\quad n\ge 1.$$

Noting that both $x+k-X_{\tau(x)}$ and ψ_{k-1} are bounded, we conclude that

$$\lim_{x\to\infty}\mathbb{E}_{x+k}\{x+k-X_{\tau(x)};\ \tau(x)<\infty\} \ =\ \mathbb{E}\psi_{k-1}.$$

Substituting this into (4.87), we find, for all $k\ge 1$,

$$\mathbb{P}_{x+k}\{\tau(x)=\infty\} \sim \frac{2\mu-b}{bx}\mathbb{E}\psi_{k-1}\quad \text{as } x\to\infty. \tag{4.88}$$

We can now use these asymptotics to study the asymptotic behaviour of the renewal mass function $h(x)$. Choose any $j_0 \in [1, J]$ such that

$$\mathbb{P}\{\xi = j_0\} > 0 \tag{4.89}$$

and consider the following up-crossing stopping times:

$$\sigma(x) := \min\{n \geq 1 : X_{n-1} \leq x,\ X_n > x\},$$
$$\gamma(x) := \min\{n \geq 1 :\ X_{n-1} \leq x,\ X_n = x + j_0\},$$

and let us evaluate the probabilities

$$p_i(x) \ := \ \mathbb{P}_{x+i}\{\gamma(x) = \infty\}$$

for $i = 1, \ldots, J$ and large values of x. For all $i \leq J$, the Markov property leads to the equation

$$\begin{aligned}
&\mathbb{P}_{x+i}\{\gamma(x) = \infty\} \\
&\quad = \mathbb{P}_{x+i}\{\sigma(x) = \infty\} + \sum_{j=1,\ j \neq j_0}^{J} \mathbb{P}_{x+i}\{\sigma(x) < \infty, X_{\sigma(x)} = x + j\}\mathbb{P}_{x+j}\{\gamma(x) = \infty\} \\
&\quad = \mathbb{P}_{x+i}\{\tau(x) = \infty\} + \sum_{j=1,\ j \neq j_0}^{J} \mathbb{P}_{x+i}\{\tau(x) < \infty, X_{\sigma(x)} = x + j\}\mathbb{P}_{x+j}\{\gamma(x) = \infty\},
\end{aligned}$$

because the transience of $\{X_n\}$ implies that

$$\mathbb{P}_{x+i}\{\tau(x) < \infty, \sigma(x) = \infty\} \ = \ 0.$$

Hence, the $(J-1)$-dimensional vector

$$p(x) := (p_1(x), \ldots, p_{j_0-1}(x), p_{j_0+1}(x), \ldots, p_J(x))^\top$$

satisfies the equation

$$p(x) = q(x) + A(x)p(x),$$

where

$$q_i(x) \ = \ \mathbb{P}_{x+i}\{\sigma(x) = \infty\} \ = \ \mathbb{P}_{x+i}\{\tau(x) = \infty\}$$

and $A(x)$ is a matrix with entries $A_{ij}(x)$, $i, j \in \{1, \ldots, j_0 - 1, j_0 + 1, \ldots, J\}$, with

$$A_{ij}(x) \ = \ \mathbb{P}_{x+i}\{\sigma(x) < \infty, X_{\sigma(x)} = x + j\}.$$

Therefore, provided the matrix $I - A(x)$ is invertible, we have

$$p(x) = (I - A(x))^{-1} q(x). \tag{4.90}$$

In view of $\mathbb{P}\{\xi = j_0\} > 0$ – see (4.89) – and because $\mathbb{Z}$ is the minimal lattice for ξ, it follows that there exists an $\varepsilon > 0$ such that

$$A_{ij_0} := \mathbb{P}\{S_\sigma = j_0 \mid S_0 = i\} > 2\varepsilon \quad \text{for all } i \le J,$$

where $\sigma := \inf\{n \ge 1 : S_{n-1} \le 0,\ S_n > 0\}$ is finite a.s. By the condition (4.78),

$$\mathbb{P}\{X_{\sigma(x)} = x + j_0, \sigma(x) < \infty \mid X_0 = x + i\} \to \mathbb{P}\{S_\sigma = j_0 \mid S_0 = i\}$$

as $x \to \infty$; hence there is an x_0 such that, for all $x \ge x_0$,

$$A_{ij_0}(x) = \mathbb{P}\{X_{\sigma(x)} = x + j_0, \sigma(x) < \infty \mid X_0 = x + i\} > \varepsilon \quad \text{for all } i \le J.$$

Thus each row of the matrix $A(x)$ sums to a number less than $1 - \varepsilon$. So the matrix $I - A(x)$ is invertible and

$$(I - A(x))^{-1} \to (I - A)^{-1} \quad \text{as } x \to \infty,$$

where

$$A_{ij} = \mathbb{P}\{S_\sigma = j \mid S_0 = i\},$$

and it follows from (4.90) and (4.88) that, as $x \to \infty$,

$$\begin{aligned} p(x) &\sim \frac{2\mu - b}{bx}(I - A)^{-1}(\mathbb{E}\psi_1, \ldots, \mathbb{E}\psi_{j_0-1}, \mathbb{E}\psi_{j_0+1}, \ldots, \mathbb{E}\psi_J)^\top \\ &=: \frac{1}{x}(c_1, \ldots, c_{j_0-1}, c_{j_0+1}, \ldots, c_J)^\top. \end{aligned}$$

Thus

$$p_{j_0}(x) - \mathbb{P}_{x+j_0}\{\gamma(x) = \infty\} \tag{4.91}$$

$$= \mathbb{P}_{x+j_0}\{\tau(x) = \infty\} + \sum_{j=1,\ j \ne j_0}^{J} A_{j_0 j}(x) p_j(x)$$

$$\sim \frac{1}{x}\Big(\frac{2\mu - b}{b}\mathbb{E}\psi_{j_0-1} + \sum_{j=1,\ j \ne j_0}^{J} A_{j_0 j} c_j\Big) \quad \text{as } x \to \infty. \tag{4.92}$$

Denote by $N(x)$ the number of visits of $\{X_n\}$ to the state x. We have

$$h(x) = \mathbb{E}\sum_{n=1}^{\gamma(x)-1} \mathbb{I}\{X_n = x\} + \mathbb{E}_{x+j_0} N(x)\mathbb{P}\{\gamma(x) < \infty\}. \tag{4.93}$$

Since the random variable

$$\sum_{n=1}^{\gamma(x)-1} \mathbb{I}\{X_n = x\}$$

is stochastically dominated by a geometric random variable with parameter $1-\mathbb{P}\{\xi(x)=j_0\}$ and $\mathbb{P}\{\xi(x)=j_0\}\to\mathbb{P}\{\xi=j_0\}>0$ as $x\to\infty$, there exists a sufficiently large $x_1\in\mathbb{Z}^+$ such that the first term on the right-hand side of (4.93) is bounded above for all $x\ge x_1$,

$$\sup_{x\ge x_1}\mathbb{E}\sum_{n=1}^{\gamma(x)-1}\mathbb{I}\{X_n=x\}<\infty. \tag{4.94}$$

In addition, since all $p_i(x)\to 0$,

$$\mathbb{P}\{\gamma(x)<\infty\}\to 1 \quad \text{as } x\to\infty. \tag{4.95}$$

Further, by the Markov property,

$$\mathbb{E}_{x+j_0}N(x)=\mathbb{E}_{x+j_0}\sum_{n=1}^{\gamma(x)-1}\mathbb{I}\{X_n=x\}+\mathbb{P}_{x+j_0}\{\gamma(x)<\infty\}\mathbb{E}_{x+j_0}N(x),$$

which yields, by (4.91),

$$\begin{aligned}\mathbb{E}_{x+j_0}N(x)&=\frac{1}{p_{j_0}(x)}\mathbb{E}_{x+j_0}\sum_{n=1}^{\gamma(x)-1}\mathbb{I}\{X_n=x\}\\&\sim cx\mathbb{E}_{x+j_0}\sum_{n=1}^{\gamma(x)-1}\mathbb{I}\{X_n=x\}.\end{aligned}$$

Taking into account that

$$\mathbb{E}_{x+j_0}\sum_{n=1}^{\gamma(x)-1}\mathbb{I}\{X_n=x\}\to\mathbb{E}\sum_{n=1}^{\gamma-1}\mathbb{I}\{S_n=0\mid S_0=j_0\}\quad\text{as } x\to\infty,$$

where

$$\gamma:=\inf\{k: S_{k-1}\le 0,\ S_k=j_0\},$$

we conclude that

$$\mathbb{E}_{x+j_0}N(x)\sim\widehat{c}x\quad\text{as } x\to\infty.$$

Substituting this, together with (4.94) and (4.95), into (4.93) we deduce that $h(x)\sim\widehat{c}x$ as $x\to\infty$. Then it follows from the integral renewal theorem, Theorem 4.12, that necessarily $\widehat{c}=2/(2\mu-b)$ and so (4.82) is proven.

To prove (4.83), let us first note that the Markov property implies

$$\begin{aligned}&\mathbb{P}\Big\{\sum_{n=1}^{\infty}\mathbb{I}\{X_n=x\}>N\Big\}\\&\quad=\mathbb{P}\{X_n=x\text{ for some } n\ge 0\}\mathbb{P}^N\{X_n=x\text{ for some } n\ge 1\mid X_0=x\}.\end{aligned}$$

We take

$$c_1(x) := \mathbb{P}\{X_n = x \text{ for some } n \ge 0\};$$

it tends to 1 as $x \to \infty$ because, by the boundedness of jumps from above, see (4.81), for $X_0 < x$, we have

$$\begin{aligned}1 &= \mathbb{P}\{X_n \in [x, x+J] \text{ for some } n\}\\ &= \mathbb{P}\{X_n = x \text{ for some } n\}\\ &\quad + \mathbb{P}\{X_n \in [x+1, x+J] \text{ for some } n,\ X_n \ne x \text{ for all } n\},\end{aligned}$$

where the second probability on the right-hand side tends to zero as $x \to \infty$. Indeed this probability is not greater than

$$\begin{aligned}&\sum_{i=1}^{J} \mathbb{P}\{X_n = x+i \text{ for some } n,\ X_n \ne x \text{ for all } n\}\\ &\qquad \le \sum_{i=1}^{J} \mathbb{P}\{X_n \ne x \text{ for all } n \mid X_0 = x+i\}\end{aligned}$$

and the ith probability on the right-hand side converges as $x \to \infty$ to

$$\mathbb{P}\{S_n \ne 0 \text{ for all } n \mid S_0 = i\} = 0,$$

since $\mathbb{E}\xi = 0$. Thus (4.83) holds with

$$c_2(x) = x\mathbb{P}\{X_n \ne x \text{ for all } n \ge 1 \mid X_0 = x\}$$

because

$$\mathbb{E}\sum_{n=1}^{\infty} \mathbb{I}\{X_n = x\} = \frac{c_1(x)}{1 - \mathbb{P}\{X_n = x \text{ for some } n \ge 1 \mid X_0 = x\}} \sim \frac{2x}{2\mu - b}$$

as $x \to \infty$, by (4.82). □

Theorem 4.15 *Let the conditions of Theorem 4.14 hold. Then, for $q(z) \ge 0$ and any distribution of X_0,*

$$\sum_{n=0}^{\infty} \mathbb{E}\{e^{-\sum_{k=0}^{n-1} q(X_k)};\ X_n = x\} \sim \frac{2x}{2\mu - b}\,\mathbb{E}e^{-\sum_{k=0}^{\infty} q(X_k)} \quad \text{as } x \to \infty.$$

Proof We may apply Lemma 4.5, all of whose conditions are satisfied by Theorem 4.14. □

We now turn to the case when (4.79) holds with $\mu = b/2$. In this case we prove the following result.

Theorem 4.16 *Let there exist an $\gamma > 0$ such that*

$$\frac{2m_1(x)}{m_2(x)} = \frac{1}{x} + \frac{1}{x\log x} + \cdots + \frac{1}{x\log x \cdots \log_{(m-1)} x} + \frac{\gamma + 1 + o(1)}{x\log x \cdots \log_{(m)} x} \tag{4.96}$$

as $x \to \infty$ and

$$\limsup_{n\to\infty} X_n = \infty \quad \textit{with probability } 1.$$

Furthermore we assume the convergence (4.78) *and that the limit ξ satisfies* (4.80)*. In addition, let the jumps $\xi(x)$ be bounded below and above by J uniformly for all $x \in \mathbb{Z}^+$, that is,*

$$|\xi(x)| \le J \qquad \textit{for all } x \in \mathbb{Z}^+.$$

Then there exists a positive constant c such that

$$h(x) := H\{x\} \sim cx\log x \ldots \log_{(m)} x \quad \textit{as } x \to \infty. \tag{4.97}$$

Moreover,

$$\mathbb{P}\Big\{\sum_{n=0}^{\infty} \mathbb{I}\{X_n = x\} > N\Big\} = c_1(x)\Big(1 - \frac{c_2(x)}{x\log x \cdots \log_{(m)} x}\Big)^N, \tag{4.98}$$

where $c_1(x) \to 1$ and $c_2(x) \to 1/c$ as $x \to \infty$; hence the family of random variables

$$\frac{1}{x\log x \cdots \log_{(m)} x} \sum_{n=0}^{\infty} \mathbb{I}\{X_n = x\}, \quad x \in \{1, 2, 3, \ldots\}, \tag{4.99}$$

is uniformly integrable.

Proof We first derive an asymptotic formula for the probability $\mathbb{P}_{x+k}\{\tau(x) = \infty\}$. We define functions $U_\pm$ by the relations

$$U_\pm(x) = \frac{1}{(\gamma \pm \delta)(\log_{(m)} x)^{\gamma\pm\delta}}.$$

It is easy to see that

$$U'_\pm(x) = -\frac{1}{x\log x \cdots \log_{(m-1)} x\, (\log_{(m)} x)^{\gamma+1\pm\delta}} \tag{4.100}$$

and

$$\frac{U''_\pm(x)}{U'_\pm(x)} = -\frac{1}{x} - \frac{1}{x\log x} - \cdots - \frac{1}{x\log x \cdots \log_{(m-1)} x} - \frac{(\gamma + 1 \pm \delta)}{x\log x \cdots \log_{(m-1)} x\, \log_{(m)} x}. \tag{4.101}$$

Let us compute the drift of $U_\pm(X_n)$. Since the jumps are bounded, using a Taylor expansion we have

$$\begin{aligned}\mathbb{E}(U_\pm(x+\xi(x)) - U_\pm(x)) &= U'_\pm(x)m_1(x) + \frac{1}{2}m_2(x)U''_\pm(x)m_2(x) + O(U'''_\pm(x)) \\ &= \frac{U'_\pm(x)m_2(x)}{2}\left(\frac{2m_1(x)}{m_2(x)} + \frac{U''_\pm(x)}{U'_\pm(x)}\right) + o\left(\frac{1}{x^3}\right).\end{aligned}$$

Taking into account (4.96), (4.100), and (4.101), we infer that

$$\mathbb{E}(U_\pm(x+\xi(x)) - U_\pm(x)) \sim \pm\frac{b\delta}{2}\frac{1}{(x\log x\cdots\log_{(m)}x)^2(\log_{(m)}x)^{\gamma+2\pm\delta}}.$$

In particular, the sequences $\{U_-(X_{n\wedge\tau(x)})\}$ and $\{U_+(X_{n\wedge\tau(x)})\}$ are super- and submartingales respectively for all sufficiently large x.

Furthermore, it follows from the definition of $U_\pm$ and from (4.100) that

$$U_\pm(x+j) - U_\pm(x) \sim -jU'_\pm(x) \sim -j(\gamma\pm\delta)\frac{U_\pm(x)}{x\log x\cdots\log_{(m)}x}.$$

Using this relation and repeating the arguments from the derivation of (4.88), we obtain, for every $k \geq 1$,

$$\mathbb{P}_{x+k}\{\tau(x) = \infty\} \sim \frac{\gamma}{x\log x\cdots\log_{(m)}x}\mathbb{E}\psi_{k-1} \quad \text{as } x\to\infty. \tag{4.102}$$

The remaining part of the proof coincides with that of Theorem 4.14. □

4.10 Comments for Chapter 4

A limit theorem for Markov chains with asymptotically zero drift was produced for the first time by Lamperti in [112], where the convergence to a Γ-distribution was proven for null-recurrent and transient Markov chains with jumps all of whose moments are finite. His proof was based on the method of moments. He also claimed that his proof combined with a truncation argument for jumps would continue to work for chains only if we assume that $\sup_x \mathbb{E}\xi^4(x) < \infty$, but no proof was provided.

This result was also proven later by Klebaner [99] for more general random sequences of the martingale type with jumps satisfying the following condition: for all $k \geq 3$, $\mathbb{E}|\xi(x)|^k = o(x^{k-2})$ as $x\to\infty$. The corresponding result is restricted to transient sequences.

Later, the convergence to a Γ-distribution was extended by Kersting [94] to martingale-type transient random sequences with jumps having moments of order $2+\delta$, bounded for some $\delta > 0$, and under some additional smoothness conditions on the drift.

The diffusion approximation by a Bessel process was originally proven by Lamperti in [112], again under the condition that absolute moments of jumps of any order are bounded. His proof was based on using the method of moments as in the proof of weak convergence to a Γ-distribution.

The convergence to the three-dimensional Bessel process for a simple symmetric random walk conditioned to stay non-negative has been known for a long time from the classic paper by Pitman [133]. Bryn-Jones and Doney [30] proved this convergence to the three-dimensional Bessel process for a general random walk on the lattice conditioned to stay non-negative under minimal moment conditions; see also Caravenna and Chaumont [32] for general results in this area. In our text, the application of functional results to random walks conditioned to stay non-negative is discussed in Section 11.1.

For a general lattice Markov chain with drift proportional to c/x and the $(2+\delta)$th moment of jumps bounded, the weak convergence to a Bessel process was proven by Bertoin and Kortchemski [15] only for a high-level initial state, $X_0 = \sqrt{n}$.

Csáki et al. [39] proved a strong approximation of a certain nearest neighbour random walk by a transient Bessel process that was constructed from the latter using stopping times.

Rosenkrantz [138] considered the following nearest neighbour Markov chain with the following special transition probabilities: $p(0,1) = 1$,

$$p(x,x-1) = \frac{1}{2}\Big(1 - \frac{\lambda}{x+\lambda}\Big), \quad p(x,x+1) = 1 - p(x,x-1), \quad x \ge 1, \tag{4.103}$$

where $\lambda > -1/2$. This Markov chain was introduced by Karlin and McGregor [86], who managed to compute n-step transition probabilities using the theory of orthogonal polynomials. Using orthogonal polynomials Rosenkrantz proved a local version of Lamperti's result on the convergence to a Γ-distribution and estimated large-deviation probabilities $\mathbb{P}\{X_n^2/nb > x\}$ in the range $x = o(\sqrt{n})$ where the Γ-tail approximation still works.

In [27], Brézis et al. considered a nearest neighbour Markov chain with transition probabilities similar to (4.103), for which they obtained a large-deviation result in the range where $x^2 - (\log n)/2 \to -\infty$, by a different technique based on an estimation of the closeness of the expected values of a smooth function of the original scaled process to those of the limiting diffusion. This result allowed them to prove the law of the iterated logarithm, that is,

$$\mathbb{P}_0\Big\{\limsup_{n\to\infty} \frac{X_n^2}{2n\log\log n} \le 1\Big\} = 1.$$

Guivarc'h et al. [76, Theorems 42 and 43] obtained weak convergence to a Γ-distribution, and also the local renewal theorem, in the transient case for a nearest neighbour chain with transition probabilities (4.103). They used the orthogonal polynomials technique of Rosenkrantz [138].

The orthogonal Laguerre polynomials technique was used by Voit in [149] for proving convergence to a Γ-distribution for critical branching processes with immigration.

A local version of Lamperti's Γ-convergence [112] was proven by Alexander in [5, Theorem 2.4] for a nearest neighbour null-recurrent Markov chain with transition probabilities

$$p(x,x-1) = 1/2 - \lambda/x + o(1/x), \quad p(x,x+1) = 1 - p(x,x-1),\ x \geq 1.$$

An integral (elementary) renewal theorem for a transient Markov chain with drift $m_1(x)$ asymptotically proportional to $1/x$ at infinity was proven in [42]; it was shown there that the renewal function then behaves as cx^2 for large values of x.

5

Limit Theorems for Transient Markov Chains with Drift Decreasing More Slowly Than $1/x$

As in the last chapter, we again assume that the first two moments of the jumps of a Markov chain $\{X_n\}$ demonstrate regular behaviour at infinity, but now we consider the case where the drift decreases at a rate slower than $1/x$, that is,

$$x\mathbb{E}\xi(x) \to \infty \quad \text{as } x \to \infty.$$

A particular example is, for some $c, b > 0$ and $\beta \in (0,1)$,

$$m_1(x) \sim c/x^{\beta},\ m_2(x) \to b > 0 \quad \text{as } x \to \infty.$$

Then clearly $\{X_n\}$ escapes to infinity at a faster rate than in the case of a drift of order $1/x$, and, in contrast with the case of convergence to a Γ-distribution, the law of large numbers holds:

$$\frac{X_n^{1+\beta}}{n} \stackrel{p}{\to} c(1+\beta) \quad \text{as } n \to \infty.$$

The asymptotic behaviour of the renewal measure is as follows:

$$\sum_{n=0}^{\infty} \mathbb{P}\{X_n \le x\} \sim \frac{x^{1+\beta}}{c(1+\beta)} \quad \text{as } x \to \infty.$$

In addition, we have the following weak convergence to a normal distribution:

$$\frac{X_n - (c(1+\beta)n)^{1/(1+\beta)}}{\sqrt{b\frac{1+\beta}{1+3\beta}n}} \Rightarrow N_{1/2+\mu/b,2} \quad \text{as } n \to \infty.$$

In this chapter, we study this type of results.

5.1 Law of Large Numbers

As seen from the results discussed in the last chapter, in the case of a drift of order $1/x$ there is no law of large numbers for X_n with a positive limit. In this section we show that a drift approaching zero more slowly than $1/x$ gives rise to a law of large numbers for X_n.

Let $v(x) > 0$ be a decreasing differentiable function such that

$$xv(x) \to \infty \quad \text{as } x \to \infty, \tag{5.1}$$

which is equivalent to $1/v(x) = o(x)$ and thus

$$V(x) \leq x/v(x) \;=\; o(x^2) \quad \text{as } x \to \infty, \tag{5.2}$$

where a convex function V is defined as

$$V(x) := \int_0^x \frac{1}{v(y)} dy \quad \text{for } x > 0;$$

$V(x) = 0$ for $x \leq 0$. In this chapter, the function $v(x)$ is responsible for the drift of the chain; it describes the asymptotic behaviour of the truncated drift function, that is,

$$m_1^{[s(x)]}(x) \sim v(x) \quad \text{as } x \to \infty. \tag{5.3}$$

In the previous chapter we considered the case where the second truncated moment $m_2^{[s(x)]}(x)$ is convergent to a positive constant, so the drift function $m_1^{[s(x)]}(x)$ and the quotient $2m_1^{[s(x)]}(x)/m_2^{[s(x)]}(x)$ are asymptotically proportional to each other. This means that $v(x)$ is typically asymptotically proportional to the reference function $r(x)$ describing the latter quotient.

In this chapter we do not assume convergence of the second moment; it is allowed to grow unboundedly as x tends to infinity, in which case the quotient $2m_1^{[s(x)]}(x)/m_2^{[s(x)]}(x)$ decays faster than the drift. We assume only that

$$\frac{2xm_1^{[s(x)]}(x)}{m_2^{[s(x)]}(x)} \to \infty \quad \text{as } x \to \infty. \tag{5.4}$$

Throughout this chapter we also assume some regular behaviour of $v(x)$. We also assume that the function $v(x)$ is differentiable and that

$$v'(x) = O(v(x)/x) \quad \text{as } x \to \infty. \tag{5.5}$$

Then the function $v(x)$ is $o(x)$-*insensitive*, that is,

$$v(x \pm o(x)) \sim v(x) \quad \text{as } x \to \infty, \tag{5.6}$$

and hence, for any function $s(x)$ of order $o(x)$ and $n \geq 1$,

$$V(x) \geq \int_{x-ns(x)}^{x} \frac{1}{v(y)} dy \geq \frac{ns(x)}{v(x - ns(x))} \sim \frac{ns(x)}{v(x)}$$

and

$$V(x \pm s(x)) = V(x) + \int_{x}^{x \pm s(x)} \frac{1}{v(y)} dy = V(x) \pm \frac{s(x)(1 + o(1))}{v(x)},$$

which yield, respectively,

$$V(x)v(x)/s(x) \to \infty \quad \text{as } x \to \infty \tag{5.7}$$

and

$$V(x \pm s(x)) \sim V(x) \quad \text{as } x \to \infty. \tag{5.8}$$

Theorem 5.1 *For some increasing function $s(x)$ of order $o(x)$ as $x \to \infty$, let the drift conditions* (5.3) *and* (5.4) *hold. Also let the following conditions hold:*

$$\mathbb{E}\{|\xi(x)|;\ \xi(x) < -s(x)\} = o(v(x)), \tag{5.9}$$

$$\mathbb{P}\{|\xi(x)| > s(x)\} \leq p(x)v(x), \tag{5.10}$$

where $p(x)$ is a decreasing function that is integrable at infinity. Assume also that

$$\limsup_{n\to\infty} X_n = \infty \quad \textit{with probability } 1. \tag{5.11}$$

Then

$$\frac{V(X_n)}{n} \xrightarrow{p} 1 \quad \textit{as } n \to \infty.$$

Since the function V is convex, its inverse V^{-1} is concave and hence

$$\frac{X_n}{V^{-1}(n)} \xrightarrow{p} 1 \quad \text{as } n \to \infty. \tag{5.12}$$

Let us give a sufficient condition for (5.9) and (5.10) to hold. If the family $\{|\xi(x)|,\ x \geq 0\}$ possesses a majorant Ξ satisfying $\mathbb{E}V(\Xi) < \infty$, that is, $|\xi(x)| \leq_{st} \Xi$ for all x, then there exists a function $s(x)$ of order $o(x)$ such that (5.9) and (5.10) hold, the second following from Lemma 2.32 with $U(x) \equiv 1$. Here Lemma 2.32 applies because $V(x)/x$ is increasing owing to the inequality $V(x) < x/v(x)$, which implies that the derivative of $V(x)/x$ is positive.

Proof By the condition (5.4), there exists a decreasing function $r(x) > 0$ such that $xr(x) \to \infty$ and, for all sufficiently large x,

$$\frac{2m_1^{[s(x)]}(x)}{m_2^{[s(x)]}(x)} \geq r(x) \tag{5.13}$$

and $s(x) = o(1/r(x))$. Owing to Lemma 2.31, we can assume that

$$r'(x) = O(r(x)/x) = o(r^2(x)) \quad \text{as } x \to \infty. \tag{5.14}$$

As in the proof of Theorem 4.8, we consider a modified Markov chain $\{\widetilde{X}_n\}$ on the same probability space as $\{X_n\}$ with truncated jumps $\widetilde{\xi}(x) = \xi(x)\mathbb{I}\{|\xi(x)| \leq s(x)\}$, and, as explained in that proof, we can assume that $\{\widetilde{X}_n\}$ satisfies the unboundedness-of-trajectories condition (5.11).

Owing to (5.13) and the convergence $xr(x) \to \infty$ we have

$$\frac{2m_1^{[s(x)]}(x)}{m_2^{[s(x)]}(x)} \geq \frac{2}{x} \quad \text{for all sufficiently large } x,$$

so the condition (2.108) is satisfied and Theorem 2.21 implies a.s. convergence $\widetilde{X}_n \to \infty$ as $n \to \infty$.

The chain $\{\widetilde{X}_n\}$ satisfies all the conditions of Theorem 4.3; hence its renewal measure satisfies

$$H_y^{\widetilde{X}}(x, x + 1/r(x)] \leq \frac{c}{v(x)r(x)} \qquad \text{for some } c < \infty.$$

Therefore, Lemma 4.1 is applicable to the chains $Y = X$ and $Z = \widetilde{X}$ with $l(x) = 1/r(x)$ and it then suffices to prove that

$$\frac{V(Z_n)}{n} \xrightarrow{p} 1 \quad \text{as } n \to \infty. \tag{5.15}$$

Let us evaluate the expectation of the increment of $V^{1+\alpha}(Z_n)$, $\alpha \geq 0$. For all sufficiently large x,

$$\begin{aligned}
&\mathbb{E}\{V^{1+\alpha}(Z_{n+1}) - V^{1+\alpha}(Z_n) \mid Z_n = x\} \\
&\quad = \mathbb{E}\{V^{1+\alpha}(x + \xi(x)) - V^{1+\alpha}(x);\ |\xi(x)| \leq s(x)\} \\
&\quad = (V^{1+\alpha})'(x)m_1^{[s(x)]}(x) + \mathbb{E}\{(V^{1+\alpha})''(x + \theta\xi(x))\xi^2(x)/2;\ |\xi(x)| \leq s(x)\} \\
&\quad = (1+\alpha)V^{\alpha}(x)\frac{1}{v(x)}m_1^{[s(x)]}(x) \\
&\qquad + (1+\alpha)\mathbb{E}\Big\{\Big(\alpha V^{\alpha-1}\frac{1}{v^2} - V^{\alpha}\frac{v'}{v^2}\Big)(x + \theta\xi(x))\xi^2(x)/2;\ |\xi(x)| \leq s(x)\Big\}.
\end{aligned} \tag{5.16}$$

Owing to the condition (5.3), the first term on the right-hand side equals

$$(1+\alpha)V^{\alpha}(x)\frac{1}{v(x)}m_1^{[s(x)]}(x) = (1+\alpha+o(1))V^{\alpha}(x) \quad \text{as } x\to\infty.$$

By (5.5) and (5.7) with $s = 1/r$,

$$\begin{aligned}
&\alpha\frac{V^{\alpha-1}(x+y)}{v^2(x+y)} - V^{\alpha}(x+y)\frac{v'(x+y)}{v^2(x+y)}\\
&= V^{\alpha}(x+y)\Big(\frac{\alpha}{V(x+y)v^2(x+y)} + o(1)\frac{r(x+y)}{v(x+y)}\Big)\\
&= o(V^{\alpha}(x+y)r(x+y)/v(x+y))\\
&= o(V^{\alpha}(x)r(x)/v(x)) \qquad (5.17)
\end{aligned}$$

as $x\to\infty$ uniformly on the set $|y|\le s(x)=o(1/r(x))$, owing to the insensitivity conditions (5.6) and (5.8). Since the drift condition (5.13) may be rearranged as

$$m_2^{[s(x)]}(x) \le \frac{2m_1^{[s(x)]}(x)}{r(x)} \sim \frac{2v(x)}{r(x)}, \qquad (5.18)$$

the relation (5.17) implies that the second term on the right-hand side of (5.16) is of order $o(V^{\alpha}(x))$ as $x\to\infty$. Substituting all this into (5.16) we finally deduce that, as $x\to\infty$,

$$\mathbb{E}\{V^{1+\alpha}(Z_{n+1}) - V^{1+\alpha}(Z_n) \mid Z_n = x\} = (1+\alpha+o(1))V^{\alpha}(x). \quad (5.19)$$

Now setting $\alpha = 0$, we get

$$\mathbb{E}\{V(Z_{n+1}) - V(Z_n) \mid Z_n = x\} \to 1 \quad \text{as } x\to\infty. \qquad (5.20)$$

Applying here the a.s. convergence $Z_n\to\infty$, we conclude that

$$\mathbb{E}V(Z_n) \sim n \quad \text{as } n\to\infty. \qquad (5.21)$$

Next take $\alpha = 1$ in (5.19). Then

$$\mathbb{E}\{V^2(Z_{n+1}) - V^2(Z_n)\} = (2+o(1))\mathbb{E}V(Z_n) \sim 2n \quad \text{as } n\to\infty.$$

Therefore

$$\mathbb{E}\Big(\frac{V(Z_n)}{n}\Big)^2 \to 1 \quad \text{as } n\to\infty.$$

Together with (5.21) this yields the convergence of the variances

$$\mathbb{V}\text{ar}\,\frac{V(Z_n)}{n} \to 0 \quad \text{as } n\to\infty,$$

which in its turn implies the desired convergence (5.15). □

5.2 Strong Law of Large Numbers

As usual, the strong law of large numbers requires stronger assumptions than the law of large numbers. Below we assume a stronger condition on $m_2^{[s(x)]}(x)$ than the drift condition (5.4) which can be seen as an upper bound on $m_2^{[s(x)]}(x)$; see (5.18).

Theorem 5.2 *Let the conditions of Theorem 5.1 hold. In addition, let*

$$m_2^{[s(x)]}(x) = o\Big(\frac{xv(x)}{f(V(x))}\Big) \quad \text{as } x \to \infty, \tag{5.22}$$

for some increasing function $f : \mathbb{R}^+ \to \mathbb{R}^+$ such that both the functions $f(x)$ and $x/f(x)$ are concave and

$$\sum_{n=1}^{\infty} \frac{1}{nf(n)} < \infty. \tag{5.23}$$

Then

$$\frac{V(X_n)}{n} \stackrel{\text{a.s.}}{\to} 1 \quad \text{as } n \to \infty.$$

As in the case of Theorem 5.1, this convergence, owing to the concavity of the inverse V^{-1}, implies that

$$\frac{X_n}{V^{-1}(n)} \stackrel{\text{a.s.}}{\to} 1 \quad \text{as } n \to \infty.$$

Corollary 5.3 *Let the condition* (5.11) *hold. Let $\mathbb{E}\xi(x) \sim c/x^\beta$, where $c > 0$ and $\beta \in [0,1)$, and let*

$$\sup_x \mathbb{E}|\xi(x)|^{1+\beta} \log^{1+\delta}(1 + |\xi(x)|) < \infty \quad \text{for some } \delta > 0.$$

Then

$$\frac{X_n^{1+\beta}}{n} \stackrel{\text{a.s.}}{\to} c(1+\beta) \quad \text{as } n \to \infty. \tag{5.24}$$

Proof of Corollary 5.3. Here $v(x) = c/x^\beta$ and $V(x) = x^{1+\beta}/c(1+\beta)$. Observe that

$$\begin{aligned} m_2^{[s(x)]}(x) &= \mathbb{E}\{\xi^2(x);\ |\xi(x)| \le s(x)\} \\ &\le \frac{s^{1-\beta}(x)}{\log^{1+\delta} s(x)} \mathbb{E}|\xi(x)|^{1+\beta} \log^{1+\delta}(1 + |\xi(x)|). \end{aligned}$$

Consider a truncation level $s(x) = x/\log^{\delta/4} x$ and a function $f(x) = \log^{1+\delta}(1 + x)$; then the conditions (5.22) and (5.23) are satisfied. The conditions (5.3), (5.13), (5.9), and (5.10) are satisfied too because

$$\begin{aligned}\mathbb{E}\{|\xi(x)|;\ |\xi(x)|>s(x)\} &\le \frac{1}{s^\beta(x)\log^{1+\delta}s(x)}\mathbb{E}|\xi(x)|^{1+\beta}\log^{1+\delta}(1+|\xi(x)|)\\ &\le c\frac{\log^{\beta\delta/4}x}{x^\beta\log^{1+\delta}x} = o(1/x^\beta) = o(v(x))\end{aligned}$$

and

$$\begin{aligned}\mathbb{P}\{|\xi(x)|>s(x)\} &\le \frac{1}{s^{1+\beta}(x)\log^{1+\delta}s(x)}\mathbb{E}|\xi(x)|^{1+\beta}\log^{1+\delta}(1+|\xi(x)|)\\ &\le c\frac{\log^{(1+\beta)\delta/4}x}{x^{1+\beta}\log^{1+\delta}x} = o(v(x))\frac{1}{x\log^{1+\delta/2}x},\end{aligned}$$

where the quotient on the right-hand side is integrable at infinity. □

Corollary 5.4 *Let* $\mathbb{E}\xi(x)\sim c(\log x)^{1+\beta}/x$, $\beta>0$, *and*

$$\sup_x \mathbb{E}\xi^2(x)<\infty.$$

Then

$$\frac{X_n^2}{(\log X_n)^{1+\beta}n}\xrightarrow{\text{a.s.}}2c \quad \text{as } n\to\infty,$$

which is equivalent to the following convergence:

$$\frac{X_n^2}{n\log^{1+\beta}n}\xrightarrow{\text{a.s.}}2^{2+\beta}c \quad \text{as } n\to\infty.$$

Proof of Corollary 5.4. Under this drift condition, $v(x) = c(\log x)^{1+\beta}/x$ for sufficiently large x and thus $V(x)\sim x^2/(2c\log^{1+\beta}x)$. Observe that the value of $m_2^{[s(x)]}(x)$ is bounded here regardless of the choice of the truncation level $s(x)$. Consider a truncation level $s(x) = x/\log^{\beta/4}x$ and a function $f(x)=\log^{1+\beta/2}(1+x)$. Then the conditions (5.22) and (5.23) are satisfied. The conditions (5.3), (5.13), (5.9), and (5.10) are satisfied too because

$$\begin{aligned}\mathbb{E}\{|\xi(x)|;\ |\xi(x)|>s(x)\} &\le \frac{1}{s(x)}\mathbb{E}\xi^2(x)\\ &\le c\frac{\log^{\beta/4}x}{x} = o((\log x)^{1+\beta}/x) = o(v(x))\end{aligned}$$

and

$$\begin{aligned}\mathbb{P}\{|\xi(x)|>s(x)\} &\le \frac{1}{s^2(x)}\mathbb{E}\xi^2(x)\\ &\le c\frac{\log^{\beta/2}x}{x^2} = o(v(x))\frac{1}{x\log^{1+\beta/3}x},\end{aligned}$$

where the quotient on the right-hand side is integrable at infinity. □

Notice that drifts such as $v(x) = (\log x)/x$ or that decrease more rapidly are excluded from consideration in Theorem 5.2 because then $xv(x) = \log x$ and $f(x)$ cannot be chosen as growing faster than $\log x$ to satisfy (5.22); thus the condition (5.23) fails. The law of large numbers, Theorem 5.1, is still applicable.

The proof of Theorem 5.2 is based on the following generalisation of the strong law of large numbers to martingales; see e.g. [80, Theorem 2.2].

Theorem 5.5 *Let $\{\mathcal{F}_n\}$ be a filtration and $\{X_n\}$ be a martingale with respect to $\{\mathcal{F}_n\}$ which is square integrable. If*

$$\sum_{n=1}^{\infty} \frac{\mathbb{E}(X_{n+1} - X_n)^2}{n^2} < \infty$$

then $X_n/n \to 0$ a.s. as $n \to \infty$.

Proof of Theorem 5.2. As in Theorem 5.1, it suffices to show that

$$\frac{V(Z_n)}{n} \to 1 \quad \text{a.s. as } n \to \infty. \tag{5.25}$$

We introduce $r(x)$ and $s(x) = o(1/r(x))$ as in the proof of Theorem 5.1. In addition, we can guarantee that the following version of the condition (5.22) holds true:

$$m_2^{[s(x)]}(x) = o\Big(\frac{v(x)}{r(x)f(V(x))}\Big) \quad \text{as } x \to \infty. \tag{5.26}$$

Denote

$$m_n^V := \mathbb{E}\{V(Z_{n+1}) - V(Z_n)|\mathcal{F}_n\},$$

where $\mathcal{F}_n = \sigma(Z_0, \dots, Z_n)$. By the Markov property, as calculated in the proof of Theorem 5.1 with $\alpha = 0$, on the event $Z_n \to \infty$ we have

$$m_n^V \to 1 \quad \text{as } n \to \infty. \tag{5.27}$$

Put

$$D_n := V(Z_{n+1}) - V(Z_n) - m_n^V,$$

so that

$$V(Z_n) = \sum_{k=0}^{n-1} m_k^V + \sum_{k=0}^{n-1} D_k.$$

By (5.27) and the convergence $Z_n \to \infty$ we have

$$\frac{1}{n}\sum_{k=0}^{n-1} m_k^V \to 1 \quad \text{a.s. as } n \to \infty,$$

and consequently the required convergence (5.25) would follow once it can be proven that

$$\frac{1}{n}\sum_{k=0}^{n-1} D_k \to 0 \quad \text{a.s. as } n \to \infty. \tag{5.28}$$

The process $\sum_{k=0}^{n-1} D_k$ constitutes a martingale with respect to the filtration $\{\mathcal{F}_{n-1}\}$. Hence the a.s. convergence (5.28) would follow by Theorem 5.5 if we could manage to prove that the increments of this martingale satisfy the condition

$$\sum_{n=1}^{\infty} \frac{\mathbb{E}D_n^2}{n^2} < \infty. \tag{5.29}$$

By the construction of D_k and the insensitivity condition (5.6), for $x \geq \widehat{x}$ we have

$$\begin{aligned}
\mathbb{E}\{D_k^2 \mid Z_k = x\} &= \mathbb{V}\text{ar}\, D_k \\
&\leq \mathbb{E}\{[V(x+\xi(x)) - V(x)]^2;\ |\xi(x)| \leq s(x)\} \\
&\leq c_1 (V'(x))^2 \mathbb{E}\{\xi^2(x);\ |\xi(x)| \leq s(x)\} \\
&= c_2 \frac{1}{v^2(x)} \frac{v(x)}{r(x) f(V(x))} \leq c_3 \frac{V(x)}{f(V(x))},
\end{aligned}$$

owing to (5.26) and (5.7). Since the function $y/f(y)$ is concave, by Jensen's inequality it holds that

$$\mathbb{E}D_k^2 \leq c_3 \frac{\mathbb{E}V(Z_k)}{f(\mathbb{E}V(Z_k))} \leq c_3 \frac{2k}{f(k/2)}$$

for sufficiently large k, as follows from (5.21). For $x < \widehat{x}$,

$$\mathbb{E}\{D_k^2 \mid Z_k = x\} \leq V^2(\widehat{x} + s(\widehat{x})) =: c_4.$$

Then it follows from the concavity of $f(y)$ that $\mathbb{E}D_k^2 \leq 2c_3 k/f(k)$, which yields

$$\sum_{k=1}^{\infty} \frac{\mathbb{E}D_k^2}{k^2} \leq \sum_{k=1}^{\infty} \left(\frac{2c_3}{kf(k)} + \frac{c_4}{k^2} \right) < \infty,$$

by the condition (5.23); hence (5.29) holds and the proof is complete. □

5.3 Integral Renewal Theorem for Transient Chain Satisfying Law of Large Numbers

In this section we discuss asymptotics of the renewal measure for $\{X_n\}$ satisfying the conditions of the law of large numbers, Theorem 5.1. Note that, in particular, we do not assume convergence of the second moment at infinity.

Theorem 5.6 *Under the conditions of the law of large numbers, Theorem 5.1, there exists an $\widehat{x}$ such that, given any distribution of X_0,*

$$H(\widehat{x},x] \sim V(x) = \int_0^x \frac{1}{v(y)}dy \quad \textit{as } x \to \infty.$$

Proof We split the proof of the asymptotics for H into two parts, consisting of the upper and lower bounds. First let us prove a proper upper bound. Let $r(x)$ be defined in Theorem 5.1. The chain $\{X_n\}$ satisfies all the conditions of Theorem 4.3. For any $A > 1$, by the Markov property and (4.32),

$$\begin{aligned} H(\widehat{x},x] \le \mathbb{E} &\sum_{n=0}^{T\left(x+A/r(x)\right)-1} \mathbb{I}\{\widehat{x} < X_n \le x\} \\ &+\mathbb{P}\Big\{X_n \le x \text{ for some } n \Big| X_0 > x + \frac{A}{r(x)}\Big\} \sup_z H_z(\widehat{x},x] \\ &\le \mathbb{E}L(\widehat{x}, T(x+A/r(x))) \\ &+ \big(\exp(\delta(R(x) - R(x+A/r(x)))) + o(1)\big) \sup_z H_z(\widehat{x},x] \quad (5.30) \end{aligned}$$

as $x \to \infty$ uniformly for all $A > 1$, where the stopping time $T(t)$ is defined by

$$T(t) := \min\{n \ge 1 : X_n > t\},$$

and L was defined in (4.13). We have

$$\begin{aligned} \exp\left(\delta(R(x) - R(x+A/r(x)))\right) &= \exp\left(-\delta \int_x^{x+A/r(x)} r(y)dy\right) \\ &\le \exp\left(-\delta A r(x+A/r(x))/r(x)\right) \le e^{-\delta A/2}, \end{aligned}$$

for all sufficiently large x, owing to the lower bound (2.8), which is applicable owing to (5.14). Applying the upper bound of Theorem 4.3 to the right-hand side of (5.30) we deduce that, for some $c < \infty$,

$$H(\widehat{x},x] \le \mathbb{E}L(\widehat{x}, T(x+A/r(x))) + \big(e^{-\delta A/2} + o(1)\big)c \int_{\widehat{x}}^{x+1/r(x)} \frac{1}{v(z)}dz$$

as $x \to \infty$ uniformly for all $A > 1$. Now applying Theorem 4.2, we deduce that

$$\mathbb{E}L(\widehat{x}, T(x + A/r(x))) \le \int_{\widehat{x}}^{x+A/r(x)+s(x+A/r(x))} \frac{1}{v(z)} dz$$

and therefore

$$\begin{aligned} H(\widehat{x}, x] &\le \left(1 + ce^{-\delta A/2} + o(1)\right) \int_{\widehat{x}}^{x+(A+1)/r(x)} \frac{1}{v(z)} dz \\ &\sim \left(1 + ce^{-\delta A/2} + o(1)\right) V(x) \quad \text{as } x \to \infty, \end{aligned} \tag{5.31}$$

for any fixed A, owing to (5.8). Now letting $A \to \infty$, we obtain the required upper bound for $H(0, x]$.

The lower bound is simpler. Indeed,

$$\begin{aligned} H(\widehat{x}, x] &= \sum_{n=0}^{\infty} \mathbb{P}\{\widehat{x} < X_n \le x\} \\ &\ge \sum_{10V(\widehat{x}) \le n \le (1-\varepsilon)V(x)} \mathbb{P}\{V(\widehat{x}) < V(X_n) \le V(x)\} \\ &\ge \sum_{10V(\widehat{x}) \le n \le (1-\varepsilon)V(x)} \mathbb{P}\left\{0.1 < \frac{V(X_n)}{n} \le \frac{1}{1-\varepsilon}\right\}, \end{aligned}$$

for any fixed $\varepsilon > 0$. Therefore, by the law of large numbers for X_n, $V(X_n)/n \to 1$; hence

$$H(\widehat{x}, x] \ge (1 - \varepsilon + o(1))V(x) \quad \text{as } x \to \infty.$$

This concludes the proof owing to the arbitrary choice of $\varepsilon > 0$. □

5.4 Central Limit Theorem

In this section we study the case where $xm_1(x) \to \infty$ as $x \to \infty$ and the strong law of large numbers holds true,

$$\frac{X_n}{V^{-1}(n)} \stackrel{a.s.}{\to} 1 \quad \text{as } n \to \infty, \tag{5.32}$$

given any distribution of X_0; for the sufficient conditions see Theorem 5.2. Then the natural next step is to study the fluctuations around the mean value. In the result below we specify additional conditions that guarantee a normal approximation to these fluctuations.

In addition to the condition (5.1), let the function $v(x)$ be regularly varying at infinity with index $-\beta \in [-1, 0]$. Then, by Karamata's theorem,

$$V(x) = \int_0^x \frac{1}{v(y)} dy \sim \frac{1}{1+\beta} \frac{x}{v(x)} \quad \text{as } x \to \infty. \tag{5.33}$$

In this section we consider the case where the second truncated moment of jumps has a positive limit at infinity, so that the drift function $m_1^{[s(x)]}(x)$ and the quotient $2m_1^{[s(x)]}(x)/m_2^{[s(x)]}(x)$ are asymptotically proportional to each other. For that reason any function $r(x)$ of order $o(v(x))$ gives a lower bound for the quotient, that is, it satisfies the drift condition (5.13).

Notice that the function $r(x) = \sqrt{v(x)/x}$ is asymptotically sandwiched between $1/x$ and $v(x)$; more precisely,

$$\frac{\sqrt{v(x)/x}}{1/x} \to \infty \quad \text{and} \quad \frac{\sqrt{v(x)/x}}{v(x)} \to 0 \quad \text{as } x \to \infty. \tag{5.34}$$

Theorem 5.7 *For some increasing function $s(x) = o\big(\sqrt{x/v(x)}\big)$, let*

$$m_1^{[s(x)]}(x) = v(x) + o(\sqrt{v(x)/x}) \ \text{ and } \ m_2^{[s(x)]}(x) \to b > 0 \tag{5.35}$$

as $x \to \infty$, and let the following conditions hold:

$$\mathbb{E}\{|\xi(x)|;\ \xi(x) \le -s(x)\} = o(v(x)) \quad \text{as } x \to \infty, \tag{5.36}$$

$$\mathbb{P}\{|\xi(x)| > s(x)\} \le p(x)v(x), \tag{5.37}$$

where $p(x)$ is a decreasing function integrable at infinity. Then

$$\frac{X_n - V^{-1}(n)}{\sqrt{b\frac{1+\beta}{1+3\beta}n}} \Rightarrow N_{0,1} \quad \text{as } n \to \infty.$$

The proof is based on the following generalisation of the central limit theorem to martingales, which goes back to [29, Theorem 2]. Let $\{\mathcal{F}_n, n \ge 1\}$ be a filtration and $\{X_n, n \ge 1\}$ be a square integrable martingale with respect to $\{\mathcal{F}_n\}$.

Theorem 5.8 *Let $\{X_n\}$ be a martingale such that*

$$\frac{\sum_{k=1}^n \mathbb{E}\{(X_{k+1} - X_k)^2 \mid \mathcal{F}_k\}}{\mathbb{E}X_n^2} \overset{p}{\to} 1 \quad \text{as } n \to \infty$$

and the conditioned Lindeberg condition holds: for all $\varepsilon > 0$,

$$\frac{1}{\mathbb{E}X_n^2} \sum_{k=1}^n \mathbb{E}\{(X_{k+1} - X_k)^2 \mathbb{I}\{|X_{k+1} - X_k| > \varepsilon\sqrt{\mathbb{E}X_n^2}\} \mid \mathcal{F}_k\} \overset{p}{\to} 0 \quad \text{as } n \to \infty.$$

Then $X_n/\sqrt{\mathbb{E}X_n^2}$ converges weakly to a standard normal distribution as $n \to \infty$.

Proof of Theorem 5.7. As in the proof of Theorem 4.8, we consider a modified Markov chain $\{\widetilde{X}_n\}$ on the same probability space as $\{X_n\}$ with jumps $\widetilde{\xi}(x) = \xi(x)\mathbb{I}\{|\xi(x)| \le s(x)\}$, and, as explained in that proof, we can assume that $\{\widetilde{X}_n\}$ satisfies the unboundedness-of-trajectories condition (5.11).

Notice that the chain $\{\widetilde{X}_n\}$ satisfies the conditions (3.56) and (3.57) with $r(x) = \sqrt{v(x)/x}$, for a sufficiently large $\widehat{x}$. The relation $s(x) = o(\sqrt{x/v(x)})$ is equivalent to $s(x) = o(1/r(x))$. Since $v(x)$ is regularly varying at infinity, it satisfies the condition (4.24). Therefore Theorem 4.3 applies to $\{\widetilde{X}_n\}$; hence

$$H_y^{\widetilde{X}}(x, x + 1/r(x)] \le \frac{c_1}{v(x)r(x)},$$

which in its turn allows us to apply Lemma 4.1 to a pair of chains $Y = X$ and $Z = \widetilde{X}$. Hence it suffices to prove the statement of the theorem for the process $\{Z_n\}$, that is, it is sufficient to prove that

$$\frac{Z_n - V^{-1}(n)}{\sqrt{b\frac{1+\beta}{1+3\beta}n}} \Rightarrow N_{0,1} \quad \text{as } n \to \infty. \tag{5.38}$$

The analogue of (5.32) for Z_n reads as

$$\frac{Z_n}{V^{-1}(n)} \stackrel{\text{a.s.}}{\to} 1 \quad \text{as } n \to \infty. \tag{5.39}$$

Denote

$$m_n^V := \mathbb{E}\{V(Z_{n+1}) - V(Z_n) \mid \mathcal{F}_n\},$$

where $\mathcal{F}_n := \sigma(Z_0, \ldots, Z_n)$. We have $m_n^V = m^V(Z_n)$ where

$$\begin{aligned} m^V(x) &:= \mathbb{E}\{V(Z_{n+1}) - V(Z_n) \mid Z_n = x\} \\ &= \mathbb{E}\{V(x + \xi(x)) - V(x);\ |\xi(x)| \le s(x)\} \\ &= V'(x)m_1^{[s(x)]}(x) + \frac{1}{2}\mathbb{E}\{V''(x + \theta\xi(x))\xi^2(x);\ |\xi(x)| \le s(x)\} \\ &= \frac{1}{v(x)}m_1^{[s(x)]}(x) - \frac{1}{2}\mathbb{E}\Big\{\frac{v'}{v^2}(x + \theta\xi(x))\xi^2(x);\ |\xi(x)| \le s(x)\Big\}. \end{aligned}$$

Then it follows from the conditions (5.35) and (5.5) that

$$\begin{aligned} m^V(x) &= 1 + o(1/\sqrt{xv(x)}) + O(v'(x)/v^2(x)) \\ &= 1 + o(1/\sqrt{xv(x)}) \quad \text{as } x \to \infty. \end{aligned} \tag{5.40}$$

Further, define

$$Q_n := \mathbb{E}\{(V(Z_{n+1}) - V(Z_n))^2 | \mathcal{F}_n\}.$$

We observe that $Q_n = Q(Z_n)$ where

$$\begin{aligned} Q(x) &:= \mathbb{E}\{(V(x+\xi(x)) - V(x))^2;\ |\xi(x)| \le s(x)\} \\ &= \mathbb{E}\{(V'(x+\theta\xi(x))\xi(x))^2;\ |\xi(x)| \le s(x)\} \\ &\sim b/v^2(x) \quad \text{as } x \to \infty, \end{aligned} \tag{5.41}$$

because $V'(x+y) = 1/v(x+y) \sim 1/v(x)$ as $x \to \infty$ uniformly for $|y| \le s(x)$.

Let us center $V(Z_n)$, that is, let us consider

$$\begin{aligned} \widetilde{Z}_n &:= V(Z_n) - \sum_{j=0}^{n-1} m_j^V \\ &= V(Z_n) - V(Z_{n-1}) - m_{n-1}^V + \widetilde{Z}_{n-1}, \end{aligned}$$

so that $\{\widetilde{Z}_n\}$ constitutes a martingale with respect to the filtration $\{\mathcal{F}_n\}$. It follows from the strong law of large numbers (5.39) and from (5.41) that

$$\begin{aligned} &v^2(V^{-1}(j))\mathbb{E}\{(\widetilde{Z}_{j+1} - \widetilde{Z}_j)^2 \mid \mathcal{F}_j\} \\ &= v^2(V^{-1}(j))\mathbb{E}\{(V(Z_{j+1}) - V(Z_j) - m_j^V)^2 \mid \mathcal{F}_j\} \\ &= v^2(V^{-1}(j))[\mathbb{E}\{(V(Z_{j+1}) - V(Z_j))^2 \mid \mathcal{F}_j\} - (m_j^V)^2] \\ &= v^2(V^{-1}(j))[\mathbb{E}\{(V(Z_{j+1}) - V(Z_j))^2 \mid \mathcal{F}_j\} + O(1)] \\ &\stackrel{\text{a.s.}}{\to} b \quad \text{as } j \to \infty, \end{aligned}$$

which implies the convergence

$$\frac{1}{\sigma_n^2} \sum_{j=0}^{n-1} \mathbb{E}\{(\widetilde{Z}_{j+1} - \widetilde{Z}_j)^2 \mid \mathcal{F}_j\} \stackrel{\text{a.s}}{\to} 1 \quad \text{as } n \to \infty, \tag{5.42}$$

where

$$\begin{aligned} \sigma_n^2 &:= b \sum_{j=0}^{n-1} \frac{1}{v^2(V^{-1}(j))} \\ &\ge b \frac{n}{2} \frac{1}{v^2(V^{-1}((n-1)/2))} \ge c_1 \frac{n}{v^2(V^{-1}(n))} \quad \text{for some } c_1 > 0. \end{aligned} \tag{5.43}$$

Since $|Z_1 - Z_0| \le s(x)$ given $Z_0 = x$,

$$\begin{aligned} |V(Z_1) - V(Z_0)| &= V'(x+\theta\xi(x))|\xi(x)|\mathbb{I}\{|\xi(x)| \le s(x)\} \\ &\le \frac{s(x)}{v(x+s(x))}. \end{aligned}$$

By the choice of $s(x) = o(\sqrt{x/v(x)})$, given $Z_0 = x + y$,

$$|V(Z_1) - V(Z_0)|^2 \le \gamma(x) x / v^3(x) \quad \text{for all } |y| \le x/2,$$

where $\gamma(x) \to 0$ as $x \to \infty$. Hence, on the event $|Z_n - V^{-1}(n)| \le V^{-1}(n/2)$ we have

$$\begin{aligned}\left|V(Z_{n+1}) - V(Z_n)\right|^2 &\le \gamma(V^{-1}(n))\frac{V^{-1}(n)}{v^3(V^{-1}(n))}\\ &\le \gamma(V^{-1}(n))\frac{c_2 n}{v^2(V^{-1}(n))},\end{aligned}$$

because $z/v(z) \le c_2 V(z)$ for some $c_2 < \infty$, by (5.33). Then, on the same event, by (5.43) we have

$$\left|V(Z_{n+1}) - V(Z_n)\right|^2 \le \gamma(V^{-1}(n))\frac{c_2}{c_1}\sigma_n^2.$$

By the strong law of large numbers (5.39),

$$\mathbb{P}\{|Z_n - V^{-1}(n)| \le V^{-1}(n/2) \text{ for all sufficiently large } n\} = 1.$$

This allows us to conclude that, for any fixed $\delta > 0$,

$$\mathbb{E}\{(\widetilde{Z}_{j+1} - \widetilde{Z}_j)^2;\ |\widetilde{Z}_{j+1} - \widetilde{Z}_j| \ge \delta\sigma_n \mid \mathcal{F}_j\} \overset{a.s.}{\to} 0 \quad \text{as } j \to \infty,\ j \le n-1;$$

hence

$$\frac{1}{\sigma_n^2}\sum_{j=0}^{n-1}\mathbb{E}\{(\widetilde{Z}_{j+1} - \widetilde{Z}_j)^2;\ |\widetilde{Z}_{j+1} - \widetilde{Z}_j| \ge \delta\sigma_n \mid \mathcal{F}_j\} \overset{\text{a.s.}}{\to} 0 \quad \text{as } n \to \infty.$$

So, the martingale $\{\widetilde{Z}_n\}$ satisfies the conditions of the central limit theorem for martingales – see Theorem 5.8 – and we conclude that

$$\frac{\widetilde{Z}_n}{\sigma_n} = \frac{V(Z_n) - \sum_{j=0}^{n-1} m_j^V}{\sigma_n} \Rightarrow N_{0,1} \quad \text{as } n \to \infty.$$

Further, as follows from the decomposition (5.40) for the mean drift of $V(Z_n)$,

$$\left|\sum_{j=0}^{n-1} m_j^V - n\right| \le c_3\sum_{j=0}^{n-1}\mathbb{I}\{Z_j \le V^{-1}(j)/2\} + \sum_{j=0}^{n-1}\frac{\psi_j}{\sqrt{V^{-1}(j)v(V^{-1}(j)/2)}},$$

where $\psi_j \to 0$ as $j \to \infty$. The first sum on the right-hand side is bounded by

$$\zeta := c_3\sum_{j=0}^{\infty}\mathbb{I}\{Z_j \le V^{-1}(j)/2\},$$

which is a proper random variable, owing to the strong law of large numbers (5.32), whereas the second sum is of order

$$o(1)\sum_{j=0}^{n-1}\frac{1}{\sqrt{V^{-1}(j)v(V^{-1}(j))}} = o\left(\frac{n}{\sqrt{V^{-1}(n)v(V^{-1}(n))}}\right) \quad \text{as } n \to \infty.$$

Since $V(z) \le z/v(z)$,

$$\frac{V^{-1}(n)}{v(V^{-1}(n))} \ge V(V^{-1}(n)) = n$$

and hence

$$\frac{n}{\sqrt{V^{-1}(n)v(V^{-1}(n))}} \le \frac{\sqrt{n}}{v(V^{-1}(n))}.$$

Combining all these results, including the lower bound (5.43) for σ_n, we get

$$\left|\sum_{j=0}^{n-1} m_j^V - n\right| \le o(\sigma_n) + \zeta \quad \text{as } n \to \infty.$$

Thus,

$$\frac{V(Z_n) - n}{\sigma_n} \Rightarrow N_{0,1} \quad \text{as } n \to \infty.$$

To conclude convergence to a normal distribution for Z_n itself, we make use of the mean-value theorem as follows:

$$\begin{aligned}\frac{Z_n - V^{-1}(n)}{\sigma_n} &= \frac{V^{-1}(V(Z_n)) - V^{-1}(n)}{\sigma_n} \\ &= (V^{-1})'(\theta_n)\frac{V(Z_n) - n}{\sigma_n}\end{aligned}$$

where θ_n is sandwiched between n and $V(Z_n)$. Therefore, owing to the equality $V' = 1/v$,

$$\frac{Z_n - V^{-1}(n)}{\sigma_n} = v(V^{-1}(\theta_n))\frac{V(Z_n) - n}{\sigma_n}.$$

By the strong law of large numbers (5.39), $\theta_n/n \to 1$ with probability 1 as $n \to \infty$. Therefore, $v(V^{-1}(\theta_n))/v(V^{-1}(n)) \to 1$ and hence

$$\frac{Z_n - V^{-1}(n)}{\widetilde{\sigma}_n} \Rightarrow N_{0,1}$$

where

$$\begin{aligned}\widetilde{\sigma}_n^2 &:= \sigma_n^2 v^2(V^{-1}(n)) \\ &= b\sum_{j=0}^{n-1} \frac{v^2(V^{-1}(n))}{v^2(V^{-1}(j))}.\end{aligned}$$

The sequence $v^2(V^{-1}(j))$ is regularly varying with index $-\dfrac{2\beta}{1+\beta}$: hence

$$\sum_{j=0}^{n-1} \frac{v^2(V^{-1}(n))}{v^2(V^{-1}(j))} \sim n^{-\frac{2\beta}{1+\beta}} \sum_{j=0}^{n-1} j^{\frac{2\beta}{1+\beta}} \sim \frac{1+\beta}{1+3\beta} n \quad \text{as } n \to \infty, \tag{5.44}$$

and the proof is complete. □

Theorem 5.9 *Let the conditions of Theorem 5.7 hold. Let*

$$\frac{\sqrt{n} v(V^{-1}(n))}{\log n} \to \infty \quad \textit{as } n \to \infty. \tag{5.45}$$

Then

$$\frac{\max_{k \le n} X_k - V^{-1}(n)}{\sqrt{b \frac{1+\beta}{1+3\beta} n}} \Rightarrow N_{0,1} \quad \textit{as } n \to \infty.$$

Since the function $v(V^{-1}(n))$ is regularly varying at infinity with index $-\beta/(1+\beta) > -1/2$, provided that $\beta \in [0,1)$, the condition (5.45) automatically holds for $\beta \in [0,1)$.

Proof It is again sufficient to prove the same result for the process $\{Z_n\}$, that is, it is sufficient to show that, for $M_n := \max_{k \le n} Z_k$,

$$\frac{M_n - V^{-1}(n)}{\sqrt{b \frac{1+\beta}{1+3\beta} n}} \Rightarrow N_{0,1} \quad \text{as } n \to \infty. \tag{5.46}$$

Since $M_n \ge Z_n$, it suffices to show that, for all $\varepsilon > 0$,

$$\mathbb{P}\{M_n \le Z_n + \varepsilon\sqrt{n}\} \to 1 \quad \text{as } n \to \infty.$$

Indeed,

$$\begin{aligned} \mathbb{P}\{M_n > Z_n + \varepsilon\sqrt{n}\} &\le \mathbb{P}\{M_n \notin [V^{-1}(n)/2, 2V^{-1}(n)]\} \\ &\quad + \mathbb{P}\{Z_n < M_n - \varepsilon\sqrt{n},\ M_n \in [V^{-1}(n)/2, 2V^{-1}(n)]\}. \end{aligned}$$

First, by the strong law of large numbers for Z_n, $M_n/V^{-1}(n) \to 1$ with probability 1, so

$$\mathbb{P}\{M_n \notin [V^{-1}(n)/2, 2V^{-1}(n)]\} \to 0 \quad \text{as } n \to \infty.$$

Second,

$$\begin{aligned}
&\mathbb{P}\{Z_n < M_n - \varepsilon\sqrt{n},\ M_n \in [V^{-1}(n)/2, 2V^{-1}(n)]\} \\
&\quad \le \sum_{k=0}^{n-1} \mathbb{P}\{M_n = Z_k,\ Z_n < Z_k - \varepsilon\sqrt{n},\ Z_k \in [V^{-1}(n)/2, 2V^{-1}(n)]\} \\
&\quad \le \sum_{k=0}^{n-1} \mathbb{P}\{Z_n < Z_k - \varepsilon\sqrt{n},\ Z_k \in [V^{-1}(n)/2, 2V^{-1}(n)]\}.
\end{aligned}$$

Therefore

$$\begin{aligned}
&\mathbb{P}\{Z_n < M_n - \varepsilon\sqrt{n},\ M_n \in [V^{-1}(n)/2, 2V^{-1}(n)]\} \\
&\quad \le \sum_{k=0}^{n-1} \int_{V^{-1}(n)/2}^{2V^{-1}(n)} \mathbb{P}\{Z_k \in dy\}\mathbb{P}\{Z_n < y - \varepsilon\sqrt{n} \mid Z_k = y\} \\
&\quad \le n \times \sup_{y \in [V^{-1}(n)/2, 2V^{-1}(n)]} \mathbb{P}\{Z_m < y - \varepsilon\sqrt{n} \text{ for some } m \ge 1 \mid Z_0 = y\}.
\end{aligned}$$

The process $\{Z_n\}$ satisfies all the conditions of Theorem 3.12 with $r(x) = v(x)/b$; thus, for all $y \in [V^{-1}(n)/2, 2V^{-1}(n)]$,

$$\begin{aligned}
\mathbb{P}\{Z_m < y - \varepsilon\sqrt{n} \mid Z_0 = y\} &\le \exp\left(-\delta \int_{y-\varepsilon\sqrt{n}}^{y} v(z)dz\right) \\
&\le \exp\left(-\delta\varepsilon\sqrt{n}v(y)\right) \\
&\le \exp\left(-\delta\varepsilon\sqrt{n}v(2V^{-1}(n))\right), \qquad (5.47)
\end{aligned}$$

because the function $v(z)$ is decreasing. Therefore, by the regular variation of v and the condition (5.45),

$$\sup_{y \in [V^{-1}(n)/2, 2V^{-1}(n)]} \mathbb{P}\{Z_m < y - \varepsilon\sqrt{n} \mid Z_0 = y\} = o(1/n) \quad \text{as } n \to \infty,$$

which yields

$$P\{Z_n < M_n - \varepsilon\sqrt{n},\ M_n \in [V^{-1}(n)/2, 2V^{-1}(n)]\} \to 0 \quad \text{as } n \to \infty.$$

The proof is complete. □

Recall that $T(x) = \min\{n : X_n > x\}$.

Corollary 5.10 *Under the conditions of Theorem 5.7 and* (5.45),

$$\frac{T(x) - V(x)}{\sqrt{b\frac{1+\beta}{1+3\beta}\frac{x}{v^3(x)}}} \Rightarrow N_{0,1} \quad \textit{as } x \to \infty.$$

Proof Since $\{T(x) \le n\} = \{\sup_{k\le n} X_k > x\}$,

$$\mathbb{P}\left\{\frac{T(x)-V(x)}{\sqrt{b\frac{1+\beta}{1+3\beta}\frac{x}{v^3(x)}}} \le u\right\} = \mathbb{P}\left\{\sup_{k\le n} X_k > x\right\},$$

where

$$n := V(x) + u\sqrt{b\frac{1+\beta}{1+3\beta}\frac{x}{v^3(x)}}.$$

Since $(V^{-1}(z))' = 1/V'(V^{-1}(z)) = v(V^{-1}(z))$ and $n \sim V(x)$, we have $(V^{-1}(n))' \sim v(x)$. Therefore

$$V^{-1}(n) = x + u\sqrt{b\frac{1+\beta}{1+3\beta}\frac{x}{v(x)}} + o(\sqrt{x/v(x)}).$$

Hence,

$$\begin{aligned}\mathbb{P}\left\{\sup_{k\le n} X_k > x\right\} &= \mathbb{P}\left\{\frac{\sup_{k\le n} X_k - V^{-1}(n)}{\sqrt{b\frac{1+\beta}{1+3\beta}n}} > \frac{x - V^{-1}(n)}{\sqrt{b\frac{1+\beta}{1+3\beta}n}}\right\}\\ &= \mathbb{P}\left\{\frac{\sup_{k\le n} X_k - V^{-1}(n)}{\sqrt{b\frac{1+\beta}{1+3\beta}n}} > -u + o(1)\right\},\end{aligned}$$

and reference to Theorem 5.9 completes the proof. □

5.5 Functional Central Limit Theorem

In the last section we proved the central limit theorem for a transient Markov chain. The key idea of the proof is the extraction of a martingale for which the central limit theorem is known from Brown [29]; see Theorem 5.8. Since this reference also contains a functional version of this result, it allows us to state and prove the following weak convergence to a Gaussian process for $\{X_n\}$.

Theorem 5.11 *Under the conditions of Theorem 5.7, the process*

$$\frac{X_{[nt]} - V^{-1}(nt)}{\sqrt{b\frac{1+\beta}{1+3\beta}n}}, \quad t \in [0,1],$$

converges weakly in $D[0,1]$ as $n \to \infty$ to the process

$$t^{-\frac{\beta}{1+\beta}} B\left(t^{\frac{1+3\beta}{1+\beta}}\right),$$

where $B(t)$ is a standard Brownian motion. The limiting process is Gaussian with zero mean and covariance function $t(t/s)^{\frac{\beta}{1+\beta}}$ for $s \geq t$.

Proof As in the proof of Theorem 4.11, it suffices to prove the same convergence for the process $Z_{[nt]}$. The weak convergence in the space $D[0,1]$ to the limiting process is equivalent to the following two statements: for any fixed $t_0 \in (0,1)$,

$$\frac{Z_{[nt]} - V^{-1}(nt)}{\sqrt{b\frac{1+\beta}{1+3\beta}n}} \Rightarrow t^{-\frac{\beta}{1+\beta}} B\left(t^{\frac{1+3\beta}{1+\beta}}\right) \quad \text{as } n \to \infty \text{ in the space } D[t_0, 1] \tag{5.48}$$

and

$$\sup_{t \leq t_0} \left| \frac{Z_{[nt]} - V^{-1}(nt)}{\sqrt{b\frac{1+\beta}{1+3\beta}n}} \right| \Rightarrow 0 \quad \text{as } n \to \infty,\ t_0 \to 0. \tag{5.49}$$

The calculations of the last section leading to the central limit theorem for $V(Z_n)$ allow us to apply the functional limit theorem for martingales obtained by Brown [29, Theorem 3] to the process in $D[0,1]$ defined as $(V(Z_k)-k)/\sigma_n$ on the interval $[\sigma_k^2/\sigma_n^2, \sigma_{k+1}^2/\sigma_n^2)$, where

$$\sigma_n^2 = b\sum_{j=0}^{n-1} \frac{1}{v^2(V^{-1}(j))} \sim b\frac{1+\beta}{1+3\beta}\frac{n}{v^2(V^{-1}(n))} \quad \text{as } n \to \infty, \tag{5.50}$$

owing to (5.44); hence σ_n^2 is a regularly varying sequence with index $\frac{1+3\beta}{1+\beta}$, because $v(x)$ is regularly varying with index $-\beta$. The process defined in this way converges weakly in the space $D[0,1]$ to a Brownian motion, that is,

$$\sum_{k=1}^{n} \frac{V(Z_k) - k}{\sigma_n} \mathbb{I}\{\sigma_k^2/\sigma_n^2 \leq t < \sigma_{k+1}^2/\sigma_n^2\} \Rightarrow B(t) \quad \text{in the space } D[0,1]. \tag{5.51}$$

The regular variation of σ_n^2 implies that

$$\frac{\sigma_{[nt]}^2}{\sigma_n^2} \to t^{\frac{1+3\beta}{1+\beta}} \quad \text{as } n \to \infty \text{ uniformly for all } t \in [t_0, 1].$$

Hence

$$\frac{V(Z_{[nt]}) - nt}{\sigma_n} \Rightarrow B\left(t^{\frac{1+3\beta}{1+\beta}}\right) = t^{\frac{\beta}{1+\beta}} B(t) \quad \text{in the space } D[t_0, 1].$$

Now we need to explain how to proceed from $V(Z_{[nt]})$ to $Z_{[nt]}$. By the mean-value theorem,

$$\frac{Z_{[nt]} - V^{-1}(nt)}{\sqrt{b\frac{1+\beta}{1+3\beta}n}} = \frac{V^{-1}(V(Z_{[nt]})) - V^{-1}(nt)}{\sqrt{b\frac{1+\beta}{1+3\beta}n}}$$

$$= \frac{\sigma_n}{\sqrt{b\frac{1+\beta}{1+3\beta}n}}(V^{-1})'(\theta)\frac{V(Z_{[nt]}) - nt}{\sigma_n}$$

where θ lies between nt and $V(Z_{[nt]})$. Therefore, by the equality $V' = 1/v$,

$$\frac{Z_{[nt]} - V^{-1}(nt)}{\sqrt{b\frac{1+\beta}{1+3\beta}n}} = \frac{\sigma_n}{\sqrt{b\frac{1+\beta}{1+3\beta}n}}v(V^{-1}(\theta))\frac{V(Z_{[nt]}) - nt}{\sigma_n}. \tag{5.52}$$

It follows from the strong law of large numbers for $V(Z_n)$ – see Theorem 5.2 – that

$$\frac{V(Z_{[nt]})}{n} \to t \quad \text{in the space } D[0,1],$$

so we also have $\theta/n \to t$ in $D[0,1]$. Then, since v is assumed regularly varying at infinity and $v(V^{-1}(nt))/v(V^{-1}(n)) \sim 1/t^{\beta/(1+\beta)}$ as $n \to \infty$,

$$t^{\frac{\beta}{1+\beta}}\frac{v(V^{-1}(\theta))}{v(V^{-1}(n))} \to 1 \quad \text{in the space } D[t_0,1].$$

Hence we may replace $v(V^{-1}(\theta))$ in (5.52) by $t^{-\frac{\beta}{1+\beta}}v(V^{-1}(n))$ on the interval $t \in [t_0,1]$. Taking into account (5.50), we deduce the first required statement, (5.48).

Further, the second statement, (5.49), may be reformulated as follows. For all $\gamma > 0$ and $\delta > 0$ there exist $t_0 > 0$ and $n_0 \in \mathbb{N}$ such that

$$\mathbb{P}\left\{\sup_{t\le t_0}\left|\frac{Z_{[nt]} - V^{-1}(nt)}{\sqrt{b\frac{1+\beta}{1+3\beta}n}}\right| > \gamma\right\} \le \delta \quad \text{for all } n > n_0. \tag{5.53}$$

Indeed, first choose t_0 such that

$$\mathbb{P}\left\{\sup_{t\le t_0}|B(t)| > \gamma\right\} \le \delta/2.$$

Then it follows from (5.51) that there exists an $n_0 \in \mathbb{N}$ such that

$$\mathbb{P}\left\{\sup_{k:\sigma_k^2/\sigma_n^2\le t_0}\left|\frac{V(Z_k) - k}{\sigma_n}\mathbb{I}\{\sigma_k^2/\sigma_n^2 \le t < \sigma_{k+1}^2/\sigma_n^2\}\right| > \gamma\right\} \le \delta$$

for all $n > n_0$. Equivalently,

$$\mathbb{P}\left\{\sup_{k:\sigma_k^2/\sigma_n^2\le t_0}\left|\frac{V(Z_k)-k}{\sigma_n}\right|>\gamma\right\}\le\delta\quad\text{for all } n>n_0.$$

If we take $k \le nt_0'$, where $t_0' = t_0^{\frac{1+3\beta}{1+\beta}}/2$, then $\sigma_k^2/\sigma_n^2 \le t_0$ for all sufficiently large n. Therefore,

$$\mathbb{P}\left\{\sup_{t\le t_0'}\left|\frac{V(Z_{[nt]})-nt}{\sigma_n}\right|>\gamma\right\}\le\delta\quad\text{for all } n>n_0.$$

Then we apply the same calculations as in (5.52) and conclude that (5.53) holds. □

5.6 Normal Approximation at High Level

In this section a version of the central limit theorem is deduced for a Markov chain starting from a high level. This type of Normal approximation is more appropriate for the purpose of proving asymptotics for the renewal measure.

As in the last two sections we consider the case where the second truncated moment of jumps has a positive limit at infinity, so that again the drift function $m_1^{[s(x)]}(x)$ and the quotient $2m_1^{[s(x)]}(x)/m_2^{[s(x)]}(x)$ are asymptotically proportional to each other. This allows us to choose a $\gamma > 0$ sufficiently small that, for

$$r(x) := \gamma v(x)$$

the condition (3.56) is fulfilled for the chain $\{X_n\}$, for a sufficiently large $\widehat{x}$. Note that $r(x)$ as defined above satisfies the condition (3.57) owing to (5.5), which now reads

$$v'(x) = o(v^2(x)) \quad \text{as } x \to \infty. \tag{5.54}$$

This, in particular, specifies the insensitivity condition (5.6) as follows, for any fixed $c < \infty$:

$$v(x \pm c/v(x)) \sim v(x) \quad \text{as } x \to \infty. \tag{5.55}$$

In the previous sections we applied a convex function V to X_n in order to obtain a chain with an asymptotically (positive) constant drift, which helps us to prove the law of large numbers and the central limit theorem. For the purposes of this section – where we are considering a normal approximation at a high x level – it is more convenient to make calculations for $\{X_n\}$ itself because the drift of $\{X_n\}$ does not change much on a time scale $O(1/v^2(x))$, owing to (5.55), provided that the drift is proportional to $v(x)$.

Theorem 5.12 *For some increasing function $s(x) = o(1/v(x))$, where the decreasing function $v(x)$ satisfies $xv(x) \to \infty$, (5.55), and (5.54), let*

$$m_1^{[s(x)]}(x) \sim v(x) \text{ and } m_2^{[s(x)]}(x) \to b > 0, \tag{5.56}$$

$$\mathbb{E}\{|\xi(x)|;\ \xi(x) \le -s(x)\} = o(v(x)) \quad \text{as } x \to \infty, \tag{5.57}$$

$$\mathbb{P}\{|\xi(x)| > s(x)\} \le p(x)v(x), \tag{5.58}$$

where the decreasing function $p(x) > 0$ is integrable at infinity. Then, for any fixed $t > 1$ and $h \in \mathbb{R}$,

$$\mathbb{P}_x\Big\{X_n - x \le \frac{h}{v(x)}\Big\} - \Phi\Big(\frac{h/v(x) - nv(x)}{\sqrt{nb}}\Big) \to 0$$

as $x, n \to \infty$ in such a way that $1/t \le nv^2(x) \le t$; in what follows Φ stands for the standard normal distribution function. Moreover,

$$\sup_{x \le y \le x + o(1/v(x))} \left| \mathbb{P}_y\Big\{X_n - x \le \frac{h}{v(x)}\Big\} - \Phi\Big(\frac{h/v(x) - nv(x)}{\sqrt{nb}}\Big) \right| \to 0.$$

We start with the following tightness result for $\{X_n\}$.

Lemma 5.13 *For some increasing function $s(x) = o(1/v(x))$, where the decreasing function $v(x)$ satisfies $xv(x) \to \infty$, (5.55), and (5.54), let*

$$\delta v(x) \le m_1^{[s(x)]}(x) \le v(x)/\delta, \tag{5.59}$$

for some $\delta > 0$ and all sufficiently large x. Assume also that

$$\sup_x m_2^{[s(x)]}(x) < \infty \tag{5.60}$$

and that the conditions (5.58) and (5.57) hold. Then, for every fixed $t > 0$ and $\varepsilon > 0$, there exists an $h < \infty$ such that

$$\mathbb{P}_x\Big\{-\frac{h}{v(x)} \le X_n - x \le \frac{h}{v(x)} \text{ for all } n \le \frac{t}{v^2(x)}\Big\} \ge 1 - \varepsilon$$

for all sufficiently large x.

Proof As in the proof of Theorem 4.8, we consider a modified Markov chain $\{\widetilde{X}_n\}$ on the same probability space as $\{X_n\}$, with jumps $\widetilde{\xi}(x) = \xi(x)\mathbb{I}\{|\xi(x)| \le s(x)\}$, and, as explained in that proof, we can assume that $\{\widetilde{X}_n\}$ satisfies the unboundedness-of-trajectories condition (5.11).

As discussed at the beginning of this section, the chain $\{\widetilde{X}_n\}$ satisfies the condition (3.56) with $r(x) := \gamma v(x)$. Therefore, Theorem 4.3 is applicable to $\{\widetilde{X}_n\}$; hence

$$H_y^{\widetilde{X}}(x, x+1/v(x)] \le \frac{c_1}{v^2(x)},$$

which in its turn allows us to apply Lemma 4.1 to a pair of chains $Y = X$ and $Z = \widetilde{X}$. Hence it suffices to prove the result of the lemma for $\{Z_n\}$. That is, it is sufficient to show that, for a sufficiently large $h > 0$,

$$\mathbb{P}_x\left\{-\frac{h}{v(x)} \le Z_n - x \le \frac{h}{v(x)} \text{ for all } n \le \frac{t}{v^2(x)}\right\} \ge 1 - \varepsilon, \quad (5.61)$$

ultimately in x.

Similarly to (5.47) we deduce that, for some $\gamma > 0$,

$$\mathbb{P}_x\left\{\min_{n\ge 0} Z_n \le x - \frac{h}{v(x)}\right\} \le e^{-\gamma h} \to 0 \quad \text{as } x,\ h \to \infty. \quad (5.62)$$

Let us center Z_n, that is, let us consider the process

$$\widetilde{Z}_n := Z_n - x - \sum_{j=0}^{n-1} m_1^{[s(Z_j)]}(Z_j), \quad (5.63)$$

which constitutes a martingale with respect to $\mathfrak{F}_n := \sigma(Z_0, \ldots, Z_n)$. By the condition (5.59) we have, for all $N \ge 1$,

$$0 < \sum_{n=0}^{N-1} m_1^{[s(Z_j)]}(Z_n) \le N\frac{1}{\delta} \max_{z > x - h/v(x)} v(z) \le N\frac{2}{\delta}v(x)$$

on the event $\min_n Z_n > x - h/v(x)$, where from (5.55) the last inequality follows for all sufficiently large x. Hence, for any $y > 0$,

$$\mathbb{P}_x\left\{\min_n Z_n > x - \frac{h}{v(x)},\ \max_{n\le N}|Z_n - x| > y\right\} \le \mathbb{P}_x\left\{\max_{n\le N}|\widetilde{Z}_n| > y - \frac{2}{\delta}Nv(x)\right\}.$$

By Doob's inequality for martingales,

$$\mathbb{P}_x\left\{\max_{n\le N}|\widetilde{Z}_n| > y - \frac{2}{\delta}Nv(x)\right\} \le \frac{\mathbb{E}_x\widetilde{Z}_N^2}{(y - 2Nv(x)/\delta)^2}.$$

The second moments of the jumps of the martingale $\{\widetilde{Z}_n\}$ are bounded by some $c < \infty$, see the condition (5.60); therefore,

$$\mathbb{P}_x\left\{\max_{n\le N}|\widetilde{Z}_n| > y - \frac{2}{\delta}Nv(x)\right\} \le \frac{Nc}{(y - 2Nv(x)/\delta)^2}.$$

Now taking $N = t/v^2(x)$ and $y = h/v(x)$, we obtain that

$$\mathbb{P}_x\left\{\max_{n\le N}|\widetilde{Z}_n| > y - \frac{2}{\delta}Nv(x)\right\} \le \frac{tc}{(h - 2t/\delta)^2} \le \frac{\varepsilon}{2},$$

for all sufficiently large h. Therefore

$$\mathbb{P}_x\left\{\min_n Z_n > x - \frac{h}{v(x)},\quad \max_{n\le t/v^2(x)} |Z_n - x| > \frac{h}{v(x)}\right\} \le \frac{\varepsilon}{2},$$

which together with (5.62) completes the proof of (5.61). □

The proof of Theorem 5.12 is based on the following generalisation of the central limit theorem to a triangular array of martingales, which goes back to [71, Theorem 4].

Theorem 5.14 *For all $j \ge 1$, let $\{\mathcal{F}_{n,j}, n \ge 1\}$ be a filtration and $\{X_{n,j}, n \ge 1\}$ be a square integrable martingale with respect to $\{\mathcal{F}_{n,j}\}$. Let $n_j \to \infty$ as $j \to \infty$, let*

$$\frac{\sum_{k=1}^{n_j} \mathbb{E}\{(X_{k+1,j} - X_{k,j})^2 \mid \mathcal{F}_{k,j}\}}{\mathbb{E}X_{n_j,j}^2} \xrightarrow{p} 1 \quad \text{as } j \to \infty,$$

and let the following conditioned Lindeberg condition hold: for all $\varepsilon > 0$,

$$\frac{1}{\mathbb{E}X_{n_j,j}^2} \sum_{k=1}^{n_j} \mathbb{E}\{(X_{k+1,j} - X_{k,j})^2 \mathbb{I}\{|X_{k+1,j} - X_{k,j}| > \varepsilon\sqrt{\mathbb{E}X_{n_j,j}^2}\} \mid \mathcal{F}_{k,j}\} \xrightarrow{p} 0$$

as $j \to \infty$. Then $X_{n_j,j}/\sqrt{\mathbb{E}X_{n_j,j}^2}$ converges weakly to a standard normal distribution as $j \to \infty$.

Proof of Theorem 5.12. As shown in Lemma 5.13, it suffices to prove the claim of Theorem 5.12 for the chain $\{Z_n\}$, that is, it is sufficient to prove that

$$\frac{Z_n - x - nv(x)}{\sqrt{nb}} \Rightarrow N_{0,1} \tag{5.64}$$

as $x, n \to \infty$ in such a way that $1/t \le nv^2(x) \le t$.

Since the chain $\{Z_n\}$ satisfies all the conditions of Lemma 5.13, for any function $h(x) \to \infty$, given $Z_0 = x$, we have

$$\mathbb{P}_x\left\{-\frac{h(x)}{v(x)} \le Z_n - x \le \frac{h(x)}{v(x)} \text{ for all } n \le \frac{t}{v^2(x)}\right\} \to 1 \quad \text{as } x \to \infty. \tag{5.65}$$

The process $\{\widetilde{Z}_n\}$ defined in (5.63) constitutes a martingale – parameterised by x – whose second moment of jumps converges to b as $x \to \infty$. Owing to the construction of the jumps of $\{Z_n\}$ and because $s(x) = o(1/v(x))$ we obtain, for any $\varepsilon > 0$,

$$\mathbb{E}\{\xi^2(x);\ s(x) \ge |\xi(x)| \ge \varepsilon\sqrt{n}\} \to 0 \quad \text{as } x, n \to \infty$$

in such a way that $1/t \le nv^2(x) \le t$. Together with (5.62) this implies that, for the same range of x and n,

$$\mathbb{E}\{(Z_{k+1}-Z_k)^2;\ |Z_{k+1}-Z_k| \ge \varepsilon\sqrt{n} \mid Z_0 \ge x\} \to 0$$

uniformly for $k \le n$. These observations guarantee that the conditioned Lindeberg condition for the central limit theorem for martingales in a triangular array setting – see Theorem 5.14 – is met for $\widetilde{Z}_n$ and n satisfying $1/t \le nv^2(x) \le t$. Then we conclude that, given $Z_0 = x$, the random sequence

$$\frac{\widetilde{Z}_n}{\sqrt{nb}} = \frac{Z_n - x - \sum_{j=0}^{n-1} m_1^{[s(Z_j)]}(Z_j)}{\sqrt{nb}}$$

converges weakly as $x, n \to \infty$ to a standard normal distribution.

Let us choose $h(x) \to \infty$ sufficiently slowly that in the spatial range $x - h(x)/v(x) \le y \le x + h(x)/v(x)$ we have

$$v(y) \sim m_1^{[s(y)]}(y) \sim m_1^{[s(x)]}(x) \sim v(x) \quad \text{as } x \to \infty,$$

which is possible owing to (5.55). Then, within the temporal range $n \le t/v^2(x)$, we deduce from (5.65) that

$$\frac{\sum_{j=0}^{n-1} m_1^{[s(Z_j)]}(Z_j)}{\sqrt{nb}} - \sqrt{n/b}\,v(x) \xrightarrow{p} 0 \quad \text{as } x \to \infty.$$

Therefore

$$\frac{Z_n - x}{\sqrt{nb}} - \sqrt{n/b}\,v(x) = \frac{Z_n - x - nv(x)}{\sqrt{nb}}$$

converges weakly as $x \to \infty$ to a standard normal distribution, and the proof of the first claim in Theorem 5.12 is complete.

For the second claim, the same arguments with minor modifications can be applied to show that, for any sequences x_k and y_k such that $y_k \ge x_k$, $y_k - x_k = o(1/v(x_k))$, the following holds true:

$$\mathbb{P}_{y_k}\left\{X_n - x_k \le \frac{h}{v(x_k)}\right\} - \Phi\left(\frac{h/v(x_k) - nv(x_k)}{\sqrt{nb}}\right) \to 0$$

as $k, n \to \infty$ in such a way that $1/t \le nv^2(x_k) \le t$. Then the second claim is immediate, by contradiction. □

5.7 Integro-Local Renewal Theorem for Transient Chain with Normal Limit

In this section we discuss the asymptotics of partial and full renewal measures for X with a normal distribution limit. We pay special attention to the fact that both cases are investigated under truncation at a level $s(x) = o(1/v(x))$.

Theorem 5.15 *Under the conditions of Theorem 5.12, for every fixed $h > 0$ and $B > 0$,*

$$\sum_{n=0}^{[B/v^2(x)]} \mathbb{P}_y\Big\{X_n \in \Big(x, x + \frac{h}{v(x)}\Big]\Big\} \sim \frac{f(h,B)}{v^2(x)}$$

as $x \to \infty$ uniformly for all $y \in [x, x + o(1/v(x))]$, where $f(h,B) \uparrow h$ as $B \to \infty$.

Proof Owing to the normal distribution approximation provided by Theorem 5.12 we conclude that, for every fixed B,

$$\begin{aligned}
&\sum_{n=0}^{[B/v^2(x)]} \mathbb{P}_y\Big\{X_n \in \Big(x, x + \frac{h}{v(x)}\Big]\Big\} \\
&\quad = \sum_{n=0}^{[B/v^2(x)]} \Big(\Phi\Big(\frac{h - nv^2(x)}{\sqrt{nbv^2(x)}}\Big) - \Phi\Big(-\frac{nv^2(x)}{\sqrt{nbv^2(x)}}\Big) + o(1)\Big)
\end{aligned}$$

as $x \to \infty$ uniformly for all $y \in [x, x + o(1/v(x))]$. Approximating the sum on the right by an integral we obtain that its value is given by

$$\frac{1}{v^2(x)} \int_0^B \Big(\Phi\Big(\frac{h-z}{\sqrt{bz}}\Big) - \Phi\Big(-\frac{z}{\sqrt{bz}}\Big)\Big)dz + o\Big(\frac{1}{v^2(x)}\Big) \quad \text{as } x \to \infty. \tag{5.66}$$

The last integral equals

$$\begin{aligned}
f(h,B) &= \int_0^B \Big(\Phi\Big(\frac{h-z}{\sqrt{bz}}\Big) - \Phi\Big(-\frac{z}{\sqrt{bz}}\Big)\Big)dz \\
&= \int_0^B \frac{dz}{\sqrt{bz}} \int_0^h \varphi\Big(\frac{u-z}{\sqrt{bz}}\Big)du.
\end{aligned}$$

Changing the order of integration and making the substitution $z = v^2/b$, we obtain the equalities

$$\frac{1}{\sqrt{2\pi}}\int_0^h du\int_0^B \frac{1}{\sqrt{bz}}e^{-(u-z)^2/2bz}dz$$
$$=\frac{1}{\sqrt{2\pi}}\int_0^h e^{u/b}du\int_0^B \frac{1}{\sqrt{bz}}\exp\left(-u^2/2bz-z/2b\right)dz$$
$$=\frac{2}{b\sqrt{2\pi}}\int_0^h e^{u/b}du\int_0^{\sqrt{bB}} \exp\left(-u^2/2v^2-v^2/2b^2\right)dv.$$

The limit of the internal integral as $B\to\infty$ is known – see, e.g [74, p. 337, 3.325] – and is simply

$$\int_0^\infty e^{-u^2/2v^2-v^2/2b^2}dv=\frac{b\sqrt{2\pi}}{2}e^{-u/b}.$$

Combining the above we deduce that

$$\int_0^\infty \left(\Phi\left(\frac{h-z}{\sqrt{bz}}\right)-\Phi\left(-\frac{z}{\sqrt{bz}}\right)\right)dz=h.$$

Together with (5.66) this implies the result. □

Now let us turn to the asymptotic behaviour of the renewal measure.

Theorem 5.16 *Under the conditions of Theorem 5.12, for every fixed $h>0$ and distribution of X_0,*

$$H\left(x,x+\frac{h}{v(x)}\right]\sim\frac{h}{v^2(x)}\qquad as\ x\to\infty.$$

Proof We consider the same function $r(x)$ as in the proof of Lemma 5.13, so the conditions (3.56) and (3.57) are satisfied for all sufficiently large x.

We split the proof of the integro local asymptotics for H into two parts, the upper and lower bounds. First let us prove an appropriate upper bound. By the Markov property it is sufficient to show that, uniformly for all $y>x$,

$$\limsup_{x\to\infty} v^2(x)H_y\left(x,x+\frac{h}{v(x)}\right]\le h. \tag{5.67}$$

The chain $\{X_n\}$ satisfies all the conditions of Theorem 4.3. Then, for any $A>h$, by the Markov property and (4.32),

$$H_y\left(x,x+\frac{h}{v(x)}\right]$$
$$\le \mathbb{E}_y\sum_{n=0}^{T\left(x+\frac{A}{v(x)}\right)-1}\mathbb{I}\left\{X_n\in\left(x,x+\frac{h}{v(x)}\right]\right\}$$
$$+\ \mathbb{P}\left\{X_n\le x+\frac{h}{v(x)}\text{ for some }n\mid X_0>x+\frac{A}{v(x)}\right\}\sup_z H_z\left(x,x+\frac{h}{v(x)}\right]$$

$$\leq \mathbb{E}_y \sum_{n=0}^{T\left(x+\frac{A}{v(x)}\right)-1} \mathbb{I}\Big\{X_n \in \Big(x, x + \frac{h}{v(x)}\Big]\Big\}$$
$$+ \Big(\exp\Big(\delta\Big(R\Big(x + \frac{h}{v(x)}\Big) - R\Big(x + \frac{A}{v(x)}\Big)\Big)\Big) + o(1)\Big) \sup_z H_z\Big(x, x + \frac{h}{v(x)}\Big] \tag{5.68}$$

as $x \to \infty$ uniformly for all $A > h$, where the stopping time T is defined as

$$T(t) := \min\{n \geq 1 : X_n > t\}.$$

We have

$$\begin{aligned}
&\exp\Big(\delta\Big(R\Big(x + \frac{h}{v(x)}\Big) - R\Big(x + \frac{A}{v(x)}\Big)\Big)\Big) \\
&\quad = \exp\Big(-\delta \int_{x+h/v(x)}^{x+A/v(x)} r(y)dy\Big) \\
&\quad \leq \exp\big(-\delta(A-h)r(x + A/v(x))/v(x)\big) \leq \exp\big(-\delta(A-h)/2\big),
\end{aligned}$$

for all sufficiently large x. Applying the upper bound of Theorem 4.3 to the right-hand side of (5.68) we deduce that, for some $c < \infty$,

$$\begin{aligned}
H_y\Big(x, x + \frac{h}{v(x)}\Big] &\leq \mathbb{E}_y \sum_{n=0}^{T\left(x+\frac{A}{v(x)}\right)-1} \mathbb{I}\Big\{X_n \in \Big(x, x + \frac{h}{v(x)}\Big]\Big\} \\
&\quad + \big(e^{-\delta(A-h)/2} + o(1)\big)\frac{c}{v^2(x)}
\end{aligned} \tag{5.69}$$

as $x \to \infty$, for all $A > h$. The mean of the sum may be estimated as follows. For $C > A$,

$$\begin{aligned}
&\mathbb{E}_y \sum_{n=0}^{T\left(x+\frac{A}{v(x)}\right)-1} \mathbb{I}\Big\{X_n \in \Big(x, x + \frac{h}{v(x)}\Big]\Big\} \\
&\quad \leq \mathbb{E}_y \sum_{n=0}^{[C/v^2(x)]} \mathbb{I}\Big\{X_n \in \Big(x, x + \frac{h}{v(x)}\Big]\Big\} \\
&\qquad + \mathbb{E}_y\Big\{\sum_{n=0}^{T\left(x+\frac{A}{v(x)}\right)-1} \mathbb{I}\{X_n > x\};\ T\Big(x + \frac{A}{v(x)}\Big) > \frac{C}{v^2(x)}\Big\}
\end{aligned}$$

$$
\begin{aligned}
&= \mathbb{E}_y \sum_{n=0}^{[C/v^2(x)]} \mathbb{I}\Big\{X_n \in \Big(x, x+\frac{h}{v(x)}\Big]\Big\} \\
&\quad + \mathbb{E}_y \Big\{L\Big(x, T\Big(x+\frac{A}{v(x)}\Big)\Big);\ T\Big(x+\frac{A}{v(x)}\Big) > \frac{C}{v^2(x)}\Big\}.
\end{aligned}
$$

For $y > x$, the second term on the right-hand side is not greater than

$$
\begin{aligned}
&\mathbb{E}_y \Big\{L\Big(x, T\Big(x+\frac{A}{v(x)}\Big)\Big);\ X_n < x - \frac{D}{v(x)} \text{ for some } n \ge 1\Big\} \\
&\quad + \mathbb{E}_y \Big\{L\Big(x, T\Big(x+\frac{A}{v(x)}\Big)\Big); \\
&\qquad X_n \ge x - \frac{D}{v(x)} \text{ for all } n \le T\Big(\frac{A}{v(x)}\Big) - 1,\ T\Big(x+\frac{A}{v(x)}\Big) > \frac{C}{v^2(x)}\Big\} \\
&\le \mathbb{E}_y \Big\{L\Big(x, T\Big(x+\frac{A}{v(x)}\Big)\Big);\ X_n < x - \frac{D}{v(x)} \text{ for some } n \ge 1\Big\} \\
&\quad + \mathbb{E}_y \Big\{L\Big(x-\frac{D}{v(x)}, T\Big(x+\frac{A}{v(x)}\Big)\Big); L\Big(x-\frac{D}{v(x)}, T\Big(x+\frac{A}{v(x)}\Big)\Big) > \frac{C}{v^2(x)}\Big\}.
\end{aligned}
$$

Fix an $\varepsilon > 0$. By Theorem 4.2, for any fixed A and D, the family of random variables

$$
v^2(x) L\Big(x - \frac{D}{v(x)}, T\Big(x+\frac{A}{v(x)}\Big)\Big), \quad x \le y,\ X_0 = y,
$$

is uniformly integrable; hence, there is a $C = C(A, D)$ such that

$$
\begin{aligned}
\sup_{y: y \ge x} v^2(x) \mathbb{E}_y \Big\{&L\Big(x - \frac{D}{v(x)}, T\Big(x+\frac{A}{v(x)}\Big)\Big); \\
&L\Big(x - \frac{D}{v(x)}, T\Big(x+\frac{A}{v(x)}\Big)\Big) > \frac{C}{v^2(x)}\Big\} \le \varepsilon,
\end{aligned}
$$

for all sufficiently large x. Since

$$
\sup_{y > x} \mathbb{P}_y \Big\{X_n < x - \frac{D}{v(x)} \text{ for some } n \ge 1\Big\} \to 0 \quad \text{as } D \to \infty,
$$

by uniform integrability there exists a $D = D(A)$ such that

$$
\sup_{y \ge x} v^2(x) \mathbb{E}_y \Big\{L\Big(x, T\Big(x+\frac{A}{v(x)}\Big)\Big);\ X_n < x - \frac{D}{v(x)} \text{ for some } n \ge 1\Big\} \le \varepsilon,
$$

for all sufficiently large x. Combining all this we conclude that, uniformly for all $y \in (x, h/v(x)]$,

$$\limsup_{x\to\infty} v^2(x)\mathbb{E}_y \sum_{n=0}^{T\left(x+\frac{A}{v(x)}\right)-1} \mathbb{I}\left\{X_n \in \left(x, x+\frac{h}{v(x)}\right]\right\}$$
$$\leq \limsup_{x\to\infty} v^2(x)\mathbb{E}_y \sum_{n=0}^{[C/v^2(x)]} \mathbb{I}\left\{X_n \in \left(x, x+\frac{h}{v(x)}\right]\right\} + 2\varepsilon,$$

which, on being substituted into (5.69), gives

$$\limsup_{x\to\infty} v^2(x)H_y\left(x, x+\frac{h}{v(x)}\right]$$
$$\leq \limsup_{x\to\infty} v^2(x)\mathbb{E}_y \sum_{n=0}^{[C/v^2(x)]} \mathbb{I}\left\{X_n \in \left(x, x+\frac{h}{v(x)}\right]\right\} + ce^{-\delta(A-h)/2} + 2\varepsilon.$$

As already shown in Theorem 5.15,

$$v^2(x) \sum_{n=0}^{[C/v^2(x)]} \mathbb{P}_y\left\{X_n \in \left(x, x+\frac{h}{v(x)}\right]\right\} \to f(h,C) \quad \text{as } x\to\infty,$$

which implies the following upper bound, for each fixed $A > 1$:

$$\limsup_{x\to\infty} v^2(x)H_y\left(x, x+\frac{h}{v(x)}\right] \leq f(h,C) + ce^{-\delta(A-h)/2} + 2\varepsilon,$$

where $C = C(A, D(A))$. Letting $A \to \infty$, we get the required upper bound (5.67).

Now let us proceed with the lower bound. First notice that, by Theorem 5.15,

$$\liminf_{x\to\infty} v^2(x)H_y\left(x, x+\frac{h}{v(x)}\right] \geq h \tag{5.70}$$

as $x \to \infty$ uniformly for all $y \in [x, x+o(1/v(x))]$. It remains to prove that (5.70) holds for any fixed y. By the Markov property, it suffices to show that the overshoot over the level x is less than $s(x)$ with high probability, that is,

$$\mathbb{P}_y\{X_{T(x)} - x > s(x)\} \to 0 \quad \text{as } x\to\infty. \tag{5.71}$$

Indeed, for any fixed $x_0 > 0$,

$$\mathbb{P}_y\{X_{T(x)} - x > s(x)\} \le \sum_{n=1}^{\infty} \int_0^x \mathbb{P}_y\{X_n \in dz\}\mathbb{P}\{z + \xi(z) > x + s(x)\}$$
$$= \left(\int_0^{x_0} + \int_{x_0}^{x}\right)\mathbb{P}\{z + \xi(z) > x + s(x)\}H_y(dz).$$

The first integral on the right-hand side is bounded by

$$\int_0^{x_0} \mathbb{P}\{\xi(z) > s(x)\}H_y(dz) \to 0 \quad \text{as } x \to \infty,$$

owing to the dominated convergence theorem. Since $x - z + s(x) \ge s(z)$ for all $z \le x$, it follows from the condition (5.58) that the second integral is dominated by

$$\int_{x_0}^{x} \mathbb{P}\{\xi(z) > s(z)\}H_y(dz) \le \int_{x_0}^{\infty} p(z)v(z)H_y(dz) \to 0 \quad \text{as } x_0 \to \infty;$$

see the calculations leading to (4.6). Combining these results yields the convergence (5.71) for the overshoot. This concludes the proof. □

Theorem 5.15 and the proof of Theorem 5.16 imply the following result.

Theorem 5.17 *Under the conditions of Theorem 5.12, for every fixed $h > 0$,*

$$\sum_{k=0}^{n} \mathbb{P}_y\Big\{X_k \in \Big(x, x + \frac{h}{v(x)}\Big]\Big\} = \frac{1}{v^2(x)} f(h, nv^2(x)) + o\Big(\frac{1}{v^2(x)}\Big)$$

as $x \to \infty$ uniformly for all $y \in [x, x + o(1/v(x))]$ and for all $n \ge 1$, where $f(h,z) \uparrow h$ as $z \to \infty$.

Theorem 5.18 *Under the conditions of Theorems 5.7 and 5.16, given any distribution of X_0 and any fixed $h > 0$,*

$$\sum_{k=0}^{n} \mathbb{P}\Big\{X_k \in \Big(x, x + \frac{h}{v(x)}\Big]\Big\} = \frac{h}{v^2(x)} \Phi\left(\frac{n - V(x)}{\sqrt{b\frac{1+\beta}{1+3\beta}\frac{x}{v^3(x)}}}\right) + o\Big(\frac{1}{v^2(x)}\Big) \tag{5.72}$$

as $x \to \infty$ uniformly for all $n \ge 1$.

Proof We have

$$\sum_{k=0}^{n} \mathbb{P}\Big\{X_k \in \Big(x, x + \frac{h}{v(x)}\Big]\Big\} = \mathbb{E} \sum_{k=T(x)}^{n} \mathbb{I}\Big\{X_k \in \Big(x, x + \frac{h}{v(x)}\Big]\Big\}.$$

As (5.71) shows, $v(x)(X_{T(x)} - x) \to 0$ in probability. This allows us to apply Theorem 5.17: as $x \to \infty$,

$$\mathbb{E} \sum_{k=T(x)}^{n} \mathbb{I}\Big\{X_k \in \Big(x, x + \frac{h}{v(x)}\Big]\Big\} = \frac{1}{v^2(x)} \mathbb{E} f\big(h, v^2(x)(n - T(x))^+\big) + o\Big(\frac{1}{v^2(x)}\Big).$$

Further, fix $u \in \mathbb{R}$ and take

$$n = V(x) + u\sqrt{b\frac{1+\beta}{1+3\beta}\frac{x}{v^3(x)}}.$$

Then we have

$$\begin{aligned} v^2(x)(n - T(x))^+ &= \sqrt{b\frac{1+\beta}{1+3\beta} x v(x)} \frac{(n - T(x))^+}{\sqrt{b\frac{1+\beta}{1+3\beta}\frac{x}{v^3(x)}}} \\ &= \sqrt{b\frac{1+\beta}{1+3\beta} x v(x)} \Bigg(u + \frac{V(x) - T(x)}{\sqrt{b\frac{1+\beta}{1+3\beta}\frac{x}{v^3(x)}}}\Bigg)^+. \end{aligned}$$

Since $xv(x) \to \infty$, the last quantity tends to infinity with probability

$$\mathbb{P}\Bigg\{\frac{V(x) - T(x)}{\sqrt{b\frac{1+\beta}{1+3\beta}\frac{x}{v^3(x)}}} > -u\Bigg\} \to \Phi(u) \quad \text{as } x \to \infty,$$

and equals zero with probability tending to $1 - \Phi(u)$, both by Corollary 5.10. Taking into account that $f(h, z) \to h$ as $z \to \infty$, we conclude that

$$\mathbb{E} f\big(h, v^2(x)(n - T(x))^+\big) \to h\Phi(u) \quad \text{as } x \to \infty,$$

which completes the proof. □

5.8 Local Renewal Theorem for Transient Chain on $\mathbb{Z}$ with Normal Limit

In this section we formulate and prove a local version of the renewal theorem in the case of convergence to a normal distribution. Following the technique developed so far, we can do this only for a lattice Markov chain. Without loss of generality, let $\mathbb{Z}$ be the minimal lattice on which X lives. Similarly to the case of convergence to a Γ-distribution, it is unlikely that the local renewal theorem would be valid if we assumed only a regular asymptotic behaviour of the moments of jumps. We believe that a local renewal theorem can be proven only if we assume weak convergence of the jumps $\xi(x)$ to some random variable ξ on $\mathbb{Z}$, that is,

$$\xi(x) \Rightarrow \xi \quad \text{as } x \to \infty. \tag{5.73}$$

Theorem 5.19 *Let $v(x)$ be a decreasing differentiable function satisfying $xv(x) \to \infty$ and $v'(x) = o(v^2(x))$ and let*

$$m_1(x) \sim v(x) \quad \text{and} \quad m_2(x) \to b > 0 \quad \text{as } x \to \infty \tag{5.74}$$

and

$$\limsup_{n\to\infty} X_n = \infty \quad \text{with probability } 1.$$

Furthermore we assume the convergence (5.73). Let $\mathbb{Z}$ be the minimal lattice for ξ, and let the limit ξ satisfy

$$\mathbb{E}\xi = 0, \quad \mathbb{E}\xi^2 = b. \tag{5.75}$$

In addition, let the jumps $\xi(x)$ be bounded below and above by J uniformly for all $x \in \mathbb{Z}^+$, that is,

$$|\xi(x)| \leq J \text{ for all } x \in \mathbb{Z}^+. \tag{5.76}$$

Then

$$h(x) := H\{x\} \sim \frac{1}{v(x)} \quad \text{as } x \to \infty. \tag{5.77}$$

Moreover,

$$\mathbb{P}\left\{\sum_{n=0}^{\infty} \mathbb{I}\{X_n = x\} > N\right\} = c_1(x)(1 - c_2(x)v(x))^N, \tag{5.78}$$

where $c_1(x), c_2(x) \to 1$ as $x \to \infty$; hence the family of random variables

$$v(x)\sum_{n=0}^{\infty} \mathbb{I}\{X_n = x\}, \quad x \in \{1,2,3,\ldots\}, \tag{5.79}$$

is uniformly integrable.

More general results are derived in Chapter 6, via a different technique based on the martingale approach.

Proof Let $\delta > 0$ and define two decreasing functions

$$U_\pm(x) := \int_x^\infty e^{-R_\pm(y)}dy, \quad x > 0,$$

where

$$R_\pm(y) := \frac{2\pm\delta}{b}\int_0^y v(z)dy.$$

By the mean-value theorem, for all x and $j\in\mathbb{Z}$ there is a $\theta=\theta(j,x)\in(0,1)$ such that

$$U_\pm(x+j)-U_\pm(x) = -je^{-R_\pm(x+\theta j)} \sim -je^{-R_\pm(x)} \quad \text{as } x\to\infty,$$

because, for any fixed $u>0$,

$$|R_\pm(x+u)-R_\pm(x)| \le \frac{2+\delta}{b}uv(x) \to 0 \quad \text{as } x\to\infty,$$

since $v(x)\to 0$. By L'Hôpital's rule,

$$\begin{aligned}\lim_{x\to\infty}\frac{U_\pm(x)}{\frac{1}{v(x)}e^{-R_\pm(x)}} &= \lim_{x\to\infty}\frac{U'_\pm(x)}{\left(\frac{1}{v(x)}e^{-R_\pm(x)}\right)'}\\ &= \lim_{x\to\infty}\frac{e^{-R_\pm(x)}}{\left(\frac{v'(x)}{v^2(x)}+\frac{2\pm\delta}{b}\right)e^{-R_\pm(x)}} = \frac{b}{2\pm\delta},\end{aligned}$$

owing to the condition $v'(x)=o(v^2(x))$. Therefore,

$$U_\pm(x+j)-U_\pm(x) \sim -j\frac{2\pm\delta}{b}v(x)U_\pm(x).$$

Then, since the $\xi(x)$ are bounded below, we obtain that, for all fixed $k\ge 1$,

$$\begin{aligned}&\mathbb{E}_{x+k}\left\{U_\pm(X_{\tau(x)})-U_\pm(x+k);\ \tau(x)<\infty\right\}\\ &\quad\sim \frac{2\pm\delta}{b}v(x)U_\pm(x+k)\mathbb{E}_{x+k}\{x+k-X_{\tau(x)};\ \tau(x)<\infty\},\end{aligned} \tag{5.80}$$

where

$$\tau(x) := \min\{n\ge 1 : X_n\le x\}.$$

Let us compute the drift of $U_\pm(X_n)$. Since the jumps are bounded, using a Taylor expansion we have

$$\begin{aligned}&\mathbb{E}(U_\pm(x+\xi(x))-U_\pm(x))\\ &\quad= U'_\pm(x)m_1(x)+\frac{1}{2}m_2(x)U''_\pm(x)m_2(x)+O(U'''_\pm(x))\\ &\quad= -e^{-R_\pm(x)}m_1(x)+\frac{1\pm\delta/2}{b}v(x)e^{-R_\pm(x)}m_2(x)+O\left(v^2(x)e^{-R_\pm(x)}\right)\\ &\quad\sim \pm(\delta/2+o(1))v(x)e^{-R_\pm(x)} \quad \text{as } x\to\infty.\end{aligned}$$

Therefore, the sequence $U_-(X_{n\wedge\tau(x)})$ is a supermartingale for all sufficiently large x. Then, by the optional stopping theorem (see e.g. [56, Section 4.7]),

$$\mathbb{E}_{x+k}\{U_-(X_{\tau(x)});\ \tau(x)<\infty\} \le U_-(x+k).$$

This is equivalent to

$$\mathbb{E}_{x+k}\left\{U_-(X_{\tau(x)}) - U_-(x+k);\ \tau(x)<\infty\right\} \le U_-(x+k)\mathbb{P}_{x+k}\{\tau(x)=\infty\}.$$

Now using (5.80), we get

$$\mathbb{P}_{x+k}\{\tau(x)=\infty\} \ge \frac{2-2\delta}{b}v(x)\mathbb{E}_{x+k}\{x+k-X_{\tau(x)};\ \tau(x)<\infty\}. \quad (5.81)$$

Since $U_+(X_{n\wedge\tau(x)})$ is a submartingale for all sufficiently large x,

$$\mathbb{E}_{x+k}\{U_+(X_{\tau(x)});\ \tau(x)<\infty\} \ge U_+(x+k).$$

This implies that

$$\mathbb{P}_{x+k}\{\tau(x)=\infty\} \le \frac{2+2\delta}{b}v(x)\mathbb{E}_{x+k}\{x+k-X_{\tau(x)};\ \tau(x)<\infty\}.$$

Combining this lower bound with (5.81), due to the arbitrary choice of $\delta>0$ we conclude that

$$\mathbb{P}_{x+k}\{\tau(x)=\infty\} = \frac{2+o(1)}{b}v(x)\mathbb{E}_{x+k}\{x+k-X_{\tau(x)};\ \tau(x)<\infty\}. \quad (5.82)$$

The rest of the proof is almost the same as that of Theorem 4.14. □

5.9 Comments for Chapter 5

The weak law of large numbers in the form of (5.12) was originally proven by Lamperti in [112, Theorem 7.1] under the condition that the fourth moment of the jumps is bounded and the drift is of order θ/x^β, $\beta\in(0,1)$. His proof was based on the method of moments, like everything else in that paper.

The strong law of large numbers in the form of (5.24) for a nearest neighbour Markov chain was proven by Voit in [148, Theorem 2.11] via an orthogonal polynomials technique.

Various laws of large numbers – both weak and strong – and central limit theorems were proven by Keller et al. [90] under a minimal moment condition on the positive parts of the jumps – namely, the existence of a square integrable majorant – and under the assumption that the jumps are bounded below. The strong law of large numbers under the minimal moment condition was proven by Kersting in [93].

In [123, Theorem 2.3], Menshikov and Wade proved the strong law of large numbers in the form of (5.24) under the assumption that the moments of jumps of order $2+2\beta+\delta$, $\delta > 0$, are bounded. In the same paper, the authors proved a central limit theorem like Theorem 5.7 for a drift proportional to $1/x^\beta$ under the assumption that the jumps have bounded moments of order

$$\max\left(2+2\beta, 1+\frac{2}{1+\beta}\right).$$

6

Asymptotics for Renewal Measure for Transient Markov Chain via Martingale Approach

For a transient Markov chain $\{X_n\}$ on $\mathbb{R}$ with asymptotically zero drift, the average time spent by $\{X_n\}$ in the interval $(x, x+1]$ is, roughly speaking, the reciprocal of the drift and tends to infinity as x grows.

In this chapter we present a general approach relying on the diffusion approximation to prove renewal theorems for Markov chains; for that reason we consider Markov chains that may be approximated by a diffusion process. Then, if we have some result of the renewal type for diffusion processes as in Section 1.5.2, we should be able to obtain a similar result for a Markov chain having similar asymptotic behaviour of the first two moments of jumps. In particular, we will see in the examples below that, provided we have the Green function for the diffusion process, we should, in principle, be able to construct an approximation for the Green function of the Markov chain and thus to derive a renewal theorem.

We will apply a martingale-type technique and show that the asymptotic behaviour of the renewal measure depends heavily on the rate at which the drift vanishes. As in the last two chapters, two main cases are distinguished: either the drift of the chain decreases as $1/x$ or it decreases much more slowly than that, say as $1/x^\alpha$ for some $\alpha \in (0,1)$. In contrast with the case of asymptotically positive drift to be considered in Chapter 10, the case of vanishing drift is quite tricky to analyse owing to the fact that the Markov chain tends to infinity rather slowly and hence one should take into account diffusion fluctuations.

6.1 Asymptotics for Renewal Measure on Increasing Intervals

Throughout this chapter we assume that the trajectories of $\{X_n\}$ are unbounded, that is,

$$\limsup_{n\to\infty} X_n = \infty \text{ a.s.} \tag{6.1}$$

This condition holds true for any irreducible Markov chain on $\mathbb{Z}^+$ because such a chain leaves any finite collection of states in finite time, with probability 1.

Theorem 6.1 *Let $\{X_n\}$ be such that* (6.1) *holds and*

$$m_1^{[s(x)]}(x) \sim \frac{\mu}{x}, \qquad m_2^{[s(x)]}(x) \to b \in (0,\infty) \quad \text{as } x \to \infty, \tag{6.2}$$

for some $\mu > b/2$ and an increasing truncation level $s(x)$ of order $o(x)$. Assume also that

$$\mathbb{P}\{|\xi(y)| \ge s(y)\} \le p(y)/y \tag{6.3}$$

for some decreasing function $p(x)$ that is integrable at infinity, and that

$$|\xi(y)|\mathbb{I}\{|\xi(y)| \le s(y)\} \le_{\rm st} \widehat{\xi} \quad \text{for all } y \ge 0, \tag{6.4}$$

where

$$\mathbb{E}\widehat{\xi}^2 < \infty. \tag{6.5}$$

Then, for every function $h(x) \uparrow \infty$ of order $o(x)$, we have

$$H(x, x+h(x)] \sim \frac{2}{2\mu - b} x h(x) \quad \text{as } x \to \infty.$$

Notice that both conditions (6.3) and (6.4) are met for some $s(x) = o(x)$ if $|\xi(y)| \le_{\rm st} \widehat{\xi}$ for all y and for some $\widehat{\xi}$ satisfying (6.5).

In the course of the proof of this and subsequent theorems we will construct a bounded non-negative supermartingale, which shows that $X_n \to \infty$ a.s. This convergence means the transience of any set bounded on the right.

We now turn to the critical case $\mu = b/2$ where the properties of the chain – particularly recurrence and transience – depend on further terms in the asymptotic expansions for the moments of increments. As the next theorem shows, this is also true for the renewal function.

Theorem 6.2 *Let $\{X_n\}$ be such that* (6.1) *holds and that there exist $m \ge 1$, $\gamma > 0$, and an increasing truncation level $s(x)$ of order $o(x)$ such that*

$$\frac{2m_1^{[s(x)]}(x)}{m_2^{[s(x)]}(x)} = \frac{1}{x} + \frac{1}{x\log x} + \cdots + \frac{1}{x\log x \cdots \log_{(m-1)} x} + \frac{\gamma + 1 + o(1)}{x\log x \cdots \log_{(m)} x}$$

and $m_2^{[s(x)]}(x) \to b > 0$ *as* $x \to \infty$. *Assume that, for some* $\varepsilon > 0$,

$$\mathbb{P}\{|\xi(x)| > s(x)\} = o(1/x^2 \log^{2+\varepsilon} x), \tag{6.6}$$

$$\mathbb{E}\{|\xi(x)|^3;\ |\xi(x)| \le s(x)\} = o(x/\log^{1+\varepsilon} x), \tag{6.7}$$

$$|\xi(y)|\mathbb{I}\{|\xi(y)| < s(y)\} \le_{\rm st} \widehat{\xi}, \tag{6.8}$$

where $\widehat{\xi}$ *satisfies* (6.5). *Then, for every function* $h(x) \uparrow \infty$ *of order* $o(x)$, *we have*

$$H(x, x + h(x)] \sim \frac{2h(x)}{b\gamma} x \log x \cdots \log_{(m)} x \quad \text{as } x \to \infty.$$

The proof of the integral renewal theorem in the case $\mu > b/2$ in Section 4.8 was based on the convergence of X_n^2/n towards a Γ-distribution. This approach is not applicable under the conditions of Theorem 6.2, although the convergence to a Γ-distribution is still valid. The reason is that some chains with $\mu = b/2$ are null-recurrent while other are transient, but this difference disappears in the weak limit. The only statement which can be obtained from weak convergence here is the following lower bound:

$$\lim_{x\to\infty} \frac{H(0,x]}{x^2} = \infty$$

In the next theorem we consider the case where the drift decreases more slowly than $1/x$, that is, $m_1(x)x \to \infty$.

Theorem 6.3 *Let* $\{X_n\}$ *be such that* (6.1) *holds and that there exist a decreasing* $v(x)$ *satisfying* $xv(x) \to \infty$ *and* $v'(x) = o(v^2(x))$ *and an increasing truncation level* $s(x) = o(1/v(x))$ *such that*

$$m_1^{[s(x)]}(x) \sim v(x), \quad m_2^{[s(x)]}(x) \to b \in (0,\infty) \quad \text{as } x \to \infty.$$

Assume also that

$$\mathbb{P}\{|\xi(y)| \ge s(y)\} \le p(y)v(y), \tag{6.9}$$

$$\xi(y)\mathbb{I}\{|\xi(y)| < s(y)\} \le_{st} \widehat{\xi} \quad \text{for all } y \ge 0, \tag{6.10}$$

where $p(x)$ *is a non-increasing non-negative function that is integrable at infinity and* $\widehat{\xi}$ *satisfies* (6.5). *Then, for every function* $h(x) \uparrow \infty$ *of order* $o(1/v(x))$, *we have*

$$H(x, x + h(x)] \sim \frac{h(x)}{v(x)} \quad \text{as } x \to \infty.$$

In the two examples considered in Subsections 1.4.2 and 1.5.2 – nearest neighbour Markov chain and a diffusion process – it is possible to construct an appropriate martingale that allows us to find the renewal measure in a closed form. For the general Markov chains considered in the last three theorems, this martingale approach does not work because it is extremely difficult to construct such a martingale. However, it is possible to construct a quantity that is almost a martingale and allows us to derive the asymptotic behaviour of the renewal measure; this is done in Section 6.2.

6.2 Proof of Integro-Local Renewal Theorem for Increasing Intervals

Let $r(x)$ be a decreasing differentiable function on $[0,\infty)$ satisfying the condition

$$r'(x) = O(r^2(x)) \quad \text{as } x \to \infty. \tag{6.11}$$

In what follows, $r(x)$ approximates the quotient $2m_1^{[s(x)]}(x)/m_2^{[s(x)]}(x)$. We shall impose assumptions on the truncated moments of Markov chains, and while doing that we shall always assume that the truncation function $s(x)$ is increasing and satisfies

$$s(x) = o\,(1/r(x)) \quad \text{as } x \to \infty.$$

Define $R(x) = 0$ for $x \le 0$ and

$$R(x) := \int_0^x r(y)dy, \quad x > 0, \quad \text{and} \quad U(x) := \int_x^\infty e^{-R(z)}dz, \quad x \in \mathbb{R}, \tag{6.12}$$

where $U(x)$ is assumed finite; compare with the function U defined in (1.31). Clearly,

$$\frac{U''(x)}{U'(x)} = -r(x).$$

Owing to (6.11),

$$r(x+y) \sim r(x), \quad R(x+y)-R(x) \to 0, \quad \text{and} \quad e^{-R(x+y)} \sim e^{-R(x)} \tag{6.13}$$

as $x \to \infty$ uniformly for $|y| \le s(x)$. Also,

$$U'''(x) = (r^2(x) - r'(x))e^{-R(x)} = O\big(r^2(x)e^{-R(x)}\big) \tag{6.14}$$

and, consequently,

$$U'''(x+y) = O\big(r^2(x)e^{-R(x)}\big) \quad \text{as } x \to \infty \text{ uniformly for } |y| \le s(x). \tag{6.15}$$

Let

$$G(y) := U(0) - U(y) = \int_0^y e^{-R(z)} dz.$$

We start with a result showing that $G(X_n)$ is almost a martingale provided that the quotient $2m_1^{[s(x)]}(x)/m_2^{[s(x)]}(x)$ is asymptotically proportional to $r(x)$.

Lemma 6.4 *Let $\theta(y)$ be a non-negative bounded function. Let*

$$\mathbb{E}\{|\xi(y)|^3;\ |\xi(y)| \le s(y)\} = o\big(m_2^{[s(y)]}(y)\theta(y)/r(y)\big) \quad \text{as } y \to \infty. \tag{6.16}$$

(i) *If*

$$\mathbb{P}\{\xi(y) < -s(y)\} = 0 \quad \text{for all } y \ge 0 \tag{6.17}$$

and

$$\frac{2m_1^{[s(y)]}(x)}{m_2^{[s(y)]}(y)} \ge (1 + \theta(y))r(y) \quad \text{for all sufficiently large } y \tag{6.18}$$

then there exists a $y^* > 0$ such that

$$\mathbb{E}\{G(y + \xi(y)) - G(y);\ \xi(y) \le s(y)\} \ge 0 \quad \text{for all } y > y^*.$$

(ii) *If*

$$\mathbb{P}\{\xi(y) > s(y)\} = 0 \quad \text{for all } y \ge 0 \tag{6.19}$$

and

$$\frac{2m_1^{[s(y)]}(x)}{m_2^{[s(y)]}(y)} \le (1 - \theta(y))r(y) \quad \text{for all sufficiently large } y \tag{6.20}$$

then there exists a $y^* > 0$ such that

$$\mathbb{E}\{G(y + \xi(y)) - G(y);\ \xi(y) \ge -s(y)\} \le 0 \quad \text{for all } y > y^*.$$

Proof (i) Since the function $G(y)$ is increasing,

$$\mathbb{E}G(y + \xi(y)) - G(y) \ge \mathbb{E}\{G(y + \xi(y)) - G(y);\ |\xi(y)| \le s(y)\},$$

owing to the condition (6.17). We have that $G'(y) = e^{-R(y)}$, $G''(y) = -r(y)e^{-R(y)}$, and $G'''(y + z) = O(r^2(y))e^{-R(y)}$ as $y \to \infty$ uniformly for all $|z| \le s(y)$ owing to the upper bound (6.15) on U''' and (6.13). Therefore the application of a Taylor expansion up to the third derivative yields that, for some $\gamma = \gamma(y, \xi(y)) \in [0, 1]$,

$$\begin{aligned}
&\mathbb{E}\{G(y+\xi(y)) - G(y);\ |\xi(y)| \le s(y)\} \\
&\quad = m_1^{[s(y)]}(y)G'(y) + \frac{1}{2}m_2^{[s(y)]}(y)G''(y) \\
&\qquad + \frac{1}{6}\mathbb{E}\{\xi^3(y)G'''(y+\gamma\xi(y));\ |\xi(y)| \le s(y)\} \\
&\quad = m_1^{[s(y)]}(y)e^{-R(y)} - \frac{1}{2}m_2^{[s(y)]}(y)r(y)e^{-R(y)} \\
&\qquad + O\Big(r^2(y)e^{-R(y)}\mathbb{E}\{|\xi^3(y)|;\ |\xi(y)| \le s(y)\}\Big) \quad \text{as } y \to \infty.
\end{aligned}$$

The sum of the first two terms on the last right-hand side gives

$$\frac{1}{2}e^{-R(y)}\big(2m_1^{[s(y)]}(y) - m_2^{[s(y)]}(y)r(y)\big) \ge \frac{1}{2}e^{-R(y)}m_2^{[s(y)]}(y)\theta(y)r(y),$$

owing to the condition (6.18). The third term on the right-hand side is of order $o\big(m_2^{[s(y)]}(y)\theta(y)r(y)e^{-R(y)}\big)$ owing to the condition (6.16). These observations conclude the proof of (i).

(ii) Since the function $G(y)$ is increasing,

$$\mathbb{E}G(y+\xi(y)) - G(y) \le \mathbb{E}\{G(y+\xi(y)) - G(y);\ |\xi(y)| \le s(y)\},$$

by the condition (6.19). The rest of the proof is very similar to that of part (i). □

6.2.1 Upper Bound for Renewal Measure

Our derivation of an upper bound for the renewal measure of $\{X_n\}$ is based on the Lyapunov function $G^{**}_{h,x}(y)$ defined below in (6.23).

For any x and $h > 0$, consider a piecewise-differentiable function

$$g^{**}_{h,x}(y) := \begin{cases} 0, & y \le x, \\ 2(y-x), & y \in (x, x+h], \\ 2h, & y \in (x+h, x+h+s(x+h)], \\ 2he^{R(x+h+s(x+h))-R(y)}, & y > x+h+s(x+h), \end{cases} \tag{6.21}$$

whose derivative satisfies

$$g^{**\prime}_{h,x}(y) \;=\; 2\mathbb{I}\{y \in [x, x+h]\} \quad \text{for all } y < x+h+s(x+h),\ y \ne x, x+h. \tag{6.22}$$

Its integral, that is, the function which originates from the key function (1.34) for diffusion processes,

$$G^{**}_{h,x}(y) := \int_0^y g^{**}_{h,x}(z)dz, \tag{6.23}$$

is an increasing bounded function, $G^{**}_{h,x}(\infty) < \infty$, because

$$g^{**}_{h,x}(y) \ \le \ 2he^{R(x+h+s(x+h))-R(y)} \quad \text{for all } y \tag{6.24}$$

and hence

$$\begin{aligned} G^{**}_{h,x}(\infty) &\le 2h \int_x^\infty e^{R(x+h+s(x+h))-R(y)} dy \\ &= 2he^{R(x+h+s(x+h))} U(x) \\ &\le 2hU(x)e^{R(x+h)+R'(x+h)s(x+h)} \\ &= 2hU(x)e^{R(x+h)+r(x+h)s(x+h)}, \end{aligned} \tag{6.25}$$

because R is concave. As $s(x) = o(1/r(x))$,

$$\begin{aligned} G^{**}_{h,x}(\infty) &\le 2hU(x)e^{R(x+h)+o(1)} \\ &\le 2hU(x)e^{R(x)+o(1)} \quad \text{as } x \to \infty, \end{aligned} \tag{6.26}$$

for $h \le s(x)$, owing to (6.13).

The function $G^{**}_{h,x}(y)$ is convex for $y \le x+h$. For $y > x+h$, it increases in a concave way with slope $2h$ at point $x+h$. Notice that, for $y > x+h+s(x+h)$ and $z > 0$,

$$G^{**}_{h,x}(y+z) - G^{**}_{h,x}(y) = 2he^{R(x+h+s(x+h))}(G(y+z) - G(y))$$

and, by (6.24), for $y > x+h+s(x+h)$ and $z \le 0$,

$$G^{**}_{h,x}(y+z) - G^{**}_{h,x}(y) \ge 2he^{R(x+h+s(x+h))}(G(y+z) - G(y)).$$

Therefore, for all $y > x+h+s(x+h)$ and $z \in \mathbb{R}$,

$$G^{**}_{h,x}(y+z) - G^{**}_{h,x}(y) \ge 2he^{R(x+h+s(x+h))}(G(y+z) - G(y)). \tag{6.27}$$

Further, for $y \in (x+h, x+h+s(x+h)]$,

$$g^{**}_{h,x}(y+z) \ \ge \ 2he^{R(y)-R(y+z)} \quad \text{for } z > 0$$

and

$$g^{**}_{h,x}(y+z) \ \le \ 2h \ \le \ 2he^{R(y)-R(y+z)} \quad \text{for } z \le 0.$$

Therefore, for $y \in (x+h, x+h+s(x+h)]$,

$$G^{**}_{h,x}(y+z) - G^{**}_{h,x}(y) \ge 2he^{R(y)}(G(y+z) - G(y)). \tag{6.28}$$

Lemma 6.5 *Assume that the conditions* (6.16)–(6.18) *hold. Then there exists an $x^* > 0$ such that, for all $x > x^*$, $y \ge 0$, $h \le s(x)$, and $t \in (0, h/2)$,*

$$\mathbb{E}G^{**}_{h,x}(y+\xi(y)) - G^{**}_{h,x}(y) \ge m_2^{[t]}(y)\mathbb{I}\{y \in [x+t, x+h-t]\}. \tag{6.29}$$

Proof Since the function $G^{**}_{h,x}(y)$ is zero for $y \le x$ and positive for $y > x$, the mean drift of $G^{**}_{h,x}$ is non-negative for all $y \in [0, x]$ and the inequality (6.29) follows, for this range of y.

Since $G^{**}_{h,x}(y)$ is increasing and since (6.17) holds,

$$\mathbb{E}G^{**}_{h,x}(y+\xi(y)) - G^{**}_{h,x}(y) \geq \mathbb{E}\{G^{**}_{h,x}(y+\xi(y)) - G^{**}_{h,x}(y);\ |\xi(y)| \leq s(y)\}$$
$$=: E.$$

The positivity of E for $y > x+h$ follows from (6.27) and (6.28), by Lemma 6.4.

It remains only to estimate E from below for $y \in [x, x+h]$. Let us apply a Taylor expansion for $G^{**}_{h,x}$ with integral remainder term,

$$E = m_1^{[s(y)]}(y)g^{**}_{h,x}(y) + \mathbb{E}\bigg\{\int_y^{y+\xi(y)} g^{**\prime}_{h,x}(z)(y+\xi(y)-z)dz;\ |\xi(y)| \leq s(y)\bigg\}. \tag{6.30}$$

Since $g^{**}_{h,x}(z) \geq 0$ and $g^{**\prime}_{h,x}(z) \geq 0$ for all $z \in [0, x+h+s(x+h)]$, we obtain for all sufficiently large x and $y \in [x, x+h]$, $t \in (0, h/2)$,

$$\begin{aligned}
E &\geq \mathbb{E}\bigg\{\int_y^{y+\xi(y)} g^{**\prime}_{h,x}(z)(y+\xi(y)-z)dz;\ |\xi(y)| \leq t\bigg\} \\
&\geq 2\mathbb{I}\{y \in [x+t, x+h-t]\}\mathbb{E}\bigg\{\int_y^{y+\xi(y)} (y+\xi(y)-z)dz;\ |\xi(y)| \leq t\bigg\} \\
&= m_2^{[t]}(y)\mathbb{I}\{y \in [x+t, x+h-t]\},
\end{aligned}$$

because $g^{**\prime}_{h,x}(z) = 2$ for all $z \in (x, x+h]$; this concludes the proof. □

Proposition 6.6 *Assume that the conditions of Lemma 6.5 hold. Then there exists an $x^* > 0$ such that, for all $x > x^*$, $h \leq s(x)$, and $t \in (0, h/2)$,*

$$H(x+t, x+h-t] \leq \frac{G^{**}_{h,x}(\infty) - \mathbb{E}G^{**}_{h,x}(X_0)}{\min_{y\in[x+t,x+h-t]} m_2^{[t]}(y)}.$$

Proof Consider the decomposition

$$G^{**}_{h,x}(X_n) = \sum_{k=0}^{n-1}(G^{**}_{h,x}(X_{k+1}) - G^{**}_{h,x}(X_k)) + G^{**}_{h,x}(X_0).$$

Since $G^{**}_{h,x}(y)$ is bounded by $G^{**}_{h,x}(\infty)$, we obtain

$$\begin{aligned}
G^{**}_{h,x}(\infty) &\geq \mathbb{E}G^{**}_{h,x}(X_n) \\
&= \mathbb{E}G^{**}_{h,x}(X_0) + \sum_{k=0}^{n-1}\mathbb{E}\{G^{**}_{h,x}(X_{k+1}) - G^{**}_{h,x}(X_k)\} \\
&\geq \mathbb{E}G^{**}_{h,x}(X_0) + \sum_{k=0}^{n-1}\mathbb{E}\{m_2^{[t]}(X_k);\ X_k \in (x+t, x+h-t]\},
\end{aligned}$$

for $x > x_*$, by Lemma 6.5. Hence, for any n,

$$\sum_{k=0}^{n-1} \mathbb{P}\{X_k \in (x+t, x+h-t]\} \le \frac{G^{**}_{h,x}(\infty) - \mathbb{E}G^{**}_{h,x}(X_0)}{\min_{y\in[x+t,x+h-t]} m_2^{[t]}(y)}.$$

Letting n tend to infinity, we arrive at the required conclusion. □

6.2.2 Lower Bound for Renewal Measure

We now turn to an accompanying lower bound for the renewal measure. To this end we consider a continuous piecewise-differentiable function

$$g^*_{h,x}(y) := \begin{cases} 0, & y \le x, \\ 2(y-x), & y \in (x, x+h], \\ 2he^{R(x+h)-R(y)}, & y > x+h, \end{cases} \tag{6.31}$$

whose derivative satisfies

$$g^{*\prime}_{h,x}(y) \le 2\mathbb{I}\{y \in [x, x+h]\} \quad \text{for all } y \ge 0,\ y \ne x,\ x+h. \tag{6.32}$$

Its integral, which similarly to (6.23), originates from the key function (1.34) for diffusion processes,

$$G^*_{h,x}(y) := \int_0^y g^*_{h,x}(z)dz, \tag{6.33}$$

is an increasing bounded function, $G^*_{h,x}(\infty) < \infty$, and

$$\begin{aligned} G^*_{h,x}(\infty) - h^2 + 2he^{R(x+h)}U(x \mid h) \\ \ge 2he^{R(x)}U(x+h). \end{aligned} \tag{6.34}$$

For $h \le s(x) = o(1/r(x))$, owing to (6.11),

$$G^*_{h,x}(\infty) \ge (2+o(1))he^{R(x)}U(x) \quad \text{as } x \to \infty. \tag{6.35}$$

Also define a concave function

$$G^{*<}_{h,x}(y) := h^2 + 2he^{R(x+h)} \int_{x+h}^y e^{-R(z)}dz, \tag{6.36}$$

whose derivative is $2he^{R(x+h)-R(y)}$ and for which $G^{*<}_{h,x}(x+h) = G^*_{h,x}(x+h)$. Observe the inequality

$$G^*_{h,x}(y) \ge G^{*<}_{h,x}(y) \quad \text{for all } y \le x+h, \tag{6.37}$$

and the equality

$$G^*_{h,x}(y) = G^{*<}_{h,x}(y) \quad \text{for all } y \ge x+h. \tag{6.38}$$

Hence, for $y > x + h$ and $z > 0$,

$$\begin{aligned} G^*_{h,x}(y-z) - G^{*<}_{h,x}(y-z) &\le G^*_{h,x}(y) - G^{*<}_{h,x}(y-z) \\ &= G^{*<}_{h,x}(y) - G^{*<}_{h,x}(y-z) \\ &= 2he^{R(x+h)}(G(y) - G(y-z)). \quad (6.39) \end{aligned}$$

Lemma 6.7 *Assume that the conditions* (6.16), (6.19), *and* (6.20) *hold. Then there exists an* $x^* > 0$ *such that, for all* $x > x^*$, $y \ge 0$, $h \le s(x)$, *and* $t \in (0, h/2)$,

$$\mathbb{E}G^*_{h,x}(y+\xi(y)) - G^*_{h,x}(y) \le \begin{cases} 0, & y \le x - s(x), \\ 2h\mathbb{E}\{\xi(y); \xi(y) \in (x-y, s(y))\}, & y \in (x-s(x), x-t], \\ (1+hr(y))m_2^{[s(y)]}(y), & y \in (x-t, x+h+t], \\ 3h\mathbb{E}\{|\xi(y)|; -s(y) < \xi(y) < x+h-y\}, & y > x+h+t. \end{cases}$$

Proof Since $G^*_{h,x}(y)$ is increasing in y, we obtain

$$\begin{aligned} \mathbb{E}G^*_{h,x}(y+\xi(y)) - G^*_{h,x}(y) &\le \mathbb{E}\{G^*_{h,x}(y+\xi(y)) - G^*_{h,x}(y);\ \xi(y) \ge -s(y)\} \\ &= \mathbb{E}\{G^*_{h,x}(y+\xi(y)) - G^*_{h,x}(y);\ |\xi(y)| \le s(y)\} \\ &=: E, \end{aligned}$$

owing to (6.19).

In the case $y \le x - s(x)$, we have $y + \xi(y) \le x - s(x) + s(y) \le x$, so $G^*_{h,x}(y+\xi(y)) = G^*_{h,x}(y) = 0$ and the conclusion of the lemma follows for $y \le x - s(x)$.

In the case $x - s(x) < y \le x - t$, it follows from the definition of $G^*_{h,x}$ that $G^*_{h,x}(x+z) \le 2hz$ for all $z > 0$, which yields $G^*_{h,x}(y+z) \le 2h(y-x+z)$ for all $y \le x$ and $z > 0$. Therefore,

$$E \le 2h\mathbb{E}\left\{\xi(y); \xi(y) \in (x-y, s(y)]\right\}, \quad (6.40)$$

and the conclusion of the lemma follows for $x - s(x) < y \le x - t$.

In the case $y \in (x-t, x+h+t]$, we proceed similarly to Lemma 6.5. By the Taylor expansion (6.30), we have

$$\begin{aligned} E &\le m_1^{[s(y)]}(y)g^*_{h,x}(y) + m_2^{[s(y)]}(y) \\ &\le \frac{1}{2}m_2^{[s(y)]}(y)r(y)g^*_{h,x}(y) + m_2^{[s(y)]}(y) \\ &\le m_2^{[s(y)]}(y)(hr(y)+1), \end{aligned}$$

by (6.20), where $\theta(y) \geq 0$, (6.32), and the inequality $g^*_{h,x}(y) \leq 2h$, for all sufficiently large y. Thus the lemma's conclusion follows for $y \in (x-t, x+h+t]$.

In the case $y > x+h+t$, since the function $G(y)$ is concave,

$$G(y) - G(y-z) \leq zG'(y-z) = ze^{-R(y-z)} \quad \text{for all } z > 0.$$

Therefore, as $y \to \infty$,

$$G(y) - G(y-z) \leq ze^{-R(y)}(1+o(1)) \quad \text{uniformly for all } z \in [0, s(y)].$$

Thus it follows from (6.39) that, as $y \to \infty$,

$$\begin{aligned} G^*_{h,x}(y-z) - G^{*<}_{h,x}(y-z) &\leq 2hze^{R(x+h)-R(y)}(1+o(1)) \\ &\leq 2hz(1+o(1)) \end{aligned} \tag{6.41}$$

uniformly for all h, $z \in [0, s(y)]$. The inequality (6.37) and the equality (6.38) allow us to conclude that, for $y > x+h$,

$$\begin{aligned} E &= \mathbb{E}\{G^{*<}_{h,x}(y+\xi(y)) - G^{*<}_{h,x}(y);\ |\xi(y)| \leq s(y)\} \\ &\quad + \mathbb{E}\{G^*_{h,x}(y+\xi(y)) - G^{*<}_{h,x}(y+\xi(y));\ |\xi(y)| \leq s(y)\} \\ &= \mathbb{E}\{G^{*<}_{h,x}(y+\xi(y)) - G^{*<}_{h,x}(y);\ |\xi(y)| \leq s(y)\} \\ &\quad + \mathbb{E}\{G^*_{h,x}(y+\xi(y)) - G^{*<}_{h,x}(y+\xi(y));\ \xi(y) \in [-s(y), x+h-y]\} \\ &\leq \mathbb{E}\{G^*_{h,x}(y+\xi(y)) - G^{*<}_{h,x}(y+\xi(y));\ \xi(y) \in [-s(y), x+h-y]\}, \end{aligned}$$

by the second statement of Lemma 6.4. Applying (6.41) we deduce that, for all sufficiently large x and $y > x+h$,

$$E \leq 3h\mathbb{E}\{|\xi(y)|;\ \xi(y) \in [-s(y), x+h-y]\}.$$

Combining these results we conclude the claim of the lemma for $y > x+h+t$. □

Proposition 6.8 *Let the assumptions of Lemma 6.7 hold. Then there exists an* $x^* > 0$ *such that, for all* $x > x^*$, $y \geq 0$, $h \leq s(x)$, *and* $t \in (0, h/2)$,

$$H(x-t, x+h+t] \geq \frac{G^*_{h,x}(\infty) - \mathbb{E}G^*_{h,x}(X_0) - \delta(x)}{\max_{y \in [x-t, x+h+t]}(1+hr(y))m_2^{[s(y)]}(y)},$$

where

$$\begin{aligned} \delta(x) = 2h &\int_{x-s(x)}^{x-t} H(dy)\mathbb{E}\{\xi(y);\ x-y < \xi(y) < s(y)\} \\ &+ 3h \int_{x+h+t}^{\infty} H(dy)\mathbb{E}\{|\xi(y)|;\ -s(y) < \xi(y) < x+h-y\}. \end{aligned}$$

Proof Consider the decomposition

$$G^*_{h,x}(X_n) = \sum_{k=0}^{n-1}(G^*_{h,x}(X_{k+1}) - G^*_{h,x}(X_k)) + G^*_{h,x}(X_0).$$

Then we can deduce from Lemma 6.7 that, for some $c < \infty$ and all $x > x_*$,

$$\begin{aligned}
&\mathbb{E}G^*_{h,x}(X_n)\\
&\quad = \mathbb{E}G^*_{h,x}(X_0) + \sum_{k=0}^{n-1}\mathbb{E}(G^*_{h,x}(X_{k+1}) - G^*_{h,x}(X_k))\\
&\quad \le \mathbb{E}G^*_{h,x}(X_0) + \sum_{k=0}^{n-1}\mathbb{E}\left\{(1+hr(X_k))m_2^{[s(X_k)]}(X_k); X_k \in (x-t, x+h+t]\right\}\\
&\qquad + 2h\sum_{k=0}^{n-1}\int_{x-s(x)}^{x-t}\mathbb{P}\{X_k \in dy\}\mathbb{E}\{\xi(y); x-y < \xi(y) < s(y)\}\\
&\qquad + 3h\sum_{k=0}^{n-1}\int_{x+h+t}^{\infty}\mathbb{P}\{X_k \in dy\}\mathbb{E}\{|\xi(y)|; -s(y) < \xi(y) < x+h-y\}.
\end{aligned}$$

Hence, for any n,

$$\sum_{k=0}^{n-1}\mathbb{P}\{X_k \in (x-t, x+h+t]\} \ge \frac{\mathbb{E}G^*_{h,x}(X_n) - \mathbb{E}G^*_{h,x}(X_0) - \delta(x)}{\max_{y\in[x-t,x+h+t]}(1+hr(y))m_2^{[s(y)]}(y)}.$$

Letting $n \to \infty$ we arrive at the required conclusion owing to the convergence $G^*_{h,x}(X_n) \to G^*_{h,x}(\infty)$, which in turn follows from Lemma 6.5 together with the martingale convergence theorem and the assumption (6.1). □

In order to get a lower bound in a closed form, we need to derive conditions under which the term $\delta(x)$ in Proposition 6.8 is of order $o(G^*_{h,x}(\infty))$ as $x \to \infty$. In the next result we demonstrate how to bound $\delta(x)$ provided that an appropriate upper bound for the renewal measure is available.

Lemma 6.9 *For some $h = h(x) \le s(x)$ and $t = t(x) \le h/2$, let*

$$\sup_{y:\, x/2\le y\le 2x} H(y, y+t] \le C_1 t U(x)e^{R(x)} \quad \textit{for some } C_1 < \infty, \tag{6.42}$$

and, for some random variable ξ with $\mathbb{E}\xi^2 < \infty$, let

$$|\xi(y)| \le_{st} \xi \quad \textit{for all } y \ge 0. \tag{6.43}$$

Then $\delta(x) \le chU(x)e^{R(x)}\mathbb{E}\{\xi^2;\ |\xi| > t\}$ for some $c < \infty$.

Proof Let us analyse the first term in $\delta(x)$. The stochastic majorisation condition (6.43) yields that

$$\int_{x-s(x)}^{x-t} H(dy)\mathbb{E}\{\xi(y);\ x-y<\xi(y)<s(y)\} \le \int_{x-s(x)}^{x-t} H(dy)\mathbb{E}\{\xi;\ \xi>x-y\}.$$

Further, using the upper bound (6.42) we deduce

$$\begin{aligned}\int_{x-s(x)}^{x-t} H(dy)\mathbb{E}\{\xi;\ \xi>x-y\} &\le \sum_{n=1}^{s(x)/t} H(x-(n+1)t, x-nt]\mathbb{E}\{\xi;\ \xi>nt\}\\ &\le C_2tU(x)e^{R(x)}\sum_{n=1}^{s(x)/t}\mathbb{E}\{\xi;\ \xi>nt\}\\ &\le C_2tU(x)e^{R(x)}\mathbb{E}\{\xi^2/t;\ \xi>t\}\\ &= C_2U(x)e^{R(x)}\mathbb{E}\{\xi^2;\ \xi>t\}.\end{aligned}$$

Hence the first term in $\delta(x)$ is not greater than $2C_2hU(x)e^{R(x)}\mathbb{E}\{\xi^2;\ \xi>t\}$, as required.

The second term in $\delta(x)$ can be bounded in the same way, namely

$$\begin{aligned}&\int_{x+h+t}^{\infty} H(dy)\mathbb{E}\{|\xi(y)|;\ -s(y)<\xi(y)<x+h-y\}\\ &\quad=\int_{x+h+t}^{x+h+s(x)} H(dy)\mathbb{E}\{|\xi(y)|;\ -s(x)<\xi(y)<x+h-y\}\\ &\quad\le\int_{x+h+t}^{x+h+s(x)} H(dy)\mathbb{E}\{|\xi|;\ \xi<x+h-y\}\\ &\quad=\int_{t}^{s(x)} H(x+h+dy)\mathbb{E}\{|\xi|;\ \xi<-y\},\end{aligned}$$

and, as above,

$$\begin{aligned}&\int_{t}^{s(x)} H(x+h+dy)\mathbb{E}\{|\xi|;\ \xi<-y\}\\ &\quad\le\sum_{n=1}^{s(x)/t} H(x+h+nt, x+h+(n+1)t]\mathbb{E}\{|\xi|;\ \xi<-nt\}\\ &\quad\le C_3tU(x)e^{R(x)}\sum_{n=1}^{s(x)/t}\mathbb{E}\{|\xi|;\ \xi<-nt\}\\ &\quad\le C_3U(x)e^{R(x)}\mathbb{E}\{\xi^2;\ \xi<-t\},\end{aligned}$$

which concludes the proof. □

6.2.3 Two Markov Chains with Asymptotically Equal Jumps

As in Section 4.1, let $\{Y_n\}$ and $\{Z_n\}$ be two Markov chains with jumps $\eta(x)$ and $\zeta(x)$ respectively. Denote by H^Y and H^Z their renewal measures.

Lemma 6.10 *Let the conditions of Lemma 4.1 hold. If there exists a non-negative function $g(x)$ such that*

$$H^Z(x,x+h(x)] \sim g(x) \quad \text{as } x \to \infty \tag{6.44}$$

for any distribution of Z_0 and

$$\sup_y H_y^Z(x,x+h(x)] = O(g(x)) \quad \text{as } x \to \infty, \tag{6.45}$$

then, for any distribution of Y_0,

$$H^Y(x,x+h(x)] \sim g(x) \quad \text{as } x \to \infty.$$

If, in addition, the family of random variables

$$\frac{1}{g(x)} \sum_{n=0}^{\infty} \mathbb{I}\{Z_n \in (x,x+h(x)]\}$$

is uniformly integrable then

$$\frac{1}{g(x)} \sum_{n=0}^{\infty} \mathbb{I}\{Y_n \in (x,x+h(x)]\}$$

is also uniformly integrable.

Proof Let us consider two sequences of independent random fields $\{\eta_n(x),\ x \in \mathbb{R}\}_{n\ge 0}$ and $\{\zeta_n(x),\ x \in \mathbb{R}\}_{n\ge 0}$ as in (4.5) and also the Markov chains $\{Y_n\}$ and $\{Z_n\}$ as found there.

Fix an $\varepsilon > 0$ and let x_ε be given by Lemma 4.1. Let $\tau := \min\{n \ge 0 : Y_n > x_\varepsilon\}$ and consider $\{Z_k\}$ with initial value $Z_0 = Y_\tau$. Define

$$\mu := \min\{k \ge 1 : Z_k \ne Y_{\tau+k}\}.$$

By Lemma 4.1, $\mathbb{P}\{\mu < \infty\} \le \varepsilon$. For $x > x_\varepsilon$,

$$\begin{aligned}\sup_y H_y^Y(x,x+h(x)] \le{}& \sup_y \mathbb{E}_y \sum_{n=\tau}^{\tau+\mu-1} \mathbb{I}\{Y_n \in (x,x+h(x)]\} \\ &+ \sup_y \mathbb{E}_y \sum_{n=\tau+\mu}^{\infty} \mathbb{I}\{Y_n \in (x,x+h(x)]\}.\end{aligned}$$

The first expectation on the right-hand side is not greater than $H_y^Z(x, x+h(x)]$ because $Y_n = Z_{n-\tau}$ between τ and $\tau+\mu-1$. The second expectation possesses the following upper bound:

$$
\begin{aligned}
&\mathbb{E}_y \sum_{n=\tau+\mu}^{\infty} \mathbb{I}\{Y_n \in (x, x+h(x)]\} \\
&\quad = \mathbb{E}_y \Big\{ \sum_{n=\tau+\mu}^{\infty} \mathbb{I}\{Y_n \in (x, x+h(x)]\} \Big| \mu < \infty \Big\} \mathbb{P}\{\mu < \infty\} \\
&\quad \le \sup_z H_z^Y(x, x+h(x)] \varepsilon.
\end{aligned}
$$

Therefore

$$
\sup_y H_y^Y(x, x+h(x)] \le \frac{1}{1-\varepsilon} \sup_y H_y^Z(x, x+h(x)]. \tag{6.46}
$$

For any distribution of Y_0 and $x > x_\varepsilon$ we have

$$
\begin{aligned}
&H^Y(x, x+h(x)] \\
&\quad = \mathbb{E} \sum_{n=\tau}^{\tau+\mu-1} \mathbb{I}\{Y_n \in (x, x+h(x)]\} + \mathbb{E} \sum_{n=\tau+\mu}^{\infty} \mathbb{I}\{Y_n \in (x, x+h(x)]\} \\
&\quad = \mathbb{E} \sum_{n=\tau}^{\tau+\mu-1} \mathbb{I}\{Z_n \in (x, x+h(x)]\} + \mathbb{E} \sum_{n=\tau+\mu}^{\infty} \mathbb{I}\{Y_n \in (x, x+h(x)]\} \\
&\quad = \mathbb{E} H_{Y_\tau}^Y(x, x+h(x)] \\
&\qquad \mathbb{E}\mathbb{E}_{Y_\tau} \sum_{n=\mu}^{\infty} \mathbb{I}\{Z_n \in (x, x+h(x)]\} + \mathbb{E} \sum_{n=\tau+\mu}^{\infty} \mathbb{I}\{Y_n \in (x, x+h(x)]\}.
\end{aligned}
$$

As we saw in the first part of the proof, for all large enough x,

$$
\begin{aligned}
\mathbb{E} \sum_{n=\tau+\mu}^{\infty} \mathbb{I}\{Y_n \in (x, x+h(x)]\} &\le \varepsilon \sup_y H_y^Y(x, x+h(x)] \\
&\le \frac{\varepsilon}{1-\varepsilon} \sup_y H_y^Z(x, x+h(x)],
\end{aligned}
$$

owing to (6.46). Similarly,

$$
\begin{aligned}
\mathbb{E}_{Z_\tau} \sum_{n=\mu}^{\infty} \mathbb{I}\{Z_n \in (x, x+h(x)]\} &\le \mathbb{E}\mathbb{P}_{Y_\tau}(\mu < \infty) \sup_y H_y^Z(x, x+h(x)] \\
&\le \varepsilon \sup_y H_y^Z(x, x+h(x)].
\end{aligned}
$$

Therefore

$$|H^Y(x,x+h(x)] - \mathbb{E}H^Z_{Y_\tau}(x,x+h(x)]| \le \frac{\varepsilon}{1-\varepsilon}\sup_y H^Z_y(x,x+h(x)].$$

Letting $\varepsilon \to 0$ and using (6.45) we conclude that

$$|H^Y(x,x+h(x)] - \mathbb{E}H^Z_{Z_\tau}(x,x+h(x)]| = o(g(x)) \quad \text{as } x \to \infty.$$

According to (6.44) and (6.45), $\mathbb{E}H^Z_{Y_\tau}(x,x+h(x)] \sim g(x)$ and this completes the proof. □

6.2.4 Proofs of Theorems 6.1, 6.2, and 6.3

Proof of Theorem 6.1. Consider a modified Markov chain $\{\widetilde{X}_n\}$ on the same probability space as $\{X_n\}$ with jumps $\widetilde{\xi}(x)$ defined as follows:

$$\widetilde{\xi}(x) = \begin{cases} \xi(x) & \text{if } |\xi(x)| \le s(x), \\ \text{any value} & \text{if } |\xi(x)| > s(x). \end{cases}$$

If $\{\widetilde{X}_n\}$ does not satisfy the unboundedness-of-trajectories condition (6.1), we can increase the value of $s(x)$ on some set bounded on the right in such a way that $\{\widetilde{X}_n\}$ does then satisfy (6.1). Indeed, it follows from the conditions (6.2), (6.4), and (6.5) that there exist a sufficiently high level x_0 and an $\varepsilon > 0$ such that $\mathbb{P}\{\xi(x) \ge \varepsilon\} \ge \varepsilon$ for all $x \ge x_0$. Then it suffices to increase $s(x)$ on the set $(-\infty, x_0]$ to ensure the condition (6.1) for $\{\widetilde{X}_n\}$.

Without loss of generality we will assume that $h(x) \le s(x)$. Let us choose a function $t(x) \uparrow \infty$ of order $o(h(x))$ as $x \to \infty$.

Fix some $c > 1$ and consider $r(x) = c/(1+x)$. Then

$$R(x) = c\log(1+x) \quad \text{and} \quad U(x) = (1+x)^{1-c}/(c-1).$$

Therefore

$$U(x)e^{R(x)} = \frac{x+1}{c-1}. \tag{6.47}$$

The chain $\{\widetilde{X}_n\}$ satisfies the condition (6.17). Fix some $c^{**} \in (1, 2\mu/b)$ and define $r^{**}(x) = c^{**}/(1+x)$, which ensures the condition (6.18) with $\theta(y) = \theta = (2\mu/bc^{**} - 1)/2 > 0$. The condition (6.16) is immediate from the upper bound

$$\mathbb{E}\{|\xi(y)|^3;\ |\xi(y)| \le s(y)\} \le s(y)m_2^{[s(y)]}(y) \tag{6.48}$$

and the relation $s(y) = o(y)$. Also,

$$m_2^{[t(x)]}(x) \to b \quad \text{as } x \to \infty,$$

by the conditions (6.4) and (6.5). As a result, by Proposition 6.6, as $x \to \infty$ we have

$$\begin{aligned}\widetilde{H}(x+t(x), x+h(x)-t(x)] &\le \frac{G^{**}_{h,x}(\infty)}{b+o(1)}\\ &\le \frac{2+o(1)}{(c^{**}-1)b}xh(x),\end{aligned}$$

owing to (6.26) and (6.47). Letting $c^{**} \to 2\mu/b$, we get

$$\widetilde{H}(x+t(x), x+h(x)-t(x)] \le \frac{2+o(1)}{2\mu-b}xh(x) \quad \text{as } x \to \infty.$$

Taking into account that $t(x) = o(h(x))$ we obtain the upper bound

$$\widetilde{H}(x, x+h(x)] \le \frac{2+o(1)}{2\mu-b}xh(x) \quad \text{as } x \to \infty. \tag{6.49}$$

The chain $\{\widetilde{X}_n\}$ satisfies the condition (6.19). Fix some $c^* > 2\mu/b$ and define $r^*(x) = c^*/(1+x)$, which ensures the condition (6.20) with $\theta(y) = \theta = (1-2\mu/bc^*)/2 > 0$. Then it follows from Proposition 6.8 that, as $x \to \infty$,

$$\begin{aligned}\widetilde{H}(x-t(x), x+h(x)+t(x)] &\ge \frac{G^*_{h,x}(\infty) - \mathbb{E}G^*_{h,x}(X_0) - \delta(x)}{b+o(1)}\\ &\ge (2+o(1))\frac{h(x)\frac{x}{c^*-1} - \delta(x)}{b+o(1)},\end{aligned}$$

owing to (6.35) and (6.47). By the condition (6.4), the chain $\{\widetilde{X}_n\}$ satisfies (6.43), and this, together with the upper bound (6.49) for the renewal measure generated by $\{\widetilde{X}_n\}$, yields the upper bound for $\delta(x)$ given by Lemma 6.9. Therefore

$$\widetilde{H}(x-t(x), x+h(x)+t(x)] \ge \frac{2+o(1)}{(c^*-1)b}xh(x),$$

owing to (6.47). Letting $c^* \to 2\mu/b$, since $t(x) = o(h(x))$ we finally arrive at

$$\widetilde{H}(x, x+h(x)] \ge \frac{2+o(1)}{2\mu-b}xh(x) \quad \text{as } x \to \infty.$$

Combining this lower bound with the upper bound (6.49), we conclude that

$$\widetilde{H}(x, x+h(x)] \sim \frac{2}{2\mu-b}xh(x) \quad \text{as } x \to \infty.$$

Together with the condition (6.3) this allows us to apply Lemma 6.10 to the two Markov chains, $Y = X$ and $Z = \widetilde{X}$; hence the asymptotics are the same as for the renewal measure generated by $\{X_n\}$. □

Proof of Theorem 6.2. As in the proof of Theorem 6.1, from the very beginning we may assume that $|\xi(y)| \le s(y)$ for all y, which implies both (6.17) and (6.19). Without loss of generality we assume that $h(x) \le s(x)$.

Fix $c > 1$ and consider

$$\begin{aligned} r(x) &= \frac{1}{x + e_{(m)}} + \frac{1}{(x + e_{(m)}) \log(x + e_{(m)})} \\ &\quad + \cdots + \frac{c}{(x + e_{(m)}) \log(x + e_{(m)}) \cdots \log_{(m)}(x + e_{(m)})}, \end{aligned}$$

where $e_{(m)} > 0$ is defined by $\log_{(m)} e_{(m)} = 1$. Therefore,

$$\begin{aligned} R(x) &= \log(x + e_{(m)}) + \log\log(x + e_{(m)}) \\ &\quad + \cdots + \log_{(m)}(x + e_{(m)}) + c \log_{(m+1)}(x + e_{(m)}) - C_m \end{aligned}$$

and

$$U(x) = \frac{e^{C_m}}{c-1} \left(\log_{(m)}(x + e_{(m)})\right)^{1-c},$$

which implies from (6.26) that, for $c^{**} < \gamma + 1$,

$$G^{**}_{h(x),x}(\infty) \le \frac{2 + o(1)}{c^{**} - 1} h(x) x \log x \cdots \log_{(m)} x \quad \text{as } x \to \infty,$$

and, from (6.35), for $c^* > \gamma + 1$,

$$G^{*}_{h(x),x}(\infty) \ge \frac{2 + o(1)}{c^{*} - 1} h(x) x \log x \cdots \log_{(m)} x \quad \text{as } x \to \infty.$$

Repeating the arguments used in the proof of Theorem 6.1, we obtain the desired result. □

Proof of Theorem 6.3. As in the proof of Theorem 6.1, from the very beginning we may assume that $|\xi(y)| \le s(y)$ for all y, which implies both (6.17) and (6.19). Without loss of generality we can assume that $h(x) \le s(x)$. Let us choose a function $t(x) \uparrow \infty$ of order $o(h(x))$ as $x \to \infty$.

Fix some $c > 0$ and consider $r(x) = cv(x)$. Then, by l'Hôpital's rule,

$$\frac{U(x)}{U'(x)} \sim \frac{1}{r(x)}.$$

Therefore, as follows from (6.26),

$$G^{**}_{h(x),x}(\infty) \le (2 + o(1)) \frac{h(x)}{r(x)} \quad \text{as } x \to \infty, \tag{6.50}$$

and, from (6.35),

$$G^{*}_{h(x),x}(\infty) \ge (2 + o(1)) \frac{h(x)}{r(x)} \quad \text{as } x \to \infty. \tag{6.51}$$

Considering $c^{**} < 2/b$ and $c^* > 2/b$ and repeating the arguments used in the proof of Theorem 6.1, we conclude the proof. □

6.3 Asymptotics for Renewal Measure on Fixed Intervals

While the asymptotic behaviour of the renewal measure on increasing intervals is derived under the assumption of regular behaviour of the first two moments only, it seems that the local renewal theorem can be proved only for Markov chains that are asymptotically homogeneous in space. The next result gives us a tool for deriving the asymptotic behaviour of the renewal measure on intervals from the results for sufficiently slowly increasing intervals. It requires the weak convergence of jumps at infinity, that is, we consider a Markov chain that is asymptotically homogeneous in space, which is defined as a Markov chain such that, for some random variable ξ,

$$\xi(x) \Rightarrow \xi \quad \text{as } x \to \infty; \tag{6.52}$$

if there is no asymptotic homogeneity in space then the asymptotic behaviour of $H(x, x+h]$ may be very different. For Markov chains on $\mathbb{Z}^+$ with bounded jumps, this was studied in Sections 4.9 and 5.8 via a careful analysis of the returning probabilities at high level.

Theorem 6.11 *Let* (6.52) *hold and the family of random variables* $\{|\xi(x)|, x \in \mathbb{R}\}$ *admit an integrable majorant* Ξ, *that is,* $\mathbb{E}\Xi < \infty$ *and*

$$|\xi(x)| \le_{st} \Xi \quad \text{for all } x \in \mathbb{R}. \tag{6.53}$$

Assume that there exist a bounded function $v(x) > 0$, *an increasing level* $\widetilde{t}(x) \uparrow \infty$, *and a constant* $C_H < \infty$ *such that, for any* $t(x) \uparrow \infty$ *satisfying* $t(x) \le \widetilde{t}(x)$,

$$\frac{v(x)H(x, x + t(x)]}{t(x)} \to C_H \quad \text{as } x \to \infty. \tag{6.54}$$

If the limiting random variable ξ *is non-lattice then* $v(x)H(x, x+h] \to C_H h$ *as* $x \to \infty$, *for all fixed* $h > 0$.

If the chain $\{X_n\}$ *is integer valued and* $\mathbb{Z}$ *is the minimal lattice for the variable* ξ *then* $v(k)H\{k\} \to C_H$ *as* $k \to \infty$, *and, in addition, the family of random variables*

$$v(k)\sum_{n=0}^{\infty}\mathbb{I}\{X_n = k\}, \quad k > 0, \tag{6.55}$$

is uniformly integrable.

Let us apply the last result to the chains considered in Theorems 6.1–6.3. In addition, under specific assumptions on the drift function we are able to generalise the uniform-integrability conclusion from lattice chains to general Markov chains.

Corollary 6.12 *Under the conditions of Theorem 6.1,* (6.52), *and* (6.53), *we have, for every* $h > 0$,

$$H(x, x+h] \sim \frac{2h}{2\mu - b} x \quad \text{as } x \to \infty$$

if the limiting random variable ξ *is non-lattice and*

$$H\{k\} \sim \frac{2}{2\mu - b} k \quad \text{as } k \to \infty$$

if the chain $\{X_n\}$ *is integer valued and* $\mathbb{Z}$ *is the minimal lattice for the variable* ξ.

In addition, for some $\widehat{x} \in \mathbb{R}$, *the family of random variables*

$$\frac{1}{x} \sum_{n=0}^{\infty} \mathbb{I}\{X_n \in (x, x+1]\}, \quad x \ge \widehat{x},$$

is uniformly integrable.

For lattice Markov chains, the last corollary is an improvement on Theorem 4.14, where the same asymptotics were proven only in the case of bounded jumps. A similar improvement on Theorem 4.16 holds true.

In Section 11.1, we will apply the last result to derive the local asymptotics of the renewal measure for a random walk conditioned to stay positive.

Corollary 6.13 *Under the conditions of Theorem 6.2,* (6.52), *and* (6.53), *we have, for every* $h > 0$,

$$H(x, x+h] \sim \frac{2h}{b\gamma} x \log x \cdots \log_{(m)} x \quad \text{as } x \to \infty,$$

if the limiting random variable ξ *is non-lattice and*

$$H\{k\} \sim \frac{2}{b\gamma} k \log k \cdots \log_{(m)} k \quad \text{as } k \to \infty,$$

if the chain $\{X_n\}$ *is integer valued and* $\mathbb{Z}$ *is the minimal lattice for the variable* ξ.

In addition, for some $\widehat{x} \in \mathbb{R}$, *the family of random variables*

$$\frac{1}{x \log x \cdots \log_{(m)} x} \sum_{n=0}^{\infty} \mathbb{I}\{X_n \in (x, x+1]\}, \quad x \ge \widehat{x},$$

is uniformly integrable.

Corollary 6.14 *Under the conditions of Theorem 6.3,* (6.52), *and* (6.53), *we have, for every* $h > 0$,

$$H(x, x+h] \sim \frac{h}{v(x)} \quad \text{as } x \to \infty,$$

if the limiting random variable ξ *is non-lattice and*

$$H\{k\} \sim \frac{1}{v(k)} \quad \text{as } k \to \infty$$

if the chain $\{X_n\}$ *is integer valued and* $\mathbb{Z}$ *is the minimal lattice for the variable* ξ.

In addition, for some $\widehat{x} \in \mathbb{R}$, *the family of random variables*

$$v(x) \sum_{n=0}^{\infty} \mathbb{I}\{X_n \in (x, x+1]\}, \quad x \ge \widehat{x},$$

is uniformly integrable.

The last result is an improvement on Theorem 5.19 and it is particularly useful for proving the local asymptotics for a random walk conditioned to stay positive, which represents one of the classical examples of chains with asymptotically zero drift; see Proposition 11.1.

6.4 Key Renewal Theorem

We now turn to the renewal equation

$$Z(B) = z(B) + \int_{\mathbb{R}} Z(dy)\, P(y, B), \quad B \subset \mathcal{B}(\mathbb{R}),$$

where $\mathcal{B}$ is a Borel algebra and z is a finite nonnegative measure on $\mathbb{R}$. This is more than sufficient to ensure that

$$Z(B) = \int_{\mathbb{R}} z(du) H_u(B), \quad B \in \mathcal{B}(\mathbb{R}),$$

is a unique locally finite solution to the renewal equation. The analysis of the preceding subsection allows us to deduce the asymptotic behaviour of the measure Z at infinity. The proof is immediate from the dominated convergence theorem.

Theorem 6.15 *Let* $B \in \mathcal{B}(\mathbb{R})$. *Assume that, for some positive function* $g(x)$ *and for all* $y \in \mathbb{R}$,

$$H_y(x + B) \sim g(x) \quad \text{as } x \to \infty,$$

and, for some $c < \infty$,

$$H_y(x+B) \le cg(x) \quad \text{for all } x,\ y \in \mathbb{R}.$$

If z *is a finite measure then*

$$Z(x+B) \sim z(\mathbb{R})g(x) \quad \text{as } x \to \infty.$$

6.5 Proof of Results of Section 6.3

In this section, our first goal is to provide an approach that allows us to reduce the proof of the asymptotic behaviour of the renewal measure on intervals of fixed length to a proof for sufficiently slowly growing intervals, that is, Theorem 6.11.

Lemma 6.16 *Assume that there exist monotone functions* $v(x) > 0$ *and* $\tilde{t}(x) \uparrow \infty$ *such that, for any* $t(x) \uparrow \infty$ *satisfying* $t(x) \le \tilde{t}(x)$,

$$\sup_{x\ge 1} \frac{v(x)H(x, x+t(x)]}{t(x)} < \infty.$$

Then

$$\sup_{x\ge 1} v(x)H(x, x+1] < \infty. \tag{6.56}$$

Proof Suppose that (6.56) does not hold. Then there exists a sequence $x_n \uparrow \infty$ such that

$$\alpha_n := v(x_n)H(x_n, x_n+1] \to \infty \quad \text{as } n \to \infty.$$

Since both α_n and $\tilde{t}(x_n)$ tend to infinity, there exists a sequence $t_n \uparrow \infty$ such that $t_n \le \tilde{t}(x_n)$ and $t_n = o(\alpha_n)$ as $n \to \infty$. Let t be defined as follows:

$$t(x) = t_n, \quad x_n \le x < x_{n+1}.$$

Clearly, $t(x) \le \tilde{t}(x)$ and $t(x) \uparrow \infty$. Then, for sufficiently large n,

$$\frac{v(x_n)H(x_n, x_n+t(x_n)]}{t(x_n)} \ge \frac{v(x_n)H(x_n, x_n+1]}{t(x_n)} = \frac{\alpha_n}{t(x_n)} \to \infty,$$

which contradicts the hypothesis. □

Proof of Theorem 6.11. By Lemma 6.16 it follows from the assumption (6.54) that the supremum in (6.56) is finite. In turn, this allows us to apply Helly's selection theorem to the family of measures $\{v(x)H(x+\cdot),\ x \in \mathbb{R}\}$ (see, for example, Theorem 2 in [63, Section VIII.6]). Hence, there exists a sequence of points $x_n \to \infty$ such that the sequence of measures $v(x_n)H(x_n+\cdot)$ converges weakly to some measure λ as $n \to \infty$. The following two results characterise λ.

Lemma 6.17 *Let F denote the distribution of ξ. The weak limit λ of the sequence of measures $v(x_n)H(x_n+\cdot)$ satisfies the identity $\lambda = \lambda * F$.*

Proof The measure λ is positive and σ-finite with necessity. Fix any smooth function $f(x)$ with a bounded support; let $A > 0$ be such that $f(x) = 0$ for $x \notin [-A, A]$. The weak convergence of measures means the following convergence of integrals:

$$\begin{aligned}\int_{-\infty}^{\infty} f(x)v(x_n)H(x_n+dx) &= \int_{-A}^{A} f(x)v(x_n)H(x_n+dx)\\ &\to \int_{-A}^{A} f(x)\lambda(dx) \quad \text{as } n\to\infty. \quad (6.57)\end{aligned}$$

On the other hand, owing to the equality $H(\cdot) = \mathbb{P}\{X_0 \in \cdot\} + H * P(\cdot)$, we have the following representation for the left-hand side of (6.57):

$$\begin{aligned}&\int_{-A}^{A} f(x)v(x_n)\mathbb{P}\{X_0 \in x_n + dx\}\\ &+ \int_{-A}^{A} f(x)\int_{-\infty}^{\infty} P(x_n+y, x_n+dx)v(x_n)H(x_n+dy). \qquad (6.58)\end{aligned}$$

Since f and v are bounded,

$$\begin{aligned}\int_{-A}^{A} f(x)v(x_n)\mathbb{P}\{X_0 \in x_n+dx\} &\le \|f\|_\infty \|v\|_\infty \mathbb{P}\{X_0 \in [x_n - A, x_n + A]\}\\ &\to 0 \quad \text{as } n\to\infty. \qquad (6.59)\end{aligned}$$

The second term in (6.58) is equal to

$$\int_{-\infty}^{\infty} v(x_n)H(x_n+dy)\int_{-A}^{A} f(x)P(x_n+y, x_n+dx). \qquad (6.60)$$

The weak convergence $P(t, t+\cdot) \Rightarrow F(\cdot)$ as $t\to\infty$ implies convergence of the inner integral in (6.60):

$$\int_{-A}^{A} f(x)P(x_n+y, x_n+dx) \to \int_{-A}^{A} f(x)F(dx-y);$$

here the rate of convergence can be estimated in the following way:

$$\begin{aligned}\Delta(n,y) &:= \left|\int_{-A}^{A} f(x)(P(x_n+y, x_n+dx) - F(dx-y))\right|\\ &= \left|\int_{-A}^{A} f'(x)(\mathbb{P}\{\xi(x_n+y)\le x-y\} - F(x-y))dx\right|\\ &\le \|f'\|_\infty \int_{-A-y}^{A-y} |\mathbb{P}\{\xi(x_n+y)\le x\} - F(x)|dx.\end{aligned}$$

Thus, the asymptotic homogeneity of the chain yields for every fixed $C > 0$ the uniform convergence

$$\sup_{y\in[-C,C]} \Delta(n,y) \to 0 \quad \text{as } n \to \infty. \tag{6.61}$$

In addition, by the majorisation condition (6.53), for all $x \in \mathbb{R}$,

$$|\mathbb{P}\{\xi(x_n+y) \le x\} - F(x)| \le 2\mathbb{P}\{\Xi > |x|\}.$$

Hence, for all y,

$$\begin{aligned}\Delta(n,y) &\le 2\|f'\|_\infty \int_{-A-y}^{A-y} \mathbb{P}\{\Xi > |x|\}\,dx \\ &\le 4A\|f'\|_\infty \mathbb{P}\{\Xi > |y| - A\}. \end{aligned} \tag{6.62}$$

We have an upper bound

$$\begin{aligned}\Delta_n &:= \left|\int_{-\infty}^{\infty} v(x_n)H(x_n+dy)\left(\int_{-A}^{A} f(x)P(x_n+y,x_n+dx) - \int_{-A}^{A} f(x)F(dx-y)\right)\right| \\ &\le \int_{-\infty}^{\infty} \Delta(n,y)v(x_n)H(x_n+dy).\end{aligned}$$

For any fixed $C > 0$, (6.61) and (6.56) imply that

$$\begin{aligned}\int_{-C}^{C} \Delta(n,y)v(x_n)H(x_n+dy) &\le \sup_{y\in[-C,C]} \Delta(n,y) \sup_n\big(v(x_n)H[x_n-C,x_n+C]\big) \\ &\to 0 \quad \text{as } n\to\infty.\end{aligned}$$

The remaining part of the integral can be estimated by (6.62):

$$\begin{aligned}&\limsup_{n\to\infty} \int_{|y|\ge C} \Delta(n,y)v(x_n)H(x_n+dy) \\ &\quad\le 4A\|f'\|_\infty \limsup_{n\to\infty} \int_{|y|\ge C} \mathbb{P}\{\Xi > |y|-A\}v(x_n)H(x_n+dy).\end{aligned}$$

Since Ξ has finite mean, the property (6.56) of the renewal measure H allows us to choose the value of C sufficiently large to make the 'lim sup' as small as we please. Therefore, $\Delta_n \to 0$ as $n \to \infty$. Hence, (6.60) has the same limit as the sequence of integrals

$$\int_{-\infty}^{\infty} v(x_n)H(x_n+dy)\int_{-A}^{A} f(x)F(dx-y).$$

Now, the weak convergence to λ implies that (6.60) has the limit

$$\int_{-\infty}^{\infty} \lambda(dy) \int_{-\infty}^{\infty} f(x)F(dx-y) = \int_{-\infty}^{\infty} f(x) \int_{-\infty}^{\infty} F(dx-y)\lambda(dy)$$
$$= \int_{-\infty}^{\infty} f(x)(F*\lambda)(dx). \qquad (6.63)$$

By (6.57)–(6.59) and (6.63), we conclude the identity

$$\int_{-\infty}^{\infty} f(x)\lambda(dx) = \int_{-\infty}^{\infty} f(x)(F*\lambda)(dx).$$

Since this identity holds for any smooth function f with a bounded support, the measures λ and $F*\lambda$ coincide and the proof is complete. □

Further, we use the following statement from Choquet and Deny [35].

Proposition 6.18 *Let F be a distribution not concentrated at 0. Let λ be a non-negative measure satisfying the equality $\lambda = \lambda * F$ and the property* $\sup_{n\in\mathbb{Z}} \lambda[n,n+1] < \infty$.

If F is non-lattice then λ is proportional to the Lebesgue measure.

If F is lattice with minimal span 1 and $\lambda(\mathbb{R}\setminus\mathbb{Z}) = 0$ then λ is proportional to the counting measure.

The concluding part of the proof of Theorem 6.11 will be carried out for the non-lattice case. Choose any sequence of points $x_n \to \infty$ such that the measure $v(x_n)H(x_n+\cdot)$ converges weakly to some measure λ as $n \to \infty$. It follows from Lemma 6.17 and Proposition 6.18 that then $\lambda(dx) = \alpha\, dx$ with some α, that is,

$$v(x_n)H(x_n+dx) \Rightarrow \alpha\, dx \quad \text{as } n \to \infty.$$

Then, for any $A > 0$ and $k \in \{0,1,2,\ldots\}$,

$$v(x_n)H(x_n+kA, x_n+(k+1)A] \to \alpha A.$$

Therefore there exists a sufficiently slowly growing sequence $t_n \uparrow \infty$ such that

$$\frac{v(x_n)H(x_n, x_n+t_n]}{t_n} \to \alpha.$$

It follows from the assumption (6.54) that $\alpha = C_H$.

We complete the proof of the local limit of the renewal measure using an argument by contradiction. Suppose there exists a sequence $\{x_n\}$ such that

$$v(x_n)H(x_n, x_n+h] \not\to C_H h \quad \text{as } n \to \infty. \qquad (6.64)$$

However, by Helly's selection theorem and the arguments above there exists a further subsequence x_{n_k} for which

$$v(x_{n_k})H(x_{n_k}, x_{n_k} + h] \to C_H h,$$

which contradicts (6.64).

Now let us show the uniform integrability in the lattice case; to prove this, let us first notice that the Markov property implies that

$$\begin{aligned}
&\mathbb{P}\Big\{\sum_{n=1}^{\infty}\mathbb{I}\{X_n = x\} > N\Big\} \\
&\quad = \mathbb{P}\{X_n = x \text{ for some } n \ge 0\}\mathbb{P}^N\{X_n = x \text{ for some } n \ge 1 \mid X_0 = x\} \\
&\quad = \mathbb{P}\{X_n = x \text{ for some } n \ge 0\}\Big(1 - \mathbb{P}\{X_n \ne x \text{ for all } n \ge 1 \mid X_0 = x\}\Big)^N.
\end{aligned} \tag{6.65}$$

We denote

$$p_1(x) \;:=\; \mathbb{P}\{X_n = x \text{ for some } n \ge 0\};$$

$p_1(x)$ tends to 1 as $x \to \infty$ for the following reason. For any fixed $\varepsilon > 0$, by the condition (6.53) on jumps, there exists a sufficiently large J such that, for all $X_0 < x$,

$$\begin{aligned}
1 - \varepsilon &\le \mathbb{P}\{X_n \in [x, x + J] \text{ for some } n\} \\
&= \mathbb{P}\{X_n = x \text{ for some } n\} \\
&\quad + \mathbb{P}\{X_n \in [x + 1, x + J] \text{ for some } n,\ X_n \ne x \text{ for all } n\}.
\end{aligned}$$

The second probability on the right-hand side tends to zero as $x \to \infty$; indeed, it is not greater than

$$\begin{aligned}
&\sum_{i=1}^{J}\mathbb{P}\{X_n = x + i \text{ for some } n,\ X_n \ne x \text{ for all } n\} \\
&\qquad \le \sum_{i=1}^{J}\mathbb{P}\{X_n \ne x \text{ for all } n \mid X_0 = x + i\}.
\end{aligned}$$

The ith probability on the right-hand side converges as $x \to \infty$ to

$$\mathbb{P}\{S_n \ne 0 \text{ for all } n \mid S_0 = i\} = 0,$$

since $\mathbb{E}\xi = 0$. Hence, owing to the arbitrary choice of ε, we have that $p_1(x) \to 1$ as $x \to \infty$.

If follows from (6.65) that

$$\begin{aligned}
&\mathbb{E}\Big\{\sum_{n=1}^{\infty}\mathbb{I}\{X_n=x\};\ \sum_{n=1}^{\infty}\mathbb{I}\{X_n=x\}>N\Big\}\\
&\quad=\sum_{k=N}^{\infty}\mathbb{P}\Big\{\sum_{n=1}^{\infty}\mathbb{I}\{X_n=x\}>k\Big\}+N\mathbb{P}\Big\{\sum_{n=1}^{\infty}\mathbb{I}\{X_n=x\}>N\Big\}\\
&\quad=p_1(x)\big(1-p_2(x)\big)^N(N+1/p_2(x)),
\end{aligned}$$

where

$$p_2(x)=\mathbb{P}\{X_n\neq x \text{ for all } n\ge 1 \mid X_0=x\}.$$

Taking into account that $p_1(x)\to 1$ and

$$\frac{p_1(x)}{p_2(x)}=\mathbb{E}\sum_{n=1}^{\infty}\mathbb{I}\{X_n=x\}\sim\frac{C_H}{v(x)}\quad \text{as } x\to\infty,$$

we derive the asymptotics

$$p_2(x)\sim p_1(x)\frac{v(x)}{C_H}\sim\frac{v(x)}{C_H}\quad \text{as } x\to\infty.$$

Thus, for all sufficiently large x,

$$\frac{v(x)}{2C_H}\le p_2(x)\le\frac{2v(x)}{C_H},$$

which yields, for all sufficiently large x,

$$\mathbb{E}\Big\{\sum_{n=1}^{\infty}\mathbb{I}\{X_n=x\};\ \sum_{n=1}^{\infty}\mathbb{I}\{X_n=x\}>N\Big\}\le\big(1-v(x)/2C_H\big)^N(N+2C_H/v(x)),$$

and hence the required uniform integrability. □

Proof of the uniform integrability of Corollary 6.12. It suffices to prove that, for some $h>0$ and $\widehat{x}\in\mathbb{R}$, the family of random variables

$$\frac{1}{x}\sum_{n=0}^{\infty}\mathbb{I}\{X_n\in(x,x+h]\},\quad x\ge\widehat{x},$$

is uniformly integrable. In turn, by Lemma 6.10, it is sufficient to prove the last result for a Markov chain $\{Y_n\}$ with jumps

$$\eta(x):=\max(\xi(x),\ -s(x)).$$

This Markov chain satisfies all the conditions of Corollary 3.13 for all $\delta \in (0, 2\mu/b - 1)$. By the Markov property,

$$\mathbb{P}\Big\{\sum_{n=1}^{\infty} \mathbb{I}\{Y_n \in (x, x+h]\} > N\Big\} \le \mathbb{P}\{Y_n \in (x, x+h] \text{ for some } n \ge 0\}$$
$$\times \sup_{y \in (x,x+h]} \mathbb{P}^N\{Y_n \in (x, x+h] \text{ for some } n \ge 1 \mid Y_0 = y\}$$
$$\le \sup_{y \in (x,x+h]} \mathbb{P}^N\{Y_n \in (x, x+h] \text{ for some } n \ge 1 \mid Y_0 = y\}.$$

Therefore

$$\mathbb{P}\Big\{\sum_{n=1}^{\infty} \mathbb{I}\{Y_n \in (x, x+h]\} > N\Big\}$$
$$\le \Big(1 - \inf_{y \in (x,x+h]} \mathbb{P}\{Y_n \notin (x, x+h] \text{ for all } n \ge 1 \mid Y_0 = y\}\Big)^N.$$

Let us choose $h > 0$ such that

$$p := \mathbb{P}\{\xi > 3h\} > 0$$

and then $\widehat{x}$ such that

$$\mathbb{P}\{\eta(x) > 2h\} \ge \mathbb{P}\{\xi > 3h\} = p > 0 \quad \text{for all } x > \widehat{x},$$

which is possible owing to the asymptotic homogeneity (6.52). Under such choices of h and $\widehat{x}$, for all $x > \widehat{x}$ and $y \in (x, x+h]$ we have

$$\mathbb{P}\{Y_n \notin (x, x+h] \text{ for all } n \ge 1 \mid Y_0 = y\}$$
$$\ge \mathbb{P}\{\eta(y) > 2h\}\mathbb{P}\{Y_n > x+h \text{ for all } n \ge 1 \mid Y_0 > x+2h\}$$
$$\ge p\mathbb{P}\{Y_n > x+h \text{ for all } n \ge 1 \mid Y_0 > x+2h\}.$$

As follows from Theorem 3.12, for all $z > x + 2h$,

$$\mathbb{P}\{Y_n > x+h \text{ for all } n \ge 1 \mid Y_0 = z\}$$
$$= 1 - \mathbb{P}\{Y_n \le x+h \text{ for some } n \ge 1 \mid Y_0 = z\}$$
$$\ge 1 - \left(\frac{x+h}{x+2h}\right)^{\delta}$$
$$\sim \delta h/x \quad \text{as } x \to \infty,$$

which in its turn implies that, for all sufficiently large x,

$$\inf_{y \in (x,x+h]} \mathbb{P}\{Y_n \notin (x, x+h] \text{ for all } n \ge 1 \mid Y_0 = y\} \ge p\delta h/2x$$

and thus

$$\mathbb{P}\Big\{\sum_{n=1}^{\infty}\mathbb{I}\{Y_n \in (x, x+h]\} > N\Big\} \le (1 - c/x)^N \quad \text{where } c = p\delta h/2 > 0,$$

which implies the required uniform integrability. □

Proofs of the uniform integrability of Corollaries 6.13 and 6.14. These proofs use the same method as the preceding proof. □

6.6 Comments for Chapter 6

The renewal theory for a random walk with positive drift – which is the simplest example of a transient Markov chain (spatially and temporally homogeneous) – has been intensively studied since the 1940s. The integral (elementary) renewal theorem for a random walk with positive jumps and finite mean goes back to Feller [62] and states that $H(0,x] \sim x/\mathbb{E}\xi_1$ as $x \to \infty$. More detailed information is available via the local renewal theorem, which was proved for lattice random variables in [58] and for non-lattice random variables in [18]. In the finite-mean case the local renewal theorem gives the following sharp asymptotics: $H(x,x+h] \to h/\mathbb{E}\xi_1$ as $x \to \infty$, for any fixed $h > 0$. Later, in [19], Blackwell extended the local renewal theorem to the case of independent and identically distributed random variables with positive mean that can take values with both signs, using the important concepts of what were called by Feller ladder heights and ladder epochs. Blackwell's original proof was considered to be quite complicated and a number of attempts were made to give an easier proof. A rather simple proof was given by Feller and Orey [64]; see also [63]. Further studies also considered the behaviour of the remainder in the local renewal theorem; see [136] and the references therein. In the infinite-mean case the asymptotics in Blackwell's theorem was not sharp. In the 1960s and 1970s a local renewal theorem was proved for regularly varying increments of index $\alpha > 1/2$; see Garsia and Lamperti [72] and Erickson [59]. Subsequently there have been various improvements on these results, but the complete answer has been obtained only recently; see Caravenna and Doney [33].

There exist a number of generalisations of the renewal theorem for various stochastic processes. A natural extension is for a non-homogeneous (in time) random walk, that is, a random walk with independent, but not necessarily identically distributed, increments. Probably the first result in this direction was that of Cox and Smith [37], where the local renewal theorem was derived from the local central limit theorem for a non-homogeneous random walk. Further extensions may be found in Smith [142], Williamson [151], and Maejima [118]. Renewal theorems for multidimensional random walks may be found in Doney [49], Nagaev [128], Guibourg and Hervé [75], and the recent paper Berger [12]; see also references therein.

The Markov setting has mostly been considered in the literature for the case of Markov-modulated random walks; see e.g. Kesten [96], Athreya et al. [10], Klüppelberg and Pergamenchtchikov [100], and Shurenkov [141]. In this setting one can normally use the Harris regeneration and split the process into independent cycles. Then the traditional setting of Blackwell's theorem can be used.

For the results cited above, it is essential that the underlying process possesses some independence structure. In the present chapter we considered transient Markov chains where the cycle structure is not available, which makes reduction to Blackwell's theorem impossible. Clearly, in order to observe regular asymptotics for the renewal process, we need to assume regular behaviour of the Markov chain at infinity. In particular, if the drift of X, $m_1(x)$, has a positive limit at infinity, say a, then the local renewal result, $H(x,x+h] \to h/a$, is known only for a Markov chain that is asymptotically homogeneous in space; see Korshunov [105]. This chapter was based on the paper by Denisov et al. [44].

7

Doob's h-Transform: Transition from Recurrent to Transient Chain and Vice Versa

This short chapter is the most conceptual part of the book. Our purpose here is to describe, without superfluous details, a change of measure strategy which allows us to transform a recurrent chain into a transient one, and vice versa. This strategy is motivated by the exponential change of measure technique, which goes back to Cramér [38], where, in the context of large deviations in collective risk theory, it allows one to transform a negatively drifted random walk into one with positive drift.

We now briefly sketch Cramér's approach. For any distribution F with finite γ-exponential moment $\varphi(\gamma) := \int e^{\gamma x} F(dx) < \infty$, we define a new distribution $F^{(\gamma)}$ by $F^{(\gamma)}(dx) = e^{\gamma x} F(dx)/\varphi(\gamma)$, which is the exponential change of measure F with the parameter γ. The remarkable fact is that the exponential change of measure preserves independence; that is, the nth convolution of F, $e^{\gamma x} F^{(*n)}(dx)/\varphi^n(\gamma)$, is equal to the nth convolution of the exponentially transformed F, $(F^{(\gamma)})^{(*n)}$.

The expectation of the transformed distribution $F^{(\gamma)}$ is equal to

$$\frac{1}{\varphi(\gamma)} \int_{\mathbb{R}} x e^{\gamma x} F(dx) = \frac{\varphi'(\gamma)}{\varphi(\gamma)} = \frac{d}{dx} \log \varphi(\gamma).$$

The right-hand side function is increasing in γ because

$$\frac{d^2}{dx^2} \log \varphi(\gamma) = \frac{\varphi''(\gamma)}{\varphi(\gamma)} - \left(\frac{\varphi'(\gamma)}{\varphi(\gamma)}\right)^2$$

represents the variance of $F^{(\gamma)}$ and hence is positive. So, by increasing the value of γ we increase the expectation of $F^{(\gamma)}$.

For a collective risk process with initial risk reserve $x > 0$, the ruin probability is described by the tail probability

$$\mathbb{P}\{M_\infty > x\} = \mathbb{P}\Big\{\sup_{n\geq 0} S_n > x\Big\},$$

where the underlying random walk S_n has a typical jump $\xi = X - c\tau$, where X represents the typical claim size, τ is the typical inter-arrival time, and c is the premium rate. As discussed in Section 1.3,

$$\mathbb{P}\{M_\infty > x\} = \lim_{n\to\infty} \mathbb{P}\{W_n > x\}.$$

Under the net-profit condition $c > \mathbb{E}X/\mathbb{E}\tau$, $\{S_n\}$ has a negative drift; hence the Markov chain W_n has an asymptotically negative drift too. Under the additional assumption that $\varphi(\beta) = \mathbb{E}e^{\beta\xi} = 1$ for some $\beta > 0$, Cramér suggested applying the β-exponential change of measure to a negatively driven random walk S_n, which leads to a random walk with positive drift equal to $\varphi'(\beta)$ and eventually to an exponential estimate for the ruin probability; see Section 1.3. Notice that $e^{\beta x}$ is a harmonic function for the spatially homogeneous process S_n, not for the chain W_n.

As Markov chains are not spatially homogeneous, we cannot apply exactly the same approach to them. Instead, historically it is Doob's transform with a superharmonic function that leads to substochastic transition probabilities. In the context of Lamperti's problem, the most natural substitution for the exponential change of measure is Doob's h-transform with an asymptotically harmonic function. Again, as for convolutions, the remarkable fact is that Doob's h-transform applied to the n-step transition kernel equals the n-step h-transformed transition kernel; see below.

A very important novelty of our approach, in comparison with the classical Doob's h-transform, consists in the fact that we introduce weight functions which are not necessarily harmonic or superharmonic; they are only asymptotically harmonic at infinity. This leads to potentially excessive transition masses, that is, the resulting transformed transition kernel is not necessarily substochastic as it is in the existing literature; see Chung and Walsh [36], Doob [52] and [53], Levin and Peres [116], Popov [134], and Woess [152].

Doob's h-transforms with an asymptotically harmonic function connect in a natural way our previous chapters on the asymptotic behaviour of transient chains with subsequent chapters devoted to recurrent chains.

The main challenge in applying Doob's h-transform to Lamperti chains is to identify such asymptotically harmonic functions under various drift scenarios; see Lemmas 8.6, 9.3, and the proof of Theorem 10.8. Then it can be applied to the proofs of Theorem 8.2, dealing with drifts proportional to $1/x$, of Theorem 9.2 where $xm_1(x) \to \infty$, and of Theorem 10.8, dealing with asymptotically negative drifts.

7.1 Doob's h-Transform for Transition Kernels

7.1.1 General Change of Measure Methodology for Transition Kernels

Let S be a measurable space with a σ-algebra $\mathcal{A}(S)$. Let $P(x,A) : S\times\mathcal{A}(S) \to \mathbb{R}^+$ be a non-negative transition kernel on S, that is, such that it is measurable in x for all fixed A and is a non-negative measure in A for all fixed x. It is not necessarily stochastic.

Let $U(x) > 0$ be a positive measurable function such that

$$\int_S U(y)P(x,dy) < \infty \quad \text{for all } x \in S; \tag{7.1}$$

such a function U is called a *weight function*. This allows us to define a new transition kernel

$$Q(x,A) := \int_A \frac{U(y)}{U(x)} P(x,dy),$$

which is just *Doob's h-transform* for P with weight function U. If U is a harmonic function for P, that is, if

$$U(x) = \int_S U(y)P(x,dy) \quad \text{for all } x \in S$$

then Q is a transition probability kernel.

In order to ensure that the powers of Q are well defined, we need to strengthen the condition (7.1) as follows:

$$c_S := \sup_{x\in S} \int_S \frac{U(y)}{U(x)} P(x,dy) < \infty. \tag{7.2}$$

Then it is legitimate to carry out the following standard calculations

$$\begin{aligned}
&Q^n(x,A)\\
&:= \int_S Q(x,dy_1)\cdots\int_S Q(y_{n-2},dy_{n-1})\int_A Q(y_{n-1},dy_n)\\
&= \int_S \frac{U(y_1)}{U(x)}P(x,dy_1)\cdots\int_S \frac{U(y_{n-1})}{U(y_{n-2})}P(y_{n-2},dy_{n-1})\int_A \frac{U(y_n)}{U(y_{n-1})}P(y_{n-1},dy_n)\\
&= \int_S \frac{U(y_n)}{U(x)}P(x,dy_1)\cdots\int_S P(y_{n-2},dy_{n-1})\int_A P(y_{n-1},dy_n)\\
&= \int_A \frac{U(y_n)}{U(x)}P^n(x,dy_n),
\end{aligned}$$

which shows that Doob's h-transform of the nth power of P, P^n, is equal to the nth power of Doob's h-transform of P, Q^n. Similarly, for any collection of sets $A_1, \ldots, A_n \in \mathcal{A}(S)$,

$$\int_{A_1} Q(x,dy_1)\cdots\int_{A_{n-1}} Q(y_{n-2},dy_{n-1})\int_{A_n} Q(y_{n-1},dy_n)$$
$$= \int_{A_1} P(x,dy_1)\cdots\int_{A_{n-1}} P(y_{n-2},dy_{n-1})\int_{A_n} \frac{U(y_n)}{U(x)}P(y_{n-1},dy_n).$$

Performing the inverse change of measure we get

$$P^n(x,dy) = \frac{U(x)}{U(y)}Q^n(x,dy) \tag{7.3}$$

and

$$\int_{A_1} P(x,dy_1)\cdots\int_{A_{n-1}} P(y_{n-2},dy_{n-1})\int_{A_n} P(y_{n-1},dy_n)$$
$$= \int_{A_1} Q(x,dy_1)\cdots\int_{A_{n-1}} Q(y_{n-2},dy_{n-1})\int_{A_n} \frac{U(x)}{U(y_n)}Q(y_{n-1},dy_n). \tag{7.4}$$

Denote

$$q(x) := -\log Q(x,S).$$

Let us consider the normalised kernel

$$\widehat{P}(x,dy) \;=\; \frac{Q(x,dy)}{Q(x,S)} \;=\; Q(x,dy)e^{q(x)}$$

and let $\{\widehat{X}_n\}$ be a Markov chain with these transition probabilities. Then

$$Q(x,dy) \;=\; \widehat{P}(x,dy)e^{-q(x)}$$

and hence, by (7.3), we arrive at the following basic equalities:

$$P^n(x,dy) = \frac{U(x)}{U(y)}\mathbb{E}_x\{e^{-\sum_{k=0}^{n-1} q(\widehat{X}_k)};\ \widehat{X}_n \in dy\} \tag{7.5}$$

and

$$\int_{A_1} P(x,y_1)\cdots\int_{A_{n-1}} P(y_{n-2},dy_{n-1})P(y_{n-1},dy_n)$$
$$= \frac{U(x)}{U(y_n)}\mathbb{E}_x\{e^{-\sum_{k=0}^{n-1} q(\widehat{X}_k)};\ \widehat{X}_1 \in A_1,\ldots,\widehat{X}_{n-1} \in A_{n-1}, \widehat{X}_n \in dy_n\}. \tag{7.6}$$

Let $\widehat{\xi}(x)$ be the jumps of $\{\widehat{X}_n\}$.

7.1.2 Application to Killed Markov Chain

In this subsection we specify how the above transformation works in the case in which we are mostly interested – the transition kernel corresponding to a Markov chain killed at entering some fixed set. Namely, let $\{X_n\}$ be a Markov chain with transition probabilities $P(\cdot,\cdot)$, let $B \subset S$ be some fixed set, and let $\tau_B := \min\{n \geq 1 : X_n \in B\}$. Consider a substochastic transition kernel

$$P_B(x, A) := P(x, A \setminus B) = \mathbb{P}_x\{X_1 \in A, \ \tau_B > 1\},$$

which is the transition kernel corresponding to $\{X_n\}$ *killed* at entering B.

Given a weight function $U(x) > 0$ for all $x \notin B$, the corresponding change of measure produces a transition kernel Q which may be rewritten as follows:

$$\begin{aligned} Q(x,dy) &:= \frac{U(y)}{U(x)}\mathbb{P}_x\{X_1 \in dy, \tau_B > 1\} \\ &= \frac{U(y)}{U(x)}\mathbb{P}_x\{X_1 \in dy, X_1 \notin B\}. \end{aligned} \tag{7.7}$$

Consequently, performing the inverse change of measure we arrive at the following basic equality:

$$\begin{aligned} \mathbb{P}_x\{X_n \in dy, \tau_B > n\} &= \frac{U(x)}{U(y)} Q^n(x,dy) \\ &= \frac{U(x)}{U(y)}\mathbb{E}_x\{e^{-\sum_{k=0}^{n-1} q(\widehat{X}_k)}; \ \widehat{X}_n \in dy\}, \end{aligned} \tag{7.8}$$

where

$$q(x) := -\log \int_{S\setminus B} \frac{U(y)}{U(x)} P(x,dy) \tag{7.9}$$

and $\{\widehat{X}_n\}$ is a Markov chain with transition probabilities

$$\widehat{P}(x, A) = \frac{Q(x, A)}{Q(x, S)} = \frac{\int_{A\setminus B} U(y)P(x,dy)}{\int_{S\setminus B} U(y)P(x,dy)}. \tag{7.10}$$

In other words, for any positive Borel function $f(y)$,

$$\begin{aligned} \mathbb{E}_x\{f(X_n); \ \tau_B > n\} &= U(x) \int_{S\setminus B} \frac{f(y)}{U(y)}\mathbb{E}_x\{e^{-\sum_{k=0}^{n-1} q(\widehat{X}_k)}; \ \widehat{X}_n \in dy\} \\ &= U(x)\mathbb{E}_x\left\{e^{-\sum_{k=0}^{n-1} q(\widehat{X}_k)} \frac{f(\widehat{X}_n)}{U(\widehat{X}_n)}\right\}. \end{aligned} \tag{7.11}$$

7.2 How to Increase Drift via Change of Measure with Weight Function Close to Harmonic Function

7.2.1 Stochastic Kernel

Let $\{X_n\}$ be a Markov chain on $\mathbb{R}$ with jumps $\xi(x)$. For some increasing function $s(x)$ and decreasing function $r(x) \downarrow 0$ as $x \to \infty$, let

$$\frac{2m_1^{[s(x)]}(x)}{m_2^{[s(x)]}(x)} \sim -r(x), \tag{7.12}$$

$$m_2^{[s(x)]}(x) \to b > 0. \tag{7.13}$$

If we want to increase the drift – say if we need to pass from a recurrent Markov chain to a transient one – then clearly an increasing weight should be applied. So, let $U(x) \geq 0$ be an increasing differentiable function such that, for some $c_U > 0$,

$$\frac{U'(x)}{U(x)} \sim c_U r(x) \quad \text{as } x \to \infty \tag{7.14}$$

with

$$U(x+y) \sim U(x) \quad \text{and} \quad U'(x+y) \sim U'(x) \tag{7.15}$$

as $x \to \infty$ uniformly for all $|y| \leq s(x)$.

We assume that U is close to a harmonic function in the following sense:

$$\mathbb{E}_x U(X_1) = \mathbb{E}U(x + \xi(x)) \sim U(x) \quad \text{as } x \to \infty. \tag{7.16}$$

This condition provides the asymptotic stochasticity of Q, that is, $Q(x, \mathbb{R}) \to 1$ as $x \to \infty$.

Let Q, $\widehat{P}(\cdot,\cdot)$, $\{\widehat{X}_n\}$, and $\widehat{\xi}(x)$ be defined for $P(\cdot,\cdot)$, with weight function U as described in Section 7.1.1.

Lemma 7.1 *Let conditions* (7.12)–(7.16) *hold. Then*

$$\mathbb{E}\{\widehat{\xi}(x);\ |\widehat{\xi}(x)| \leq s(x)\} \sim (c_U - 1/2)br(x), \tag{7.17}$$

$$\mathbb{E}\{(\widehat{\xi}(x))^2;\ |\widehat{\xi}(x)| \leq s(x)\} \to b \tag{7.18}$$

as $x \to \infty$, hence

$$\frac{2\widehat{m}_1^{[s(x)]}(x)}{\widehat{m}_2^{[s(x)]}(x)} \sim (2c_U - 1)r(x).$$

In addition,

$$\mathbb{P}\{\widehat{\xi}(x) < -s(x)\} \le (1+o(1))\mathbb{P}\{\xi(x) < -s(x)\}, \tag{7.19}$$
$$\mathbb{E}\{|\widehat{\xi}(x)|;\ \widehat{\xi}(x) < -s(x)\} \le (1+o(1))\mathbb{E}\{|\xi(x)|;\ \xi(x) < -s(x)\}, \tag{7.20}$$
$$\mathbb{P}\{\widehat{\xi}(x) > s(x)\} \le (1+o(1))\frac{\mathbb{E}\big\{U(x+\xi(x));\ \xi(x) > s(x)\big\}}{U(x)}. \tag{7.21}$$

Proof By the construction of $\{\widehat{X}_n\}$ and the condition (7.16),

$$\mathbb{E}\{\widehat{\xi}(x);\ |\widehat{\xi}(x)| \le s(x)\} = \frac{\mathbb{E}\{U(x+\xi(x))\xi(x);\ |\xi(x)| \le s(x)\}}{\mathbb{E}U(x+\xi(x))}$$
$$\sim \frac{\mathbb{E}\{U(x+\xi(x))\xi(x);\ |\xi(x)| \le s(x)\}}{U(x)}.$$

By Taylor's theorem,

$$\mathbb{E}\{U(x+\xi(x))\xi(x);\ |\xi(x)| \le s(x)\}$$
$$= U(x)\mathbb{E}\{\xi(x);\ |\xi(x)| \le s(x)\} + \mathbb{E}\{U'(x+\theta\xi(x))\xi^2(x);\ |\xi(x)| \le s(x)\},$$

where $\theta = \theta(x,\xi(x)) \in (0,1)$. The first term on the right-hand side is equivalent to $-bU(x)r(x)/2$, as follows from (7.12) and (7.13). By the condition (7.15), $U'(x+\theta\xi(x)) \sim U'(x)$ as $x \to \infty$ uniformly for all $|\xi(x)| \le s(x)$, which implies, as $x \to \infty$,

$$\mathbb{E}\{U'(x+\theta\xi(x))\xi^2(x);\ |\xi(x)| \le s(x)\} \sim U'(x)\mathbb{E}\{\xi^2(x);\ |\xi(x)| \le s(x)\}$$
$$\sim U'(x)b \sim c_U br(x)U(x),$$

owing to the conditions (7.13) and (7.14). Altogether these expressions yield

$$\mathbb{E}\{U(x+\xi(x))\xi(x);\ |\xi(x)| \le s(x)\} \sim (c_U - 1/2)br(x)U(x) \quad \text{as } x \to \infty,$$

and (7.17) follows. The second result, (7.18), follows if we apply (7.13), (7.15), and (7.16) to the right-hand side of

$$\mathbb{E}\{\widehat{\xi}^2(x);\ |\widehat{\xi}(x)| \le s(x)\} = \frac{\mathbb{E}\{U(x+\xi(x))\xi^2(x);\ |\xi(x)| \le s(x)\}}{\mathbb{E}U(x+\xi(x))}.$$

Using (7.16) and recalling that U is increasing, we also get

$$\mathbb{P}\{\widehat{\xi}(x) < -s(x)\} = \frac{\mathbb{E}\{U(x+\xi(x)); \xi(x) < -s(x)\}}{\mathbb{E}U(x+\xi(x))}$$
$$\sim \frac{\mathbb{E}\{U(x+\xi(x)); \xi(x) < -s(x)\}}{U(x)}$$
$$\le \mathbb{P}\{\xi(x) < -s(x)\},$$

and similarly for (7.20). The last assertion, (7.21), follows by (7.16), and the proof is complete. □

7.2.2 Killed Markov Chain

Let $\widehat{x} \in \mathbb{R}^+$ be some level. For an $\{X_n\}$ killed at entering $B := (-\infty, \widehat{x}\,]$, let us perform the change of measure with an increasing weight function U and consider the corresponding kernel Q,

$$Q(x, A) = \frac{\mathbb{E}\{U(x + \xi(x));\ x + \xi(x) \in A \cap (\widehat{x}, \infty)\}}{U(x)}, \tag{7.22}$$

and the embedded Markov chain $\{\widehat{X}_n\}$ with transition probabilities

$$\widehat{P}(x, A) = \frac{\mathbb{E}\{U(x + \xi(x));\ x + \xi(x) \in A \cap (\widehat{x}, \infty)\}}{\mathbb{E}\{U(x + \xi(x));\ x + \xi(x) > \widehat{x}\}}, \tag{7.23}$$

if $\mathbb{P}\{x + \xi(x) > \widehat{x}\} > 0$ and $\widehat{P}(x, A) = \mathbb{I}\{x \in A\}$ otherwise. Let $\widehat{\xi}(x)$ be the jumps of $\{\widehat{X}_n\}$.

The following result is almost immediate from Lemma 7.1.

Lemma 7.2 *Let the conditions* (7.12)–(7.16) *hold for some* $s(x) \le x/2$ *and let*

$$\mathbb{P}\{x + \xi(x) \le \widehat{x}\} \to 0 \quad \textit{as } x \to \infty. \tag{7.24}$$

Then the claims (7.17)–(7.21) *hold.*

If, in addition, $c_U > 1/2$ *and* (7.24) *holds for any* $\widehat{x}$, *then there exists an* $\widehat{x}$ *sufficiently large that*

$$\mathbb{E}\{\widehat{\xi}(x);\ |\widehat{\xi}(x)| \le s(x)\} \ge \frac{c_U - 1/2}{2} br(x) \quad \textit{for all } x \ge \widehat{x}. \tag{7.25}$$

Proof By the condition (7.24),

$$\mathbb{E}\{U(x + \xi(x));\ x + \xi(x) \le \widehat{x}\} \le U(\widehat{x})\mathbb{P}\{x + \xi(x) \le \widehat{x}\} \to 0$$

as $x \to \infty$; hence

$$\mathbb{E}\{U(x + \xi(x));\ x + \xi(x) > \widehat{x}\} \sim \mathbb{E}U(x + \xi(x)) \sim U(x),$$

owing to (7.16). Thus (7.17)–(7.21) follow from Lemma 7.1 since $s(x) \le x/2$. □

7.3 How to Decrease Drift via Change of Measure with Weight Function Close to Harmonic Function

7.3.1 Stochastic Kernel

In this section let $\{X_n\}$ be a Markov chain on $\mathbb{R}$ such that, for some increasing function $s(x)$ and decreasing function $r(x) \to 0$ as $x \to \infty$, we have

$$\frac{2m_1^{[s(x)]}(x)}{m_2^{[s(x)]}(x)} \sim r(x), \tag{7.26}$$

$$m_2^{[s(x)]}(x) \to b > 0. \tag{7.27}$$

If we want to decrease the drift – say if we need to pass from a transient Markov chain to a recurrent one – then clearly a decreasing weight should be applied. So, let $U(x) > 0$ be a decreasing differentiable function such that (7.14) for some $c_U < 0$ and (7.15) hold. As in the previous section, we assume that U is close to a harmonic function in the sense (7.16).

In the same way as Lemma 7.1, the following result follows.

Lemma 7.3 *Let conditions* (7.26), (7.27), *and* (7.14)–(7.16) *hold. Then*

$$\mathbb{E}\{\widehat{\xi}(x);\ |\widehat{\xi}(x)| \le s(x)\} \sim (c_U + 1/2)br(x), \tag{7.28}$$

$$\mathbb{E}\{(\widehat{\xi}(x))^2;\ |\widehat{\xi}(x)| \le s(x)\} \to b \tag{7.29}$$

as $x \to \infty$, hence

$$\frac{2\widehat{m}_1^{[s(x)]}(x)}{\widehat{m}_2^{[s(x)]}(x)} \sim (2c_U + 1)r(x). \tag{7.30}$$

In addition,

$$\mathbb{P}\{\widehat{\xi}(x) > s(x)\} \le (1 + o(1))\mathbb{P}\{\xi(x) > s(x)\}, \tag{7.31}$$

$$\mathbb{P}\{\widehat{\xi}(x) < -s(x)\} \le (1 + o(1))\frac{\mathbb{E}\big\{U(x + \xi(x));\ \xi(x) < -s(x)\big\}}{U(x)}. \tag{7.32}$$

7.3.2 Killed Markov Chain

Let $\widehat{x} \in \mathbb{R}^+$ be some level. For an $\{X_n\}$ killed at entering $B := (-\infty, \widehat{x}]$, let us perform the change of measure with a decreasing weight function U and consider the corresponding kernel Q and the embedded Markov chain $\{\widehat{X}_n\}$.

Then, similarly to Lemma 7.2, we get the following result.

Lemma 7.4 *Let the conditions* (7.26), (7.27), *and* (7.14)–(7.16) *hold. Then the conclusions* (7.28)–(7.32) *follow.*

7.4 Cycle Structure of Markov Chain and Doob's Transform

Let a Markov chain $\{X_n\}$ on $\mathbb{R}$ be recurrent in the sense that, for some $\widehat{x} \in \mathbb{R}$, the set $(-\infty, \widehat{x}\,]$ is recurrent, that is,

$$\mathbb{P}_x\{\tau_{(-\infty,\widehat{x}]} < \infty\} = 1 \quad \text{for all } x > \widehat{x}. \tag{7.33}$$

Let $\{X_n\}$ possess a sigma-finite non-negative invariant measure π, that is, a measure π that solves the equation

$$\pi(A) = \int_{\mathbb{R}} P(x, A)\pi(dx) \quad \text{for all } A \in \mathcal{B}(\mathbb{R});$$

we do not assume that this invariant measure is unique. It follows from (7.33) that

$$\pi(-\infty, \widehat{x}\,] > 0. \tag{7.34}$$

The case where π is finite corresponds to positive recurrence, while the case where π is infinite corresponds to null recurrence.

In addition, assume that

$$\pi(-\infty, \widehat{x}\,] < \infty. \tag{7.35}$$

The conditions (7.34) and (7.35) allow us to construct an *aggregated* Markov chain $\{X_n^*\}$ on $[\widehat{x}, \infty)$ with the following transition probabilities: for $x > \widehat{x}$,

$$P^*(x, A) = \begin{cases} P(x, A) & \text{for } A \subseteq (\widehat{x}, \infty), \\ P(x, (-\infty, \widehat{x}\,]) & \text{for } A = \{\widehat{x}\} \end{cases} \tag{7.36}$$

and

$$P^*(\widehat{x}, A) = \begin{cases} \displaystyle\int_{(-\infty,\widehat{x}\,]} \frac{P(y, A)}{\pi(-\infty, \widehat{x}\,]}\pi(dy) & \text{for } A \subseteq (\widehat{x}, \infty), \\ \displaystyle\int_{(-\infty,\widehat{x}]} \frac{P(y, (-\infty, \widehat{x}\,])}{\pi(-\infty, \widehat{x}\,]}\pi(dy) & \text{for } A = \{\widehat{x}\}. \end{cases} \tag{7.37}$$

Then the measure π^* which aggregates states from $(-\infty, \widehat{x}\,]$ to an $\widehat{x}$, that is, $\pi^*\{\widehat{x}\} = \pi(-\infty, \widehat{x}\,] > 0$ and $\pi^*(A) = \pi(A)$ for all $A \subseteq (\widehat{x}, \infty)$, is an invariant measure for $\{X_n^*\}$. Indeed,

$$\begin{aligned} \int_{[\widehat{x},\infty)} P^*(y, \{\widehat{x}\})\pi^*(dy) &= \pi^*(\widehat{x})P^*(\widehat{x}, \{\widehat{x}\}) + \int_{(\widehat{x},\infty)} P^*(y, \{\widehat{x}\})\pi^*(dy) \\ &= \int_{(-\infty,\widehat{x}]} P(y, (-\infty, \widehat{x}\,])\pi(dy) + \int_{(\widehat{x},\infty)} P(y, (-\infty, \widehat{x}\,])\pi(dy) \\ &= \pi(-\infty, \widehat{x}\,] \;=\; \pi^*\{\widehat{x}\}, \end{aligned}$$

as π is invariant for P, and, for $A \subseteq (\widehat{x}, \infty)$,

$$\begin{aligned}\int_{[\widehat{x},\infty)} P^*(y, A)\pi^*(dy) &= \pi^*(\widehat{x})P^*(\widehat{x}, A) + \int_{(\widehat{x},\infty)} P^*(y, A)\pi^*(dy)\\ &= \int_{(-\infty,\widehat{x}]} P(y, A)\pi(dy) + \int_{(\widehat{x},\infty)} P(y, A)\pi(dy)\\ &= \pi(A) \;=\; \pi^*(A).\end{aligned}$$

We assume that the atom $\widehat{x}$ is non-degenerate, that is,

$$P^*(\widehat{x}, \{\widehat{x}\}) < 1; \tag{7.38}$$

a sufficient condition for this to hold is given next.

Lemma 7.5 *Let the conditions* (7.33) *and* (7.35) *hold true and let*

(i) *either π be a probability measure and*

$$\pi(\widehat{x}, \infty) > 0 \tag{7.39}$$

(ii) *or π be sigma-finite and, for all initial states,*

$$\mathbb{P}\big\{\limsup_{n\to\infty} X_n > \widehat{x}\big\} = 1. \tag{7.40}$$

Then (7.38) *follows.*

Proof (i) Consider a stationary Markov chain $\{X_n\}$ having distribution π for all n. If $P^*(\widehat{x}, \{\widehat{x}\}) = 1$ in (7.37) then

$$\int_{(-\infty,\widehat{x}]} P(y, (\widehat{x}, \infty))\mathbb{P}\{X_0 \in dy\} = 0$$

and hence

$$\begin{aligned}\mathbb{P}\{X_1 > \widehat{x}\} &= \int_{(\widehat{x},\infty)} P(y, (\widehat{x}, \infty))\mathbb{P}\{X_0 \in dy\}\\ &= \mathbb{P}\{X_0 > \widehat{x}, X_1 > \widehat{x}\}.\end{aligned}$$

By induction,

$$\mathbb{P}\{X_n > \widehat{x}\} = \mathbb{P}\{X_0 > \widehat{x}, \ldots, X_n > \widehat{x}\};$$

hence the recurrence of the set $(-\infty, \widehat{x}\,]$ implies the convergence $\mathbb{P}\{X_n > \widehat{x}\} \to 0$ as $n \to \infty$, which contradicts the stationarity of $\{X_n\}$ and (7.39).

(ii) The condition (7.35) allows us to consider a Markov chain $\{X_n\}$ with initial distribution concentrated on $(-\infty, \widehat{x}\,]$,

$$\mathbb{P}\{X_0 \in dy\} = \frac{\pi(dy)}{\pi(-\infty, \widehat{x}\,]}, \quad y \le \widehat{x}.$$

If $P^*(\widehat{x},\{\widehat{x}\}) = 1$ then

$$\begin{aligned}\mathbb{P}\{X_1 \le \widehat{x}\} &= \int_{(-\infty,\widehat{x}\,]} P(y,(-\infty,\widehat{x}\,])\mathbb{P}\{X_0 \in dy\} \\ &= \int_{(-\infty,\widehat{x}\,]} \frac{P(y,(-\infty,\widehat{x}\,])}{\pi(-\infty,\widehat{x}\,]}\pi(dy) \\ &= P^*(\widehat{x},\{\widehat{x}\}) \;=\; 1.\end{aligned}$$

By induction, $\mathbb{P}\{X_n \le \widehat{x}\} = 1$ for all n, which contradicts (7.40). □

So, under the conditions (7.33), (7.35), and (7.38) the aggregated Markov chain $\{X_n^*\}$ on $[\widehat{x},\infty)$ is Harris recurrent (see the start of Chapter 8) with a non-degenerate atom at state $\widehat{x}$ – for a definition of Harris recurrence see [126] – regardless of whether π is finite. Then the following representation for the invariant measure π^* via a cycle structure (generated by the atom $\widehat{x}$) of the Markov chain $\{X_n^*\}$ is well known; see e.g. [126, Theorem 10.4.9]:

$$\begin{aligned}\pi^*(dy) &= \pi^*(\widehat{x})\mathbb{E}_{\widehat{x}} \sum_{n=1}^{\tau^*_{\widehat{x}}-1} \mathbb{I}\{X_n^* \in dy\} \\ &= \pi^*(\widehat{x}) \sum_{n=1}^{\infty} \mathbb{P}_{\widehat{x}}\{X_n^* \in dy;\ \tau^*_{\widehat{x}} > n\}, \quad y > \widehat{x},\end{aligned} \tag{7.41}$$

where $\tau^*_{\widehat{x}} = \min\{n \ge 1 : X_n^* = \widehat{x}\}$. This is equivalent to the following representation for the invariant measure π of $\{X_n\}$:

$$\pi(dy) = \int_B \pi(dz) \sum_{n=1}^{\infty} \mathbb{P}_z\{X_n \in dy;\ \tau_B > n\}, \quad y > \widehat{x}, \tag{7.42}$$

where $B = (-\infty,\widehat{x}\,]$. By the Markov property,

$$\mathbb{P}_z\{X_n \in dy,\ \tau_B > n\} = \int_{\widehat{x}}^{\infty} \mathbb{P}_z\{X_1 \in dx\}\mathbb{P}_x\{X_{n-1} \in dy,\ \tau_B > n-1\}.$$

Therefore, for $y > \widehat{x}$,

$$\begin{aligned}\pi(dy) &= \int_B \pi(dz) \int_{\widehat{x}}^{\infty} \mathbb{P}_z\{X_1 \in dx\} \sum_{n=0}^{\infty} \mathbb{P}_x\{X_n \in dy,\ \tau_B > n\} \\ &= \int_{\widehat{x}}^{\infty} \mu(dx) \sum_{n=0}^{\infty} \mathbb{P}_x\{X_n \in dy,\ \tau_B > n\},\end{aligned}$$

where

$$\begin{aligned}\mu(dx) &:= \int_B \pi(dz)\mathbb{P}_z\{X_1 \in dx\}\\ &= \int_B \pi(dz)P(z,dx)\end{aligned} \tag{7.43}$$

is a measure on $(\widehat{x},\infty)$. Substituting (7.8), we get

$$\pi(dy) = \frac{1}{U(y)}\int_{\widehat{x}}^{\infty} \mu(dx)U(x)\sum_{n=0}^{\infty}\mathbb{E}_x\{e^{-\sum_{k=0}^{n-1}q(\widehat{X}_k)};\ \widehat{X}_n \in dy\}.$$

Consider the chain $\{\widehat{X}_n\}$ with initial distribution

$$\mathbb{P}\{\widehat{X}_0 \in dz\} = \frac{\mu(dz)U(z)}{c^*}, \quad z \in (\widehat{x},\infty), \tag{7.44}$$

where c^* is a normalising constant,

$$\begin{aligned}c^* &:= \int_{\widehat{x}}^{\infty} \mu(dx)U(x)\\ &= \int_B \pi(dz)\int_{\widehat{x}}^{\infty} U(x)P(z,dx).\end{aligned}$$

Then we have

$$\pi(dy) = \frac{\widehat{H}^{(q)}(dy)}{U(y)}c^*,$$

where the weighted renewal measure $\widehat{H}^{(q)}$ for $\{\widehat{X}_n\}$ is defined by

$$\widehat{H}^{(q)}(dy) = \sum_{n=0}^{\infty}\mathbb{E}\{e^{-\sum_{k=0}^{n-1}q(\widehat{X}_k)};\ \widehat{X}_n \in dy\}. \tag{7.45}$$

The constant c^* is finite if

$$\sup_{z\in B}\int_{\widehat{x}}^{\infty} U(x)P(z,dx) < \infty.$$

Provided the condition (7.2) holds, the constant c^* possesses the following upper bound:

$$c^* \le c_S\int_B U(z)\pi(dz), \tag{7.46}$$

which is not greater than $c_S U(\widehat{x})\pi(-\infty,\widehat{x}]$ if the function $U(x)$ is increasing.

The above calculations imply, in particular, that

$$\pi(x_1,x_2] = c^*\int_{x_1}^{x_2}\frac{\widehat{H}^{(q)}(dy)}{U(y)}. \tag{7.47}$$

So, the main method for investigating an invariant measure is to identify an increasing test function $U(x)$ which is sufficiently close to a harmonic function, in the sense that its drift is sufficiently small for large x to imply small values of $q(x)$. We also need to choose $U(x)$ in such a way that the chain $\{\widehat{X}_n\}$ is transient. Then the factorisation result for the renewal function $\widehat{H}^{(q)}$, see Section 4.4, and an integro-local renewal theorem for $\{\widehat{X}_n\}$ allow us to derive asymptotics for the tail distribution of the invariant measure π.

7.5 Last-Visit Decomposition and Doob's Transform

For the pre-stationary distribution of X_n, we follow the last-visit decomposition approach. Let $\widehat{x} \in \mathbb{R}$ and set $B := (-\infty, \widehat{x}]$. Regardless of the recurrence or transience of $\{X_n\}$, splitting the set of all trajectories of $\{X_n\}$ by the last visit to B, we get, for $y > \widehat{x}$,

$$\begin{aligned}\mathbb{P}\{X_n \in dy\} &= \sum_{j=1}^{n} \mathbb{P}\{X_{n-j} \in B, X_{n-j+1}, \ldots, X_{n-1} \notin B, X_n \in dy\}\\ &= \sum_{j=1}^{n} \int_B \mathbb{P}\{X_{n-j} \in dz\} \int_{\widehat{x}}^{\infty} P(z,du)\mathbb{P}_u\{X_{j-1} \in dy, \tau_B > j-1\}.\end{aligned}$$

Substituting (7.8), we obtain the following equality:

$$\begin{aligned}&\mathbb{P}\{X_n \in dy\}\\ &= \sum_{j=1}^{n} \int_B \mathbb{P}\{X_{n-j} \in dz\} \int_{\widehat{x}}^{\infty} P(z,du)\frac{U(u)}{U(y)}\mathbb{E}_u\Big\{e^{-\sum_{k=0}^{j-2} q(\widehat{X}_k)};\ \widehat{X}_{j-1} \in dy\Big\},\end{aligned} \tag{7.48}$$

where $q(x)$ and $\{\widehat{X}_n\}$ are defined in (7.9) and (7.10) respectively. Equivalently, for all $x > \widehat{x}$ and $h > 0$,

$$\begin{aligned}&\mathbb{P}\{X_n \in (x,x+h]\}\\ &= \sum_{j=1}^{n} \int_B \mathbb{P}\{X_{n-j} \in dz\} \int_{\widehat{x}}^{\infty} P(z,du)U(u)\mathbb{E}_u\bigg\{\frac{e^{-\sum_{k=0}^{j-2} q(\widehat{X}_k)}}{U(\widehat{X}_{j-1})};\ \widehat{X}_{j-1} \in (x,x+h]\bigg\}.\end{aligned} \tag{7.49}$$

The last representation allows us to study the tail distribution of a positive recurrent chain $\{X_n\}$ by considering a suitable increasing test function $U(x)$ that makes the chain $\{\widehat{X}_n\}$ transient. Then the factorisation result for the renewal function $H^{(q)}$ with weights, see Section 4.4, an integro-local renewal theorem for $\{\widehat{X}_n\}$, and the convergence in total variation of X_n to π allow us to derive asymptotics for the tail distribution of X_n.

7.6 Comments for Chapter 7

Doob introduced this kind of transform in [52] for Brownian motion, with a superharmonic function U for which the resulting transition kernel is substochastic; this approach was further developed in [53].

In [152, Section 7.6], Woess used Doob's h-processes to exhibit two general criteria that are useful for recognising the minimal harmonic functions.

A more systematic exposition of Doob's h-transform as generated by a superharmonic function can be found in Chung and Walsh [36, Chapter 11].

Doob's h-transform as based on harmonic functions was briefly discussed by Levin and Peres [116, Section 17.6] in the context of hitting absorbing states.

In the context of simple random walks in one and two dimensions, Doob's h-transform was used by Popov in [134, Section 4.1], where it led to a Lamperti-type Markov chain with drift of order $1/x$; see Section 1.4.3.

Doob's h-transform with a harmonic function applied to a recurrent random walk killed at entering $(-\infty, 0)$ appears in Bertoin and Doney [13] where they consider the construction of a recurrent random walk conditioned to stay non-negative.

8

Tail Analysis for Recurrent Markov Chains with Drift Proportional to $1/x$

In this chapter we consider a recurrent Markov chain $\{X_n\}$ possessing an invariant measure which is either probabilistic, in the case of positive recurrence, or σ-finite, in the case of null recurrence. We denote this measure by π.

If we consider an irreducible aperiodic Markov chain on $\mathbb{Z}$ then the existence of a probabilistic invariant measure is equivalent to the finiteness of $\mathbb{E}_0\tau_0$, where $\tau_0 := \min\{n \ge 1 : X_n = 0\}$. The case of null recurrence corresponds to an a.s. finite τ_0 with infinite mean, $\mathbb{E}\tau_0 = \infty$. For the state space $\mathbb{R}$, a standard notion of recurrence is that of Harris recurrence; see [126] for the related definitions. Harris recurrence guarantees that an invariant measure is unique up to a constant multiplier.

We consider the case where π has right unbounded support, that is, $\pi(x, \infty) > 0$ for all x. Our main aim is to describe the asymptotic behaviour of its tail, $\pi(x, \infty)$, for a class of Markov chains with asymptotically zero drift.

Here we consider a Markov chain $\{X_n\}$ such that the first two truncated moments of jumps satisfy the following conditions:

$$m_2^{[s(x)]}(x) \to b > 0 \quad \text{and} \quad m_1^{[s(x)]}(x)x \to -\mu \in \mathbb{R} \qquad \text{as } x \to \infty,$$

where the function $s(x) = o(x)$ is increasing and $\mu > -b/2$. In this case the tail of π typically decays as a regularly varying function with index $-2\mu/b + 1$; see Corollary 8.4 below. We have already observed this effect for chains with jumps ± 1 and 0 in Example 1.31, for the positive recurrent case $\mu > b/2$.

As follows from Example 1.44, a stationary density of a diffusion with the same drift and diffusion coefficients is asymptotically equivalent to $c/x^{2\mu/b}$ as $x \to \infty$.

8.1 Markov Chains with Asymptotically Zero Drift: Heavy-Tailedness of Invariant Measure

We start with a result which states that a typical stationary Markov chain with asymptotically zero drift generates a heavy-tailed invariant distribution that is very different from the case of a Markov chains with asymptotically negative drift that is bounded away from zero.

Theorem 8.1 *Let a Markov chain $\{X_n\}$ on $\mathbb{R}$ have asymptotically zero drift, i.e. $m_1(x) \to 0$ as $x \to \infty$ and, in addition,*

$$\liminf_{x\to\infty} \mathbb{E}\{\xi^2(x);\ \xi(x) > 0\} > 0. \tag{8.1}$$

Then any right unbounded invariant distribution π of $\{X_n\}$ is heavy-tailed, that is,

$$\int e^{\lambda y}\pi(dy) = \infty \quad \text{for all } \lambda > 0.$$

Proof Assume that on the contrary the invariant distribution π is right unbounded with finite exponential moment of some order $\lambda > 0$. Let $\{X_n\}$ be stationary with distribution π. Then, for any x_0,

$$\mathbb{E}(V(X_1) - V(X_0)) = 0, \tag{8.2}$$

where $V(x) := \max(e^{\lambda x}, e^{\lambda x_0})$. Since

$$\mathbb{E}(V(X_1) - V(X_0)) \ge \mathbb{E}\{V(X_1) - V(X_0);\ X_0 > x_0\}$$

and since X_0 has right unbounded support, it would be a contradiction with (8.2) if we could prove that, for some x_0,

$$v(x) := \mathbb{E}\{V(X_1) - V(X_0) \mid X_0 = x\} > 0 \quad \text{for all } x > x_0. \tag{8.3}$$

For all $x > x_0$,

$$v(x) \ge \mathbb{E}e^{\lambda(x+\xi(x))} - e^{\lambda x} = e^{\lambda x}(\mathbb{E}e^{\lambda\xi(x)} - 1).$$

Since $e^y \ge 1 + y$ for all y and $e^y \ge 1 + y + y^2/2$ for all $y > 0$,

$$\mathbb{E}e^{\lambda\xi(x)} - 1 \ge \lambda m_1(x) + \frac{\lambda^2}{2}\mathbb{E}\{\xi^2(x);\ \xi(x) > 0\}.$$

Since $\lambda m_1(x) \to 0$ as $x \to \infty$ and we have the condition (8.1), there exists a sufficiently large x_0 that the sum on the right-hand side of the last inequality is positive for all $x > x_0$, which proves (8.3) and hence the claim of the theorem. □

Let us show by an example that the condition (8.1), which expresses a kind of non-degeneracy of jumps, is essential for the theorem to hold. Consider the skip-free Markov chain $\{X_n\}$ on $\mathbb{Z}^+$ described in Section 1.4, that is, $\xi(x)$ takes the values -1, 1, and 0 only, with probabilities $p_-(x)$, $p_+(x)$, and $p_0(x)$ respectively; $p_-(0) = 0$. The invariant probabilities $\pi(x)$, $x \in \mathbb{Z}^+$, are computed in (1.7),

$$\pi(x) = \pi(0) \prod_{k=1}^{x} \frac{p_+(k-1)}{p_-(k)}.$$

Consider the case where $p_+(x) := 1/(2(x+1))$ and $p_-(x) := 1/(x+1)$. In this case the drift is asymptotically zero but the stationary probabilities are asymptotically equivalent to $cx/2^x$, so the invariant distribution is light-tailed. Clearly, here the condition (8.1) fails.

8.2 Stationary Measure of Recurrent Chains: Power-Like Asymptotics

This section is devoted to the precise asymptotic behaviour of the invariant measure in the case where the drift behaves asymptotically like c/x.

As discussed in Sections 1.4 and 1.5.1, there are two types of Markov chains for which the invariant measure is explicitly calculable. Both are related to skip-free processes, either on a lattice $\mathbb{Z}^+$ or on a continuous state space $\mathbb{R}^+$.

The first type, for which the stationary distribution is explicitly known, is a Markov chain on $\mathbb{Z}^+$ with $\xi(x)$ taking values -1, 1, and 0 only, with probabilities $p_-(x)$, $p_+(x)$, and $p_0(x)$ respectively, and for which $p_-(0) = 0$; see Section 1.4. The second type is that of diffusion processes on $\mathbb{R}^+$ (slotted in time if we wanted just a Markov chain); see Section 1.5. For both types we observe a power tail behaviour of the invariant probabilities in the case where the drift is asymptotically proportional to $-\mu/x$ as $x \to \infty$.

In this chapter we consider a recurrent Markov chain $\{X_n\}$ on $\mathbb{R}$ whose jumps are such that

$$m_2^{[s(x)]}(x) \to b > 0 \text{ and } m_1^{[s(x)]}(x)x \to -\mu \in \mathbb{R} \quad \text{as } x \to \infty, \tag{8.4}$$

where the function $s(x) = o(x)$ is increasing and $\mu > -b/2$.

- The case $\mu \in (-b/2, b/2)$ usually corresponds to the null recurrence of $\{X_n\}$; see Corollary 2.16.
- The case $\mu > b/2$ corresponds to positive recurrence; see Corollary 2.5.

- In the case $\mu = b/2$ either null or positive recurrence can happen; see Corollaries 2.6 and 2.17.

In addition, we assume that

$$\frac{2m_1^{[s(x)]}(x)}{m_2^{[s(x)]}(x)} = -r(x) + o(p(x)) \quad \text{as } x \to \infty \tag{8.5}$$

for some monotone function $r(x) \to 0$ satisfying $r(x)x \to 2\mu/b > -1$ as $x \to \infty$ and some decreasing function $p(x) \geq 0$ that is integrable at infinity. Since $p(x)$ is decreasing and integrable, $p(x)x \to 0$ as $x \to \infty$. We also assume that

$$r'(x) = O(1/x^2) \quad \text{and} \quad p'(x) = O(1/x^2). \tag{8.6}$$

As follows from Lemma 2.30, the second relation can always be satisfied by choosing a more slowly decreasing integrable function $p(x)$.

Given (8.4), an equivalent way to state the assumption (8.5) is as follows:

$$m_1^{[s(x)]}(x) + \frac{m_2^{[s(x)]}(x)}{2} r(x) = o(p(x)) \quad \text{as } x \to \infty. \tag{8.7}$$

Define a monotone function

$$R(x) := \int_0^x r(y)dy, \quad x > 0, \tag{8.8}$$

with $R(x) = 0$ for $x \leq 0$. Since $xr(x) \to 2\mu/b > -1$,

$$\frac{R(x)}{\log x} \to \frac{2\mu}{b} > -1 \quad \text{as } x \to \infty.$$

Define the following increasing function, which plays the most important rôle in our analysis of recurrent Markov chains: $U(x) = 0$ for $x \leq 0$ and, for $x > 0$,

$$U(x) := \int_0^x e^{R(y)}dy \to \infty \quad \text{as } x \to \infty, \tag{8.9}$$

again owing to $2\mu/b > -1$; in what follows we show that the function $U(x)$ is very close to being a harmonic function for large values of x. Note that $U(x)$ solves the equation $U'' - rU' = 0$ for $x > 0$.

According to our assumptions,

$$r(x) = \frac{2\mu}{b}\frac{1}{x} + \frac{\varepsilon(x)}{x},$$

where $\varepsilon(x) \to 0$ as $x \to \infty$. In view of the representation theorem for slowly varying functions, see e.g. [17, Theorem 1.3.1], there exists a function $\ell(x)$, slowly varying at infinity, which is such that

$$e^{R(x)} = x^{\rho-1}\ell(x) \quad \text{and} \quad U(x) \sim x^{\rho}\ell(x)/\rho \quad \text{where } \rho = 2\mu/b + 1 > 0.$$

The main result in this section is the following theorem, which provides exact asymptotics for the stationary measure of the recurrent Markov chain with asymptotically zero drift described above.

Theorem 8.2 *Let* $\{X_n\}$ *be a recurrent Markov chain and let* $\pi(\cdot)$ *be its stationary measure. Let* $\pi(-\infty, x] < \infty$ *for all* x *and, for any initial state, let*

$$\mathbb{P}\Big\{\limsup_{n\to\infty} X_n = \infty\Big\} = 1. \tag{8.10}$$

Let the first two truncated moments of jumps satisfy the conditions (8.4) *and* (8.5), *where* $r(x)$ *and* $p(x)$ *satisfy the regularity condition* (8.6). *Assume that the following integrability conditions hold:*

$$\sup_{x\in\mathbb{R}} \frac{\mathbb{E}U(\xi(x))}{1+U(x)} < \infty \tag{8.11}$$

and, as $x \to \infty$,

$$\mathbb{P}\{|\xi(x)| > s(x)\} = o(p(x)/x), \tag{8.12}$$

$$\mathbb{E}\big\{|\xi(x)|^3;\ |\xi(x)| \le s(x)\big\} = o(x^2 p(x)). \tag{8.13}$$

In addition, let

$$\mathbb{E}\big\{U(\xi(x));\ \xi(x) > s(x)\big\} = o(p(x)e^{R(x)}). \tag{8.14}$$

Then, for some $c > 0$,

$$\pi(x_1, x_2] \sim c \int_{x_1}^{x_2} \frac{y}{U(y)} dy$$

as $x_1, x_2 \to \infty$ *in such a way that* $\liminf x_2/x_1 > 1$.

It follows from the condition (8.10) that π has right unbounded support, that is, $\pi(x, \infty) > 0$ for all x.

Regarding applications, in Theorem 11.2 we apply the last result above to derive asymptotics of the invariant measure for a reflected random walk with zero drift, at the end of Section 11.3.3 for a branching process with migration, and in Theorem 11.17 for a stochastic difference equation.

Corollary 8.3 *If* $2\mu > b$, $\{X_n\}$ *is positive recurrent, and the conditions of Theorem 8.2 hold then*

$$\pi(x, \infty) \sim \frac{c}{\rho - 2} \frac{x^2}{U(x)} \quad \text{as } x \to \infty.$$

If $2\mu \in (-b, b)$, $\{X_n\}$ is null recurrent, and the conditions of Theorem 8.2 hold then

$$\pi(-\infty, x) \sim \frac{c}{2-\rho}\frac{x^2}{U(x)} \quad \text{as } x \to \infty.$$

Corollary 8.4 *In addition, let $r(x) = 2\mu/bx$. If $2\mu > b$ and $\{X_n\}$ is positive recurrent then*

$$\pi(x, \infty) \sim \frac{c\rho}{\rho-2}\frac{1}{x^{2\mu/b-1}} \quad \text{as } x \to \infty.$$

If $2\mu \in (-b, b)$ and $\{X_n\}$ is null recurrent then

$$\pi(-\infty, x) \sim \frac{c\rho}{2-\rho}x^{1-2\mu/b} \quad \text{as } x \to \infty.$$

In the case $2\mu = b$, we have the following result.

Corollary 8.5 *In addition, for some $m \ge 1$ and $\gamma \ne 0$, let*

$$r(x) = \frac{1}{x} + \frac{1}{x\log x} + \cdots + \frac{1}{x\log x \cdots \log_{(m-1)} x} + \frac{1+\gamma}{x\log x \cdots \log_{(m)} x}.$$

If $\gamma > 0$ and $\{X_n\}$ is positive recurrent then

$$\pi(x, \infty) \sim \frac{2c}{\gamma}\frac{1}{\log_{(m)}^{\gamma} x} \quad \text{as } x \to \infty.$$

If $\gamma < 0$ and $\{X_n\}$ is null recurrent then

$$\pi(-\infty, x) \sim \frac{2c}{-\gamma}\log_{(m)}^{-\gamma} x \quad \text{as } x \to \infty.$$

Before proving Theorem 8.2 let us formulate and prove some auxiliary results. First we construct the necessary Lyapunov function. Consider the function $r_p(x) := r(x) - p(x)$ and define $R_p(x) = U_p(x) = 0$ for $x \le 0$ and

$$R_p(x) := \int_0^x r_p(y)dy, \quad U_p(x) := \int_0^x e^{R_p(y)}dy \quad \text{for } x > 0. \tag{8.15}$$

We have $r_p(x) \le r(x)$, $R_p(x) \le R(x)$, and $U_p(x) \le U(x)$ for $x \ge 0$. Since

$$C_p := \int_0^\infty p(y)dy \quad \text{is finite,}$$

we have

$$R_p(x) = R(x) - C_p + o(1) \quad \text{as } x \to \infty. \tag{8.16}$$

Therefore

$$U_p(x) \sim e^{-C_p}U(x) \to \infty \quad \text{as } x \to \infty, \tag{8.17}$$

because $U(x) \to \infty$. Further, since $xr_p(x) = xr(x) - xp(x) \to 2\mu/b$,

$$\frac{U_p'(x)}{(xe^{R_p(x)})'} = \frac{e^{R_p(x)}}{(1+xr_p(x))e^{R_p(x)}} \to \frac{b}{2\mu+b} \quad \text{as } x \to \infty.$$

Then L'Hôpital's rule yields

$$U_p(x) \sim \frac{b}{2\mu+b} xe^{R_p(x)} \sim \frac{be^{-C_p}}{2\mu+b} xe^{R(x)} \quad \text{as } x \to \infty. \tag{8.18}$$

In what follows we need to know the asymptotic behaviour of the drift of $U_p(X_n)$.

Lemma 8.6 *Assume that* (8.4)–(8.6) *and* (8.12)–(8.14) *hold. Then*

$$\begin{aligned} \mathbb{E}U_p(x+\xi(x)) - U_p(x) &\sim -\frac{b}{2}p(x)e^{R_p(x)} \\ &\sim -\frac{2\mu+b}{2}\frac{p(x)}{x}U_p(x) \quad \text{as } x \to \infty, \end{aligned} \tag{8.19}$$

where the last equivalence is due to (8.18).

Proof We start with the following decomposition:

$$\begin{aligned} \mathbb{E}U_p(x+\xi(x)) - U_p(x) = {} & \mathbb{E}\{U_p(x+\xi(x)) - U_p(x);\ \xi(x) < -s(x)\} \\ & + \mathbb{E}\{U_p(x+\xi(x)) - U_p(x);\ |\xi(x)| \le s(x)\} \\ & + \mathbb{E}\{U_p(x+\xi(x)) - U_p(x);\ \xi(x) > s(x)\}. \end{aligned} \tag{8.20}$$

Here the first term on the right-hand side is negative and may be bounded below as follows:

$$\begin{aligned} \mathbb{E}\{U_p(x+\xi(x)) - U_p(x);\ \xi(x) < -s(x)\} &\ge -U_p(x)\mathbb{P}\{\xi(x) < -s(x)\} \\ &= o(p(x)/x)U_p(x) \\ &= o(p(x)e^{R_p(x)}), \end{aligned} \tag{8.21}$$

by the condition (8.12) and the equivalence (8.18). Furthermore, the third term on the right-hand side of (8.20) is positive and may be bounded in the following way:

$$\begin{aligned} &\mathbb{E}\{U_p(x+\xi(x)) - U_p(x);\ \xi(x) > s(x)\} \\ &\quad \le \mathbb{E}\{U_p(x+\xi(x));\ \xi(x) > s(x)\} \\ &\quad \le \mathbb{E}\{U_p(2x) + U_p(2\xi(x));\ \xi(x) > s(x)\} \\ &\quad \le c\big(U_p(x)\mathbb{P}\{\xi(x) > s(x)\} + \mathbb{E}\{U_p(\xi(x));\ \xi(x) > s(x)\}\big), \end{aligned}$$

owing to the regular variation of U_p at infinity. Hence

$$\mathbb{E}\{U_p(x+\xi(x)) - U_p(x);\ \xi(x) > s(x)\} = o(p(x)e^{R_p(x)}), \tag{8.22}$$

owing to the conditions (8.12) and (8.14). To estimate the second term on the right-hand side of (8.20), we make use of a Taylor expansion:

$$\begin{aligned}
&\mathbb{E}\{U_p(x+\xi(x)) - U_p(x);\ |\xi(x)| \le s(x)\} \\
&\quad = U_p'(x)\mathbb{E}\{\xi(x); |\xi(x)| \le s(x)\} + \frac{1}{2}U_p''(x)\mathbb{E}\{\xi^2(x); |\xi(x)| \le s(x)\} \\
&\qquad + \frac{1}{6}\mathbb{E}\big\{U_p'''(x+\theta\xi(x))\xi^3(x); |\xi(x)| \le s(x)\big\},
\end{aligned} \tag{8.23}$$

where $0 \le \theta = \theta(x,\xi(x)) \le 1$. By the construction of U_p we have

$$U_p'(x) = e^{R_p(x)} \quad \text{and} \quad U_p''(x) = r_p(x)e^{R_p(x)} = (r(x)-p(x))e^{R_p(x)}. \tag{8.24}$$

Then it follows that

$$\begin{aligned}
&U_p'(x)m_1^{[s(x)]}(x) + \frac{1}{2}U_p''(x)m_2^{[s(x)]}(x) \\
&\quad = e^{R_p(x)}\left(m_1^{[s(x)]}(x) + (r(x)-p(x))\frac{m_2^{[s(x)]}(x)}{2}\right) \\
&\quad = \frac{m_2^{[s(x)]}(x)}{2}e^{R_p(x)}\left(\frac{2m_1^{[s(x)]}(x)}{m_2^{[s(x)]}(x)} + r(x) - p(x)\right) \\
&\quad = -\frac{m_2^{[s(x)]}(x)}{2}e^{R_p(x)}p(x)(1+o(1)) \\
&\quad \sim -\frac{b}{2}e^{R_p(x)}p(x),
\end{aligned} \tag{8.25}$$

by the condition (8.5).

Finally, let us estimate the last term in (8.23). Notice that, by the condition (8.6) on the derivatives of $r(x)$ and $p(x)$,

$$U_p'''(x) = \big(r'(x) - p'(x) + (r(x)-p(x))^2\big)e^{R_p(x)} = O(1/x^2)e^{R_p(x)},$$

and hence

$$\begin{aligned}
&\big|\mathbb{E}\big\{U_p'''(x+\theta\xi(x))\xi^3(x); |\xi(x)| \le s(x)\big\}\big| \\
&\quad \le \frac{c_1}{x^2}\mathbb{E}\big\{|\xi^3(x)|;\ |\xi(x)| \le s(x)\big\}e^{R_p(x)},
\end{aligned}$$

because $s(x) = o(x)$ and the function $e^{R_p(x)}$ is regularly varying at infinity. Then, in view of (8.13),

$$\big|\mathbb{E}\big\{U_p'''(x+\theta\xi(x))\xi^3(x);\ |\xi(x)| \le s(x)\big\}\big| = o(p(x)e^{R_p(x)}). \tag{8.26}$$

Therefore it follows from (8.23), (8.25), and (8.26) that

$$\mathbb{E}\{U_p(x+\xi(x)) - U_p(x);\ |\xi(x)| \le s(x)\} = -\frac{b}{2}p(x)e^{R_p(x)} + o(p(x)e^{R_p(x)}). \tag{8.27}$$

Substituting (8.21), (8.22), and (8.27) into (8.20), we finally get the desired expression for $\mathbb{E}U_p(x+\xi(x)) - U_p(x)$. This completes the proof of the lemma. □

Fix an $\widehat{x} > 0$. Define a transition kernel Q on $S = (\widehat{x}, \infty)$ via the following change of measure:

$$Q(x,dy) := \frac{U_p(y)}{U_p(x)} P(x,dy), \quad x,\ y > \widehat{x}.$$

Since

$$\frac{\mathbb{E}U(x+\xi(x))}{1+U(x)} \le \frac{U(2x)}{1+U(x)} + \frac{\mathbb{E}U(2\xi(x))}{1+U(x)}$$

and the function U is regularly varying at infinity, the condition (8.11) implies that

$$\sup_{x\in\mathbb{R}} \frac{\mathbb{E}U(x+\xi(x))}{1+U(x)} < \infty. \tag{8.28}$$

Then it follows that the kernel Q satisfies the condition (7.2), which allows us to apply the machinery developed in Chapter 7. We have

$$Q(x,\mathbb{R}) = \frac{\mathbb{E}\{U_p(x+\xi(x));\ x+\xi(x) > \widehat{x}\}}{U_p(x)}. \tag{8.29}$$

Lemma 8.6 yields the following result.

Corollary 8.7 *Under the conditions of Lemma 8.6, there exists an $\widehat{x}$ such that*

$$-(2\mu+b)\frac{p(x)}{x}U_p(x) \le \mathbb{E}U_p(x+\xi(x)) - U_p(x) \le 0 \quad \textit{for all } x > \widehat{x}.$$

Everywhere, in what follows, $\widehat{x}$ is any level guaranteed by Corollary 8.7, $B = (-\infty, \widehat{x}]$ and $\tau_B := \min\{n \ge 1 : X_n \in B\}$. Then the definition of the transition kernel Q may be rewritten as follows:

$$\begin{aligned} Q(x,dy) &= \frac{U_p(y)}{U_p(x)}\mathbb{P}_x\{X_1 \in dy, \tau_B > 1\} \\ &= \frac{U_p(y)}{U_p(x)}\mathbb{P}_x\{X_1 \in dy, X_1 > \widehat{x}\}. \end{aligned} \tag{8.30}$$

We have from the upper bound in Corollary 8.7 that

$$Q(x,\mathbb{R}) = \frac{\mathbb{E}\{U_p(x+\xi(x)); \tau_B > 1\}}{U_p(x)} \le \frac{\mathbb{E}U_p(x+\xi(x))}{U_p(x)} \le 1 \quad \text{for all } x > \widehat{x}.$$

In other words, Q restricted to $(\widehat{x},\infty)$ is a substochastic kernel. It follows from (8.12) that

$$\begin{aligned}\mathbb{E}\{U_p(x+\xi(x));\tau_B=1\} &= \mathbb{E}_x\{U_p(X_1);\ X_1\le\widehat{x}\}\\ &\le U_p(\widehat{x})\mathbb{P}\{x+\xi(x)\le\widehat{x}\}\\ &= o(p(x)/x).\end{aligned}\tag{8.31}$$

Combining this with the lower bound in Corollary 8.7 we obtain that

$$q(x)\ :=\ -\log Q(x,\mathbb{R})=O(p(x)/x).\tag{8.32}$$

Let us consider the following normalised kernel:

$$\widehat{P}(x,dy)\ :=\ \frac{Q(x,dy)}{Q(x,\mathbb{R})}$$

and let $\{\widehat{X}_n\}$ be a Markov chain with this transition probability; let $\widehat{\xi}(x)$ be its jump from the state x. Consequently, by (7.8),

$$\mathbb{P}_x\{X_n\in dy,\tau_B>n\}=\frac{U_p(x)}{U_p(y)}\mathbb{E}_x\big\{e^{-\sum_{k=0}^{n-1}q(\widehat{X}_k)};\ \widehat{X}_n\in dy\big\}.\tag{8.33}$$

Lemma 8.8 *Under the conditions of Lemma 8.6, as $x\to\infty$,*

$$\mathbb{E}\{\widehat{\xi}(x);\ |\widehat{\xi}(x)|\le s(x)\}\ \sim\ \frac{\mu+b}{x},\tag{8.34}$$

$$\mathbb{E}\{(\widehat{\xi}(x))^2;\ |\widehat{\xi}(x)|\le s(x)\}\to b,\tag{8.35}$$

$$\mathbb{P}\{|\widehat{\xi}(x)|>s(x)\}\ =\ o(p(x)/x),\tag{8.36}$$

$$\mathbb{E}\{|\widehat{\xi}(x)|;\ \widehat{\xi}(x)<-s(x)\}\ =\ o(p(x)),\tag{8.37}$$

for some decreasing function $p(x)$ that is integrable at infinity. Moreover, there exists an $\widehat{x}$ sufficiently large that

$$\mathbb{E}\{\widehat{\xi}(x);\ \widehat{\xi}(x)\le s(x)\}\ge\frac{\mu+b}{2x}\quad \textit{for all } x\ge\widehat{x}.\tag{8.38}$$

Proof It follows from (8.18) that

$$\frac{U_p'(x)}{U_p(x)}=\frac{e^{R_p(x)}}{U_p(x)}\sim\frac{2\mu+b}{bx}\quad\text{as } x\to\infty.$$

So the function U_p satisfies the condition (7.14) with $r(x)=1/x$ and $c_U=1+2\mu/b$. Also, U_p satisfies (7.15) for any $s(x)=o(x)$ because

$$\frac{U_p'(x+y)}{U_p'(x)}=\frac{e^{R_p(x+y)}}{e^{R_p(x)}}\sim e^{R(x+y)-R(x)}=e^{\int_x^{x+y}r(z)dz}=e^{O(s(x)/x)}=e^{o(1)}$$

as $x \to \infty$ uniformly for all $|y| \le s(x)$, and, by (8.18),

$$\frac{U_p(x+y)}{U_p(x)} \sim \frac{x+y}{x}\frac{e^{R(x+y)}}{e^{R(x)}} \sim e^{R(x+y)-R(x)} \to 1.$$

The function U_p satisfies (7.16) by Lemma 8.6. Finally, the condition (7.24) follows from (8.36). Thus all the conditions of Lemma 7.2 are met and (8.34)–(8.38) follow. □

Therefore the chain $\{\widehat{X}_n\}$ satisfies the conditions (4.45)–(4.47) of Theorem 4.8 with $\widehat{\mu} = \mu + b$ and $\widehat{b} = b$, so that $\widehat{\mu} > \widehat{b}/2$. Further, the lower bound (8.38) for the drift of $\{\widehat{X}_n\}$ allows us to apply Theorem 4.2 to $\{\widehat{X}_n\}$ and to conclude that, for $\widehat{T}(t) = \min\{n \ge 1 : \widehat{X}_n > t\}$,

$$\mathbb{E}_y \widehat{T}(t) = \mathbb{E}_y \widehat{L}(\widehat{x}, \widehat{T}(t)) < \infty \quad \text{for all } t > y;$$

hence, for any initial state $\widehat{X}_0 = y$,

$$\mathbb{P}_y\left\{\limsup_{n\to\infty} \widehat{X}_n = \infty\right\} = 1.$$

In turn, it follows from Theorem 2.21 that $\widehat{X}_n \to \infty$ with probability 1.

So, Theorem 4.8 is applicable to $\{\widehat{X}_n\}$, which implies the weak convergence of $(\widehat{X}_n)^2/n$ to a Γ-distribution with mean $2\mu + 3b = (2+\rho)b$ and variance $(2\mu+3b)2b = (2+\rho)2b^2$, where $\rho = 1 + 2\mu/b$, that is, a Γ-distribution with probability density function

$$\gamma(u) = \frac{1}{(2b)^{1+\rho/2}\Gamma(1+\rho/2)} u^{\rho/2} e^{-u/2b}. \tag{8.39}$$

Furthermore, by Theorem 4.3, there exists a $c < \infty$ such that

$$\widehat{H}_y(x) := \sum_{n=0}^{\infty} \mathbb{P}_y\{\widehat{X}_n \le x\} \le c(1+x^2) \quad \text{for all } x,\ y. \tag{8.40}$$

Having proven this estimate proven we now deduce the following result.

Lemma 8.9 *Under the conditions of Lemma 8.6,*

$$h(z) := \lim_{n\to\infty} \mathbb{E}_z e^{-\sum_{k=0}^{n} q(\widehat{X}_k)} > 0 \quad \textit{for all } z, \tag{8.41}$$

where q is defined in (8.32). *Moreover, $h(z) \to 1$ as $z \to \infty$.*

Proof The existence of $h(z)$ is immediate because $e^{-\sum_{k=0}^{n} q(\widehat{X}_k)}$ is decreasing in n. Since the function e^{-x} is convex, by Jensen's inequality we have

$$\mathbb{E}_z e^{-\sum_{k=0}^{n} q(\widehat{X}_k)} \ge e^{-\mathbb{E}_z \sum_{k=0}^{n} q(\widehat{X}_k)}. \tag{8.42}$$

Thus, to show positivity it suffices to prove that

$$\mathbb{E}_z \sum_{k=1}^{\infty} q(\widehat{X}_k) < \infty, \quad z > \widehat{x}. \tag{8.43}$$

Note that

$$\mathbb{E}_z \sum_{k=1}^{\infty} q(\widehat{X}_k) \le \int_{\widehat{x}}^{\infty} q(y)\widehat{H}_z(dy) \le c \int_{\widehat{x}}^{\infty} \frac{p(y)}{y} \widehat{H}_z(dy),$$

because $q(y) = O(p(y)/y)$; see (8.32). But it has been already shown in the proof of Lemma 4.1 that the last integral is finite.

To prove that $h(z) \to 1$, we note that Theorem 2.21 implies, for every fixed $N > 0$,

$$\mathbb{P}_z\{\widehat{X}_n > N \text{ for all } n \ge 1\} \to 1 \quad \text{as } z \to \infty,$$

so that

$$\widehat{H}_z(N) \to 0 \quad \text{as } z \to \infty.$$

Then, for every fixed N,

$$\lim_{z\to\infty} \mathbb{E}_z \sum_{k=0}^{\infty} q(\widehat{X}_k) \le \sup_{z>\widehat{x}} \int_N^{\infty} q(y)\widehat{H}_z(dy).$$

According to (4.6),

$$\lim_{N\to\infty} \sup_{z>\widehat{x}} \int_N^{\infty} q(y)\widehat{H}_z(dy) = 0.$$

Therefore, we infer that

$$\lim_{z\to\infty} \mathbb{E}_z \sum_{k=0}^{\infty} q(\widehat{X}_k) = 0$$

and finally conclude that $\lim_{z\to\infty} h(z) = 1$, again from (8.42). □

For the tail asymptotics of the recurrence times derived below in Section 8.5, we need the following two assertions.

Corollary 8.10 *Assume that the conditions of Lemma 8.6 are valid. Then $h(x)$ is a harmonic function for the kernel Q, that is,*

$$h(x) = \int_{\widehat{x}}^{\infty} h(y)Q(x,dy) \quad \textit{for all } x > \widehat{x}.$$

Furthermore,

$$W_p(x) := h(x)U_p(x) \tag{8.44}$$

is a harmonic function for the Markov chain $\{X_n\}$ *killed at the time of the first visit to* $(-\infty,\widehat{x}\,]$*:*

$$W_p(x) = \mathbb{E}_x\{W_p(X_1);\ X_1 > \widehat{x}\} \quad \textit{for all } x > \widehat{x}.$$

Proof By the Markov property,

$$\mathbb{E}_x e^{-\sum_{k=0}^{n} q(\widehat{X}_k)} = e^{-q(x)} \int_{\widehat{x}}^{\infty} \widehat{P}(x,dy)\mathbb{E}_y e^{-\sum_{k=0}^{n-1} q(\widehat{X}_k)}.$$

Letting $n \to \infty$ and using the dominated convergence theorem, we get

$$h(x) = e^{-q(x)} \int_{\widehat{x}}^{\infty} \widehat{P}(x,dy)h(y).$$

Recalling now that $e^{-q(x)}\widehat{P}(x,dy) = Q(x,dy)$, we arrive at the first statement of the corollary.

Noting also that $Q(x,dy) = \frac{U_p(y)}{U_p(x)}\mathbb{P}_x\{X_1 \in dy, X_1 > \widehat{x}\}$ for all $x,\ y > \widehat{x}$, we conclude that $h(x)U_p(x)$ is harmonic for the $\{X_n\}$ killed at entering $(-\infty,\widehat{x}\,]$, and the proof is complete. □

It turns out that, as it is formally defined via the function $U_p(x)$, the harmonic function $W_p(x)$ does not depend essentially on the choice of a function $p(x)$ that is decreasing and is integrable at infinity, since this function contributes only a constant multiplier. This observation follows from the following result.

Lemma 8.11 *Let* $V(x)$ *be a positive harmonic function for the Markov chain* $\{X_n\}$ *killed at the first visit to* $B := (-\infty,\widehat{x}\,]$*, that is,*

$$V(x) = \mathbb{E}_x\{V(X_1);\ \tau_B > 1\} \quad \textit{for all } x > \widehat{x}. \tag{8.45}$$

If, for some $C_V > 0$*,*

$$V(x) \sim C_V U(x) \quad \textit{as } x \to \infty \tag{8.46}$$

then

$$V(x) = C_V \lim_{n\to\infty} \mathbb{E}_x\{U(X_n);\ \tau_B > n\} \quad \textit{for all } x > \widehat{x}.$$

Proof It follows from (8.45) that, for all $n \ge 1$,

$$V(x) = \mathbb{E}_x\{V(X_n);\ \tau_B > n\} \quad \text{for all } x > \widehat{x}. \tag{8.47}$$

Fix an $\varepsilon > 0$. Given to the assumption (8.46), there exists an x_ε such that

$$(1-\varepsilon)V(y) \le C_V U(y) \le (1+\varepsilon)V(y) \quad \text{for all } y > x_\varepsilon.$$

Therefore

$$(1-\varepsilon)\mathbb{E}_x\{V(X_n);\ \tau_B>n, X_n>x_\varepsilon\} \tag{8.48}$$
$$\le C_V\mathbb{E}_x\{U(X_n);\ \tau_B>n, X_n>x_\varepsilon\}$$
$$\le (1+\varepsilon)\mathbb{E}_x\{V(X_n);\ \tau_B>n, X_n>x_\varepsilon\}. \tag{8.49}$$

On the other hand, by the definition of $\{\widehat{X}_n\}$, see (7.11),

$$\begin{aligned}\mathbb{E}_x\{V(X_n);\tau_B>n, X_n\le x_\varepsilon\} &= U_p(x)\mathbb{E}_x\left\{e^{-\sum_{k=0}^{n-1}q(\widehat{X}_k)}\frac{V(\widehat{X}_n)}{U_p(\widehat{X}_n)};\ \widehat{X}_n\le x_\varepsilon\right\}\\ &\le U_p(x)\mathbb{E}_x\left\{\frac{V(\widehat{X}_n)}{U_p(\widehat{X}_n)};\ \widehat{X}_n\le x_\varepsilon\right\},\end{aligned}$$

since $q(y)$ is non-negative. Recalling that the chain $\{\widehat{X}_n\}$ is transient, we conclude the convergence

$$\mathbb{E}_x\{V(X_n);\ \tau_B>n, X_n\le x_\varepsilon\}\to 0 \quad \text{as } n\to\infty. \tag{8.50}$$

By the same argument,

$$\mathbb{E}_x\{U(X_n);\ \tau_B>n, X_n\le x_\varepsilon\}\to 0 \quad \text{as } n\to\infty. \tag{8.51}$$

Combining (8.49), (8.51), and (8.47), we obtain

$$\begin{aligned}C_V\mathbb{E}_x\{U(X_n);\ \tau_B>n\} &\le (1+\varepsilon)\mathbb{E}_x\{V(X_n);\ \tau_B>n, X_n>x_\varepsilon\}+o(1)\\ &\le (1+\varepsilon)V(x)+o(1) \quad \text{as } n\to\infty.\end{aligned}$$

Combining (8.48), (8.50), and (8.47), we obtain

$$\begin{aligned}C_V\mathbb{E}_x\{U(X_n);\ \tau_B>n\} &\ge (1-\varepsilon)\mathbb{E}_x\{V(X_n);\ \tau_B>n, X_n>x_\varepsilon\}\\ &= (1-\varepsilon)V(x)+o(1) \quad \text{as } n\to\infty.\end{aligned}$$

Therefore, for any fixed $\varepsilon>0$,

$$\begin{aligned}\frac{1-\varepsilon}{C_V}V(x) &\le \liminf_{n\to\infty}\mathbb{E}_x\{U(X_n);\ \tau_B>n\}\\ &\le \limsup_{n\to\infty}\mathbb{E}_x\{U(X_n);\ \tau_B>n\} \le \frac{1+\varepsilon}{C_V}V(x).\end{aligned}$$

Letting $\varepsilon\to 0$ we conclude the existence of the limit of $\mathbb{E}_x\{U(X_n);\ \tau_B>n\}$ as $n\to\infty$, which equals $V(x)/C_V$. □

Set

$$W(x) := \lim_{n\to\infty}\mathbb{E}_x\{U(X_n);\ \tau_B>n\}. \tag{8.52}$$

According to Corollary 8.10, $W_p(x) = h(x)U_p(x)$ is harmonic and $W_p(x) \sim e^{-C_p}U(x)$. Then, by Lemma 8.11,

$$W_p(x) = e^{-C_p}W(x). \tag{8.53}$$

Consider the weighted renewal measure

$$\widehat{H}_z^{(q)}(dx) = \sum_{j=0}^{\infty} \mathbb{E}_z\{e^{-\sum_{k=0}^{j-1} q(\widehat{X}_k)};\ \widehat{X}_j \in dx\} \tag{8.54}$$

and its finite-time-horizon version,

$$\widehat{H}_{z,n}^{(q)}(dx) = \sum_{j=0}^{n} \mathbb{E}_z\{e^{-\sum_{k=0}^{j-1} q(\widehat{X}_k)};\ \widehat{X}_j \in dx\}. \tag{8.55}$$

Applying Lemma 4.5 and Theorem 4.12 to $\widehat{X}$ and taking into account Lemma 8.9, we get the following result.

Corollary 8.12 *Assume that the conditions of Lemma 8.6 are valid. Then*

$$\widehat{H}_{z,n}^{(q)}(\widehat{x},x] = h(z)\widehat{H}_{z,n}(\widehat{x},x] + o(x^2) = h(z)(\widehat{I}(n/x^2) + o(1))x^2$$

as $x \to \infty$ uniformly for all n, where $\widehat{I}$ is the function defined in Theorem 4.12 with $\widehat{\mu} = \mu + b$ and $\widehat{b} = b$. In particular,

$$\widehat{H}_z^{(q)}(\widehat{x},x] \sim h(z)\widehat{H}_z(\widehat{x},x] \sim h(z)\frac{x^2}{2\mu + b} \quad \textit{as } x \to \infty.$$

Now we are ready to prove the main result of this section.

Proof of Theorem 8.2. Since the function $y/U(y)$ is regularly varying at infinity, and $\liminf x_2/x_1 > 1$, it suffices to consider the case where $x_2 = (1+h)x_1$, $h > 0$.

Lemma 7.5 is applicable to the chain $\{X_n\}$, so it is legitimate to use the cycle representation (7.42). As follows from the representation (7.47) applied to U_p,

$$\begin{aligned}\pi(x,(1+h)x] &= c^* \int_x^{(1+h)x} \frac{H^{(q)}(dy)}{U_p(y)} \\ &\sim c^* e^{C_p} \int_x^{(1+h)x} \frac{H^{(q)}(dy)}{U(y)} \quad \text{as } x \to \infty,\end{aligned} \tag{8.56}$$

because $U_p(y) \sim e^{-C_p}U(y)$, see (8.17); $H^{(q)}$ is defined in (7.45).

Fix an $\varepsilon > 0$ and $n \in \mathbb{N}$. Let $x_k = (1 + hk/n)x$, $k = 0, \ldots, n$. Then, since the function U is increasing,

$$\sum_{k=0}^{n-1} \frac{\widehat{H}^{(q)}(x_k, x_{k+1})}{U(x_{k+1})} \le \int_x^{(1+h)x} \frac{\widehat{H}^{(q)}(dy)}{U(y)} \le \sum_{k=0}^{n-1} \frac{\widehat{H}^{(q)}(x_k, x_{k+1})}{U(x_k)}.$$

Now, according to Corollary 8.12,

$$\begin{aligned}\widehat{H}^{(q)}(x_k, x_{k+1}] &= \int_{\mathbb{R}} \widehat{H}_z^{(q)}(x_k, x_{k+1}]\mathbb{P}\{\widehat{X}_0 \in dz\} \\ &= c_q(x_{k+1}^2 - x_k^2) + o(x^2) \quad \text{as } x \to \infty \text{ uniformly for all } k \le n-1,\end{aligned}$$

where $c_q := \mathbb{E}h(\widehat{X}_0)/(2\mu + b)$. Consequently, for all sufficiently large x,

$$(c_q - \varepsilon)(x_{k+1}^2 - x_k^2) \le \widehat{H}^{(q)}(x_k, x_{k+1}] \le (c_q + \varepsilon)(x_{k+1}^2 - x_k^2)$$

for all $k \le n - 1$, which yields

$$(c_q - \varepsilon)\sum_{k=0}^{n-1} \frac{x_{k+1}^2 - x_k^2}{U(x_{k+1})} \le \int_x^{(1+h)x} \frac{\widehat{H}^{(q)}(dy)}{U(y)} \le (c_q + \varepsilon)\sum_{k=0}^{n-1} \frac{x_{k+1}^2 - x_k^2}{U(x_k)};$$

hence

$$(c_q - \varepsilon)\frac{2h}{n}\sum_{k=0}^{n-1} \frac{x_{k+1} + x_k}{U(x_{k+1})} \le \int_x^{(1+h)x} \frac{\widehat{H}^{(q)}(dy)}{U(y)} \le (c_q + \varepsilon)\frac{2h}{n}\sum_{k=0}^{n-1} \frac{x_{k+1} + x_k}{U(x_k)}.$$

Letting $n \to \infty$ and taking into account that the function $y/U(y)$ is regularly varying at infinity we derive that

$$\int_x^{(1+h)x} \frac{\widehat{H}^{(q)}(dy)}{U(y)} \sim 2c_q \int_x^{(1+h)x} \frac{y}{U(y)} dy,$$

which together with (8.56) concludes the proof. □

Corollary 8.13 *Assume that the conditions of Theorem 8.2 are valid. Then integrability of the function $y/U(y)$ at infinity is necessary and sufficient for the Markov chain $\{X_n\}$ on $\mathbb{R}^+$ to be positive recurrent.*

8.3 Local Asymptotics of Stationary Probabilities

In this section we derive sharp local asymptotics for a stationary measure π of a recurrent irreducible Markov chain with asymptotically zero drift of order $1/x$ at infinity. Following Section 6.3, we assume that the jumps $\xi(x)$ converge weakly to some random variable ξ on $\mathbb{R}$, that is, the asymptotic homogeneity condition (6.52) holds.

Theorem 8.14 *Let a recurrent Markov chain $\{X_n\}$ with invariant measure $\pi(\cdot)$ satisfy the conditions of Theorem 8.2. In addition, let $\xi(x) \Rightarrow \xi$ as $x \to \infty$ where $\mathbb{E}\xi = 0$, $\mathbb{E}\xi^2 = b$, and*

$$|\xi(y)|\mathbb{I}\{|\xi(y)| \le s(y)\} \le_{st} \Xi \quad \textit{for all } y \ge 0, \tag{8.57}$$

where $\mathbb{E}\Xi^2 < \infty$. Then, in the lattice case,

$$\pi(x) \sim c\frac{x}{U(x)} \quad \textit{as } x \to \infty \tag{8.58}$$

for some $c > 0$. In the non-lattice case, for any $h > 0$,

$$\pi(x, x+h] = ch\frac{x}{U(x)} \quad \textit{as } x \to \infty. \tag{8.59}$$

As far as applications are concerned, in Section 11.2 we apply the last result to derive the local asymptotic behaviour of the invariant measure for a reflected random walk with zero drift.

Corollary 8.15 *In addition to the conditions of Theorem 8.14, let $r(x) = 2\mu/bx$ and either $2\mu/b \in (-1, 1)$ or $2\mu/b > 1$, so that either null or positive recurrence holds respectively. Then, in the lattice case,*

$$\pi(x) \sim c\rho x^{-2\mu/b} \quad \textit{as } x \to \infty,$$

which agrees with the global asymptotics given in Corollary 8.4; $\rho = 2\mu/b + 1 > 0$. In the non-lattice case, for any $h > 0$,

$$\pi(x, x+h] \sim ch\rho x^{-2\mu/b} \quad \textit{as } x \to \infty.$$

In the case $2\mu/b = 1$, we have the following result.

Corollary 8.16 *In addition to the conditions of Theorem 8.14, for some $m \ge 1$ and $\gamma \ne 0$, let*

$$r(x) = \frac{1}{x} + \frac{1}{x\log x} + \cdots + \frac{1}{x\log x \cdots \log_{(m-1)} x} + \frac{1+\gamma}{x\log x \cdots \log_{(m)} x}$$

as $x \to \infty$, where $\gamma < 0$ corresponds to null recurrence while $\gamma > 0$ corresponds to positive recurrence. Then, in the lattice case,

$$\pi(x) \sim \frac{2c}{x\log x \cdots \log_{(m-1)} x \log_{(m)}^{1+\gamma} x} \quad \textit{as } x \to \infty,$$

which agrees with the global asymptotics given in Corollary 8.5. In the non-lattice case, for any $h > 0$,

$$\pi(x, x+h] \sim \frac{2ch}{x\log x \cdots \log_{(m-1)} x \log_{(m)}^{1+\gamma} x} \quad \textit{as } x \to \infty.$$

Proof of Theorem 8.14. As in the proof of Theorem 8.2, it follows from the representation (7.47) that

$$\pi(x,x+h] = c^* \int_x^{x+h} \frac{\widehat{H}^{(q)}(dy)}{U_p(y)},$$

where

$$\widehat{H}^{(q)}(dy) := \sum_{n=0}^{\infty} \mathbb{P}\{\widehat{X}_n \in dy\}.$$

Since the function $U_p(y)$ is regularly varying at infinity (and hence long-tailed at infinity),

$$\pi(x,x+h] \sim c^* \frac{\widehat{H}^{(q)}(x,x+h]}{U_p(x)} \quad \text{as } x \to \infty.$$

The Markov chain $\{\widehat{X}_n\}$ satisfies all the conditions of Corollary 6.12 with $\widehat{\mu} = \mu + b$ and $\widehat{b} = b$, so $2\widehat{\mu} - \widehat{b} = 2\mu + b > 0$. Indeed, the conditions (8.34)–(8.36) were checked in Lemma 8.8 and the condition (6.1) was checked directly after that. The weak convergence (6.52) for $\widehat{\xi}(x)$, that is, $\widehat{\xi}(x) \Rightarrow \xi$, follows from that for the original jumps $\xi(x)$ because $U_p(x+y)/U_p(x) \to 1$ as $x \to \infty$, for any fixed $y \in \mathbb{R}$. Finally, the majorisation condition (6.53) holds with a square integrable majorant, since it follows from (8.32) and (8.30) that, for all sufficiently large x,

$$\begin{aligned}
&\mathbb{P}\{\widehat{\xi}(x)\mathbb{I}\{|\widehat{\xi}(x)| \le s(x)|\} > y\} \\
&\quad = \frac{Q(x,(x+y,x+s(x)])}{Q(x,\mathbb{R})} \\
&\quad \le 2\frac{\mathbb{E}\{U_p(x+\xi(x));\ \xi(x)\mathbb{I}\{|\xi(x)| \le s(x)|\} > y\}}{U_p(x)} \\
&\quad \le 2\frac{U_p(2x)}{U_p(x)}\mathbb{P}\{\xi(x)\mathbb{I}\{|\xi(x)| \le s(x)|\} > y\} \\
&\qquad + 2\frac{\mathbb{E}\{U_p(2\xi(x));\ \xi(x)\mathbb{I}\{|\xi(x)| \le s(x)|\} > y\}}{U_p(x)} \\
&\quad \le c_1\mathbb{P}\{\Xi > y\} + \frac{c_1}{U(x)}\mathbb{E}\{U(\Xi);\ y < \Xi \le s(x)\},
\end{aligned}$$

owing to the regular variation of the function U_p and the conditions (8.18) and (8.57). Since $s(x) \le x$,

$$\mathbb{P}\{\widehat{\xi}(x)\mathbb{I}\{|\widehat{\xi}(x)| \le s(x)|\} > y\} \le 2c_1\mathbb{P}\{\Xi > y\},$$

which implies that

$$\widehat{\xi}(x)\mathbb{I}\{|\widehat{\xi}(x)| \le s(x)|\} \le_{st} \widehat{\Xi},$$

where $\mathbb{E}\widehat{\Xi}^2 < \infty$ owing to the assumption $\mathbb{E}\Xi^2 < \infty$. In addition,

$$\begin{aligned}
&\mathbb{P}\{\widehat{\xi}(x)\mathbb{I}\{|\widehat{\xi}(x)| \le s(x)|\} < -y\} \\
&\quad = \frac{Q(x,[x-s(x),x-y))}{Q(x,\mathbb{R})} \\
&\quad \le 2\frac{\mathbb{E}\{U_p(x+\xi(x));\ \xi(x)\mathbb{I}\{|\xi(x)| \le s(x)|\} < -y\}}{U_p(x)} \\
&\quad \le 2\mathbb{P}\{\xi(x)\mathbb{I}\{|\xi(x)| \le s(x)|\} < -y\} \\
&\quad \le 2\mathbb{P}\{\Xi > y\},
\end{aligned}$$

which implies that $\widehat{\xi}(x) \ge_{st} -\widehat{\Xi}$; thus the proof of the existence of a square integrable majorant for the family of $\widehat{\xi}(x)\mathbb{I}\{|\widehat{\xi}(x)| \le s(x)|\}$ is complete.

Hence, by Corollary 6.12 and Lemma 4.5 applied to the Markov chain $\{\widehat{X}_n\}$, we deduce that

$$\widehat{H}^{(q)}(x,x+h] \sim c_q \frac{h+o(1)}{2\widehat{\mu}-\widehat{b}} x \quad \text{as } x \to \infty,$$

which concludes the proof of Theorem 8.14 because $U_p(x) \sim c_3 U(x)$ as $x \to \infty$; see (8.17). □

8.4 Pre-Stationary Distribution of Positive Recurrent Chain with Power-Like Stationary Measure

In this section we assume that the distribution of X_n converges in total-variation distance to a unique invariant distribution π as $n \to \infty$, that is,

$$\sup_{A\in\mathcal{B}(\mathbb{R})} |\mathbb{P}\{X_n \in A\} - \pi(A)| \to 0 \quad \text{as } n \to \infty; \tag{8.60}$$

for a countable Markov chain $\{X_n\}$ this condition holds automatically provided that the chain is irreducible, aperiodic, and positive recurrent. For a real-valued chain it is related to the Harris ergodicity; see e.g. [126].

Theorem 8.17 *Assume that all the conditions of Theorem 8.2 are valid and that $\{X_n\}$ is positive recurrent and satisfies* (8.60). *Then*

$$\mathbb{P}\{X_n > x\} = (F(n/x^2) + o(1))\pi(x,\infty)$$

as $x \to \infty$ uniformly for all n, where

$$F(u) := \frac{1}{\widehat{\Gamma}(\infty)} \int_0^u \widehat{\Gamma}(1/z)\Big[1 - \frac{\rho}{2}\Big(\frac{z}{u}\Big)^{\rho/2-1}\Big)\Big]dz$$

is a continuous distribution function and $\widehat{I}$ and $\widehat{\Gamma}$ are functions defined in Theorem 4.12 with $\widehat{\mu} = \mu + b$ and $\widehat{b} = b$.

The last result shows that $\{X_n\}$ enters the stationary regime at time n of order x^2; that is, if $n = o(x^2)$ then

$$\mathbb{P}\{X_n > x\} = o(\pi(x,\infty));$$

if $n/x^2 \to u \in (0,\infty)$ then

$$\mathbb{P}\{X_n > x\} \sim F(u)\pi(x,\infty);$$

and if $n/x^2 \to \infty$ then

$$\mathbb{P}\{X_n > x\} \sim \pi(x,\infty).$$

Proof Splitting all the paths according to the time of the last visit of $\{X_n\}$ to $B = (-\infty, \widehat{x}]$, see (7.49), we get, for $x > \widehat{x}$,

$$\begin{aligned}
&\mathbb{P}\{X_n > x\} \\
&= \sum_{j=1}^{n} \int_B \mathbb{P}\{X_{n-j} \in dz\} \int_{\widehat{x}}^{\infty} P(z,du)U_p(u)\mathbb{E}_u\left\{\frac{e^{-\sum_{k=0}^{j-2} q(\widehat{X}_k)}}{U_p(\widehat{X}_{j-1})}; \widehat{X}_{j-1} > x\right\},
\end{aligned} \tag{8.61}$$

where $q(x) \ge 0$ and $\{\widehat{X}_n\}$ were defined in (7.9) and (7.10) respectively.

Fix a sequence $N_x \to \infty$ such that $N_x = o(x^2)$. Then, since $q \ge 0$ and U_p is increasing,

$$\begin{aligned}
&\sum_{j=n-N_x+1}^{n} \int_B \mathbb{P}\{X_{n-j} \in dz\} \int_{\widehat{x}}^{\infty} P(z,du)U_p(u)\mathbb{E}_u\left\{\frac{e^{-\sum_{k=0}^{j-2} q(\widehat{X}_k)}}{U_p(\widehat{X}_{j-1})}; \widehat{X}_{j-1} > x\right\} \\
&\le N_x \frac{1}{U_p(x)} \sup_{z\in B} \int_{\widehat{x}}^{\infty} P(z,du)U_p(u) \\
&\le N_x \frac{c}{U_p(x)} \sup_{z\in B}(1 + U_p(z)) \\
&= o(x^2/U_p(x)),
\end{aligned} \tag{8.62}$$

where the second bound follows from (8.28). Furthermore, the distribution of X_{n-j} converges in total variation to π uniformly for all $j \le n - N_x$; see (8.60). Therefore, as $x \to \infty$,

$$\sum_{j=1}^{n-N_x}\int_B \mathbb{P}\{X_{n-j}\in dz\}\int_{\widehat{x}}^\infty P(z,du)U_p(u)\mathbb{E}_u\left\{\frac{e^{-\sum_{k=0}^{j-2}q(\widehat{X}_k)}}{U_p(\widehat{X}_{j-1})};\ \widehat{X}_{j-1}>x\right\}$$
$$\sim\sum_{j=1}^{n-N_x}\int_B \pi(dz)\int_{\widehat{x}}^\infty P(z,du)U_p(u)\mathbb{E}_u\left\{\frac{e^{-\sum_{k=0}^{j-2}q(\widehat{X}_k)}}{U_p(\widehat{X}_{j-1})};\ \widehat{X}_{j-1}>x\right\}. \tag{8.63}$$

Similarly to (8.62), we find that

$$\sum_{j=n-N_x+1}^{n}\int_B \pi(dz)\int_{\widehat{x}}^\infty P(z,du)U_p(u)\mathbb{E}_u\left\{\frac{e^{-\sum_{k=0}^{j-2}q(\widehat{X}_k)}}{U_p(\widehat{X}_{j-1})};\ \widehat{X}_{j-1}>x\right\}$$
$$=o(x^2/U_p(x)). \tag{8.64}$$

Combining (8.61)—(8.64), we obtain

$$\begin{aligned}
&\mathbb{P}\{X_n>x\}\\
&=(1+o(1))\sum_{j=1}^{n}\int_B \pi(dz)\int_{\widehat{x}}^\infty P(z,du)U_p(u)\\
&\quad\times\mathbb{E}_u\left\{\frac{e^{-\sum_{k=0}^{j-2}q(\widehat{X}_k)}}{U_p(\widehat{X}_{j-1})};\ \widehat{X}_{j-1}>x\right\}+o\left(\frac{x^2}{U_p(x)}\right)\\
&=(1+o(1))\int_B \pi(dz)\int_{\widehat{x}}^\infty P(z,du)U_p(u)\\
&\quad\times\sum_{j=1}^{n}\int_x^\infty \mathbb{E}_u\left\{\frac{e^{-\sum_{k=0}^{j-2}q(\widehat{X}_k)}}{U_p(y)};\ \widehat{X}_{j-1}\in dy\right\}+o\left(\frac{x^2}{U_p(x)}\right)\\
&=(1+o(1))\int_{\widehat{x}}^\infty \mu(du)U_p(u)\int_x^\infty\frac{\widehat{H}^{(q)}_{u,n}(dy)}{U_p(y)}+o\left(\frac{x^2}{U_p(x)}\right)
\end{aligned} \tag{8.65}$$

as $x\to\infty$, where

$$\mu(du)=\int_B \pi(dz)P(z,du)$$

is a measure on $(\widehat{x},\infty)$, see (7.43), and

$$\widehat{H}^{(q)}_{u,n}(A):=\sum_{j=1}^{n}\mathbb{E}_u\left\{e^{-\sum_{k=0}^{j-2}q(\widehat{X}_k)};\ \widehat{X}_{j-1}\in A\right\}$$

is also a measure on $(\widehat{x},\infty)$.

For any fixed $u>\widehat{x}$, owing to Corollary 8.12,

$$\widehat{H}^{(q)}_{u,n}(\widehat{x},y]\sim h(u)(\widehat{I}(n/y^2)+o(1))y^2 \quad \text{as } y\to\infty \text{ uniformly for all } n. \tag{8.66}$$

In addition, since $q \ge 0$,

$$\sup_{u>\widehat{x}} \widehat{H}^{(q)}_{u,n}(\widehat{x}, y] \le \sup_{u>\widehat{x}} \sum_{j=1}^{n} \mathbb{P}_u\{\widehat{X}_{j-1} \in (\widehat{x}, y]\} \le c_1 y^2 \quad \text{for all } y \text{ and } n, \tag{8.67}$$

for some $c_1 < \infty$, as follows from the integral renewal theorem for $\{\widehat{X}_n\}$. Integration by parts together with (8.66) implies that, for any fixed $u > \widehat{x}$,

$$\begin{aligned}\int_x^\infty \frac{\widehat{H}^{(q)}_{u,n}(dz)}{U_p(z)} &= -\frac{\widehat{H}^{(q)}_{u,n}(\widehat{x}, x]}{U_p(x)} - \int_x^\infty \widehat{H}^{(q)}_{u,n}(\widehat{x}, z] d\frac{1}{U_p(z)} \\ &= h(u)\Big[-\frac{\widehat{I}(n/x^2)x^2}{U_p(x)} - \int_x^\infty \widehat{I}(n/z^2)z^2 d\frac{1}{U_p(z)}\Big] + o\Big(\frac{x^2}{U_p(x)}\Big)\end{aligned}$$

as $x \to \infty$ uniformly for all n. Taking into account that

$$-\frac{d}{dz}\frac{1}{U_p(z)} = \frac{U'_p(z)}{U_p^2(z)} = \frac{e^{R_p(z)}}{U_p^2(z)} \sim \frac{2\mu + b}{b}\frac{1}{zU_p(z)} \quad \text{as } z \to \infty,$$

owing to (8.18) we deduce that

$$\begin{aligned}&\int_x^\infty \frac{\widehat{H}^{(q)}_{u,n}(dz)}{U_p(z)} \\ &= h(u)\Big[-\frac{\widehat{I}(n/x^2)x^2}{U_p(x)} + \frac{2\mu + b}{b}\int_x^\infty \frac{\widehat{I}(n/z^2)z}{U_p(z)}dz\Big] + o\Big(\frac{x^2}{U_p(x)}\Big) \\ &= h(u)\Big[-\frac{\widehat{I}(n/x^2)x^2}{U_p(x)} + \frac{2\mu + b}{b}x^2\int_1^\infty \frac{\widehat{I}(n/x^2z^2)z}{U_p(xz)}dz\Big] + o\Big(\frac{x^2}{U_p(x)}\Big).\end{aligned}$$

Since the function U_p is regularly varying at infinity with index $\rho = 2\mu/b + 1 > 2$, $U_p(xz)/U_p(x) \to z^\rho$ as $x \to \infty$. Therefore

$$\begin{aligned}\int_x^\infty \frac{\widehat{H}^{(q)}_{u,n}(dz)}{U_p(z)} &= h(u)\frac{x^2}{U_p(x)}\Big[-\widehat{I}(n/x^2) + \rho\int_1^\infty \frac{\widehat{I}(n/x^2z^2)}{z^{\rho-1}}dz\Big] + o\Big(\frac{x^2}{U_p(x)}\Big) \\ &= h(u)\frac{x^2}{U_p(x)}\widehat{F}(n/x^2) + o\Big(\frac{x^2}{U_p(x)}\Big)\end{aligned} \tag{8.68}$$

as $x \to \infty$ uniformly for all n, where

$$\begin{aligned}\widehat{F}(t) &:= -\widehat{I}(t) + \rho\int_1^\infty \frac{\widehat{I}(t/z^2)}{z^{\rho-1}}dz \\ &= -\widehat{I}(t) - \frac{\rho}{\rho - 2}\int_1^\infty \widehat{I}(t/z^2)d\frac{1}{z^{\rho-2}} \\ &= \widehat{I}(t)\frac{2}{\rho - 2} + \frac{\rho}{\rho - 2}\int_1^\infty \frac{1}{z^{\rho-2}}d\widehat{I}(t/z^2).\end{aligned}$$

Thus we have

$$\begin{aligned}
\widehat{F}(t) &= \widehat{I}(t)\frac{2}{\rho-2} - \frac{2\rho t}{\rho-2}\int_1^\infty \frac{1}{z^{\rho+1}}\widehat{\Gamma}(z^2/t)dz \\
&= \widehat{I}(t)\frac{2}{\rho-2} - \frac{\rho t^{1-\rho/2}}{\rho-2}\int_0^t u^{\rho/2-1}\widehat{\Gamma}(1/u)du \\
&= \frac{2}{\rho-2}\int_0^t \widehat{\Gamma}(1/u)\left(1-\frac{\rho}{2}\left(\frac{u}{t}\right)^{\rho/2-1}\right)du.
\end{aligned}$$

Similarly, it follows from (8.67) that

$$\sup_{u>\widehat{x}}\int_x^\infty \frac{\widehat{H}_{u,n}^{(q)}(dz)}{U_p(z)} \le c_2\frac{x^2}{U_p(x)}.$$

In addition,

$$\begin{aligned}
\widehat{c} &= \int_{\widehat{x}}^\infty h(u)U_p(u)\mu(du) \\
&= \int_B \pi(dz)\int_{\widehat{x}}^\infty h(u)U_p(u)P(z,du) \;<\; \infty,
\end{aligned}$$

as follows from (8.28). Hence the dominated convergence theorem is applicable to (8.65), so, substituting (8.68) into (8.65), we obtain

$$\mathbb{P}\{X_n > x\} = \widehat{c}\frac{x^2}{U_p(x)}(\widehat{F}(n/x^2)+o(1)) \tag{8.69}$$

as $x\to\infty$ uniformly for all n. In particular, letting $n\to\infty$ we obtain

$$\pi(x,\infty) \;=\; \lim_{n\to\infty}\mathbb{P}\{X_n > x\} \sim \widehat{c}\frac{x^2}{U_p(x)}\widehat{F}(\infty) \;=\; \widehat{c}\frac{x^2}{U_p(x)}2\frac{\rho-1}{\rho-2}\widehat{I}(\infty),$$

which concludes the proof. □

8.5 Tail Asymptotics for Recurrence Times of Positive and Null Recurrent Markov Chains

In this section we study the tail behaviour of the stopping time

$$\tau_{\widehat{x}} := \inf\{n \ge 1 :\; X_n \le \widehat{x}\},$$

in the case where $\tau_{\widehat{x}}$ is a proper random variable, that is, $\{X_n\}$ is either positive or null recurrent with respect to the set $(-\infty,\widehat{x}]$.

Theorem 8.18 *Let the conditions of Theorem 8.2 hold. Let $\widehat{x}$ be chosen as in Corollary 8.7 and Lemma 8.8. Then there exists a constant $c < \infty$ such that*

$$\mathbb{P}_x\{\tau_{\widehat{x}} > n\} \le c\frac{U(x)}{U(\sqrt{n})} \quad \text{for all } n \text{ and } x > \widehat{x}. \tag{8.70}$$

Further, for any fixed $x > \widehat{x}$,

$$\mathbb{P}_x\{\tau_{\widehat{x}} > n\} \sim \frac{1}{(2b)^{\rho/2}\Gamma(1+\rho/2)}\frac{W(x)}{U(\sqrt{n})} \quad \text{as } n \to \infty, \tag{8.71}$$

where $W(x)$ is the harmonic function defined in (8.52).

In addition, if $X_0 > \widehat{x}$ a.s. and $\mathbb{E}U(X_0) < \infty$ then

$$\mathbb{P}\{\tau_{\widehat{x}} > n\} \sim \frac{\mathbb{E}W(X_0)}{(2b)^{\rho/2}\Gamma(1+\rho/2)}\frac{1}{U(\sqrt{n})} \quad \text{as } n \to \infty. \tag{8.72}$$

Notice that

$$\frac{W(x)}{U(\sqrt{n})} \sim \frac{U(x)}{U(\sqrt{n})} \quad \text{as } n,\ x \to \infty,$$

owing to Lemma 8.9 and the equivalence (8.17).

Regarding applications, in Section 11.2 we apply the last result to derive down-crossing probabilities for a reflected random walk with zero drift.

In order to prove the upper bound (8.70) for the tail of $\tau_{\widehat{x}}$ we need a couple of preliminary results. In Theorem 3.12 we constructed a function of a transient Markov chain which was a bounded supermartingale. It turns out that for a Markov chain $\{\widehat{X}_n\}$ which is specially constructed a similar result is valid under weaker conditions on the left tail distribution. Recall the definition of the function U_p in (8.15).

Lemma 8.19 *For any $\varepsilon \in (0,1)$ and a number $a > 1/\varepsilon$, there exists an $x_* > \widehat{x}$ such that*

$$\min\left(\frac{U_{ap}^{\varepsilon}(\widehat{X}_n)}{U_p(\widehat{X}_n)}, \frac{U_{ap}^{\varepsilon}(x_*)}{U_p(x_*)}\right)$$

is a positive supermartingale.

Proof By the definition of the chain $\{\widehat{X}_n\}$ and Jensen's inequality,

$$\begin{aligned}
\mathbb{E}\frac{U_{ap}^{\varepsilon}(x+\widehat{\xi}(x))}{U_p(x+\widehat{\xi}(x))} &= \int_{\widehat{x}}^{\infty}\frac{U_{ap}^{\varepsilon}(y)}{U_p(y)}\frac{Q(x,dy)}{Q(x,\mathbb{R})} \\
&= \frac{1}{\int_{\widehat{x}}^{\infty}U_p(y)P(x,dy)}\int_{\widehat{x}}^{\infty}U_{ap}^{\varepsilon}(y)P(x,dy) \\
&\le \frac{1}{\int_{\widehat{x}}^{\infty}U_p(y)P(x,dy)}\left(\int_{\widehat{x}}^{\infty}U_{ap}(y)P(x,dy)\right)^{\varepsilon}.
\end{aligned} \tag{8.73}$$

Owing to Lemma 8.6 and (8.31), as $x \to \infty$ we have

$$\int_{\widehat{x}}^{\infty} U_p(y)P(x,dy) = U_p(x)\Big(1 - \frac{2\mu + b}{2}\frac{p(x)}{x} + o\Big(\frac{p(x)}{x}\Big)\Big) \quad (8.74)$$

and

$$\int_{\widehat{x}}^{\infty} U_{ap}(y)P(x,dy) = U_{ap}(x)\Big(1 - a\frac{2\mu + b}{2}\frac{p(x)}{x} + o\Big(\frac{p(x)}{x}\Big)\Big).$$

Then

$$\Big(\int_{\widehat{x}}^{\infty} U_{ap}(y)P(x,dy)\Big)^{\varepsilon} = U_{ap}^{\varepsilon}(x)\Big(1 - a\varepsilon\frac{2\mu + b}{2}\frac{p(x)}{x} + o\Big(\frac{p(x)}{x}\Big)\Big)$$

and it follows from $a\varepsilon > 1$ that

$$\frac{1}{\int_{\widehat{x}}^{\infty} U_p(y)P(x,dy)}\Big(\int_{\widehat{x}}^{\infty} U_{ap}(y)P(x,dy)\Big)^{\varepsilon} \le \frac{U_{ap}^{\varepsilon}(x)}{U_p(x)}$$

for all sufficiently large x, which completes the proof. □

Lemma 8.20 *For*

$$\widehat{T}(z) := \min\{n \ge 0 : \widehat{X}_n > z\}, \quad z > \widehat{x},$$

there exists a $\gamma > 0$ *such that, for all n and z,*

$$\sup_x \mathbb{P}_x\{\widehat{T}(z) > n\} \le ce^{-\gamma n/z^2}.$$

Proof It follows from the definition of the chain $\{\widehat{X}_n\}$ that it can only visit $(-\infty, \widehat{x}\,]$ at time 0. Therefore,

$$\widehat{T}(z) \le 1 + \sum_{k=1}^{\widehat{T}(z)-1} \mathbb{I}\{\widehat{X}_k > \widehat{x}\}.$$

Then, by Theorem 4.2 with $v(x) \ge c_1/x$,

$$\begin{aligned}\mathbb{E}_x\widehat{T}(z) &\le \int_x^{z+s(z)} \frac{y}{c_1}dy \\ &\le \frac{1}{2c_1}(z + s(z))^2 \le c_2z^2 \quad \text{uniformly for all } x \text{ and } z. \quad (8.75)\end{aligned}$$

Next, by the Markov property, for all t and $s > 0$,

$$\begin{aligned}\mathbb{P}_x\{\widehat{T}(z) > t + s\} &= \int_0^z \mathbb{P}_x\{\widehat{T}(z) > t, X_t \in du\}\mathbb{P}_u\{\widehat{T}(z) > s\} \\ &\le \mathbb{P}_x\{\widehat{T}(z) > t\} \sup_{u \le z} \mathbb{P}_u\{\widehat{T}(z) > s\}.\end{aligned}$$

Therefore, the decreasing function $g(t) := \sup_{u\le z} \mathbb{P}_u\{\widehat{T}(z) > tz^2\}$ satisfies the inequality $g(t+s) \ge g(t)g(s)$ and $g(0) = 1$. Then the increasing function $g_0(t) := \log(1/g(t))$ is convex since $g_0(t+s) \le g_0(t)+g_0(s)$ and $g_0(0) = 0$. By the bound (8.75) and Markov's inequality, there exists a t_0 such that $g(t_0) < 1$, so that $g(t_0) = e^{-\gamma}$ with $\gamma > 0$ and $g_0(t_0) = \gamma > 0$. Then, by $g_0(0) = 0$ and by the convexity of g_0, $g_0(t) \ge \gamma(t-t_0)$ for $t \ge t_0$, which implies that $g(t) \le e^{-\gamma(t-t_0)}$, which is equivalent to the claim of the lemma. □

Lemma 8.21 *For any fixed $\varepsilon \in (0,\rho)$, $\rho = 2\mu/b+1$, there exists a constant $c(\varepsilon)$ such that, for all n, x, and $y \in (x_*, \sqrt{n}\,]$,*

$$\mathbb{P}_x\{\widehat{X}_k \le y \textit{ for some } k \in [n+1,2n]\} \le c(\varepsilon)\Big(\frac{y}{\sqrt{n}}\Big)^{\rho-\varepsilon}.$$

Proof For any $z > y$, the event whose probability we need to bound can only occur if either the chain $\{\widehat{X}_n\}$ does not exceed the level z by time n or it does exceed this level and then falls below y. Therefore, by the Markov property, the corresponding probability is not greater than the sum

$$\mathbb{P}_x\{T(z) > n\} + \sup_{u\ge z} \mathbb{P}_u\{\widehat{X}_k \le y \text{ for some } k \ge 1\}, \tag{8.76}$$

where the first term may be bounded above by Lemma 8.20. For the second term, by Lemma 8.19 we can apply the Doob inequality for supermartingales, which guarantees that there exists a constant $c_1(\varepsilon)$ such that, for all $u \ge z$,

$$\mathbb{P}_u\{\widehat{X}_k \le y \text{ for some } k \ge 1\} \le c_1(\varepsilon)\frac{U_p(y)}{U_p(u)}\frac{U_{ap}^{\varepsilon/2\rho}(u)}{U_{ap}^{\varepsilon/2\rho}(y)} \quad \text{for all } y \in (x_*, u].$$

Hence the equivalence (8.17) implies the existence of $c_2(\varepsilon)$ such that

$$\mathbb{P}_u\{\widehat{X}_k \le y \text{ for some } k \ge 1\} \le c_2(\varepsilon)\bigg(\frac{U(y)}{U(u)}\bigg)^{1-\varepsilon/2\rho} \quad \text{for all } y \in (x_*, u].$$

Since U is regularly varying at infinity with index ρ, by Potter's bounds, see e.g. [17, Theorem 1.5.6], there exists a constant $c_3(\varepsilon)$ such that

$$\frac{1}{c_3(\varepsilon)}\left(\frac{y}{u}\right)^{\rho+\varepsilon/2} \le \frac{U(y)}{U(u)} \le c_3(\varepsilon)\left(\frac{y}{u}\right)^{\rho-\varepsilon/2} \quad \text{for all } y \in (x_*, u]. \tag{8.77}$$

Consequently,

$$\sup_{u\ge z} \mathbb{P}_u\{\widehat{X}_k \le y \text{ for some } k \ge 1\} \le c_4(\varepsilon)\left(\frac{y}{z}\right)^{\rho-\varepsilon}. \tag{8.78}$$

Therefore, the estimates for each term in the upper bound (8.76) give

$$\mathbb{P}_x\{\widehat{X}_k \le y \text{ for some } k \in [n+1,2n]\} \le c_5\Big(e^{-\gamma n/z^2} + \Big(\frac{y}{z}\Big)^{\rho-\varepsilon}\Big).$$

Optimisation of the right-hand side with respect to z is not solvable in terms of elementary functions, so we choose

$$z := \sqrt{\frac{\gamma n}{\log((\sqrt{n}/y)^{\rho-\varepsilon})}},$$

which is close to the optimal value. Then

$$\begin{aligned}&\mathbb{P}_x\{\widehat{X}_k \le y \text{ for some } k \in [n+1, 2n]\}\\ &\quad\le c_5\Big(\Big(\frac{y}{\sqrt{n}}\Big)^{\rho-\varepsilon} + \Big(\frac{y}{\sqrt{\gamma n}}\Big)^{\rho-\varepsilon}\Big((\rho-\varepsilon)\log\frac{\sqrt{n}}{y}\Big)^{\frac{\rho-\varepsilon}{2}}\Big),\end{aligned}$$

which implies the claim of the lemma if we take $\varepsilon/2$ instead of ε on the right-hand side. □

Proof of Theorem 8.18. We start with the upper bound (8.70), which is the most difficult part of the theorem. It follows from (8.33) that

$$\begin{aligned}\mathbb{P}_x\{\tau_{\widehat{x}} > n\} &= U_p(x)\int_{\widehat{x}}^{\infty}\frac{1}{U_p(y)}Q^n(x,dy)\\ &= U_p(x)\int_{\widehat{x}}^{\infty}\frac{1}{U_p(y)}\mathbb{E}_x\big\{e^{-\sum_{k=0}^{n-1}q(\widehat{X}_k)};\ \widehat{X}_n \in dy\big\}. \end{aligned}\tag{8.79}$$

Since $q(x) \ge 0$,

$$\begin{aligned}\mathbb{P}_x\{\tau_{\widehat{x}} > n\} &\le U_p(x)\mathbb{E}_x\frac{1}{U_p(\widehat{X}_n)}\\ &\le c_1U(x)\mathbb{E}_x\frac{1}{U(\widehat{X}_n)},\end{aligned}\tag{8.80}$$

owing to (8.17). Summing up n successive probabilities we get

$$\begin{aligned}\sum_{k=n+1}^{2n}\mathbb{P}_x\{\tau_{\widehat{x}} > k\} &\le c_1U(x)\int_{\widehat{x}}^{\infty}\frac{1}{U(y)}\widehat{H}_{x,n}(dy)\\ &= c_1U(x)\Big(\int_{\widehat{x}}^{\sqrt{n}} + \int_{\sqrt{n}}^{\infty}\Big)\frac{1}{U(y)}\widehat{H}_{x,n}(dy),\end{aligned}\tag{8.81}$$

where

$$\widehat{H}_{x,n}(A) := \sum_{k=n+1}^{2n}\mathbb{P}_x\{\widehat{X}_k \in A\}.$$

The function U is increasing, so

$$\int_{\sqrt{n}}^{\infty}\frac{1}{U(y)}\widehat{H}_{x,n}(dy) \le \frac{n}{U(\sqrt{n})}\quad\text{for all } x \text{ and } n. \tag{8.82}$$

Further, integrating by parts, we obtain

$$\int_{\widehat{x}}^{\sqrt{n}} \frac{1}{U(y)} \widehat{H}_{x,n}(dy) = \frac{\widehat{H}_{x,n}(\widehat{x}, \sqrt{n}\,]}{U(\sqrt{n})} + \int_{\widehat{x}}^{\sqrt{n}} \frac{U'(y)\widehat{H}_{x,n}(\widehat{x}, y]}{U^2(y)} dy$$
$$\le \frac{n}{U(\sqrt{n})} + \int_{\widehat{x}}^{\sqrt{n}} \frac{e^{R(y)}\widehat{H}_{x,n}(\widehat{x}, y]}{U^2(y)} dy,$$

owing to $U' = e^R$. Combining this with (8.81) and (8.82) and noting that $e^{R(y)} \sim \rho U(y)/y$, we conclude that

$$\sum_{k=n+1}^{2n} \mathbb{P}_x\{\tau_{\widehat{x}} > k\} \le 2c_1 U(x) \frac{n}{U(\sqrt{n})} + c_2 U(x) \int_{\widehat{x}}^{\sqrt{n}} \frac{\widehat{H}_{x,n}(\widehat{x}, y]}{yU(y)} dy. \quad (8.83)$$

Next we derive an upper bound for $\widehat{H}_{x,n}$. It is clear that

$$\widehat{H}_{x,n}(\widehat{x}, y] = \mathbb{E}_x \sum_{k=n+1}^{2n} \mathbb{I}\{\widehat{X}_k \in (\widehat{x}, y]\}$$
$$\le \mathbb{P}_x\{\widehat{X}_k \in (\widehat{x}, y] \text{ for some } k \in [n+1, 2n]\} \sup_{s \le y} \sum_{k=0}^{\infty} \mathbb{P}_s\{\widehat{X}_k \in (\widehat{x}, y]\}$$
$$\le \sup_s \widehat{H}_s(\widehat{x}, y]\, \mathbb{P}_x\{\widehat{X}_k \in (\widehat{x}, y] \text{ for some } k \in [n+1, 2n]\}.$$

Applying Theorem 4.3 and Lemma 8.21, we get

$$\widehat{H}_{x,n}(\widehat{x}, y] \le c_3 y^2 \Big(\frac{y}{\sqrt{n}}\Big)^{\rho-\varepsilon}.$$

Therefore

$$\int_{\widehat{x}}^{\sqrt{n}} \frac{\widehat{H}_{x,n}(y)}{yU(y)} dy \le c_4 \int_{\widehat{x}}^{\sqrt{n}} \frac{y}{U(y)} \Big(\frac{y}{\sqrt{n}}\Big)^{\rho-\varepsilon} dy.$$

Substituting $y = u\sqrt{n}$ leads to the following expression for the last integral:

$$\frac{n}{U(\sqrt{n})} \int_{\widehat{x}/\sqrt{n}}^{1} \frac{U(\sqrt{n})}{U(u\sqrt{n})} u^{1+\rho-\varepsilon} du.$$

Applying the left-hand side inequality from (8.77) we get the upper bound

$$\int_{\widehat{x}}^{\sqrt{n}} \frac{\widehat{H}_{x,n}(y)}{yU(y)} dy \le c_5 \frac{n}{U(\sqrt{n})} \int_0^1 u^{1-3\varepsilon/2} du = c_6 \frac{n}{U(\sqrt{n})},$$

provided that $\varepsilon < 1$. Substituting this upper bound into (8.83) we obtain

$$\sum_{k=n+1}^{2n} \mathbb{P}_x\{\tau_{\widehat{x}} > k\} \le CU(x) \frac{n}{U(\sqrt{n})}.$$

Thus we have

$$\mathbb{P}_x\{\tau_{\widehat{x}} > 2n\} \le C\frac{U(x)}{U(\sqrt{n})}.$$

Since U is regularly varying at infinity, this completes the proof of the upper bound (8.70).

Now let us prove the tail asymptotics for $\tau_{\widehat{x}}$. Fix an $\varepsilon > 0$ and split the integral (8.79) into two parts:

$$\mathbb{P}_x\{\tau_{\widehat{x}} > n\} = U_p(x)\biggl(\int_{\widehat{x}}^{\varepsilon\sqrt{n}} + \int_{\varepsilon\sqrt{n}}^{\infty}\biggr)\frac{1}{U_p(y)}Q^n(x,dy). \tag{8.84}$$

The asymptotic behaviour of the second integral follows relatively easily from the weak convergence to a Γ-distribution and the dominated convergence theorem. Indeed,

$$\int_{\varepsilon\sqrt{n}}^{\infty}\frac{1}{U_p(y)}Q^n(x,dy) = \frac{1}{U_p(\sqrt{n})}\int_{\varepsilon\sqrt{n}}^{\infty}\frac{U_p(\sqrt{n})}{U_p(y)}Q^n(x,dy). \tag{8.85}$$

The monotonicity of U_p implies the following upper bound for the integrand on the right-hand side:

$$\sup_{n,\ y>\varepsilon\sqrt{n}}\frac{U_p(\sqrt{n})}{U_p(y)} \le \sup_n \frac{U_p(\sqrt{n})}{U_p(\varepsilon\sqrt{n})} < \infty, \tag{8.86}$$

because U_p is regularly varying at infinity, which also implies the convergence

$$\frac{U_p(\sqrt{n})}{U_p(u\sqrt{n})} \to \frac{1}{u^\rho} \quad \text{as } n \to \infty. \tag{8.87}$$

It follows from Theorem 4.8 that $\widehat{X}_n^2/n$ converges weakly to a Γ-distribution with probability density function $\gamma(u)$; see (8.39). Then, by Lemma 4.7, the substochastic measure $Q^n(x,\sqrt{n}\,du)$ converges weakly as $n \to \infty$ to a measure with density function $h(x)2u\gamma(u^2)$. The relations (8.86) and (8.87) allow us to apply the dominated convergence theorem and to conclude that, as $n \to \infty$,

$$\begin{aligned}
\int_{\varepsilon}^{\infty}\frac{U_p(\sqrt{n})}{U_p(u\sqrt{n})}Q^n(x,\sqrt{n}\,du) &\to h(x)\int_{\varepsilon}^{\infty}\frac{2u}{u^\rho}\gamma(u^2)du \\
&= h(x)\int_{\varepsilon^2}^{\infty}\frac{1}{u^{\rho/2}}\gamma(u)du \\
&= h(x)\frac{e^{-\varepsilon^2/2b}}{(2b)^{\rho/2}\Gamma(1+\rho/2)}.
\end{aligned}$$

Hence, (8.85) and (8.44) finally imply that

$$U_p(x)\int_{\varepsilon\sqrt{n}}^{\infty}\frac{1}{U_p(y)}Q^n(x,dy)\sim\frac{h(x)U_p(x)}{U_p(\sqrt{n})}\frac{e^{-\varepsilon^2/2b}}{(2b)^{\rho/2}\Gamma(1+\rho/2)}$$
$$=\frac{W_p(x)}{U_p(\sqrt{n})}\frac{e^{-\varepsilon^2/2b}}{(2b)^{\rho/2}\Gamma(1+\rho/2)}$$
$$=\frac{W(x)}{U(\sqrt{n})}\frac{e^{-\varepsilon^2/2b}}{(2b)^{\rho/2}\Gamma(1+\rho/2)},\tag{8.88}$$

owing to (8.53) and (8.18). Letting $\varepsilon\downarrow 0$ we conclude the following lower bound:

$$\liminf_{n\to\infty}U(\sqrt{n})\mathbb{P}_x\{\tau_{\widehat{x}}>n\}\geq\frac{W(x)}{(2b)^{\rho/2}\Gamma(1+\rho/2)},\tag{8.89}$$

which also follows by Fatou's lemma; however, (8.88) is still needed in what follows.

Fix some $\delta>0$. By the Markov property,

$$\mathbb{P}_x\{\tau_{\widehat{x}}>n\}=\int_{\widehat{x}}^{\infty}\mathbb{P}_x\{X_{[(1-\delta)n]}\in dy,\tau_{\widehat{x}}>(1-\delta)n\}\mathbb{P}_y\{\tau_{\widehat{x}}>\delta n\}.\tag{8.90}$$

It follows from the upper bound (8.70) that

$$\int_{\widehat{x}}^{\varepsilon\sqrt{n}}\mathbb{P}_x\{X_{[(1-\delta)n]}\in dy,\tau_{\widehat{x}}>(1-\delta)n\}\mathbb{P}_y\{\tau_{\widehat{x}}>\delta n\}$$
$$\leq\frac{C}{U_p(\sqrt{\delta n})}\int_{\widehat{x}}^{\varepsilon\sqrt{n}}U_p(y)\mathbb{P}_x\{X_{[(1-\delta)n]}\in dy,\tau_{\widehat{x}}>(1-\delta)n\}$$
$$=\frac{CU_p(x)}{U_p(\sqrt{\delta n})}\int_{\widehat{x}}^{\varepsilon\sqrt{n}}Q^{[(1-\delta)n]}(x,dy)$$
$$\leq\frac{CU_p(x)}{U_p(\sqrt{\delta n})}\mathbb{P}_x\{\widehat{X}_{[(1-\delta)n]}\leq\varepsilon\sqrt{n}\},$$

since Q is substochastic. The function U_p is regularly varying at infinity with index ρ; hence $U_p(\sqrt{\delta n})/U_p(\sqrt{n})\to\delta^{\rho/2}$ as $n\to\infty$. Together with the weak convergence of $\widehat{X}_n^2/n$ to a Γ-distribution, it implies that, for all $\delta>0$,

$$\lim_{\varepsilon\to 0}\limsup_{n\to\infty}U_p(\sqrt{n})\int_{\widehat{x}}^{\varepsilon\sqrt{n}}\mathbb{P}_x\{X_{[(1-\delta)n]}\in dy,\tau_{\widehat{x}}>(1-\delta)n\}\mathbb{P}_y\{\tau_{\widehat{x}}>\delta n\}=0.\tag{8.91}$$

Further,

$$\begin{aligned}
\int_{\varepsilon\sqrt{n}}^{\infty} & \mathbb{P}_x\{X_{[(1-\delta)n]} \in dy, \tau_{\widehat{x}} > (1-\delta)n\}\mathbb{P}_y\{\tau_{\widehat{x}} > \delta n\} \\
&\le \int_{\varepsilon\sqrt{n}}^{\infty} \mathbb{P}_x\{X_{[(1-\delta)n]} \in dy, \tau_{\widehat{x}} > (1-\delta)n\} \\
&= U_p(x)\int_{\varepsilon\sqrt{n}}^{\infty} \frac{1}{U_p(y)} Q^{[(1-\delta)n]}(x,dy).
\end{aligned}$$

As proven in (8.88),

$$U_p(x)\int_{\varepsilon\sqrt{n}}^{\infty} \frac{1}{U_p(y)} Q^{[(1-\delta)n]}(x,dy) \sim \frac{W(x)}{U(\sqrt{(1-\delta)n})}\frac{e^{-\varepsilon^2/2b(1-\delta)}}{(2b)^{\rho/2}\Gamma(1+\rho/2)}. \tag{8.92}$$

The substitution of (8.91) and (8.92) into (8.90) leads to

$$\limsup_{n\to\infty} U(\sqrt{n})\mathbb{P}_x\{\tau_{\widehat{x}} > n\} \le \limsup_{n\to\infty} \frac{W(x)U(\sqrt{n})}{U(\sqrt{(1-\delta)n})}\frac{1}{(2b)^{\rho/2}\Gamma(1+\rho/2)}.$$

Since $U(\sqrt{n})/U(\sqrt{(1-\delta)n}) \to (1-\delta)^{-\rho/2}$ and since $\delta > 0$ may be chosen as small as we please, we obtain the upper bound

$$\limsup_{n\to\infty} U(\sqrt{n})\mathbb{P}_x\{\tau_{\widehat{x}} > n\} \le \frac{W(x)}{(2b)^{\rho/2}\Gamma(1+\rho/2)},$$

which together with the lower bound (8.89) completes the proof of the asymptotics (8.71).

Conditioning on X_0, we conclude that (8.72) holds by the dominated convergence theorem owing to (8.71) and (8.70). □

Corollary 8.22 *Under the conditions of Theorem 8.2, for any initial distribution such that $\mathbb{E}U(X_0) < \infty$,*

$$\mathbb{P}\{\tau_{\widehat{x}} > n\} \sim \frac{\mathbb{E}\{W(X_1);\ X_1 > \widehat{x}\}}{(2b)^{\rho/2}\Gamma(1+\rho/2)}\frac{1}{U(\sqrt{n})} \quad \textit{as } n\to\infty.$$

In Theorem 11.9, we apply this result to derive asymptotic estimates for the non-extinction probability of branching processes with migration.

Proof We have

$$\begin{aligned}
&\mathbb{P}\{\tau_{\widehat{x}} > n\} \\
&= \int_{\widehat{x}}^{\infty} \mathbb{P}_y\{\tau_{\widehat{x}} > n\}\mathbb{P}\{X_0 \in dy\} + \int_{-\infty}^{\widehat{x}} \mathbb{P}\{X_0 \in dy\}\int_{\widehat{x}}^{\infty} P(y,dz)\mathbb{P}_z\{\tau_{\widehat{x}} > n-1\} \\
&\sim \frac{\mathbb{E}\{W(X_0); X_0 > \widehat{x}\} + \mathbb{E}\{W(X_1); X_0 \le \widehat{x}, X_1 > \widehat{x}\}}{(2b)^{\rho/2}\Gamma(1+\rho/2)}\frac{1}{U(\sqrt{n})},
\end{aligned}$$

by Theorem 8.18, and the result then follows from the harmonicity of W. □

Next let us discuss an implication for a discrete state space according to which it is possible to extend the results of the last theorem to the hitting time of any finite subset D of the state space:

$$\tau_D := \min\{n \geq 1 : X_n \in D\}.$$

Theorem 8.23 *Assume that $\{X_n\}$ is a countable Markov chain on a state space $\{z_0 < z_1 < z_2 < \cdots\}$ satisfying the conditions of Theorem 8.2. Then, for any finite subset D of the state space, there exists a $c = c(D) < \infty$ such that*

$$\mathbb{P}_x\{\tau_D > n\} \leq c\frac{U(x)}{U(\sqrt{n})} \quad \textit{for all n and x}. \tag{8.93}$$

In addition, for any fixed initial state x,

$$\mathbb{P}_x\{\tau_D > n\} \sim \frac{C(x,D)}{U(\sqrt{n})} \quad \textit{as } n \to \infty, \tag{8.94}$$

where

$$C(x,D) := \frac{1}{(2b)^{\rho/2}\Gamma(1+\rho/2)} \sum_{j=0}^{\infty} \mathbb{E}_x\{W(X_{j+1}); X_{j+1} > \widehat{x}, \tau_D > j\} \in (0,\infty);$$

here $\widehat{x}$ is any truncation level guaranteed by Corollary 8.7 and such that $D \subseteq B := [z_0, \widehat{x}]$.

Proof Owing to the upper bound (8.70) provided by Theorem 8.18 and by the Markov property, it is enough to prove (8.93) for $x \leq \widehat{x}$. To start, consider the case where $B \setminus D$ is a singleton, say z_1. Given $X_0 = z_1$, the distribution of the hitting time τ_B may be decomposed as the following mixture of distributions, according to the position of the chain at time $\tau_{\widehat{x}}$:

$$\mathbb{P}_{z_1}\{\tau_B > n\} = p\mathbb{P}_{z_1}\{\eta > n\} + (1-p)\mathbb{P}_{z_1}\{\theta > n\},$$

where $p = \mathbb{P}_{z_1}\{X_{\tau_B} = z_1\}$, $1 - p = \mathbb{P}_{z_1}\{X_{\tau_B} \in D\}$, the distribution of the random variable η is the conditional distribution of τ_B given $X_{\tau_B} = z_1$, and the distribution of θ is the conditional distribution of τ_B given $X_{\tau_B} \in D$. Since the chain may visit z_1 several times before hitting D, we get

$$\mathbb{P}_{z_1}\{\tau_D > n\} = (1-p)\sum_{k=0}^{\infty} p^k \mathbb{P}\{\eta_1 + \cdots + \eta_k + \theta > n\},$$

where the η_k are independent copies of η. By Theorem 8.18, the distribution of τ_B is regularly varying, so the tail distributions of both of η and θ possess regularly varying upper bounds of order $c/U(\sqrt{n})$, is known to be of subexponential type. Thus, Kesten's bound – see e.g. [67, Section 3.10] – shows that a random sum possesses the same regularly varying upper bound, and the proof of (8.93) for the case $|B \setminus D| = 1$ follows. Since we have used only the upper bound for the tail of τ_B in our proof of the upper bound for the tail of τ_D, we may apply the same argument to the case of an arbitrary number of states in $B \setminus D$, by induction on this number.

Now let us prove (8.94). For any $N < n/2$,

$$\begin{aligned}\mathbb{P}_x\{\tau_D > n\} &= \mathbb{P}_x\{\tau_D > n, X_j \le \widehat{x} \text{ for some } j \in [N,n]\}\\ &\quad + \mathbb{P}_x\{\tau_D > n, X_j > \widehat{x} \text{ for all } j \in [N,n]\}\\ &=: P_1 + P_2. \end{aligned} \tag{8.95}$$

Let us show that the first probability becomes negligible when N increases. Indeed,

$$\begin{aligned}P_1 &\le \mathbb{P}_x\{\tau_D > n, X_j \le \widehat{x} \text{ for some } j \in [N,n/2]\}\\ &\quad + \mathbb{P}_x\{\tau_D > n, X_j \le \widehat{x} \text{ for some } j \in (n/2,n-N]\}\\ &\quad + \mathbb{P}_x\{\tau_D > n, X_j > \widehat{x} \text{ for all } j \in [N,n-N],\\ &\qquad\qquad X_j \le \widehat{x} \text{ for some } j \in (n-N,n]\}\\ &=: P_{11} + P_{12} + P_{13}.\end{aligned}$$

As proven in Theorem 8.18, the tail of τ_B is regularly varying; hence

$$\mathbb{E}\mathbb{P}_{X_N}\{\tau_B = n+k\} = o(\mathbb{P}\{\tau_B > n\}) \quad \text{as } n \to \infty$$

for any fixed $k \in \mathbb{Z}$. Thus, so that, for any fixed N,

$$\begin{aligned}P_{13} &\le \mathbb{E}\mathbb{P}_{X_N}\{X_j > \widehat{x} \text{ for all } k \in [N,n-N], X_j \le \widehat{x} \text{ for some } j \in (n-N,n]\}\\ &= \mathbb{E}\mathbb{P}_{X_N}\{\tau_B \in (n-2N,n-N]\}\\ &= o(\mathbb{P}\{\tau_B > n\}) \quad \text{as } n \to \infty. \end{aligned} \tag{8.96}$$

By the Markov property,

$$P_{11} \le \mathbb{P}_x\{\tau_D > N\} \max_{y \le \widehat{x}} \mathbb{P}_y\{\tau_D > n/2\} \le \frac{c_1}{U(\sqrt{N})U(\sqrt{n/2})}, \tag{8.97}$$

by (8.93), because there is only a finite number of states in $[0,\widehat{x}\,]$ and

$$P_{12} \le \mathbb{P}_x\{\tau_D > n/2\} \max_{y \le \widehat{x}} \mathbb{P}_y\{\tau_D > N\} \le \frac{c_2}{U(\sqrt{n/2})U(\sqrt{N})}. \tag{8.98}$$

It follows from the inequalities (8.96)–(8.98) and the regular variation of U that

$$\lim_{N\to\infty} \limsup_{n\to\infty} U(\sqrt{n})P_1 = 0. \tag{8.99}$$

Further, decomposing all the trajectories according to the time of the last visit to $[z_0, \widehat{x}]$, we obtain, by the Markov property,

$$\begin{aligned}
P_2 &= \sum_{j=0}^{N-1} \sum_{y\in B\setminus D} \mathbb{P}_x\{X_j = y, \tau_D > j\}\mathbb{P}_y\{\tau_B > n-j\} \\
&\sim \frac{1}{(2b)^{\rho/2}\Gamma(1+\rho/2)} \frac{1}{U(\sqrt{n})} \\
&\quad \times \sum_{j=1}^{N-1} \sum_{y\in B\setminus D} \mathbb{P}_x\{X_j = y, \tau_D > j\}\mathbb{E}_y\{W(X_1); X_1 > \widehat{x}\}
\end{aligned}$$

as $n\to\infty$, by Corollary 8.22, because $U(\sqrt{n})$ is regularly varying. Summing over y we get, as $n\to\infty$,

$$P_2 \sim \frac{1}{(2b)^{\rho/2}\Gamma(1+\rho/2)} \frac{1}{U(\sqrt{n})} \sum_{j=0}^{N-1} \mathbb{P}_x\{W(X_{j+1}), X_{j+1} > \widehat{x}, \tau_D > j\},$$

which being substituted into (8.95) together with (8.99) gives the required result if we let $N\to\infty$. □

8.6 Limit Theorems for Conditioned Markov Chains

In this section we prove limit theorems for positive and null recurrent Markov chains $\{X_n\}$ conditioned on the event

$$\{X_1 > \widehat{x}, \ldots, X_n > \widehat{x}\}.$$

Theorem 8.24 *Let the conditions of Theorem 8.2 hold; in particular, let $2\mu > -b$. Let $\mathbb{E}U(X_0) < \infty$. Then, for all $u > 0$,*

$$\mathbb{P}\left\{\frac{X_n^2}{nb} > u \,\middle|\, \tau_{\widehat{x}} > n\right\} \to e^{-u/2} \quad \text{as } n\to\infty.$$

In Section 11.2 we will apply the last result to a reflected random walk with zero drift and in Section 11.3 to branching processes.

Proof For any fixed initial state $x > \widehat{x}$, by a change of measure we have

$$\begin{aligned}
\mathbb{P}_x\left\{\frac{X_n^2}{nb} > u, \tau_{\widehat{x}} > n\right\} &= U_p(x)\int_{\sqrt{unb}}^{\infty} \frac{1}{U_p(y)} Q_n(x,dy) \\
&\sim \frac{W(x)}{U(\sqrt{n})} \frac{e^{-u/2}}{(2b)^{\rho/2}\Gamma(1+\rho/2)},
\end{aligned}$$

as shown in (8.88). Combining this with the tail asymptotics for $\tau_{\widehat{x}}$ given in Theorem 8.18, we arrive at the required result for $x > \widehat{x}$. Then we follow the same arguments as in Corollary 8.22. □

Corollary 8.25 *Assume that* $\{X_n\}$ *is a countable Markov chain on a state space* $\{z_0 < z_1 < z_2 < \cdots\}$. *Then, for any finite subset D of the state space and for all* $u > 0$,

$$\mathbb{P}\left\{\frac{X_n^2}{nb} > u \,\middle|\, \tau_D > n\right\} \to e^{-u/2} \quad \textit{as } n \to \infty.$$

In Theorem 11.9 we apply this result to derive the limit behaviour of branching processes with migration given that they do not become extinct.

Proof Fix an $N \geq 1$. It follows from (8.99) and the asymptotic tail behaviour of τ_D – see Theorem 8.23 – that

$$\lim_{N\to\infty} \limsup_{n\to\infty} \mathbb{P}\left\{\frac{X_n^2}{nb} > u, X_j \leq \widehat{x} \text{ for some } j \in [N,n] \middle| \tau_D > n\right\} = 0. \quad (8.100)$$

Further, by the Markov property,

$$\begin{aligned}
&\mathbb{P}\left\{\frac{X_n^2}{nb} > u, \tau_D > n, X_j > \widehat{x} \text{ for all } j \in [N,n]\right\} \\
&\quad = \sum_{y>\widehat{x}} \mathbb{P}\{X_N = y, \tau_D > N\}\mathbb{P}_y\left\{\frac{X_{n-N}^2}{nb} > u, \tau_B > n - N\right\}.
\end{aligned}$$

Since $\mathbb{E}U(X_0) < \infty$ and U is regularly varying, $\mathbb{E}U(X_N)$ is also finite. Now applying Theorem 8.24, we get

$$\begin{aligned}
&\mathbb{P}\left\{\frac{X_n^2}{nb} > u, \tau_D > n, X_j > \widehat{x} \text{ for all } j \in [N,n]\right\} \\
&\qquad \sim e^{-u/2} \sum_{y>\widehat{x}} \mathbb{P}\{X_N = y, \tau_D > N\}\mathbb{P}_y\{\tau_B > n - N\} \\
&\qquad = e^{-u/2}\mathbb{P}\{\tau_D > n, X_j > \widehat{x} \text{ for all } j \in [N,n]\} \\
&\qquad = e^{-u/2} P_2,
\end{aligned}$$

where P_2 is defined in (8.95), which in combination with (8.100) yields the required limit behaviour. □

Theorem 8.26 *Let the conditions of Theorem 8.2 hold. Then, for any* $x > \widehat{x}$,

$$\mathbb{P}_x\left\{\max_{n\leq\tau_{\widehat{x}}} X_n > y\right\} \sim \frac{W(x)}{W(y)} \quad \textit{as } y \to \infty,$$

where W is the harmonic function for the chain $\{X_n\}$ *killed at the time of its first visit to* $(-\infty,\widehat{x}\,]$, *see Corollary 8.10.*

Proof First notice that

$$\mathbb{P}_x\left\{\max_{n\le\tau_{\widehat{x}}} X_n > y\right\} = \mathbb{P}_x\{\tau_{\widehat{x}} > T(y)\}.$$

The harmonicity of W_p implies that the sequence $W_p(X_n)\mathbb{I}\{\tau_{\widehat{x}} > n\}$ is a martingale. Applying the optional stopping theorem (see e.g. [56, Section 4.7]) to this martingale and to the stopping time $\tau_{\widehat{x}} \wedge T(y)$, we obtain

$$W_p(x) = \mathbb{E}_x\{W_p(X_{T(y)});\ \tau_{\widehat{x}} > T(y)\}.$$

Since $W_p(z) \sim U_p(z)$ as $z \to \infty$, we have

$$\mathbb{E}_x\{U_p(X_{T(y)});\ \tau_{\widehat{x}} > T(y)\} \to W_p(x) \quad \text{as } y \to \infty. \tag{8.101}$$

Let us split the expectation into two parts:

$$\begin{aligned}
&\mathbb{E}_x\{U_p(X_{T(y)});\ \tau_{\widehat{x}} > T(y)\} \\
&\quad = \mathbb{E}_x\{U_p(X_{T(y)});\ \tau_{\widehat{x}} > T(y), X_{T(y)} \le y + s(y)\} \\
&\qquad + \mathbb{E}_x\{U_p(X_{T(y)});\ \tau_{\widehat{x}} > T(y), X_{T(y)} > y + s(y)\}.
\end{aligned} \tag{8.102}$$

Since $s(y) = o(y)$ and U_p is a regularly varying function, $U_p(y + s(y)) \sim U_p(y)$ as $y \to \infty$, so that

$$\begin{aligned}
&\mathbb{E}_x\{U_p(X_{T(y)});\ \tau_{\widehat{x}} > T(y), X_{T(y)} \le y + s(y)\} \\
&\quad \sim U_p(y)\mathbb{P}_x\{\tau_{\widehat{x}} > T(y), X_{T(y)} \le y + s(y)\}.
\end{aligned} \tag{8.103}$$

By a change of measure with function U_p and the fact that the resulting kernel Q is substochastic, we have

$$\mathbb{E}_x\{U_p(X_{T(y)}), \tau_{\widehat{x}} > T(y), X_{T(y)} > y + s(y)\} \le U_p(x)\mathbb{P}_x\{\widehat{X}_{\widehat{T}(y)} > y + s(y)\}.$$

By the formula for total probability,

$$\begin{aligned}
\mathbb{P}_x\{\widehat{X}_{\widehat{T}(y)} > y + s(y)\} &= \sum_{n=0}^{\infty}\int_{\widehat{x}}^{y} \mathbb{P}_x\{\widehat{X}_n \in dz, \widehat{T}(y) > n\}\mathbb{P}\{\widehat{\xi}(z) > y + s(y) - z\} \\
&\le \int_{\widehat{x}}^{y} \mathbb{P}\{\widehat{\xi}(z) > s(y)\}\widehat{H}_x(dz).
\end{aligned}$$

According to (8.36), $\mathbb{P}\{\widehat{\xi}(z) > s(z)\} = o(p(z)/z)$. Then, similarly to the integral estimation in the proof of Lemma 4.1, we conclude that

$$\int_{\widehat{x}}^{\infty} \mathbb{P}\{\widehat{\xi}(z) > s(z)\}\widehat{H}_x(dz) < \infty.$$

Consequently,

$$\int_{\widehat{x}}^{y} \mathbb{P}\{\widehat{\xi}(z) > s(y)\}\widehat{H}_x(dz) \to 0 \quad \text{as } y \to \infty.$$

As a result,

$$\mathbb{E}_x\{U_p(X_{T(y)});\ \tau_{\widehat{x}} > T(y), X_{T(y)} > y + s(y)\} \to 0 \quad \text{as } y \to \infty \quad (8.104)$$

and hence

$$U_p(y)\mathbb{P}_x\{\tau_{\widehat{x}} > T(y), X_{T(y)} > y + s(y)\} \to 0 \quad \text{as } y \to \infty. \quad (8.105)$$

Applying (8.105) to (8.103) we get

$$\begin{aligned}&\mathbb{E}_x\{U_p(X_{T(y)});\ \tau_{\widehat{x}} > T(y), X_{T(y)} \le y + s(y)\}\\ &\quad = (1 + o(1))U_p(y)\mathbb{P}_x\{\tau_{\widehat{x}} > T(y)\} + o(1).\end{aligned}$$

Combining this with (8.104), we obtain from (8.102) the following equality, as $y \to \infty$,

$$\mathbb{E}_x\{U_p(X_{T(y)});\ \tau_{\widehat{x}} > T(y)\} = (1 + o(1))U_p(y)\mathbb{P}_x\{\tau_{\widehat{x}} > T(y)\} + o(1).$$

substituting this into (8.101) gives

$$U_p(y)\mathbb{P}_x\{\tau_{\widehat{x}} > T(y)\} \to W_p(x) \quad \text{as } y \to \infty,$$

which completes the proof since $U_p(y) \sim W_p(y) = e^{-c_p}W(y)$. □

Now we turn to functional limit theorems for a recurrent chain $\{X_n\}$ conditioned on $\{\tau_{\widehat{x}} > n\}$. Durrett [55] suggested a method for deriving functional limit theorems for conditional distributions of null recurrent Markov chains from the corresponding limit theorems for unconditioned chains. His approach is applicable in the case $\mu \in (-b/2, b/2)$. It immediately follows from Theorems 4.11 and 8.24 that the conditions of [55, Theorem 3.9] are satisfied. Therefore, the finite-dimensional distributions of $\{X_{[nt]}/\sqrt{bn}\}$ conditioned on $\{\tau_{\widehat{x}} > n\}$ converge to the finite-dimensional distributions of a (time-inhomogeneous) Markov process $X^+(t)$, which may be described in terms of the limiting Bessel process $Bes(t)$ in Theorem 4.11 – with drift μ/bx and diffusion coefficient 1 – and in terms of its first hitting time of the origin, $T_0 = \min\{t : Bes(t) = 0\}$. The process $X^+(t)$, $0 \le t \le 1$, is a Markov process on $\mathbb{R}^+$ starting at the origin, with entrance law (transition probabilities from the origin)

$$\mathbb{P}\{X^+(t) \in dy\} = \frac{y}{t^{\rho/2+1/2}}e^{-y^2/2t}\mathbb{P}\{T_0 > 1 - t \mid Bes(0) = y\}dy,$$

for $t \in (0, 1]$, where $\rho = 1 + 2\mu/b$ and with transition kernel, for $t > s$,

$$\begin{aligned}&\mathbb{P}\{X^+(t) \in dy \mid X^+(s) = x\}\\ &\quad = \frac{\mathbb{P}\{T_0 > 1 - t \mid Bes(0) = y\}}{\mathbb{P}\{T_0 > 1 - s \mid Bes(0) = x\}}\mathbb{P}\{Bes(t - s) \in dy, T_0 > t - s \mid Bes(0) = x\}.\end{aligned}$$

It is easy to see that $X^+(t)$ converges in probability to zero as $t \to 0$. Then, again using Theorem 3.9 in [55], we conclude that the sequence of conditional distributions is tight in $D[0,1]$. Therefore, we get weak convergence in the space $D[0,1]$.

We follow a different strategy for positive recurrent Markov chains, which allows us to avoid proving a functional limit theorem for unconditioned positive recurrent chains $\{X_{[nt]}\}$ with a starting state of order $\sqrt{n}$. To the best of our knowledge, such a functional limit is known only for chains on $\mathbb{Z}^+$; see [15, Theorem 5].

Below we suggest an alternative approach which is based on the change of measure technique and uses functional limit theorems for transient chains. As in the discussion after Theorem 4.11 in Section 4.7, we define $\{X^{(n)}(t)\}$ as a continuous piecewise-linear process whose trajectories connect points $(k/n, X_k/\sqrt{bn})$ by segments. The limiting process $X^+(t)$ may be equivalently defined, via the values of $\mathbb{E}g(X^+)$ for all bounded continuous functionals g on the space $C[0,1]$, as follows: as above, starting with the Bessel process $Bes(t)$, now with drift $(\mu+b)/bx$ and diffusion coefficient 1, we define a Markov process X^+ starting at the origin and such that

$$\mathbb{E}g(X^+) \;=\; 2^{\rho/2}\Gamma(1+\rho/2)\mathbb{E}\frac{g(Bes)}{Bes^{\rho}(1)} \;=\; \frac{\mathbb{E}g(Bes)Bes^{-\rho}(1)}{\mathbb{E}Bes^{-\rho}(1)}.$$

Theorem 8.27 *Let the conditions of Theorem 8.2 hold: in particular, let $2\mu > -b$. Then the process $X^{(n)}$ conditioned on $\{\tau_{\widehat{x}} > n\}$ converges weakly to $X^+(t)$ in the space $C[0,1]$ as $n \to \infty$.*

Proof It suffices to prove this weak convergence for the case $X_0 > \widehat{x}$. Let g be a bounded continuous functional on the space $C[0,1]$. We need to show that, for all $x > \widehat{x}$,

$$\mathbb{E}_x\{g(X^{(n)}) \mid \tau_{\widehat{x}} > n\} \;\to\; \mathbb{E}g(X^+) \quad \text{as } n \to \infty. \tag{8.106}$$

Our strategy is to represent the expectation on the left hand side as a functional of a transient Markov chain. So, we consider the process $\widehat{X}^{(n)}(t)$, $t \in [0,1]$, constructed as a continuous piecewise-linear process whose trajectories connect points $(k/n, \widehat{X}_k/\sqrt{bn})$ by segments; here $\{\widehat{X}_k\}$ is the transient Markov chain constructed in Section 7.1 as a Doob's h-transform with function U_p of the original Markov chain $\{X_k\}$. Then it follows from (7.6) that, for any bounded functional g on the space $C[0,1]$,

$$\mathbb{E}_x\{g(X^{(n)}) \mid \tau_{\widehat{x}} > n\} = \frac{U_p(x)}{\mathbb{P}_x\{\tau_{\widehat{x}} > n\}}\mathbb{E}_x\left\{\frac{e^{-\sum_{k=0}^{n-1} q(\widehat{X}_k)}}{U_p(\widehat{X}_n)} g(\widehat{X}^{(n)})\right\}.$$

By Theorem 8.18 and definitions (8.41) and (8.44),

$$\begin{aligned}\mathbb{P}_x\{\tau_{\widehat{x}} > n\} &\sim \frac{1}{(2b)^{\rho/2}\Gamma(1+\rho/2)}\frac{W_p(x)}{U_p(\sqrt{n})}\\ &\sim \frac{1}{(2b)^{\rho/2}\Gamma(1+\rho/2)}\frac{U_p(x)}{U_p(\sqrt{n})}\mathbb{E}_x e^{-\sum_{k=0}^{\infty} q(\widehat{X}_k)} \quad \text{as } n\to\infty.\end{aligned}$$

Therefore

$$\mathbb{E}_x\{g(X^{(n)}) \mid \tau_{\widehat{x}} > n\} \sim \frac{(2b)^{\rho/2}\Gamma(1+\rho/2)}{\mathbb{E}_x e^{-\sum_{k=0}^{\infty} q(\widehat{X}_k)}}\mathbb{E}_x\left\{\frac{e^{-\sum_{k=0}^{n-1} q(\widehat{X}_k)}}{U_p(\widehat{X}_n)/U_p(\sqrt{n})}g(\widehat{X}^{(n)})\right\}. \tag{8.107}$$

Fix a $\delta > 0$. Since g is bounded, it follows from Theorem 8.24 that, for all $\delta > 0$ and n,

$$\begin{aligned}\left|\mathbb{E}\{g(X^{(n)})\mathbb{I}\{X_n \le \delta\sqrt{nb}\} \mid \tau_{\widehat{x}} > n\}\right| &\le \|g\|_\infty \mathbb{P}\{X_n \le \delta\sqrt{nb} \mid \tau_{\widehat{x}} > n\}\\ &\le C\delta.\end{aligned} \tag{8.108}$$

Applying (8.107) to $g(X^{(n)})\,\mathbb{I}\{X^{(n)}(1) > \delta\}$ we get

$$\begin{aligned}&\mathbb{E}_x\{g(X^{(n)})\,\mathbb{I}\{X_n > \delta\sqrt{nb}\} \mid \tau_{\widehat{x}} > n\}\\ &\quad\sim \frac{(2b)^{\rho/2}\Gamma(1+\rho/2)}{\mathbb{E}_x e^{-\sum_{k=0}^{\infty} q(\widehat{X}_k)}}\mathbb{E}_x\left\{\frac{e^{-\sum_{k=0}^{n-1} q(\widehat{X}_k)}}{U_p(\widehat{X}_n)/U_p(\sqrt{n})}g(\widehat{X}^{(n)})\,\mathbb{I}\{\widehat{X}^{(n)}(1) > \delta\}\right\}.\end{aligned}$$

Owing to the regular variation at infinity of the function U_p, we have the convergence

$$\frac{U_p(\widehat{X}_n)}{U_p(\sqrt{n})} = \frac{U_p(\widehat{X}^{(n)}(1)\sqrt{nb})}{U_p(\sqrt{n})} \to (\widehat{X}^{(n)}(1))^\rho b^{\rho/2} \quad \text{as } n\to\infty$$

uniformly on the event $\{\widehat{X}^{(n)}(1) > \delta\} = \{\widehat{X}_n/\sqrt{nb} > \delta\}$. Hence,

$$\begin{aligned}&\mathbb{E}_x\{g(X^{(n)})\,\mathbb{I}\{X_n > \delta\sqrt{nb}\} \mid \tau_{\widehat{x}} > n\}\\ &\quad\sim \frac{2^{\rho/2}\Gamma(1+\rho/2)}{\mathbb{E}_x e^{-\sum_{k=0}^{\infty} q(\widehat{X}_k)}}\mathbb{E}_x\left\{\frac{e^{-\sum_{k=0}^{n-1} q(\widehat{X}_k)}}{(\widehat{X}^{(n)}(1))^\rho}g(\widehat{X}^{(n)})\,\mathbb{I}\{\widehat{X}^{(n)}(1) > \delta\}\right\}.\end{aligned} \tag{8.109}$$

The Bessel approximation proven in Theorem 4.11 still holds if a certain number of first values of the Markov chain are fixed; hence we conclude that, for any bounded continuous functional g on the space $C[0,1]$,

$$\begin{aligned}&\mathbb{E}\left\{\frac{g(\widehat{X}^{(n)})}{(\widehat{X}^{(n)}(1))^\rho}\,\mathbb{I}\{\widehat{X}^{(n)}(1) > \delta\} \;\middle|\; X_0 = z_0,\ldots,X_N = z_N\right\}\\ &\quad\to \mathbb{E}\left\{\frac{g(Bes)}{(Bes(1))^\rho}\,\mathbb{I}\{Bes(1) > \delta\}\right\},\end{aligned}$$

for all N and $z_0, \dots, z_N$. This makes it possible to apply Lemma 4.6; hence we have

$$\mathbb{E}_x\left\{\frac{e^{-\sum_{k=0}^{n-1} q(\widehat{X}_k)}}{(\widehat{X}^{(n)}(1))^\rho} g(\widehat{X}^{(n)})\, \mathbb{I}\{\widehat{X}^{(n)}(1) > \delta\}\right\}$$
$$\to \mathbb{E} e^{-\sum_{k=0}^{\infty} q(\widehat{X}_k)} \mathbb{E}\left\{\frac{g(Bes)}{(Bes(1))^\rho}\, \mathbb{I}\{Bes(1) > \delta\}\right\} \quad \text{as } n \to \infty.$$

From this estimate and (8.109) we obtain

$$\mathbb{E}_x\{g(X^{(n)})\, \mathbb{I}\{X_n > \delta\sqrt{nb}\} \mid \tau_{\widehat{x}} > n\}$$
$$\sim 2^{\rho/2}\Gamma(1+\rho/2)\mathbb{E}\left\{\frac{g(Bes)}{(Bes(1))^\rho}\, \mathbb{I}\{Bes(1) > \delta\}\right\} \quad \text{as } n \to \infty.$$

Combining this with (8.108) and the upper bound

$$\mathbb{E}\left\{(Bes(1))^{-\rho};\ Bes(1) < \delta\right\} = O(\delta)$$

and letting $\delta \to 0$, we get

$$\mathbb{E}\{g(X^{(n)}) \mid \tau_{\widehat{x}} > n\} \to 2^{\rho/2}\Gamma(1+\rho/2)\mathbb{E}\frac{g(Bes)}{(Bes(1))^\rho} \quad \text{as } n \to \infty,$$

and hence the desired convergence. □

Corollary 8.28 *Assume that $\{X_n\}$ is a countable Markov chain on a state space $\{z_0 < z_1 < z_2 < \cdots\}$. Then, under the conditions of Theorem 8.2, for any finite subset D of the state space the process $X^{(n)}$ conditioned on $\{\tau_D > n\}$ converges weakly to $X^+(t)$ in the space $C[0,1]$, as $n \to \infty$.*

Proof Fix $N \ge 1$. It follows from (8.99) and the asymptotic tail behaviour of τ_D – see Theorem 8.23 – that

$$\lim_{N\to\infty} \limsup_{n\to\infty} \mathbb{E}\big\{g(X^{(n)}); X_j \le \widehat{x} \text{ for some } j \in [N,n] \big| \tau_D > n\big\} = 0. \quad (8.110)$$

By the Markov property,

$$\mathbb{E}\big\{g(X^{(n)}); \tau_D > n, X_j > \widehat{x} \text{ for all } j \in [N,n]\big\}$$
$$= \sum_{y_1,\dots,y_N \notin D} \mathbb{P}\{X_1 = y_1, \dots, X_N = y_N\}$$
$$\times \mathbb{E}\{g(X^{(n)}); \tau_B > n \mid X_1 = y_1, \dots, X_N = y_N\}.$$

Now applying Theorem 8.27, which is still valid for the conditional expectations on the right-hand side, we get

$$\begin{aligned}
&\mathbb{E}\{g(X^{(n)}); \tau_D > n, X_j > \widehat{x} \text{ for all } j \in [N,n]\} \\
&\quad \sim \mathbb{E}g(X^+) \sum_{y_1,\ldots,y_N \notin D} \mathbb{P}\{X_1 = y_1, \ldots, X_N = y_N\} \\
&\qquad\qquad \times \mathbb{P}\{\tau_B > n \mid X_1 = y_1, \ldots, X_N = y_N\} \\
&\quad = \mathbb{E}g(X^+)\mathbb{P}\{\tau_D > n, X_j > \widehat{x} \text{ for all } j \in [N,n]\} \\
&\quad = \mathbb{E}g(X^+)P_2,
\end{aligned}$$

where P_2 is defined in (8.95), which in combination with (8.110) yields the required limiting behaviour. □

8.7 Limit Theorem in Critical Case $2\mu = b$

In the critical case $\mu = b/2$ we have a different type of limit behaviour. It may be described in terms of the function

$$G(x) := \int_{\widehat{x}}^{x} \frac{y}{U(y)} dy,$$

which is slowly varying at infinity because U is regularly varying with index $\rho = 2\mu/b + 1 = 2$.

Theorem 8.29 *Let $\{X_n\}$ be a Markov chain on a countable set $\{z_0 < z_1 < z_2 < \cdots\}$. Let the conditions of Theorem 8.2 hold with $\mu = b/2$. If $G(x) \to \infty$ as $x \to \infty$ then $G(X_n)/G(\sqrt{n})$ converges weakly as $n \to \infty$ to a uniform distribution on the interval* $[0, 1]$.

At the end of Section 11.3.3 we apply the last result to derive asymptotics for a branching process with migration.

Corollary 8.30 *In the particular case where, for some $m \geq 1$ and $\gamma > 0$,*

$$r(x) = \frac{1}{x} + \frac{1}{x\log x} + \cdots + \frac{1}{x\log x\cdots\log_{(m-1)} x} + \frac{1-\gamma}{x\log x\cdots\log_{(m)} x},$$

we have

$$R(x) = \log x + \log_{(2)} x + \cdots + \log_{(m)} x + (1-\gamma)\log_{(m+1)} x,$$

$$U(x) \sim \frac{x^2}{2}\log x\ \log_{(2)} x\cdots\log_{(m-1)} x\ \log_{(m)}^{1-\gamma} x,$$

$$\frac{x}{U(x)} \sim \frac{2}{\gamma x \log x \, \log_{(2)} x \cdots \log_{(m-1)} x \, \log_{(m)}^{1-\gamma} x},$$
$$G(x) \sim \frac{2}{\gamma} \log_{(m)}^{\gamma} x \quad \text{as } x \to \infty.$$

Then the following weak convergence holds true:

$$\left(\frac{\log_{(m)} X_n}{\log_{(m)} \sqrt{n}} \right)^{\gamma} \Rightarrow U[0,1] \quad \text{as } n \to \infty.$$

Proof of Theorem 8.29. According to Corollary 8.13, the assumption $G(x) \to \infty$ implies the null recurrence of $\{X_n\}$. Furthermore, by Theorem 8.23,

$$U(\sqrt{n})\mathbb{P}_x\{\tau_{z_0} > n\} \to C(x, z_0) \quad \text{as } n \to \infty.$$

Let the random variables T_k be the time intervals between consequent visits of $\{X_n\}$ to the state z_0. These random variables are independent. Moreover, T_2, $T_3, \ldots$ are identically distributed and, for every $k \geq 2$,

$$\mathbb{P}\{T_k > n\} \sim \frac{C(z_0, z_0)}{U(\sqrt{n})} \quad \text{as } n \to \infty.$$

Let θ_n denote the corresponding renewal process, that is,

$$\theta_n := \max\{k \geq 1 : T_1 + T_2 + \cdots + T_k \leq n\}.$$

Let us also introduce a sequence of undershoots:

$$O_n := n - (T_1 + T_2 + \cdots + T_{\theta_n}), \quad n \geq 1.$$

It is clear from the definition of θ_n that

$$\mathbb{P}\{O_n = j\} = \mathbb{P}\{X_{n-j} = z_0\}\mathbb{P}\{T_2 > j\} \quad \text{for } 0 \leq j \leq n-1$$

and

$$\mathbb{P}\{O_n = n\} = \mathbb{P}\{T_1 > n\}.$$

Then, for every $z > z_0$, we have

$$\begin{aligned}
\mathbb{P}\{X_n > z\} &= \sum_{j=1}^{n} \mathbb{P}\{X_{n-j} = z_0\}\mathbb{P}_{z_0}\{X_j > z, \tau_{z_0} > j\} \\
&= \sum_{j=1}^{n} \mathbb{P}\{O_n = j\}\mathbb{P}\{X_j > z \mid \tau_{z_0} > j\}.
\end{aligned}$$

According to Theorem 8.24,

$$\mathbb{P}_{z_0}\{X_j > z \mid \tau_{z_0} > j\} = e^{-z^2/2bj} + o(1) \quad \text{as } j \to \infty$$

uniformly for all z. In addition, for any fixed j,

$$\mathbb{P}_{z_0}\{X_j > z \mid \tau_{z_0} > j\} \to 0 \quad \text{as } z \to \infty.$$

Therefore

$$\mathbb{P}_{z_0}\{X_j > z \mid \tau_{z_0} > j\} = e^{-z^2/2bj} + o(1) \quad \text{as } z \to \infty$$

uniformly for all j. Hence,

$$\mathbb{P}\{X_n > z\} = \mathbb{E}\exp\left\{-\frac{z^2}{2bO_n}\right\} + o(1) \quad \text{as } z \to \infty,$$

which implies the following relation, as $n \to \infty$:

$$\begin{aligned}\mathbb{P}\left\{\frac{G(X_n)}{G(\sqrt{n})} > y\right\} &= \mathbb{P}\{X_n > G^{-1}(yG(\sqrt{n}))\}\\ &= \mathbb{E}\exp\left\{-\frac{1}{2b}\left(\frac{G^{-1}(yG(\sqrt{n}))}{\sqrt{O_n}}\right)^2\right\} + o(1). \qquad (8.111)\end{aligned}$$

Since $\mathbb{P}\{T_2 > n\} \sim C(z_0)/U(\sqrt{n})$, we obtain, as $x \to \infty$,

$$\begin{aligned}\int_0^x \mathbb{P}\{T_2 > y\}dy &\sim C(z_0)\int_{\widehat{x}^2}^x \frac{1}{U(\sqrt{y})}dy\\ &= 2C(z_0)\int_{\widehat{x}}^{\sqrt{x}} \frac{u}{U(u)}du\\ &= 2C(z_0)G(\sqrt{x}).\end{aligned}$$

Let us recall the following result of Erickson [59, Theorem 6].

Theorem 8.31 *Let $S_n = \xi_1 + \cdots + \xi_n$ be a random walk with positive jumps such that the distribution F of ξ has infinite mean and $\overline{F}(t) = L(t)/t$, $t > 0$, where L is slowly varying at infinity. Let $N(t) = \max\{n : S_n \le t\}$ be the renewal process generated by the random walk, let $Y(t) = t - S_{N(t)}$ be the undershoot, and let $Z(t) = S_{N(t)+1} - t$ be the overshoot. Then, for $0 < y \le 1$, $z > 0$,*

$$\mathbb{P}\left\{\frac{m(Y(t))}{m(t)} \le y,\ \frac{m(Z(t))}{m(t)} \le z\right\} \to \min\{y, z\} \quad \textit{as } t \to \infty,$$

where

$$m(t) = \int_0^t \overline{F}(u)du.$$

Applying this result, we conclude, for all $y \in [0, 1]$, that

$$\mathbb{P}\left\{\frac{G(\sqrt{O_n})}{G(\sqrt{n})} \le y\right\} \to y \quad \text{as } n \to \infty,$$

or, in other words,

$$\mathbb{P}\{\sqrt{O_n} \le G^{-1}(yG(\sqrt{n}))\} \to y \quad \text{as } n \to \infty. \tag{8.112}$$

Since G is a slowly varying function, the inverse function satisfies the relation

$$G^{-1}(tu) = o(G^{-1}(u)) \quad \text{as } u \to \infty,$$

for any fixed $0 < t < 1$; thus it follows from (8.112) that

$$\frac{\sqrt{O_n}}{G^{-1}(yG(\sqrt{n}))} \to 0 \quad \text{as } n \to \infty \text{ with probability } y$$

and

$$\frac{\sqrt{O_n}}{G^{-1}(yG(\sqrt{n}))} \to \infty \quad \text{as } n \to \infty \text{ with probability } 1 - y.$$

Therefore

$$\mathbb{E}\exp\Big\{-\frac{1}{2b}\Big(\frac{G^{-1}(yG(\sqrt{n}))}{\sqrt{O_n}}\Big)^2\Big\} \to 1 - y \quad \text{as } n \to \infty,$$

which completes the proof, by (8.111). □

8.8 Comments for Chapter 8

In the paper [122], Menshikov and Popov investigated the behaviour of the invariant distribution $\{\pi(x), x \in \mathbb{Z}^+\}$ for countable Markov chains with asymptotically zero drift and with bounded jumps. Some rough theorems for the local probabilities $\pi(x)$ were proven; if the condition (8.4) holds then for every $\varepsilon > 0$ there exist constants $c_- = c_-(\varepsilon) > 0$ and $c_+ = c_+(\varepsilon) < \infty$ such that

$$c_- x^{-2\mu/b-\varepsilon} \le \pi(x) \le c_+ x^{-2\mu/b+\varepsilon}.$$

The same bounds were obtained by Aspandiiarov and Iasnogorodski in [9]; their results also cover null-recurrent chains with $\mu > 0$.

The paper [106] by Korshunov is devoted to the existence and non-existence of the moments of invariant distributions. In particular, it was proven there that if (8.4) holds and the families of random variables $\{(\xi^+(x))^{2+\gamma}, x \ge 0\}$ for some $\gamma > 0$ and $\{(\xi^-(x))^2, x \ge 0\}$ are uniformly integrable then the moment of order γ of the invariant distribution π is finite if $\gamma < 2\mu/b - 1$ and infinite if π has unbounded support and $\gamma > 2\mu/b - 1$. This result implies that for every $\varepsilon > 0$ there exists a $c(\varepsilon)$ such that

$$\pi(x, \infty) \le c(\varepsilon) x^{-2\mu/b+1+\varepsilon}. \tag{8.113}$$

In [42] the present authors found the asymptotic behaviour of $\pi(x,\infty)$ for positive recurrent chains under more restrictive moment conditions. In particular, it was assumed there that the third moments of jumps converge at infinity.

Concerning Theorem 8.18, Huillet [84] and Dette [47] obtained exact formulas for recurrence times for very special chains. They used the orthogonal polynomials technique, which was suggested by Karlin and McGregor in [86].

Alexander [5] considered recurrence times for Markov chain with steps ± 1. Using the standard embedding of such a random walk into the corresponding Bessel process, he found exact asymptotics for $\mathbb{P}_x\{\tau_0 = n\}$ for all $\rho > 0$. Unfortunately, this method applies only to a skip-free chain.

From the results in Hryniv et al. [83, Theorem 2.4] one obtains the bounds

$$n^{-\rho/2}\log^{-\varepsilon} n \le \mathbb{P}_0\{\tau_0 > n\} \le n^{-\rho/2}\log^{\rho+1+\varepsilon}$$

for chains satisfying conditions similar to that of Theorem 8.18 with $r(x) = 2\mu/bx + o(1/x\log x)$.

Our Theorem 8.26 improves Theorem 2.3 of Hryniv, Menshikov, and Wade [83], where lower and upper bounds were given with an extra logarithmic term.

9

Tail Analysis for Positive Recurrent Markov Chains with Drift Going to Zero More Slowly Than $1/x$

In this chapter we consider a Markov chain $\{X_n\}$ which possesses a stationary (invariant) probability distribution π and which is such that the first two truncated moments of jumps satisfy the following conditions:

$$m_2^{[s(x)]}(x) \to b > 0 \quad \text{and} \quad m_1^{[s(x)]}(x)x \to -\infty \qquad \text{as } x \to \infty. \tag{9.1}$$

In this case the tail of π typically decays faster than any power function, as may be guessed from Corollary 8.4; it is usually of Weibullian type as seen below.

We have already observed this effect for chains with jumps ± 1 and 0 in Example 1.32. Let us consider such chains in more detail. Fix positive numbers $a_+ > a_-$, $\alpha \in (0, 1)$ and consider a chain $\{X_n\}$ on $\mathbb{Z}^+$ with transition probabilities up and down

$$p_+(x) = \frac{1}{2}\left(1 - \frac{a_+}{(x+1)^\alpha}\right), \quad p_-(x) = \frac{1}{2}\left(1 + \frac{a_-}{(x+1)^\alpha}\right), \quad x > 1,$$

with $p_0(0) + p_+(0) = 1$, $p_+(0) > 0$. Then, according to (1.7),

$$\pi(x) = \pi(0)\exp\left\{\sum_{k=1}^{x} \log \frac{p_+(k-1)}{p_-(k)}\right\}.$$

From the definition of $p_\pm$ we get

$$\log \frac{p_+(k-1)}{p_-(k)} = \log\left(1 - \frac{a_+}{k^\alpha}\right) - \log\left(1 + \frac{a_-}{(k+1)^\alpha}\right).$$

Set $d_\alpha := \max\{j : j\alpha \le 1\}$. Then, by a Taylor expansion of the logarithm function,

$$\log \frac{p_+(k-1)}{p_-(k)} = -\sum_{j=1}^{d_\alpha} \frac{a_+^j - (-a_-)^j}{j} k^{-j\alpha} + O(k^{-(d_\alpha+1)\alpha}) \quad \text{as } k \to \infty.$$

Therefore

$$\pi(x) \sim C\exp\left\{-\sum_{j=1}^{d_\alpha-1}\frac{a_+^j-(-a_-)^j}{j(1-j\alpha)}x^{1-j\alpha}-\frac{a_+^{d_\alpha}-(-a_-)^{d_\alpha}}{d_\alpha}\sum_{k=1}^{x}k^{-d_\alpha\alpha}\right\}, \tag{9.2}$$

owing to Proposition 1.30. If $d_\alpha < 1/\alpha$ then

$$\pi(x) \sim C\exp\left\{-\sum_{j=1}^{d_\alpha}\frac{a_+^j-(-a_-)^j}{j(1-j\alpha)}x^{1-j\alpha}\right\},$$

and if $d_\alpha = 1/\alpha$ then

$$\pi(x) \sim Cx^q\exp\left\{-\sum_{j=1}^{1/\alpha-1}\frac{a_+^j-(-a_-)^j}{j(1-j\alpha)}x^{1-j\alpha}\right\},$$

where $q = -\alpha(a_+^{1/\alpha}-(-a_-)^{1/\alpha})$. In this example we have

$$m_1(x) = -\frac{a_++a_-}{2(x+1)^\alpha} \quad\text{and}\quad m_2(x) = 1-\frac{a_+-a_-}{2(x+1)^\alpha}.$$

As follows from (1.22), the stationary density of a diffusion with the same drift and diffusion coefficients is asymptotically equivalent to, as $x\to\infty$,

$$\begin{aligned}
&C\exp\left\{-(a_++a_-)\int_0^x\frac{1}{(y+1)^\alpha-(a_+-a_-)/2}dy\right\}\\
&\sim C\exp\left\{-(a_++a_-)\left(\sum_{j=1}^{d_\alpha-1}\frac{(a_+-a_-)^{j-1}}{2^{j-1}(1-j\alpha)}x^{1-j\alpha}\right.\right.\\
&\qquad\left.\left.+\frac{(a_+-a_-)^{d_\alpha-1}}{2^{d_\alpha-1}}\int_1^x y^{-d_\alpha\alpha}dy\right)\right\}.
\end{aligned}$$

Comparing this expression with (9.2), we see that the main term is the same but all the correction terms have different coefficients. Since the correction terms play a rôle in the case $\alpha \le 1/2$ ($d_\alpha \ge 2$), we conclude that the densities are asymptotically equivalent for $\alpha > 1/2$ only. We also see that if $\alpha \le 1/2$ then it is not sufficient to know the asymptotic behaviour of only the first and second moments to conclude the precise asymptotic behaviour of the tail of π; as we will see later on, higher moments also play a rôle if $\alpha \le 1/2$.

9.1 Stationary Measure of Positive Recurrent Chains: Weibullian-Type Asymptotics

Our first result concerns the case where, roughly speaking, $m_1(x) = o(1/\sqrt{x})$ as $x \to \infty$. More precisely, we assume that

$$\frac{2m_1^{[s(x)]}(x)}{m_2^{[s(x)]}(x)} = -r(x) + o(p(x)) \quad \text{as } x \to \infty, \tag{9.3}$$

where a decreasing differentiable function $r(x) > 0$ satisfies $r(x)x \to \infty$ as $x \to \infty$ and

$$r^2(x) = o(p(x)) \quad \text{as } x \to \infty. \tag{9.4}$$

Here $p(x) \in [0, r(x)]$ is a decreasing differentiable function which is assumed $r(x)$-insensitive, that is, $p(x \pm 1/r(x)) \sim p(x)$, and integrable at infinity,

$$\int_0^\infty p(x)dx < \infty. \tag{9.5}$$

The increasing function $s(x)$ is assumed to be of order $o(1/r(x))$. In view of (9.1), the condition (9.3) is equivalent to

$$m_1^{[s(x)]}(x) + \frac{m_2^{[s(x)]}(x)}{2} r(x) = o(p(x)) \quad \text{as } x \to \infty. \tag{9.6}$$

We also assume that

$$|p'(x)| < |r'(x)| \text{ for all } x, \qquad |r'(x)| = o(r^2(x)) \text{ as } x \to \infty. \tag{9.7}$$

Compare the second part of this condition with (2.7) or (3.3); it is valid for functions $r(x)$ such as $x^{-\beta}$, $x^{-\beta}\log^\alpha x$ with $\beta \in (0,1)$, and $\log^\alpha x/x$ with $\alpha > 0$ and excludes the function $r(x) = 1/x$.

Define

$$R(x) := \int_0^x r(y)dy, \quad x \geq 0, \tag{9.8}$$

with $R(x) = 0$ for $x < 0$. Since $xr(x) \to \infty$, $R(x) \to \infty$ as $x \to \infty$. The function $R(x)$ is concave because $r(x)$ is decreasing. As shown in Section 2.1, $1/r(x)$ is a natural x-step responsible for the constant increase of the function $R(x)$; see (2.11) and (2.12). Under the condition (9.7), which is stronger than (2.7), we can derive an asymptotic version of the inequalities (2.8) and (2.9) as follows: for all $h > 0$,

$$\frac{1}{r(x)} - \frac{1}{r(x+h/r(x))} = \int_x^{x+h/r(x)} \frac{r'(y)}{r^2(y)}dy = o(1/r(x)) \quad \text{as } x \to \infty,$$

which implies the equivalence

$$r(x+h/r(x)) \sim r(x) \quad \text{as } x \to \infty. \tag{9.9}$$

Therefore, for any fixed $h \in \mathbb{R}$,

$$R\Big(x + \frac{h}{r(x)}\Big) = R(x) + h + o(1) \quad \text{as } x \to \infty. \tag{9.10}$$

Consider the following function:

$$U(x) := \int_0^x e^{R(y)} dy, \quad x \ge 0, \tag{9.11}$$

with $U(x) = 0$ for $x < 0$. Note that the function U solves the equation $U'' - rU' = 0$. It can be seen that the function $U(x)$ is convex. Since

$$\frac{U'(x)}{\left(\frac{1}{r(x)}e^{R(x)}\right)'} = \frac{e^{R(x)}}{\left(1 - \frac{r'(x)}{r^2(x)}\right)e^{R(x)}}$$

and $|r'(x)| = o(r^2(x))$ by (9.7), L'Hôpital's rule yields that

$$U(x) \sim \frac{1}{r(x)} e^{R(x)} \quad \text{as } x \to \infty. \tag{9.12}$$

The condition (9.4) applies to functions $r(x)$ of order $o(1/\sqrt{x})$, where the tail asymptotics of the invariant measure is determined by the functions r and U, which are defined via the asymptotic behaviour of the first two truncated moments of jumps.

Theorem 9.1 *Let* $\{X_n\}$ *be a positive recurrent Markov chain on* $\mathbb{R}$ *and let* $\pi(\cdot)$ *be its invariant probability measure. Let* π *have right unbounded support, that is,* $\pi(x,\infty) > 0$ *for all* x.

Let the first two moments of jumps truncated at some increasing level $s(x) = o(1/r(x))$ *satisfy the conditions* (9.1) *and* (9.3), *where the functions* $r(x)$ *and* $p(x)$ *satisfy* (9.4) *and* (9.7). *Let the following integrability conditions hold:*

$$\sup_{x\in\mathbb{R}} \frac{\mathbb{E}U(x+\xi(x))}{1+U(x)} < \infty \tag{9.13}$$

and, as $x \to \infty$,

$$\mathbb{P}\{|\xi(x)| > s(x)\} = o(r(x)p(x)), \tag{9.14}$$

$$\mathbb{E}\big\{U(\xi(x));\ \xi(x) > s(x)\big\} = o(p(x)), \tag{9.15}$$

$$\sup_x \mathbb{E}\big\{|\xi(x)|^3;\ |\xi(x)| \le s(x)\big\} < \infty. \tag{9.16}$$

Then there exists a $c > 0$ such that, for any fixed $h > 0$,

$$\pi\left(x, x + \frac{h}{r(x)}\right] \sim c\frac{1 - e^{-h}}{r^2(x)U(x)} \quad \text{as } x \to \infty.$$

In particular,

$$\pi(x, \infty) \sim \frac{c}{r^2(x)U(x)} \quad \text{as } x \to \infty.$$

Notice that the condition (9.4) excludes any function $r(x)$ which decreases as $1/\sqrt{x}$ or more slowly. In cases where the absolute value of the first moment decreases more slowly than $1/\sqrt{x}$, the conclusion of Theorem 9.1 fails, in general. In such case the answer depends heavily on the asymptotic properties of higher moments of the chain jumps.

In order to present the tail asymptotics for the invariant measure in the general case we need the following set of conditions.

Fix some $\gamma \in \{2, 3, 4, \dots\}$ and a decreasing function $p(x) \in C^{\gamma-1}(\mathbb{R}^+)$ that is integrable at infinity. Assume that there exists a decreasing function $r(x)$ satisfying

$$r^{\gamma}(x) = o(p(x)) \quad \text{as } x \to \infty \tag{9.17}$$

and such that

$$\frac{2m_1^{[s(x)]}(x)}{m_2^{[s(x)]}(x)} \sim -r(x) \quad \text{as } x \to \infty.$$

We further assume that the following condition – which involves all truncated moments of order up to γ – holds:

$$m_1^{[s(x)]}(x) + \sum_{j=2}^{\gamma} \frac{m_j^{[s(x)]}(x)}{j!} r^{j-1}(x) = o(p(x)) \quad \text{as } x \to \infty. \tag{9.18}$$

We also assume that the conditions (9.5) and (9.7) hold, and that, as $x \to \infty$,

$$r^{(k)}(x) = o(p(x)), \quad p^{(k)}(x) = o(p(x)) \quad \text{for all } 2 \le k \le \gamma - 1. \tag{9.19}$$

As follows from Lemma 2.30, the second relation can be always satisfied by choosing a more slowly decreasing integrable function $p(x)$.

Define $R(x)$ as in (9.8) and $U(x)$ as in (9.11).

Theorem 9.2 *Let $\{X_n\}$ be a positive recurrent Markov chain on $\mathbb{R}$ and let $\pi(\cdot)$ be its invariant probability measure. Let π have right unbounded support, that is, $\pi(x, \infty) > 0$ for all x.*

Let $\gamma \in \{2, 3, \dots\}$. Let the first γ moments of jumps truncated at some increasing level $s(x) = o(1/r(x))$ satisfy the conditions (9.1) and (9.18), with *functions $r(x)$ and $p(x)$ satisfying* (9.17), (9.7), *and* (9.19). *Let the following integrability conditions hold:*

$$\sup_{x\in\mathbb{R}} \frac{\mathbb{E}U(x+\xi(x))}{1+U(x)} < \infty \tag{9.20}$$

and, as $x\to\infty$,

$$\mathbb{P}\{|\xi(x)| > s(x)\} = o(r(x)p(x)), \tag{9.21}$$

$$\mathbb{E}\{U(\xi(x));\ \xi(x) > s(x)\} = o(p(x)), \tag{9.22}$$

$$\sup_x \mathbb{E}\{|\xi(x)|^{\gamma+1};\ |\xi(x)| \le s(x)\} < \infty. \tag{9.23}$$

Then there exists a $c > 0$ such that, for any fixed $h > 0$,

$$\pi\left(x, x+\frac{h}{r(x)}\right] \sim c\frac{1-e^{-h}}{r^2(x)U(x)} \quad \text{as } x\to\infty.$$

In particular,

$$\pi(x,\infty) \sim \frac{c}{r^2(x)U(x)} \quad \text{as } x\to\infty.$$

Let us demonstrate how the function $r(x)$ may be constructed under some regularity conditions. Assume that $m_1^{[s(x)]}(x)$ possesses the following decomposition with respect to some non-negative decreasing function $t(x) \in C^\gamma(\mathbb{R}^+)$,

$$m_1^{[s(x)]} = -t(x) + \sum_{j=2}^{\gamma-1} a_{1,j}t^j(x) + o(p(x)), \tag{9.24}$$

and that, for all $k = 2, 3, \ldots, \gamma$,

$$m_k^{[s(x)]}(x) = \sum_{j=0}^{\gamma-k} a_{k,j}t^j(x) + o(t^{1-k}(x)p(x)), \tag{9.25}$$

where the function $t(x)$ satisfies the conditions (9.7) and (9.19) for $r(x)$. Then there exists – see Lemma 9.8 below – a solution to the equation (9.18) which may be represented as

$$r(x) = \sum_{j=1}^{\gamma-1} r_j t^j(x), \tag{9.26}$$

for some reals $r_1, \ldots, r_{\gamma-1}$. The function $r(x)$ satisfies the conditions (9.7) and (9.19). In addition, since its derivative,

$$r'(x) = t'(x)(1+O(t(x))) = t'(x)(1+o(1)),$$

is non-positive for sufficiently large x, we may redefine the function $t(x)$ on a compact set in such a way that the function $r(x)$ becomes decreasing.

Theorems 9.1 and 9.2 give, at first glance, the same answer:

$$\pi\left(x, x+\frac{h}{r(x)}\right] \sim c\frac{1-e^{-h}}{r^2(x)U(x)}.$$

The difference consists in the choice of the function $r(x)$. In Theorem 9.1 this function should satisfy (9.6), while in Theorem 9.2 we use (9.18) instead of (9.6). In order to explain the difference between (9.18) and (9.6) we consider the case where the first moment behaves regularly at infinity. We first assume that (9.6) holds with $r(x) = x^{-\beta}\ell(x)$, $\beta \in (0, 1)$. Owing to the condition (9.4) we may apply Theorem 9.1 for $\beta > 1/2$ only. In this case

$$R(x) = \int_0^x y^{-\beta}\ell(y)dy \sim \frac{1}{1-\beta}x^{1-\beta}\ell(x) \quad \text{as } x \to \infty.$$

Recalling that $U(x) \sim \frac{1}{r(x)}e^{R(x)}$, we then obtain

$$\pi(x, \infty) \sim c\frac{x^{\beta}}{\ell(x)}\exp\left\{-\int_0^x y^{-\beta}\ell(y)dy\right\} \quad \text{as } x \to \infty \tag{9.27}$$

and, in particular,

$$\log \pi(x, \infty) \sim -\frac{1}{1-\beta}x^{1-\beta}\ell(x) \quad \text{as } x \to \infty. \tag{9.28}$$

If $\beta \le 1/2$ then we have to use (9.18) with $\gamma = \min\{k \in \mathbb{Z} : k\beta > 1\}$. This choice of γ follows from (9.17). In order to have a simpler representation for the answer, we assume that (9.24) and (9.25) are valid with $t(x) = x^{-\beta}\ell(x)$. As mentioned above, then

$$r(x) = x^{-\beta}\ell(x) + \sum_{j=2}^{\gamma} r_j x^{-j\beta}\ell^j(x).$$

Consequently,

$$R(x) = \int_0^x y^{-\beta}\ell(y)dy + \sum_{j=2}^{\gamma} r_j \int_0^x y^{-j\beta}\ell^j(y)dy$$

and

$$\pi(x, \infty) \sim c\frac{x^{\beta}}{\ell(x)}\exp\left\{-\int_0^x y^{-\beta}\ell(y)dy + \sum_{j=2}^{\gamma} r_j \int_0^x y^{-j\beta}\ell^j(y)dy\right\}. \tag{9.29}$$

Taking the logarithm and comparing with (9.28), we see that the logarithmic asymptotics are the same for all $\beta \in (0,1)$; however, the exact asymptotics are different. If, for example, $\beta \in (1/3, 1/2]$ and $\ell(x) \equiv 1$ then we get from (9.29) that

$$\pi(x,\infty) \sim cx^{\beta} \exp\left\{-\frac{1}{1-\beta}x^{1-\beta} + \frac{r_2}{1-2\beta}x^{1-2\beta}\right\}.$$

For $\beta > 1/2$ we have only the first summand in the exponent. Finally, in the borderline case $\beta = 1/2$ we have

$$\pi(x,\infty) \sim cx^{\beta+r_2} \exp\left\{-\frac{1}{1-\beta}x^{1-\beta}\right\},$$

which again differs from the case $\beta > 1/2$.

Finally, let us discuss the case $\beta = 1$, for which $r(x) = \ell(x)/x$ and $\ell(x) \to \infty$. Let us consider a special case where $\ell(x) = c \log x$, $c > 0$. Then

$$R(x) = \frac{c}{2}\log^2 x + c_1 + o(1);$$

$$U(x) \sim c_2 \frac{x}{\log x} e^{c(\log^2 x)/2} \quad \text{as } x \to \infty,$$

which, owing to Theorem 9.1, gives rise to a log-normal type of tail behaviour of the invariant measure:

$$\pi(x,\infty) \sim c_3 \frac{x}{\log x} e^{-c(\log^2 x)/2} \quad \text{as } x \to \infty.$$

9.2 Lyapunov Function and Corresponding Change of Measure

In this section we construct a Lyapunov function that will be used to derive exact asymptotics in Theorems 9.1 and 9.2.

Consider a function $r_p(x) := r(x) - p(x)$. We have $0 \le r_p(x) \le r(x)$; the function $r_p(x)$ is decreasing because

$$r_p'(x) = r'(x) - p'(x) < 0,$$

by the condition (9.7). Define $R_p(x) = U_p(x) = 0$ for $x \le 0$ and, for $x > 0$,

$$R_p(x) := \int_0^x r_p(y)dy, \qquad 0 \le R_p(x) \le R(x),$$

$$U_p(x) := \int_0^x e^{R_p(y)}dy, \qquad 0 < U_p(x) \le U(x).$$

Since the function $r_p(x)$ is decreasing, the function $R_p(x)$ is concave. Since

$$\int_0^\infty r(y)dy = \infty \text{ and } C_p := \int_0^\infty p(y)dy < \infty,$$

we have that

$$R_p(x) = R(x) - C_p + o(1) \quad \text{as } x \to \infty. \tag{9.30}$$

Therefore

$$U_p(x) \sim e^{-C_p} U(x) \quad \text{as } x \to \infty, \tag{9.31}$$

and, by (9.12),

$$U_p(x) \sim \frac{1}{r(x)} e^{R(x)-C_p} \sim \frac{1}{r_p(x)} e^{R_p(x)} \quad \text{as } x \to \infty. \tag{9.32}$$

Notice that the increments of the function U_p obey the following useful upper bound, for all $x, y > 0$,

$$\begin{aligned} U_p(x+y) - U_p(x) &= \int_0^y e^{R_p(x+z)} dz \\ &\le \int_0^y e^{R(x+z)} dz \\ &\le e^{R(x)} \int_0^y e^{R(z)} dz = e^{R(x)} U(y), \end{aligned} \tag{9.33}$$

provided that the function $r(x)$ is decreasing, because then the function R is concave since it is the integral of a decreasing function r.

Lemma 9.3 *Under the conditions of Theorem 9.2, as $x \to \infty$,*

$$\mathbb{E}U_p(x+\xi(x)) - U_p(x) = -p(x)r(x)U_p(x)\Big(\frac{m_2^{[s(x)]}(x)}{2} + o(1)\Big). \tag{9.34}$$

Proof We start with the following decomposition:

$$\begin{aligned} \mathbb{E}U_p(x+\xi(x)) - U_p(x) = \mathbb{E}\{&U_p(x+\xi(x)) - U_p(x);\ \xi(x) < -s(x)\} \\ +\mathbb{E}\{&U_p(x+\xi(x)) - U_p(x);\ |\xi(x)| \le s(x)\} \\ +\mathbb{E}\{&U_p(x+\xi(x)) - U_p(x);\ \xi(x) > s(x)\}. \end{aligned} \tag{9.35}$$

Since the function $U_p(x)$ is increasing, the first term on the right-hand side may be bounded as follows:

$$\begin{aligned} \big|\mathbb{E}\{U_p(x+\xi(x)) - U_p(x);\ \xi(x) < -s(x)\}\big| &\le U_p(x)\mathbb{P}\{\xi(x) < -s(x)\} \\ &= o(p(x)r(x))U_p(x), \end{aligned} \tag{9.36}$$

owing to the condition (9.21). To estimate the second term on the right-hand side of (9.35), we make use of a Taylor expansion:

$$\begin{aligned}
&\mathbb{E}\{U_p(x+\xi(x)) - U_p(x);\ |\xi(x)| \le s(x)\} \\
&\quad = \sum_{k=1}^{\gamma} \frac{U_p^{(k)}(x)}{k!} m_k^{[s(x)]}(x) + \mathbb{E}\Big\{\frac{U_p^{(\gamma+1)}(x+\theta\xi(x))}{(\gamma+1)!}\xi^{\gamma+1}(x);\ |\xi(x)| \le s(x)\Big\},
\end{aligned} \tag{9.37}$$

where $0 \le \theta = \theta(x, \xi(x)) \le 1$. By the construction of U_p,

$$U_p'(x) = e^{R_p(x)}, \quad U_p''(x) = r_p(x)e^{R_p(x)} = (r(x) - p(x))e^{R_p(x)}, \tag{9.38}$$

and, for $k = 3, \dots, \gamma + 1$,

$$U_p^{(k)}(x) = (e^{R_p(x)})^{(k-1)} = \big(r_p^{k-1}(x) + o(p(x))\big)e^{R_p(x)} \quad \text{as } x \to \infty,$$

where the remainder terms in the parentheses on the right are of order $o(p(x))$, by the conditions (9.7) and (9.19). By the definition of $r_p(x)$, for $k \ge 3$,

$$r_p^{k-1}(x) = (r(x) - p(x))^{k-1} = r^{k-1}(x) + o(p(x)),$$

which implies the relation

$$U_p^{(k)}(x) = \big(r^{k-1}(x) + o(p(x))\big)e^{R_p(x)} \quad \text{as } x \to \infty. \tag{9.39}$$

It follows from the equalities (9.38) and (9.39) that

$$\begin{aligned}
&\sum_{k=1}^{\gamma} \frac{U_p^{(k)}(x)}{k!} m_k^{[s(x)]}(x) \\
&\quad = e^{R_p(x)}\Bigg(\sum_{k=1}^{\gamma} \frac{r^{k-1}(x)}{k!} m_k^{[s(x)]}(x) + o(p(x)) - p(x)\frac{m_2^{[s(x)]}(x)}{2}\Bigg) \\
&\quad = e^{R_p(x)}\Bigg(o(p(x)) - p(x)\frac{m_2^{[s(x)]}(x)}{2}\Bigg),
\end{aligned}$$

by the conditions (9.18). Hence, the equivalence (9.32) yields

$$\sum_{k=1}^{\gamma} \frac{U_p^{(k)}(x)}{k!} m_k^{[s(x)]}(x) = -r(x)p(x)\frac{m_2^{[s(x)]}(x)}{2}U_p(x) + o(r(x)p(x))U_p(x). \tag{9.40}$$

Owing to the conditions (9.19) for $\gamma \ge 3$ and (9.7) for $\gamma = 2$ on the derivatives of $r(x)$ and to the condition (9.17), we have

$$\begin{aligned}
U_p^{(\gamma+1)}(x) &= (r^{\gamma}(x) + o(p(x)))e^{R_p(x)} \\
&= o(p(x))e^{R_p(x)} = o(p(x)r(x))U_p(x).
\end{aligned}$$

Then, owing to (9.9), (9.10), and (9.12), the last term in (9.37) possesses the following bound:

$$\begin{aligned}&\left|\mathbb{E}\left\{\frac{U_p^{(\gamma+1)}(x+\theta\xi(x))}{(\gamma+1)!}\xi^{\gamma+1}(x);\ |\xi(x)|\le s(x)\right\}\right|\\&\le o(p(x)r(x))U_p(x))\mathbb{E}\big\{|\xi(x)|^{\gamma+1};\ |\xi(x)|\le s(x)\big\}\\&=o(p(x)r(x))U_p(x),\end{aligned}$$

by the condition (9.23). Therefore, it follows from (9.37) and (9.40) that

$$\begin{aligned}&\mathbb{E}\{U_p(x+\xi(x))-U_p(x);\ |\xi(x)|\le s(x)\}\\&=-r(x)p(x)\frac{m_2^{[s(x)]}(x)}{2}U_p(x)+o(p(x)r(x))U_p(x).\end{aligned}\tag{9.41}$$

Finally, the last term in (9.35) is of order $o(p(x)r(x))U_p(x)$ owing to the upper bound (9.33), the equivalence (9.32), and the condition (9.22). Substituting this term, together with (9.36) and (9.41), into (9.35) we arrive at the claim of the lemma. □

Corollary 9.4 *Let the conditions of Theorem 9.2 hold true. Then there exists an $\widehat{x}$ such that the mean drift of the function $U_p(x)$ is 'sandwiched' as follows:*

$$-bp(x)r(x)U_p(x)\ \le\ \mathbb{E}U_p(x+\xi(x))-U_p(x)\ \le\ 0\quad \textit{for all } x>\widehat{x}.$$

9.3 Proof of Theorem 9.2

Let us define a new transition kernel via the following change of measure:

$$Q(x,dy):=\frac{U_p(y)}{U_p(x)}\mathbb{P}_x\{X_1\in dy,\tau_B>1\},\tag{9.42}$$

where $B=(-\infty,\widehat{x}\,]$, $\widehat{x}$ is defined in Corollary 9.4, and

$$\tau_B:=\min\{n\ge 1: X_n\in B\}.$$

It follows from the upper bound in Corollary 9.4 that

$$Q(x,\mathbb{R})=\frac{\mathbb{E}\{U_p(x+\xi(x)),\tau_B>1\}}{U_p(x)}\le\frac{\mathbb{E}U_p(x+\xi(x))}{U_p(x)}\le 1$$

for all $x>\widehat{x}$. In other words, Q is a substochastic kernel on $(\widehat{x},\infty)$. Furthermore, combining the lower bound in Corollary 9.4 with the estimate, from (9.21),

$$\mathbb{E}\{U_p(x+\xi(x));\tau_B=1\}\le U_p(\widehat{x})\mathbb{P}\{x+\xi(x)\le\widehat{x}\}=o(p(x)r(x)),$$

we obtain that

$$q(x) := -\log Q(x,\mathbb{R}) = O(p(x)r(x)). \tag{9.43}$$

Let us consider the normalised kernel

$$\widehat{P}(x,dy) = \frac{Q(x,dy)}{Q(x,\mathbb{R})}$$

and let $\{\widehat{X}_n\}$ denote the corresponding Markov chain; let $\widehat{\xi}(x)$ be its jump from the state x. Consequently, performing an inverse change of measure, we arrive at the following basic equality, see (7.8):

$$\mathbb{P}_x\{X_n \in dy, \tau_B > n\} = \frac{U_p(x)}{U_p(y)}\mathbb{E}_x\{e^{-\sum_{k=0}^{n-1} q(\widehat{X}_k)};\ \widehat{X}_n \in dy\}. \tag{9.44}$$

Lemma 9.5 *Under the conditions of Theorem 9.2, as $x \to \infty$, we have*

$$\mathbb{E}\{\widehat{\xi}(x);\ |\widehat{\xi}(x)| \le s(x)\} \sim \frac{b}{2}r(x), \tag{9.45}$$

$$\mathbb{E}\{(\widehat{\xi}(x))^2;\ |\widehat{\xi}(x)| \le s(x)\} \to b, \tag{9.46}$$

$$\mathbb{P}\{|\widehat{\xi}(x)| > s(x)\} = o(p(x)r(x)). \tag{9.47}$$

Moreover, there exists an $\widehat{x}$ sufficiently large that

$$\mathbb{E}\{\widehat{\xi}(x);\ \widehat{\xi}(x) \le s(x)\} \ge \frac{b}{4}r(x) \quad \textit{for all } x \ge \widehat{x}. \tag{9.48}$$

Proof We are going to apply Lemma 7.2, so we need to check its conditions. The conditions (7.12) and (7.13) are met by the conditions (9.1) and (9.3). The condition (7.24) is met because of (9.21). Further, it follows from (9.30) and (9.32) that

$$\frac{U_p'(x)}{U_p(x)} = \frac{e^{R_p(x)}}{U_p(x)} \sim \frac{e^{R(x)-C_p}}{\frac{1}{r(x)}e^{R(x)-C_p}} = r(x).$$

So, the function U_p satisfies the condition (7.14) with $c_U = 1$. Also, U_p satisfies (7.15) for any $s(x) = o(1/r(x))$ because

$$\frac{U_p'(x+y)}{U_p'(x)} = \frac{e^{R_p(x+y)}}{e^{R_p(x)}} \sim e^{R(x+y)-R(x)} = e^{\int_x^{x+y} r(z)dz} = e^{O(s(x)r(x))} = e^{o(1)}$$

as $x \to \infty$ uniformly for all $|y| \le s(x)$, and, by (9.32),

$$\frac{U_p(x+y)}{U_p(x)} \sim \frac{r(x)}{r(x+y)}\frac{e^{R(x+y)}}{e^{R(x)}} \sim e^{R(x+y)-R(x)} \to 1.$$

Finally, U_p satisfies (7.16) by Corollary 9.4. So, all the conditions of Lemma 7.2 are met and (9.45)–(9.48) follow. □

Owing to (9.48), the chain $\{\widehat{X}_n\}$ satisfies the condition (4.15) in Theorem 4.2 with $\widehat{v}(x) = br(x)/4$; hence we conclude that, for $\widehat{T}(t) = \min\{n \geq 1 : \widehat{X}_n > t\}$,

$$\mathbb{E}_y \widehat{T}(t) = \mathbb{E}_y \widehat{L}(\widehat{x}, \widehat{T}(t)) < \infty \quad \text{for all } t > y.$$

Thus, for any initial state $\widehat{X}_0 = y$,

$$\mathbb{P}\Big\{\limsup_{n\to\infty} \widehat{X}_n = \infty\Big\} = 1.$$

In turn, it follows from Theorem 2.21 that $\widehat{X}_n \to \infty$ with probability 1.

Further, the function $\widehat{v}(x)$ introduced above satisfies the condition (4.24) owing to (9.9). Therefore, Theorem 4.3 is applicable to the chain $\{\widehat{X}_n\}$, and there exists a $c < \infty$ such that

$$\begin{aligned} \widehat{H}_y(x, x + 1/r(x)] &:= \sum_{n=0}^{\infty} \mathbb{P}_y\{\widehat{X}_n \in (x, x + 1/r(x)]\} \\ &\leq \frac{c}{r^2(x)} \quad \text{for all } x, y > 0. \end{aligned} \tag{9.49}$$

Having this estimate we can now prove the following result.

Lemma 9.6 *Under the conditions of Theorem 9.2,*

$$h(z) := \lim_{n\to\infty} \mathbb{E}_z e^{-\sum_{k=0}^{n} q(\widehat{X}_k)} > 0, \quad z > \widehat{x}.$$

Moreover, $h(z) \to 1$ as $z \to \infty$.

Proof The existence of $h(z)$ as a limit is immediate from the monotonicity of the sequence $e^{-\sum_{k=0}^{n} q(\widehat{X}_k)}$ in n. By the convexity of the function e^{-x}, to show positivity it suffices to prove that

$$\mathbb{E}_z \sum_{k=0}^{\infty} q(\widehat{X}_k) < \infty, \quad z > \widehat{x}. \tag{9.50}$$

Note that

$$\mathbb{E}_z \sum_{k=0}^{\infty} q(\widehat{X}_k) = \int_{\widehat{x}}^{\infty} q(y) \widehat{H}_z(dy) \leq c \int_{\widehat{x}}^{\infty} p(y) r(y) \widehat{H}_z(dy),$$

because $q(y) = O(p(y)r(y))$, see (9.43). But it has been already shown in the proof of Lemma 4.1 that the last integral is finite under (9.49); thus the first claim of the lemma is proven.

To prove the second claim we note that it follows from Theorem 2.21 that, for every fixed $N > 0$,

$$\mathbb{P}_z\{\widehat{X}_n > N \text{ for all } n \geq 1\} \to 1 \quad \text{as } z \to \infty;$$

hence

$$\widehat{H}_z(N) \to 0 \quad \text{as } z \to \infty.$$

Then, for any fixed N,

$$\lim_{z\to\infty} \mathbb{E}_z \sum_{k=0}^{\infty} q(\widehat{X}_k) \le \sup_{z>\widehat{x}} \int_N^{\infty} q(y)\widehat{H}_z(dy).$$

According to (4.6),

$$\lim_{N\to\infty} \sup_{z>\widehat{x}} \int_N^{\infty} q(y)\widehat{H}_z(dy) = 0.$$

Therefore, we infer that

$$\lim_{z\to\infty} \mathbb{E}_z \sum_{k=0}^{\infty} q(\widehat{X}_k) = 0.$$

From this relation and Jensen's inequality we finally conclude that $\lim_{z\to\infty} h(z) = 1$. □

Consider the following weighted renewal measure on $(\widehat{x},\infty)$:

$$\widehat{H}_z^{(q)}(dx) \;=\; \sum_{j=0}^{\infty} \mathbb{E}_z\{e^{-\sum_{k=0}^{j-1} q(\widehat{X}_k)};\ \widehat{X}_j \in dx\}, \tag{9.51}$$

and its finite-time-horizon version,

$$\widehat{H}_{z,n}^{(q)}(dx) \;=\; \sum_{j=0}^{n} \mathbb{E}_z\{e^{-\sum_{k=0}^{j-1} q(\widehat{X}_k)};\ \widehat{X}_j \in dx\}. \tag{9.52}$$

Corollary 9.7 *Under the conditions of Theorem 9.2, for every fixed* $z \ge \widehat{x}$ *and* $h > 0$,

$$\widehat{H}_z^{(q)}\Big(x, x+\frac{h}{r(x)}\Big] \sim h(z)\widehat{H}_z\Big(x, x+\frac{h}{r(x)}\Big] \;\sim\; h(z)\frac{h}{r^2(x)} \quad \textit{as } x \to \infty.$$

Proof The claim follows from Lemma 4.5, which applies to $\{\widehat{X}_n\}$ because of to Theorem 5.16 and Lemmas 9.5 and 9.6. □

We again use the representation (7.47) applied to the test function U_p, which reads

$$\pi(x, x+h/r(x)] = c^* \int_x^{x+h/r(x)} \frac{\widehat{H}^{(q)}(dy)}{U_p(y)},$$

where $\widehat{H}^{(q)}$ is defined in (7.45), with initial distribution (7.44). We proceed by splitting the interval $(x, x+h/r(x))$ into equal small subintervals. So, let us fix a large $m \in \mathbb{Z}^+$ and consider points

$$x_k(m) = x + \frac{k-1}{m}\frac{h}{r(x)}, \quad k \in \{1, 2, \ldots, m+1\}.$$

Then

$$\int_x^{x+h/r(x)} \frac{\widehat{H}^{(q)}(dy)}{U_p(y)} = \sum_{k=1}^{m} \int_{x_k(m)}^{x_{k+1}(m)} \frac{\widehat{H}^{(q)}(dy)}{U_p(y)}.$$

Since the function $U_p(y)$ is increasing, we have the following lower and upper bounds:

$$\frac{\widehat{H}^{(q)}(x_k(m), x_{k+1}(m)]}{U_p(x_{k+1}(m))} \le \int_{x_k(m)}^{x_{k+1}(m)} \frac{\widehat{H}^{(q)}(dy)}{U_p(y)} \le \frac{\widehat{H}^{(q)}(x_k(m), x_{k+1}(m)]}{U_p(x_k(m))}.$$

For every fixed m, it follows from Corollary 9.7 that, as $x \to \infty$,

$$\begin{aligned}\widehat{H}^{(q)}\big(x_k(m), x_{k+1}(m)\big] &\sim \widehat{H}\big(x_k(m), x_{k+1}(m)\big] \int_B h(z)\mathbb{P}\{\widehat{X}_0 \in dz\} \\ &= \widehat{H}\big(x_k(m), x_{k+1}(m)\big] \frac{\int_B h(z)U_p(z)\mu(dz)}{\int_B U_p(z)\mu(dz)},\end{aligned}$$

where the measure μ is defined in (7.43). In turn, Theorem 5.16 yields the following asymptotics:

$$\widehat{H}^{(q)}\big(x_k(m), x_{k+1}(m)\big] \sim c\frac{h}{mr^2(x_k(m))} \quad \text{as } x \to \infty,$$

because

$$x_{k+1}(m) - x_k(m) = \frac{h}{mr(x)} \sim \frac{h}{mr(x_k(m))},$$

where

$$c := \frac{\int_B h(z)U_p(z)\mu(dz)}{\int_B U_p(z)\mu(dz)}.$$

This implies the following asymptotic upper bound:

$$\int_x^{x+h/r(x)} \frac{\widehat{H}^{(q)}(dy)}{U_p(y)} \le (c+o(1))\frac{h}{m}\sum_{k=1}^{\infty} \frac{1}{r^2(x_k(m))U_p(x_k(m))}.$$

Substituting the asymptotic relation (9.32) for U_p, we arrive at the following upper bound:

$$\int_x^{x+h/r(x)} \frac{\widehat{H}^{(q)}(dy)}{U_p(y)} \le (c+o(1))\frac{h}{m}\sum_{k=1}^{m} \frac{e^{-R(x_k(m))}}{r(x_k(m))} \quad \text{as } x \to \infty.$$

Letting $m \to \infty$ we can approximate the sum on the right, multiplied by h/m, by the integral

$$\begin{aligned} r(x)\int_x^{x+h/r(x)} \frac{e^{-R(y)}}{r(y)}dy &\sim \int_0^{h/r(x)} e^{-R(x+y)}dy \\ &= \frac{1}{r(x)}\int_0^h e^{-R(x+y/r(x))}dy \\ &\sim \frac{1}{r(x)}e^{-R(x)}\int_0^h e^{-y}dy = \frac{1-e^{-h}}{r(x)}e^{-R(x)} \end{aligned}$$

as $x \to \infty$, where we make use of (9.10). In this way the upper bound of Theorem 9.2 is found.

The corresponding lower bound may be derived in the same way, and the proof of Theorem 9.2 is then complete.

9.4 Sufficient Condition for Existence of $r(x)$ Satisfying (9.18)

Lemma 9.8 *Let $\gamma \in \{2, 3, \ldots\}$. Assume that $m_1^{[s(x)]}(x)$ possesses the following decomposition with respect to some non-negative decreasing function $t(x) \in C^\gamma(\mathbb{R}^+)$ satisfying the conditions* (9.7) *and* (9.19) *on $r(x)$,*

$$m_1^{[s(x)]}(x) = -t(x) + \sum_{j=2}^{\gamma-1} a_{1,j}t^j(x) + o(p(x)), \tag{9.53}$$

and that, for every $k = 2, 3, \ldots, \gamma$,

$$m_k^{[s(x)]}(x) = \sum_{j=0}^{\gamma-k} a_{k,j}t^j(x) + o(t^{1-k}(x)p(x)). \tag{9.54}$$

Then there exists a solution to equation (9.18) *which possesses the following decomposition:*

$$r(x) = \sum_{j=1}^{\gamma-1} r_j t^j(x), \tag{9.55}$$

for some reals $r_1, \ldots, r_{\gamma-1}$.

Proof It is sufficient to find an $r(x)$ satisfying the equality

$$m_1^{[s(x)]}(x) + \sum_{j=2}^{\gamma} \frac{1}{j!}m_j^{[s(x)]}(x)r^{j-1}(x) = o(p(x)). \tag{9.56}$$

In order to find the coefficients r_j, let us substitute (9.53), (9.54), and (9.55) into (9.56). Then we arrive at the following equality:

$$\left(-t(x)+\sum_{j=2}^{\gamma-1} a_{1,j}t^j(x)\right)+\sum_{j=2}^{\gamma}\frac{1}{j!}\left(\sum_{k=0}^{\gamma-j} a_{j,k}t^k(x)\right)\left(\sum_{k=1}^{\gamma-1} r_k t^k(x)\right)^{j-1} = o(p(x)).$$

The coefficient of t equals $-1+r_1a_{2,0}/2$, which implies that

$$r_1 = 2/a_{2,0}.$$

The coefficient of t^2 equals $a_{1,2}+\frac{1}{2}(a_{2,0}r_2+a_{2,1}r_1)+\frac{1}{6}a_{3,0}r_1^2$, which implies that

$$r_2 = -\frac{2a_{1,2}+a_{2,1}r_1+a_{3,0}r_1^2/3}{a_{2,0}}.$$

All further coefficients may be evaluated in recursive way. □

9.5 Local Asymptotics of Stationary Probabilities

Similarly to the case of $m_1(x) \sim -\mu/x$, in this section we derive sharp local asymptotics for the stationary measure π of a recurrent irreducible Markov chain with asymptotically zero drift of order $r(x)$, $xr(x) \to \infty$. Following Section 6.3, we assume that the jumps $\xi(x)$ converge weakly to some random variable ξ on $\mathbb{R}$, that is, the condition (6.52) holds.

Theorem 9.9 *Let $\{X_n\}$ be a positive recurrent Markov chain on $\mathbb{R}$ and $\pi(\cdot)$ be its invariant probabilistic measure. Let π have right unbounded support, that is, $\pi(x,\infty) > 0$ for all x.*

Let $\gamma \in \{2,3,\ldots\}$. Let the first γ moments of jumps truncated at some increasing level $s(x) = o(1/r(x))$ satisfy the conditions (9.1) and (9.18), with functions $r(x)$ and $p(x)$ satisfying (9.17), (9.7), and (9.19). Let the following integrability conditions hold:

$$\sup_{x\in\mathbb{R}} \frac{\mathbb{E}U(x+\xi(x))}{1+U(x)} < \infty \tag{9.57}$$

and, as $x \to \infty$,

$$\mathbb{P}\{|\xi(x)| > s(x)\} = o(r(x)p(x)), \tag{9.58}$$

$$\mathbb{E}\big\{U(\xi(x));\ \xi(x) > s(x)\big\} = o(p(x)), \tag{9.59}$$

$$\sup_x \mathbb{E}\big\{|\xi(x)|^{\gamma+1};\ |\xi(x)| \le s(x)\big\} < \infty. \tag{9.60}$$

Furthermore, assume the convergence $\xi(x) \Rightarrow \xi$ *and that* $\mathbb{E}\xi = 0$ *and* $\mathbb{E}\xi^2 = b$. *In addition, let*

$$-\Xi_- \le_{st} \xi(x) \le_{st} \Xi_+ \quad \textit{for all } x, \tag{9.61}$$

where $\Xi_-^2 < \infty$ *and* $\mathbb{E}U(\Xi_+)\Xi_+^2 < \infty$. *Then, in the lattice case,*

$$\pi(x) \sim ce^{-\int_0^x r(y)dy} \quad \textit{as } x \to \infty,$$

for some $c > 0$. *In the non-lattice case, for any* $h > 0$,

$$\pi(x, x+h] \sim che^{-\int_0^x r(y)dy} \quad \textit{as } x \to \infty. \tag{9.62}$$

Corollary 9.10 *In addition to the conditions and assumptions In Theorem 9.9, let,* $r(x) = \gamma/x^\beta$ *where* $\beta \in (1/2, 1)$ *and* $\gamma > 0$. *Then, in the lattice case,*

$$\pi(x) \sim ce^{-\frac{\gamma}{1-\beta}x^{1-\beta}} \quad \textit{as } x \to \infty,$$

which agrees with the global asymptotics given in (9.27). *In the non-lattice case,*

$$\pi(x, x+h] = e^{-\frac{\gamma}{1-\beta}x^{1-\beta}} \quad \textit{as } x \to \infty.$$

Proof of Theorem 9.9. This proof is very similar to that of Theorem 8.14. In particular, as shown there,

$$\pi(x, x+h] \sim c^* \frac{\widehat{H}^{(q)}(x, x+h]}{U_p(x)} \quad \text{as } x \to \infty.$$

The Markov chain $\{\widehat{X}_n\}$ satisfies all the conditions of Corollary 6.14 with $\widehat{v}(x) = r(x)b/2$ and $\widehat{b} = b$. Indeed, the drift conditions and (6.9) were checked in Lemma 9.5 and (6.1) was checked right after that. The weak convergence (6.52) for $\widehat{\xi}(x)$, that is $\widehat{\xi}(x) \Rightarrow \xi$, follows from that for the original jumps $\xi(x)$ because $U_p(x+y)/U_p(x) \to 1$ as $x \to \infty$, for any fixed $y \in \mathbb{R}$. Finally, the majorisation condition (6.53) holds for a square integrable majorant, since from (9.43), (9.42), and (9.33) it follows that, for all sufficiently large x,

$$\begin{aligned}
\mathbb{P}\{\widehat{\xi}(x) > y\} &= \frac{Q(x, (x+y, \infty))}{Q(x, \mathbb{R})} \\
&\le 2\frac{\mathbb{E}\{U_p(x+\xi(x));\ \xi(x) > y\}}{U_p(x)} \\
&\le 2\mathbb{P}\{\xi(x) > y\} + 2e^{R(x)}\frac{\mathbb{E}\{U(\xi(x));\ \xi(x) > y\}}{U_p(x)} \\
&\le 2\mathbb{P}\{\xi(x) > y\} + c_1\mathbb{E}\{U(\Xi_+);\ \Xi_+ > y\},
\end{aligned}$$

owing to (9.32) and (9.61). Therefore, owing to the condition $\mathbb{E}U(\Xi_+)\Xi_+^2 < \infty$, there exists a random variable $\widehat{\Xi}_+$ such that $\widehat{\xi}(x) \le_{st} \widehat{\Xi}_+$ and $\mathbb{E}\widehat{\Xi}_+^2 < \infty$. In addition,

$$\begin{aligned}\mathbb{P}\{\widehat{\xi}(x) < -y\} &= \frac{Q(x,(-\infty,x-y))}{Q(x,\mathbb{R})}\\ &\le 2\frac{\mathbb{E}\{U_p(x+\xi(x));\ \xi(x)<-y\}}{U_p(x)}\\ &\le 2\mathbb{P}\{\xi(x)<-y\} \le 2\mathbb{P}\{\Xi_->y\},\end{aligned}$$

which implies that $\widehat{\xi}(x) \ge_{st} -\widehat{\Xi}_-$, where $\mathbb{E}\widehat{\Xi}_-^2 < \infty$ owing to the condition $\mathbb{E}\Xi_-^2 < \infty$, and the proof of the existence of a square integrable majorant for the family of $\widehat{\xi}(x)$ is complete.

Hence, by Corollary 6.14 and Lemma 4.5 applied to the Markov chain $\{\widehat{X}_n\}$, we deduce that

$$\widehat{H}^{(q)}(x,x+h] \sim c_q\frac{h}{\widehat{v}(x)} \sim c_q\frac{2h}{br(x)} \quad \text{as } x\to\infty,$$

which concludes the proof because $U_p(x) \sim c_3U(x)$ as $x\to\infty$; see (9.31). □

9.6 Pre-Stationary Distributions

In this section we assume that the distribution of X_n converges to π in total-variation distance; see (8.60).

Theorem 9.11 *Assume that all the conditions of Theorem 9.2 are valid. If $r(x)$ is regularly varying at infinity with index $-\beta\in[-1,0]$ and satisfies $r'(x) = O(r(x)/x)$ then, for any fixed $h>0$,*

$$\frac{\mathbb{P}\{X_n\in(x,x+h/r(x)]\}}{\pi(x,x+h/r(x)]} = \Phi\left(\frac{n-V(x)}{\sqrt{b\frac{1+\beta}{1+3\beta}\frac{x}{r^3(x)}}}\right)+o(1)$$

as $x\to\infty$ uniformly for all n, where the function $V(x)$ is given by

$$V(x) = \int_0^x\left(\sum_{k=2}^{\gamma}\frac{m_k^{[s(y)]}(y)}{(k-2)!\,k}r^{k-1}(y)\right)^{-1}dy.$$

Proof Splitting all the paths according to the time of the last visit of $\{X_n\}$ to $B=(-\infty,\widehat{x}]$, see (7.48), we get, for $x>\widehat{x}$,

$$\begin{aligned}&\mathbb{P}\{X_n\in(x,x+h/r(x)]\}\\ &=\sum_{j=1}^{n}\int_B\mathbb{P}\{X_{n-j}\in dz\}\int_{\widehat{x}}^{\infty}P(z,du)U_p(u)\\ &\quad\times\mathbb{E}_u\left\{\frac{e^{-\sum_{k=0}^{j-2}q(\widehat{X}_k)}}{U_p(\widehat{X}_{j-1})};\ x<\widehat{X}_{j-1}\le x+\frac{h}{r(x)}\right\},\end{aligned} \tag{9.63}$$

where $q(x)\ge 0$ and $\{\widehat{X}_n\}$ are defined in (7.9) and (7.10) respectively.

Fix a sequence $N_x \to \infty$ of order $o(1/r^2(x))$. Then, since $q \ge 0$ and U_p is increasing,

$$\begin{aligned}\sum_{j=n-N_x+1}^{n} & \int_B \mathbb{P}\{X_{n-j}\in dz\} \int_{\widehat{x}}^{\infty} P(z,du)U_p(u) \\ & \times \mathbb{E}_u \left\{ \frac{e^{-\sum_{k=0}^{j-2} q(\widehat{X}_k)}}{U_p(\widehat{X}_{j-1})}; \ x<\widehat{X}_{j-1}\le x+\frac{h}{r(x)} \right\} \\ & \le N_x \frac{1}{U_p(x)} \sup_{z\in B} \int_{\widehat{x}}^{\infty} P(z,du)U_p(u) \\ & \le N_x \frac{c}{U_p(x)} \sup_{z\in B}(1+U_p(z)) \\ & = o(1/r^2(x)U_p(x)), \end{aligned} \tag{9.64}$$

where the second bound follows from the condition (9.20). Furthermore, the distribution of X_{n-j} converges in total variation to π uniformly for all $j \le n - N_x$; see (8.60). Therefore,

$$\begin{aligned}\sum_{j=1}^{n-N_x} & \int_B \mathbb{P}\{X_{n-j}\in dz\} \int_{\widehat{x}}^{\infty} P(z,du)U_p(u) \\ & \times \mathbb{E}_u \left\{ \frac{e^{-\sum_{k=0}^{j-2} q(\widehat{X}_k)}}{U_p(\widehat{X}_{j-1})}; \ x<\widehat{X}_{j-1}\le x+\frac{h}{r(x)} \right\} \\ & \sim \sum_{j=1}^{n-N_x} \int_B \pi(dz) \int_{\widehat{x}}^{\infty} P(z,du)U_p(u) \\ & \qquad \times \mathbb{E}_u \left\{ \frac{e^{-\sum_{k=0}^{j-2} q(\widehat{X}_k)}}{U_p(\widehat{X}_{j-1})}; \ x<\widehat{X}_{j-1}\le x+\frac{h}{r(x)} \right\}. \end{aligned} \tag{9.65}$$

Similarly to (9.64),

$$\begin{aligned}\sum_{j=n-N_x+1}^{n} & \int_B \pi(dz) \int_{\widehat{x}}^{\infty} P(z,du)U_p(u) \\ & \times \mathbb{E}_u \left\{ \frac{e^{-\sum_{k=0}^{j-2} q(\widehat{X}_k)}}{U_p(\widehat{X}_{j-1})}; \ x<\widehat{X}_{j-1}\le x+\frac{h}{r(x)} \right\} \\ & = o(1/r^2(x)U_p(x)). \end{aligned} \tag{9.66}$$

Combining (9.63)–(9.66), we obtain

$$\begin{aligned}
&\mathbb{P}\{X_n \in (x, x+h/r(x)]\} \\
&\quad = \sum_{j=1}^{n} \int_B \pi(dz) \int_{\widehat{x}}^{\infty} P(z,du)U_p(u) \\
&\qquad \times \mathbb{E}_u\left\{ \frac{e^{-\sum_{k=0}^{j-2} q(\widehat{X}_k)}}{U_p(\widehat{X}_{j-1})};\ x<\widehat{X}_{j-1}\le x+\frac{h}{r(x)} \right\} + o(1/r^2(x)U_p(x)) \\
&\quad = \int_B \pi(dz) \int_{\widehat{x}}^{\infty} P(z,du)U_p(u) \\
&\qquad \times \sum_{j=1}^{n} \int_x^{x+h/r(x)} \mathbb{E}_u\left\{ \frac{e^{-\sum_{k=0}^{j-2} q(\widehat{X}_k)}}{U_p(y)};\ \widehat{X}_{j-1}\in dy \right\} + o(1/r^2(x)U_p(x)) \\
&\quad = \int_{\widehat{x}}^{\infty} \mu(du)U_p(u) \int_x^{x+h/r(x)} \frac{\widehat{H}^{(q)}_{u,n}(dy)}{U_p(y)} + o(1/r^2(x)U_p(x)) \qquad (9.67)
\end{aligned}$$

as $x \to \infty$, where

$$\mu(du) = \int_B \pi(dz)P(z,du)$$

is a measure on $(\widehat{x}, \infty)$, see (7.43), and

$$\widehat{H}^{(q)}_{u,n}(A) := \sum_{j=1}^{n} \mathbb{E}_u\left\{ e^{-\sum_{k=0}^{j-2} q(\widehat{X}_k)};\ \widehat{X}_{j-1}\in A \right\}$$

is also a measure on $(\widehat{x}, \infty)$.

Lemma 9.12 *Under the conditions of Theorem 9.11,*

$$\begin{aligned}
\widehat{H}^{(q)}_{z,n}(x, x+h/r(x)] &= h(z)\widehat{H}_{z,n}(x, x+h/r(x)] + o(1/r^2(x)) \\
&= h(z)\frac{h}{r^2(x)}\Phi\left(\frac{n-V(x)}{\sqrt{b\frac{1+\beta}{1+3\beta}\frac{x}{r^3(x)}}} \right) + o\left(\frac{1}{r^2(x)}\right)
\end{aligned}$$

as $x \to \infty$ uniformly for all n, where Φ is the standard normal distribution function.

Proof We will apply Lemma 4.5 and Theorem 5.18 to $\{\widehat{X}_n\}$, keeping in mind Lemma 9.6.

In order to apply Theorem 5.18 we need to identify a regularly varying decreasing function $v(x)$ such that $v'(x) = O(v(x)/x)$ and

$$\widehat{m}_1^{[s(x)]}(x) := \mathbb{E}\{\widehat{\xi}(x);\ |\widehat{\xi}(x)| \le s(x)\} = v(x) + o(\sqrt{v(x)/x}). \qquad (9.68)$$

By the definition of $\widehat{\xi}(x)$,

$$\widehat{m}_1^{[s(x)]}(x) = \frac{\mathbb{E}\{U_p(x+\xi(x))\xi(x); |\xi(x)| \le s(x)\}}{Q(x,\mathbb{R}^+)U_p(x)}. \tag{9.69}$$

Using a Taylor expansion,

$$\begin{aligned}&\mathbb{E}\{U_p(x+\xi(x))\xi(x);\ |\xi(x)| \le s(x)\}\\ &= \sum_{k=1}^{\gamma} \frac{U_p^{(k-1)}(x)}{(k-1)!} m_k^{[s(x)]}(x) + \mathbb{E}\Big\{\frac{U_p^{(\gamma)}(x+\theta\xi(x))}{\gamma!}\xi^{\gamma+1}(x);\ |\xi(x)| \le s(x)\Big\}.\end{aligned}$$

It is clear that the assumption (9.23) implies the boundedness of the functions $m_k^{[s(x)]}(x)$ for all $k \le \gamma+1$. From this fact and from (9.38) and (9.39) we infer that

$$\begin{aligned}&\mathbb{E}\{U_p(x+\xi(x))\xi(x); |\xi(x)| \le s(x)\}\\ &\quad = U_p(x)m_1^{[s(x)]}(x) + U_p'(x)\sum_{k=2}^{\gamma}\frac{m_k^{[s(x)]}(x)}{(k-1)!}r^{k-2}(x)\\ &\qquad + O(p(x)r(x))U_p(x) + O(r^{\gamma}(x))U_p(x).\end{aligned}$$

By (9.17),

$$\begin{aligned}&\mathbb{E}\{U_p(x+\xi(x))\xi(x); |\xi(x)| \le s(x)\}\\ &\quad = U_p(x)m_1^{[s(x)]}(x) + U_p'(x)\sum_{k=2}^{\gamma}\frac{m_k^{[s(x)]}(x)}{(k-1)!}r^{k-2}(x) + O(p(x))U_p(x).\end{aligned}$$

Substituting this relation into (9.69) and using $Q(x,\mathbb{R}^+) = 1 + O(r(x)p(x))$, which is immediate from (9.43), we conclude that

$$\widehat{m}_1^{[s(x)]}(x) = m_1^{[s(x)]}(x) + \frac{U_p'(x)}{r_p(x)U_p(x)}\sum_{k=2}^{\gamma}\frac{m_k^{[s(x)]}(x)}{(k-1)!}r^{k-1}(x) + O(p(x)). \tag{9.70}$$

Recalling that $U_p'(x) = e^{R_p(x)}$ and using $r_p'(x) = O(r_p(x)/x)$, we obtain

$$(U_p'(x) - r_p(x)U_p(x))' = -r_p'(x)U_p(x) = O\Big(\frac{U_p(x)r_p(x)}{x}\Big).$$

Since $r_p(x)U_p(x) \sim e^{R_p(x)}$, we have

$$|U_p'(x) - r_p(x)U_p(x)| \le c_1\int_1^x \frac{e^{R_p(y)}}{y}dy \quad \text{for some } c_1 < \infty.$$

The derivative of $U_p(x)/x$ is asymptotically equivalent to $e^{R_p(x)}/x$ because $r(x)x \to \infty$. Therefore, by L'Hôpital's rule,

$$|U'_p(x) - r_p(x)U_p(x)| \;=\; O(U_p(x)/x),$$

or, in other words,

$$\frac{U'_p(x)}{r_p(x)U_p(x)} = 1 + O(1/xr(x)).$$

Substituting this into (9.70), we obtain

$$\widehat{m}_1^{[s(x)]}(x) \;=\; m_1^{[s(x)]}(x) + \sum_{k=2}^{\gamma} \frac{m_k^{[s(x)]}(x)}{(k-1)!} r^{k-1}(x) + O(1/x).$$

According to (9.18),

$$m_1^{[s(x)]}(x) = -\sum_{k=2}^{\gamma} \frac{m_k^{[s(x)]}(x)}{k!} r^{k-1}(x) + o(p(x)).$$

As a result we have the following asymptotic expansion for the expectation of the truncation at levels $\pm s(x)$ of jumps for the chain $\{\widehat{X}_n\}$:

$$\widehat{m}_1^{[s(x)]}(x) \;=\; \sum_{k=2}^{\gamma} \frac{m_k^{[s(x)]}(x)}{(k-2)!\,k} r^{k-1}(x) + O(1/x).$$

Now it is clear that (9.68) is valid with

$$v(x) = \sum_{k=2}^{\gamma} \frac{m_k^{[s(x)]}(x)}{(k-2)!\,k} r^{k-1}(x),$$

because, for some $c_2 > 0$,

$$x\sqrt{v(x)/x} \;=\; \sqrt{v(x)x} \;\geq\; c_2\sqrt{r(x)x} \;\to\; \infty \quad \text{as } x \to \infty,$$

and so

$$1/x \;=\; o(\sqrt{v(x)/x}) \quad \text{as } x \to \infty.$$

The function $v(x)$ is regularly varying at infinity since

$$\frac{v(x)}{r(x)} \;\sim\; \frac{m_2^{[s(x)]}(x)}{2} \;\to\; \frac{b}{2},$$

and the claim of the lemma follows. □

Since U_p is increasing, we deduce the following lower and upper bounds:

$$\frac{\widehat{H}^{(q)}_{u,n}(x,x+h/r(x)]}{U_p(x+h/r(x))} \le \int_x^{x+h/r(x)} \frac{\widehat{H}^{(q)}_{u,n}(dz)}{U_p(z)} \le \frac{H^{(q)}_{u,n}(x,x+h/r(x)]}{U_p(x)}. \tag{9.71}$$

For any fixed $u > \widehat{x}$, owing to Lemma 9.12,

$$\widehat{H}^{(q)}_{u,n}\Big(x,x+\frac{h}{r(x)}\Big] = h(u)\frac{h}{r^2(x)}\Phi\Bigg(\frac{n-V(x)}{\sqrt{b\frac{1+\beta}{1+3\beta}\frac{x}{r^3(x)}}}\Bigg)+o\Big(\frac{1}{r^2(x)}\Big) \tag{9.72}$$

as $y \to \infty$ uniformly for all n. In addition, since $q \ge 0$,

$$\begin{aligned}\sup_{u>\widehat{x}} \widehat{H}^{(q)}_{u,n}\Big(x,x+\frac{h}{r(x)}\Big] &\le \sup_{u>\widehat{x}} \sum_{j=1}^{n} \mathbb{P}_u\Big\{\widehat{X}_{j-1} \in \Big(x,x+\frac{h}{r(x)}\Big]\Big\}\\ &\le c_1/r^2(x)\end{aligned} \tag{9.73}$$

for all x and n and for some $c_1 < \infty$, as follows from (9.49).

From the estimate (9.72) and the equivalence $U_p(x+h/r(x)) \sim e^h U_p(x)$ – as follows from (9.10) – we infer from (9.71) that, for any fixed $u > \widehat{x}$,

$$U_p(x)r^2(x)\int_x^{x+h/r(x)} \frac{\widehat{H}^{(q)}_{u,n}(dz)}{U_p(z)} \le h(u)h\Phi\Bigg(\frac{n-V(x)}{\sqrt{b\frac{1+\beta}{1+3\beta}\frac{x}{r^3(x)}}}\Bigg)+o(1)$$

and

$$U_p(x)r^2(x)\int_x^{x+h/r(x)} \frac{\widehat{H}^{(q)}_{u,n}(dz)}{U_p(z)} \ge h(u)he^{-h}\Phi\Bigg(\frac{n-V(x)}{\sqrt{b\frac{1+\beta}{1+3\beta}\frac{x}{r^3(x)}}}\Bigg)+o(1).$$

Splitting the interval $(x,x+h/r(x)]$ into smaller intervals as in Theorem 9.2, we can justify the following asymptotics:

$$U_p(x)r^2(x)\int_x^{x+h/r(x)} \frac{\widehat{H}^{(q)}_{u,n}(dz)}{U_p(z)} \sim h(u)(1-e^{-h})\Phi\Bigg(\frac{n-V(x)}{\sqrt{b\frac{1+\beta}{1+3\beta}\frac{x}{r^3(x)}}}\Bigg)+o(1) \tag{9.74}$$

as $x \to \infty$ uniformly for all n.

Similarly, it follows from (9.73) that

$$\sup_{u>\widehat{x}}\int_x^{x+h/r(x)} \frac{\widehat{H}^{(q)}_{u,n}(dz)}{U_p(z)} \le \frac{c_2}{r^2(x)U_p(x)}.$$

In addition,

$$\begin{aligned}\widehat{c} &= \int_{\widehat{x}}^{\infty} h(u)U_p(u)\mu(du)\\ &= \int_B \pi(dz)\int_{\widehat{x}}^{\infty} h(u)U_p(u)P(z,du) < \infty,\end{aligned}$$

as follows from the condition (9.13). Hence the dominated convergence theorem is applicable to (9.67), so substituting (9.74) into (9.67) we obtain

$$\mathbb{P}\{X_n \in (x, x+h/r(x)]\} = \widehat{c}\frac{1-e^{-h}}{r^2(x)U_p(x)}\Phi\Bigl(\frac{n-V(x)}{\sqrt{b\frac{1+\beta}{1+3\beta}\frac{x}{r^3(x)}}}\Bigr) + o\Bigl(\frac{1}{r^2(x)U_p(x)}\Bigr)$$

as $x \to \infty$ uniformly for all n, and the proof of Theorem 9.11 is complete. □

9.7 Comments for Chapter 9

Markov chains with drifts satisfying $xm_1(x) \to \infty$ were considered by Menshikov and Popov in [122] as well as drifts of order $1/x$. They derived rough asymptotics for $\{\pi(x), x \in \mathbb{Z}^+\}$ for countable Markov chains with asymptotically zero drift and bounded jumps. Some rough theorems for the local probabilities $\pi(x)$ were proven; if

$$2m_1(x)/m_2(x) \sim -\theta/x^\beta \quad \text{as } x \to \infty,$$

then for every $\varepsilon > 0$ there exist constants $c_- = c_-(\varepsilon) > 0$ and $c_+ = c_+(\varepsilon) < \infty$ such that

$$c_- \exp\left(-(\theta/(1-\beta)+\varepsilon)x^{1-\beta}\right) \le \pi(x) \le c_+ \exp\left(-(\theta/(1-\beta)-\varepsilon)x^{1-\beta}\right).$$

The same bounds were obtained by Aspandiiarov and Iasnogorodski in [9].

The paper [106] by Korshunov was devoted to the existence and non-existence of the moments of invariant distributions. In particular, it was proven there that if $m_1(x) \sim -\mu/x^\beta$ and $b(x) \to b$ hold and the families of random variables

$$\{\exp\left(\frac{\gamma}{2}\log^2(1+\xi^+(x))\right)(\xi^+(x))^2,\ x \ge 0\} \quad \text{for some } \gamma > 0$$

and $\{(\xi^-(x))^2, x \ge 0\}$ are uniformly integrable, then, for an X_0 having the invariant distribution π,

- $\mathbb{E}e^{\gamma X_0^{1-\beta}} < \infty$ for $\gamma < 2\mu/(1-\beta)b$;
- $\mathbb{E}e^{\gamma X_0^{1-\beta}} = \infty$ if π has unbounded support and $\gamma > 2\mu/(1-\beta)b$.

This result implies that for every $\varepsilon > 0$ there exists a $c(\varepsilon)$ such that

$$\pi(x,\infty) \le c(\varepsilon)\exp\left(-(2\mu/(1-\beta)b-\varepsilon)x^{1-\beta}\right). \tag{9.75}$$

In the paper [106] there is also some analysis for $\gamma = 2\mu/(1-\beta)b$.

10

Markov Chains with Asymptotically Non-Zero Drift in Cramér Case

In this chapter we consider Markov chains with asymptotically constant (non-zero) drift. As shown in the previous chapter, the more slowly $m_1(x)$ tends to zero the higher must be the moments that behave regularly at infinity, in order to make it possible to describe the asymptotic tail behaviour of the invariant measure. Therefore, it is not surprising that in the case of an asymptotically negative drift bounded away from zero we need to assume that the distribution of jumps $\xi(x)$ converges weakly as x tends to infinity. This corresponds, roughly speaking, to the assumption that *all* moments behave regularly at infinity. In this chapter we slightly extend the notion of an asymptotically homogeneous Markov chain by allowing extended limiting random variables.

Definition 10.1 We say that $\{X_n\}$ is *asymptotically homogeneous in space* if

$$\xi(x) \Rightarrow \xi \quad \text{as } x \to \infty, \tag{10.1}$$

where ξ is an extended random variable taking values in $\mathbb{R} \cup \{-\infty\}$.

The class of asymptotically homogeneous chains is larger than the class of additive Markov chains, which was introduced by Aldous [4], and in which $\xi(x)$ was assumed convergent in the total-variation norm.

One of the simplest and most important examples of asymptotically homogeneous Markov chains is a *random walk with delay at zero* (*a Lindley recursion*):

$$W_{n+1} = (W_n + \xi_{n+1})^+, \quad n \ge 0, \tag{10.2}$$

where $\{\xi_n\}$ are independent copies of ξ. In this example we can observe convergence in total variation. The process $\{W_n\}$ describes the waiting-time process in a single-server queue, which is a basic model in queueing theory.

Another popular class of models closely related to asymptotically homogeneous chains originates from the stochastic recursions

$$R_n = A_n R_{n-1} + B_n, \quad n \ge 0,$$

where $\{(A_n, B_n)\}$ are independent and identically distributed random vectors in $\mathbb{R}^+ \times \mathbb{R}$. The sequence R_n does not satisfy (10.1), but some function of it is an asymptotically homogeneous Markov chain; for the details see Goldie [73, Section 2] or Section 11.5 below.

10.1 Local Renewal Theorem

In this section we assume that (10.1) holds and that the mean of the limiting variable ξ is positive. Our aim is to study the asymptotic behaviour of the renewal measure

$$H(B) := \sum_{n=0}^{\infty} \mathbb{P}\{X_n \in B\}.$$

In contrast with the case of asymptotically zero drift, one can derive a renewal theorem for an asymptotically homogeneous chain $\{X_n\}$ without the use of limit theorems for X_n. Instead, we will apply some ideas from the operator approach proposed by Feller [63].

Theorem 10.2 *Let $\xi(x) \Rightarrow \xi$ as $x \to \infty$ and $\mathbb{E}\xi > 0$. Let the family of random variables $\{|\xi(x)|, x \in \mathbb{R}\}$ admit an integrable majorant Ξ, that is, $\mathbb{E}\Xi < \infty$ and*

$$|\xi(x)| \le_{\rm st} \Xi \quad \text{for all } x \in \mathbb{R}. \tag{10.3}$$

Assume that $X_n \to \infty$ with probability 1 as $n \to \infty$ and, moreover, that its renewal measure satisfies

$$\sup_{x \in \mathbb{R}} H(x, x+1] < \infty. \tag{10.4}$$

If the limiting random variable ξ is non-lattice then $H(x, x+h] \to h/\mathbb{E}\xi$ as $x \to \infty$, for all fixed $h > 0$.
If the chain $\{X_n\}$ is integer-valued and $\mathbb{Z}$ is the minimal lattice for the variable ξ then $H\{n\} \to 1/\mathbb{E}\xi$ as $n \to \infty$.

The condition $\mathbb{E}\xi > 0$ excludes the possibility of finding an atom of ξ at point $-\infty$. The condition (10.3) and the dominated convergence theorem imply $|\xi| \le_{\rm st} \Xi$, $\mathbb{E}|\xi| < \infty$, and $\mathbb{E}\xi(x) \to \mathbb{E}\xi$ as $x \to \infty$; in particular, the chain $\{X_n\}$ has an asymptotically space-homogeneous drift.

Proof The condition (10.4) allows us to apply Helly's selection theorem to the family of measures $\{H(x+\cdot),\ x\in\mathbb{R}^+\}$ (see, for example, Theorem 2 in [63, Section VIII.6]). Hence, there exists a sequence of points $t_n\to\infty$ such that the sequence of measures $H(t_n+\cdot)$ converges weakly to some measure λ as $n\to\infty$. The following result characterises λ; it follows from Lemma 6.17 with $v(x)\equiv 1$.

Lemma 10.3 *Let F denote the distribution of ξ. The weak limit λ of the sequence of measures $H(t_n+\cdot)$ satisfies the identity $\lambda=\lambda*F$.*

In what follows the following auxiliary result is useful.

Lemma 10.4 *Let λ_n be a sequence of measures on $\mathbb{R}$ that is weakly convergent to an absolutely continuous σ-finite measure λ. Let $F_n:\mathbb{R}\to\mathbb{R}$ be a sequence of increasing functions weakly convergent to an increasing function F (that is, $F_n(n)$ is convergent at all points of continuity of $F(x)$). Then, for any $A>0$,*

$$\int_0^A F_n(x)\lambda_n(dx)\to\int_0^A F(x)\lambda(dx)\quad \textit{as } n\to\infty.$$

Proof First, by Fubini's theorem, as $n\to\infty$,

$$\begin{aligned}\int_0^A (F_n(x)-F_n(0))(\lambda_n-\lambda)(dx)&=\int_0^A(\lambda_n-\lambda)(dx)\int_0^x F_n(du)\\&=\int_0^A(\lambda_n-\lambda)(u,A]F_n(du)\ \to\ 0,\end{aligned}$$

because $\lambda_n(u,A]\to\lambda(u,A]$ as $n\to\infty$ uniformly for all $u\in[0,A]$, owing to the weak convergence $\lambda_n\Rightarrow\lambda$ and the absolute continuity of the measure λ. Therefore,

$$\int_0^A F_n(x)(\lambda_n-\lambda)(dx)\to 0\quad \text{as } n\to\infty.$$

Second,

$$\int_0^A F_n(x)\lambda(dx)\to\int_0^A F(x)\lambda(dx)\quad \text{as } n\to\infty,$$

by the dominated convergence theorem, because $F_n(x)\to F(x)$ almost everywhere, owing to the weak convergence of F_n and the monotonicity of $F_n(x)$ and $F(x)$. Taken altogether this implies the desired convergence of integrals. □

The concluding part of the proof of Theorem 10.2 will be carried out for the non-lattice case. Choose any sequence of points $t_n\to\infty$ such that the measure

$H(t_n + \cdot)$ converges weakly to some measure λ as $n \to \infty$. It follows from Lemma 10.3 and Proposition 6.18 that then $\lambda(dx) = \alpha\, dx$ with some α, i.e.,

$$H(t_n + dx) \Rightarrow \alpha\, dx \quad \text{as } n \to \infty.$$

Now, it suffices to prove that $\alpha = 1/\mathbb{E}\xi$ for any sequence t_n such that the measure $H(t_n + \cdot)$ is weakly convergent. Fix some $k \in \mathbb{N}$. Set

$$\begin{aligned} H^{(k)}(\cdot) &:= \sum_{j=k}^{\infty} \mathbb{P}\{X_j \in \cdot\} \\ &= H(\cdot) - \sum_{j=0}^{k-1} \mathbb{P}\{X_j \in \cdot\}. \end{aligned}$$

Then, owing to the weak convergence $\mathbb{P}\{X_j \in t_n + \cdot\} \Rightarrow 0$ for all j, we have

$$H^{(k)}(t_n + dx) \Rightarrow \alpha\, dx \quad \text{as } n \to \infty. \tag{10.5}$$

Consider the measure $H^{(k)} - H^{(k+1)}$; by the definition of the renewal measure, it equals the distribution of X_k, that is, for any *bounded* Borel set B, $H^{(k)}(B) - H^{(k+1)}(B) = \mathbb{P}\{X_k \in B\}$ (the equality may fail for unbounded sets, say, for $(x, \infty])$. In particular,

$$(H^{(k)} - H^{(k+1)})(0, x] = \mathbb{P}\{X_k \in (0, x]\} \to \mathbb{P}\{X_k > 0\} \quad \text{as } x \to \infty. \tag{10.6}$$

On the other hand,

$$\begin{aligned} &(H^{(k)} - H^{(k+1)})(0, x] \\ &= \int_{-\infty}^{\infty} (I - P)(y, (0, x]) H^{(k)}(dy) \\ &= -\int_{-\infty}^{0} P(y, (0, x]) H^{(k)}(dy) + \int_{0}^{x} P(y, (-\infty, 0]) H^{(k)}(dy) \\ &\quad + \int_{0}^{x} P(y, (x, \infty)) H^{(k)}(dy) - \int_{x}^{\infty} P(y, (0, x]) H^{(k)}(dy). \end{aligned} \tag{10.7}$$

By Lemma 10.4, the asymptotic homogeneity of the chain and the weak convergence (10.5) imply the following convergences of the integrals, for any fixed $A > 0$:

$$\int_{t_n - A}^{t_n} P(y, (t_n, \infty)) H^{(k)}(dy) \to \alpha \int_{0}^{A} \mathbb{P}\{\xi > z\}\, dz \tag{10.8}$$

as $n \to \infty$ and

$$\int_{t_n}^{t_n + A} P(y, (0, t_n]) H^{(k)}(dy) \to \alpha \int_{0}^{A} \mathbb{P}\{\xi \le -z\}\, dz. \tag{10.9}$$

The majorisation condition (10.3) allows us to estimate the tails of the integrals:

$$\int_0^{t_n-A} P(y,(t_n,\infty))H^{(k)}(dy) \le -\int_A^\infty \mathbb{P}\{\Xi > z\}H(t_n - dz) \tag{10.10}$$

and

$$\int_{t_n+A}^\infty P(y,(0,t_n])H^{(k)}(dy) \le \int_A^\infty \mathbb{P}\{\Xi \ge z\}H(t_n + dz). \tag{10.11}$$

Since the majorant Ξ is integrable, the condition (10.4) guarantees that the right-hand sides of the inequalities (10.10) and (10.11) can be made as small as we please by the choice of a sufficiently large A. For these reasons we conclude from (10.7)–(10.9) that

$$\begin{aligned}(H^{(k)} &- H^{(k+1)})(0,t_n] \\ &\to -\int_{-\infty}^0 P(y,(0,\infty))H^{(k)}(dy) + \int_0^\infty P(y,(-\infty,0])H^{(k)}(dy) \\ &\quad + \alpha\int_0^\infty \mathbb{P}\{\xi > z\}dz - \alpha\int_0^\infty \mathbb{P}\{\xi \le -z\}dz \quad \text{as } n\to\infty.\end{aligned}$$

Together with (10.6) this implies the following equality, for any fixed k:

$$\mathbb{P}\{X_k > 0\} = -\int_{-\infty}^0 P(y,(0,\infty))H^{(k)}(dy) + \int_0^\infty P(y,(-\infty,0])H^{(k)}(dy) + \alpha\mathbb{E}\xi. \tag{10.12}$$

Now let $k\to\infty$; then both integrals go to zero. For example, the first integral can be estimated as follows, for all $A > 0$:

$$\int_{-\infty}^0 P(y,(0,\infty))H^{(k)}(dy) \le \int_{-\infty}^{-A} \mathbb{P}\{\Xi > -y\}H(dy) + H^{(k)}(-A,0].$$

Here, for any fixed A, $H^{(k)}(-A,0]\to 0$ as $k\to\infty$, owing to (10.4). Therefore, (10.6) and (10.12) imply that $1 = \alpha\mathbb{E}\xi$ and the proof is complete. □

In the next theorem we provide some simple conditions sufficient for the condition (10.4) to hold, that is, for local compactness of the renewal measure. Denote $a\wedge b = \min\{a,b\}$.

Theorem 10.5 *Suppose that there exist $A > 0$ and $\varepsilon > 0$ such that*

$$\mathbb{E}(\xi(x)\wedge A) \ge \varepsilon \quad \textit{for all } x\in\mathbb{R}. \tag{10.13}$$

In addition, let

$$\mathbb{P}\{X_n > x \textit{ for all } n\ge 1 | X_0 = x\} \ge \delta > 0 \quad \textit{for all } x\in\mathbb{R}. \tag{10.14}$$

Then $H(x,x+h] \le (A+h)/\varepsilon\delta$ *for all* $x \in \mathbb{R}$ *and* $h > 0$*; in particular,* (10.4) *holds.*

Proof By the Markov property, it suffices to show that

$$H_y(x,x+h] \le (A+h)/\varepsilon\delta \tag{10.15}$$

for all $y \in (x,x+h]$. Given $X_0 \in (x,x+h]$, consider the stopping time

$$T(x+h) = \min\{n \ge 1 : X_n > x+h\}.$$

Since $X_{T(x+h)} \wedge (x+h+A) - X_0 \le A+h$ with probability 1,

$$\begin{aligned} A+h &\ge \mathbb{E}(X_{T(x+h)} \wedge (x+h+A) - X_0) \\ &= \sum_{n=1}^{\infty} \mathbb{E}\big([X_n \wedge (x+h+A) - X_{n-1} \wedge (x+h+A)]\,\mathbb{I}\{T(x+h) \ge n\}\big). \end{aligned}$$

Hence, the definition of $T(x+h)$ implies that

$$\begin{aligned} A+h &\ge \sum_{n=1}^{\infty} \mathbb{E}\{X_n \wedge (x+h+A) - X_{n-1} \wedge (x+h+A);\ T(x+h) \ge n\} \\ &= \sum_{n=1}^{\infty} \mathbb{E}\{X_n \wedge (x+h+A) - X_{n-1} \mid T(x+h) \ge n\}\mathbb{P}\{T(x+h) \ge n\}. \end{aligned}$$

The Markov property and condition (10.13) yield

$$\mathbb{E}\{X_n \wedge (x+h+A) - X_{n-1} \mid T(x+h) \ge n\} \ge \mathbb{E}(\xi(X_{n-1}) \wedge A) \ge \varepsilon$$

for all n. Therefore

$$A+h \ge \varepsilon \sum_{n=1}^{\infty} \mathbb{P}\{T(x+h) \ge n\} = \varepsilon \mathbb{E}T(x+h).$$

Thus the expected number of visits to the interval $(x,x+h]$ until the first exit from $(-\infty,x+h]$ does not exceed $(A+h)/\varepsilon$, independently of the initial state $X_0 \in (x,x+h]$. By the condition (10.14), after exiting $(-\infty,x+h]$ the chain is above the level $X_T(x+h)$ forever, with probability at least δ; in particular, with probability at least δ it does not visit the interval $(x,x+h]$ again. With probability at most $1-\delta$ the chain does visit this interval again, and so on. Concluding, we find that the expected number of visits to the interval $(x,x+h]$ cannot exceed the value of

$$\frac{A+h}{\varepsilon}\sum_{n=0}^{\infty}(1-\delta)^n = \frac{A+h}{\varepsilon\delta},$$

and (10.15) is proven. The proof of Theorem 10.5 is complete. □

Corollary 10.6 *Let the family of jumps $\{\xi(x), x \in \mathbb{R}\}$ possess an integrable minorant with a positive mean, that is, there exists a random variable ζ such that $\mathbb{E}\zeta > 0$ and $\xi(x) \geq_{\rm st} \zeta$ for all $x \in \mathbb{R}$. Then*

$$H(x, x+h] \leq (A+h)A/\varepsilon^2$$

for all $A > 0$ such that $\varepsilon \equiv \mathbb{E}(\zeta \wedge A) > 0$; in particular, (10.4) *holds.*

Proof Consider the partial sums $Z_n = \zeta_1 + \cdots + \zeta_n$ of independent copies of ζ. Denote the first ascending-ladder epoch by $\eta = \min\{n \geq 1 : Z_n > 0\}$. It is well known (see e.g. Theorem 2.3(c) in [8, Chapter VIII]) that

$$\mathbb{P}\{Z_n > 0 \text{ for all } n \geq 1\} = 1/\mathbb{E}\eta.$$

Since

$$\mathbb{P}\{X_n > x \text{ for all } n \geq 1 \mid X_0 = x\} \geq \mathbb{P}\{Z_n > 0 \text{ for all } n \geq 1\}$$

by the minorisation condition, the value of δ in Theorem 10.5 is at least $1/\mathbb{E}\eta$. Since $Z_{A,\eta_A} \leq A$, where $Z_{A,n} := \zeta_1 \wedge A + \cdots + \zeta_n \wedge A$ and $\eta_A := \min\{n \geq 1 : Z_{A,n} > 0\}$, we obtain $\mathbb{E}\eta_A \leq A/\varepsilon$ by Wald's equality $\mathbb{E}Z_{\eta_A} = \mathbb{E}\eta_A \mathbb{E}\zeta_1 \wedge A$. Then it follows from $\eta \leq \eta_A$ that $\mathbb{E}\eta \leq A/\varepsilon$, which yields $\delta \geq \varepsilon/A$, and the claim of the corollary follows. □

10.2 Large-Deviation Principle for Stationary Distribution

We now turn to the asymptotic behaviour of the stationary distribution of an asymptotically homogeneous chain, that is, we assume that (10.1) holds for an extended limiting variable ξ. We shall also assume that the limiting variable ξ satisfies Cramér's condition:

$$\text{there exists a } \beta > 0 \text{ such that } \mathbb{E}e^{\beta\xi} = 1. \tag{10.16}$$

As is well known, the stationary measure of the random walk $\{W_n\}$ delayed at the origin, defined in (10.2), say π_W, coincides with the distribution of $\sup_{n\geq 0}\sum_{k=1}^n \xi_k$ where the ξ_k are independent copies of ξ. Then, by the classical Cramér – Lundberg approximation, for some $c > 0$ we have

$$\pi_W(x,\infty) \sim ce^{-\beta x} \quad \text{as } x \to \infty, \tag{10.17}$$

under the additional assumption $\mathbb{E}\xi e^{\beta\xi} < \infty$, in the non-lattice case; in the lattice case x is restricted to the lattice values. Since the jumps of the chains $\{X_n\}$ and $\{W_n\}$ are asymptotically equivalent, one might expect that the stationary tail distributions of $\{X_n\}$ and $\{W_n\}$ are asymptotically equivalent. It turns out to be true on a logarithmic scale only.

Theorem 10.7 *Assume the asymptotic homogeneity* (10.1) *and Cramér's condition* (10.16). *If an invariant measure π has right unbounded support then the following lower bound holds:*

$$\liminf_{x\to\infty} \frac{\log \pi(x,\infty)}{x} \ge -\beta. \tag{10.18}$$

If, in addition,

$$\sup_{x>0} \mathbb{E}e^{\lambda\xi(x)} < \infty, \quad \sup_{x\le 0} \mathbb{E}e^{\lambda(x+\xi(x))} < \infty \quad \textit{for all } \lambda \in [0,\beta) \tag{10.19}$$

then

$$\limsup_{x\to\infty} \frac{\log \pi(x,\infty)}{x} \le -\beta. \tag{10.20}$$

In Theorem 11.15 we apply this result to derive the logarithmic asymptotics of stationary distributions of positive recurrent stochastic difference equations.

Proof Fix some $\widehat{x} \in \mathbb{R}$ and consider an aggregated Markov chain $\{X_n^*\}$ on $[\widehat{x},\infty)$ with transition probabilities defined in (7.36) and (7.37). As mentioned in the text there, the measure π^* that aggregates states from $(-\infty,\widehat{x}]$ to $\widehat{x}$, that is, $\pi^*\{\widehat{x}\} = \pi(-\infty,\widehat{x}]$ and $\pi^*(B) = \pi(B)$ for all $B \subseteq (\widehat{x},\infty)$ is an invariant measure for $\{X_n^*\}$.

First we derive the lower bound (10.18) via a comparison of $\{X_n^*\}$ with a random walk delayed at zero; we choose $\widehat{x}$ sufficiently large as follows. For any u consider a random variable $\eta(u)$ with tail distribution

$$\mathbb{P}\{\eta(u) > z\} = \inf_{v\ge u-1/u} \mathbb{P}\{\xi(v) > z + 2/u\}.$$

Then $\eta(u)$ stochastically increases as u grows and $\xi(v) \ge_{\rm st} \eta(u)$ for all $v \ge u - 1/u$. For any $A > 0$, define $\eta_A(u) := \min\{\eta(u), A\}$. Since the chain $\{X_n\}$ is asymptotically homogeneous, we have $\eta_A(u) \le_{\rm st} \xi$ for all u and $\eta_A(u) \Rightarrow \xi$ as A, $u \to \infty$. Hence, for all sufficiently large A and u, there exists a unique solution $\beta_A(u)$ to the equation $\mathbb{E}e^{\beta_A(u)\eta_A(u)} = 1$, which is always not less than β. In addition, $\beta_A(u)$ decreases as A and u grow, and

$$\beta_A(u) \downarrow \beta \quad \text{as } A,\ u \to \infty.$$

Fix an $\varepsilon \in (0,1)$ and choose a sufficiently large A and $\widehat{x}$ that $\beta_A(\widehat{x}) \in [\beta,\beta+\varepsilon]$ and $\pi(\widehat{x} - 1/\widehat{x},\widehat{x}] > 0$, which is possible because π has right unbounded support. Denote $\widehat{\eta} := \eta_A(\widehat{x})$. It follows from (7.37) that, for $u > \widehat{x}$,

$$\begin{aligned} P^*(\widehat{x},(u,\infty)) &\geq \frac{1}{\pi(-\infty,\widehat{x}]}\int_{\widehat{x}-1/\widehat{x}}^{\widehat{x}+0} P(z,(u,\infty))\pi(dz) \\ &\geq \frac{\pi(\widehat{x}-1/\widehat{x}]}{\pi(-\infty,\widehat{x}]}\mathbb{P}\{\widehat{\eta}>u-\widehat{x}\}. \end{aligned} \tag{10.21}$$

Consider a random walk $\{\widehat{W}_n\}$ delayed at $\widehat{x}$, that is,

$$\widehat{W}_n = \max(\widehat{x},\ \widehat{W}_{n-1}+\widehat{\eta}_n),$$

where the $\widehat{\eta}_n$ are independent copies of $\widehat{\eta}$. By the construction of $\widehat{\eta}=\eta_A(\widehat{x})$, $\{\widehat{W}_n\}$ is dominated by $\{X_n\}$ above $\widehat{x}$. More precisely, the following inequality is valid for all $u>\widehat{x}$, $y>\widehat{x}$ and m:

$$\begin{aligned} &\mathbb{P}\{X_k>\widehat{x} \text{ for all } k<m,\ X_m>y \mid X_0=u\} \\ &\quad\geq \mathbb{P}\{\widehat{W}_k>\widehat{x} \text{ for all } k<m,\ \widehat{W}_m>y \mid \widehat{W}_0=u\}. \end{aligned} \tag{10.22}$$

Consider a stationary version of $\{X_n\}$, that is, X_n has a distribution π for all $n\geq 0$. Then the distribution of $\max(\widehat{x},X_n)$ on $(\widehat{x},\infty)$ is the same as that of X_n^*, given that X_0^* has distribution π^*. Thus, at any time n, the decomposition of all trajectories with respect to the last visit of $\{X_k^*\}$ to the state $\widehat{x}$ gives the following lower bound, for $y>\widehat{x}$:

$$\begin{aligned} \pi(y,\infty) &= \mathbb{P}\{X_n^*>y\} \\ &\geq \sum_{j=0}^{n-1}\mathbb{P}\{X_j^*=\widehat{x}\}\int_{\widehat{x}+0}^{\infty} P^*(\widehat{x},du)\mathbb{P}\{X_k^*>\widehat{x},k\in[j+2,n-1],X_n^*>y \mid X_{j+1}^*=u\} \\ &= \pi(-\infty,\widehat{x}]\sum_{j=0}^{n-1}\int_{\widehat{x}+0}^{\infty} P^*(\widehat{x},du)\mathbb{P}\{X_k>\widehat{x},k\in[j+2,n-1],X_n>y \mid X_{j+1}=u\} \\ &\geq \pi(-\infty,\widehat{x}]\sum_{j=0}^{n-1}\int_{\widehat{x}+0}^{\infty} P^*(\widehat{x},du)\mathbb{P}\{\widehat{W}_k>\widehat{x},k\in[j+2,n-1],\widehat{W}_n>y \mid \widehat{W}_{j+1}=u\}, \end{aligned}$$

owing to (10.22). Since the probability $\mathbb{P}\{\widehat{W}_k>\widehat{x},k\in[j+2,n-1], \widehat{W}_n>y \mid \widehat{W}_{j+1}=u\}$ is increasing in u, the stochastic domination condition (10.21) yields

$$\begin{aligned} &\int_{\widehat{x}+0}^{\infty} P^*(\widehat{x},du)\mathbb{P}\{\widehat{W}_k>\widehat{x},k\in[j+2,n-1],\ \widehat{W}_n>y \mid \widehat{W}_{j+1}=u\} \\ &\quad\geq \frac{\pi(\widehat{x}-1/\widehat{x},\widehat{x}]}{\pi(-\infty,\widehat{x}]}\int_{\widehat{x}+0}^{\infty}\mathbb{P}\{\widehat{x}+\widehat{\eta}\in du\} \\ &\qquad\qquad\times\mathbb{P}\{\widehat{W}_k>\widehat{x},k\in[j+2,n-1],\ \widehat{W}_n>y \mid \widehat{W}_{j+1}=u\}. \end{aligned}$$

Therefore

$$\pi(y,\infty) \geq \pi(\widehat{x}-1/\widehat{x},\widehat{x}] \sum_{j=0}^{n-1} \int_{\widehat{x}+0}^{\infty} \mathbb{P}\{\widehat{x}+\widehat{\eta} \in du\} \times \mathbb{P}\{\widehat{W}_k > \widehat{x}, k \in [j+2, n-1],\ \widehat{W}_n > y \mid \widehat{W}_{j+1} = u\}. \tag{10.23}$$

On the other hand, applying the decomposition all trajectories of $\{\widehat{W}_n\}$ with respect to the last visit of $\{\widehat{W}_n\}$ to the state $\widehat{x}$ we deduce that, for $\widehat{W}_0 = \widehat{x}$ and $y > \widehat{x}$,

$$\begin{aligned}
\mathbb{P}\{\widehat{W}_n > y\}
&= \sum_{j=0}^{n-1} \mathbb{P}\{\widehat{W}_j = \widehat{x}\} \int_{\widehat{x}+0}^{\infty} \mathbb{P}\{\widehat{W}_{j+1} \in du \mid \widehat{W}_j = \widehat{x}\} \\
&\qquad \times \mathbb{P}\{\widehat{W}_k > \widehat{x}, k \in [j+2, n-1],\ \widehat{W}_n > y \mid \widehat{W}_{j+1} = u\} \\
&\leq \sum_{j=0}^{n-1} \int_{\widehat{x}+0}^{\infty} \mathbb{P}\{\widehat{x}+\widehat{\eta} \in du\} \mathbb{P}\{\widehat{W}_k > \widehat{x}, k \in [j+2, n-1],\ \widehat{W}_n > y \mid \widehat{W}_{j+1} = u\}.
\end{aligned}$$

Together with (10.23) this implies the following lower bound:

$$\pi(y,\infty) \geq \pi(\widehat{x}-1/\widehat{x},\widehat{x}]\mathbb{P}\{\widehat{W}_n > y\} \quad \text{for } y > \widehat{x}.$$

As $\widehat{W}_n - \widehat{x}$ is a Lindley recursion, the Cramér–Lundberg approximation (10.17) yields

$$\lim_{y\to\infty} \lim_{n\to\infty} \frac{\log \mathbb{P}\{\widehat{W}_n > y\}}{y} = -\beta_A(\widehat{x});$$

hence

$$\liminf_{y\to\infty} \frac{\log \pi(y,\infty)}{y} \geq -\beta_A(\widehat{x}).$$

Letting $\varepsilon \downarrow 0$ we conclude that the assertion (10.18) holds because $\beta_A(\widehat{x}) \leq \beta + \varepsilon$.

Let us now prove the upper bound (10.20). Fix any $\lambda < \beta$. Then the boundedness (10.19) of the exponential moments of jumps of order $(\beta + \lambda)/2 \in (\lambda, \beta)$ and the weak convergence $\xi(x) \Rightarrow \xi$ imply convergence of the exponential moments of order λ:

$$\mathbb{E}e^{\lambda\xi(x)} \to \mathbb{E}e^{\lambda\xi} < \mathbb{E}e^{\beta\xi} = 1 \quad \text{as } x \to \infty;$$

hence there exist $\widehat{x} \in \mathbb{R}$ and $\varepsilon > 0$ such that

$$\mathbb{E}e^{\lambda\xi(x)} \leq 1 - \varepsilon \quad \text{for all } x \geq \widehat{x}. \tag{10.24}$$

Fix an $A > \widehat{x}$ and consider the function $g(x) = \min(e^{\lambda x}, e^{\lambda A})$. Let $\{X_n\}$ be in the stationary regime, that is, let X_n have the distribution π for all n. Since g is bounded above by $e^{\lambda A}$, we have

$$\begin{aligned} 0 &= \mathbb{E}(g(X_1) - g(X_0)) \\ &= \Big(\int_{-\infty}^{\widehat{x}} + \int_{\widehat{x}}^{A} + \int_{A}^{\infty}\Big)(\mathbb{E}g(x+\xi(x)) - g(x))\pi(dx). \end{aligned} \tag{10.25}$$

The third integral on the right-hand side is non-positive because the increasing function $g(x)$ is constant for $x \ge A$. The first integral is bounded above by

$$c_1 := \sup_{x\le\widehat{x}} \mathbb{E}\{g(X_1) - g(x) \mid X_0 = x\} \;\le\; \sup_{x\le\widehat{x}} \mathbb{E}e^{\lambda(x+\xi(x))},$$

which is finite owing to the condition (10.19). The second integral satisfies

$$\begin{aligned} \int_{\widehat{x}}^{A} (\mathbb{E}g(x+\xi(x)) - g(x))\pi(dx) &\le \int_{\widehat{x}}^{A} (\mathbb{E}e^{\lambda(x+\xi(x))} - e^{\lambda x})\pi(dx) \\ &\le -\varepsilon \int_{\widehat{x}}^{A} e^{\lambda x}\pi(dx), \end{aligned}$$

by (10.24). Therefore, it follows from (10.25) that

$$0 \le c_1 - \varepsilon \int_{\widehat{x}}^{A} e^{\lambda x}\pi(dx).$$

Owing to the arbitrary choice of A, we obtain

$$\int_{\widehat{x}}^{\infty} e^{\lambda x}\pi(dx) \le c_1/\varepsilon,$$

which implies that $\pi(x,\infty) \le c_1 e^{-\lambda x}/\varepsilon$ for all $x \ge \widehat{x}$. Now the upper bound (10.20) follows, because we may choose $\lambda < \beta$ as close to β as we please. □

10.3 Sharp Asymptotics for Stationary Distribution

While the logarithmic asymptotic law is universal for stationary distributions of Markov chains that are asymptotically homogeneous in space, it turns out that the exact asymptotic tail behaviour of π depends not only on the distribution of ξ, but also on the speed of convergence in (10.1).

The next result describes the case where this convergence is so fast that the measure π has a tail that is asymptotically proportional to the stationary measure of $\{W_n\}$.

Theorem 10.8 *Assume the asymptotic homogeneity* (10.1) *and Cramér's condition* (10.16). *Let π have right unbounded support. Suppose that*

$$\xi(x) \le_{\mathrm{st}} \Xi, \quad x \in \mathbb{R}, \tag{10.26}$$

for some random variable Ξ such that $\mathbb{E}\Xi e^{\beta\Xi} < \infty$ and

$$|\mathbb{E}e^{\beta\xi(x)} - 1| \le \gamma(x) \tag{10.27}$$

for some decreasing function $\gamma(x)$ that is integrable at infinity.

If the distribution of ξ is non-lattice then there exists a positive constant c such that

$$\pi(x,\infty) \sim ce^{-\beta x} \quad \text{as } x \to \infty. \tag{10.28}$$

If $\{X_n\}$ takes values on $\mathbb{Z}$, and $\mathbb{Z}$ is the minimal lattice for ξ, then (10.28) *holds with x restricted to the integers.*

In Theorem 11.15 we apply this result to derive precise asymptotics for a stationary distribution of positive recurrent stochastic difference equations.

The condition (10.27) is quite close to being optimal. If, for example, all values of $\mathbb{E}e^{\beta\xi(x)} - 1$ are of the same sign and not summable then $\pi(x)e^{\beta x}$ converges either to zero or to infinity; see Corollary 10.12 below. Thus, if (10.27) is violated, then $\pi(x,\infty)$ may have exponential asymptotics such as (10.28) only in the case where $\mathbb{E}e^{\beta\xi(x)} - 1$ changes sign infinitely often.

Example 10.9 Consider a Markov chain $\{X_n\}$ on $\mathbb{Z}^+$ with jumps to nearest neighbours only:

$$\mathbb{P}\{\xi(i) = 1\} = 1 - \mathbb{P}\{\xi(i) = -1\} = p + \varphi(i).$$

Assume that, as $i \to \infty$,

$$\varphi(i) \sim \begin{cases} i^{-\gamma}, & i = 2k, \\ -i^{-\gamma}, & i = 2k+1, \end{cases}$$

for some $\gamma \in (1/2, 1)$. Clearly, then the asymptotic homogeneity (10.1) and Cramér's condition (10.16) hold true while the condition (10.27) fails.

Let us have a look at values of $\{X_n\}$ at even time epochs, that is, let us consider the chain

$$Y_k = X_{2k}, \quad k \ge 0.$$

Then we have

$$\begin{aligned}
\mathbb{P}_i\{Y_1 - i = -2\} &= (q - \varphi(i))(q - \varphi(i-1)), \\
\mathbb{P}_i\{Y_1 - i = 0\} &= (q - \varphi(i))(p + \varphi(i-1)) + (p + \varphi(i))(q - \varphi(i+1)), \\
\mathbb{P}_i\{Y_1 - i = 2\} &= (p + \varphi(i))(p + \varphi(i+1)),
\end{aligned}$$

where $q := 1 - p$. From these equalities we obtain

$$\begin{aligned}
\mathbb{E}_i\left(\frac{q}{p}\right)^{Y_1-i} - 1 &= \left(\frac{p^2}{q^2}-1\right)\mathbb{P}_i\{Y_1 - i = -2\} + \left(\frac{q^2}{p^2}-1\right)\mathbb{P}_i\{Y_1 - i = 2\} \\
&= \left(\frac{p^2}{q^2}-1\right)(q-\varphi(i))(q-\varphi(i-1)) + \left(\frac{q^2}{p^2}-1\right)(p+\varphi(i))(p+\varphi(i+1)) \\
&= -q\left(\frac{p^2}{q^2}-1\right)(\varphi(i)+\varphi(i-1)) + p\left(\frac{q^2}{p^2}-1\right)(\varphi(i)+\varphi(i+1)) + O(i^{-2\gamma}).
\end{aligned}$$

Noting that $\varphi(i) + \varphi(i+1) = O(i^{-\gamma-1})$ as $i \to \infty$, we conclude that the sequence $|\mathbb{E}_i(q/p)^{Y_1-i} - 1|$ is summable and, consequently, we may apply Theorem 10.8. Since π is also stationary for Y, we obtain $\pi(i) \sim c(p/q)^i$ as $i \to \infty$.

Proof of Theorem 10.8. We start, as usual, with the construction of an appropriate Lyapunov function that is sufficiently close to a harmonic function. Let p be a bounded decreasing function $p(x) : \mathbb{R} \to \mathbb{R}^+$ that is regularly varying at infinity with index -1 and integrable at infinity. Set

$$g(x) := \min\Big(1, \int_x^\infty p(y)dy\Big) \tag{10.29}$$

and consider

$$U_p(x) := e^{\beta x}(1 + g(x)). \tag{10.30}$$

We want to show that there exists a $p(x)$ such that

$$\mathbb{E}U_p(x + \xi(x)) - U_p(x) = -e^{\beta x}p(x)(\mathbb{E}\xi e^{\beta\xi} + o(1)) \quad \text{as } x \to \infty. \tag{10.31}$$

By the definition of $U_p(x)$,

$$\begin{aligned}
&\mathbb{E}U_p(x+\xi(x)) - U_p(x) \\
&\quad = e^{\beta x}\big(\mathbb{E}e^{\beta\xi(x)}(1 + g(x+\xi(x))) - 1 - g(x)\big) \\
&\quad = e^{\beta x}(1+g(x))(\mathbb{E}e^{\beta\xi(x)} - 1) + e^{\beta x}\mathbb{E}(g(x+\xi(x)) - g(x))e^{\beta\xi(x)}.
\end{aligned} \tag{10.32}$$

Owing to Lemma 2.29, the assumption (10.27) yields the existence of a $p(x)$ satisfying the conditions above and such that

$$|\mathbb{E}e^{\beta\xi(x)} - 1| = o(p(x)). \tag{10.33}$$

Fix some increasing function $s(x) = o(x)$ and split the second term on the right-hand side of (10.32) into three parts:

$$\begin{aligned}\mathbb{E}(g(x+\xi(x))-g(x))e^{\beta\xi(x)} &= \mathbb{E}\{(g(x+\xi(x))-g(x))e^{\beta\xi(x)};\ \xi(x)<-s(x)\}\\ &\quad+\mathbb{E}\{(g(x+\xi(x))-g(x))e^{\beta\xi(x)};\ |\xi(x)|\le s(x)\}\\ &\quad+\mathbb{E}\{(g(x+\xi(x))-g(x))e^{\beta\xi(x)};\ \xi(x)>s(x)\}.\end{aligned}$$

Because g is decreasing and is bounded by 1, we have

$$0 \le \mathbb{E}\{(g(x+\xi(x))-g(x))e^{\beta\xi(x)};\ \xi(x)<-s(x)\} \le e^{-\beta s(x)}. \tag{10.34}$$

Since $p(x)$ is assumed to be regularly varying at infinity, we have the equivalence $g(x+\xi(x))-g(x) \sim -p(x)\xi(x)$ as $x\to\infty$ uniformly on the set $|\xi(x)|\le s(x)$. Therefore

$$\begin{aligned}&\mathbb{E}\{(g(x+\xi(x))-g(x))e^{\beta\xi(x)};\ |\xi(x)|\le s(x)\}\\ &\sim -p(x)\mathbb{E}\{\xi(x)e^{\beta\xi(x)};\ |\xi(x)|\le s(x)\}.\end{aligned}$$

Recalling that the family $\xi(x)$ possesses a majorant Ξ with $\mathbb{E}\Xi e^{\beta\Xi}<\infty$, we infer that

$$\mathbb{E}\{\xi(x)e^{\beta\xi(x)};\ |\xi(x)|\le s(x)\} \to \mathbb{E}\xi e^{\beta\xi} \quad \text{as } x\to\infty.$$

As a result,

$$\mathbb{E}\{(g(x+\xi(x))-g(x))e^{\beta\xi(x)};\ |\xi(x)|\le s(x)\} \sim -p(x)\mathbb{E}\xi e^{\beta\xi}. \tag{10.35}$$

The existence of Ξ implies also that the function $\mathbb{E}\{e^{\beta\xi(x)};\ \xi(x)>s(x)\}$ is dominated by $\mathbb{E}\{e^{\beta\Xi};\ \Xi>s(x)\}$. Since $\mathbb{E}\Xi e^{\beta\Xi}$ is finite, the function $\mathbb{E}\{e^{\beta\Xi};\ \Xi>s(x)\}$ is decreasing and summable provided that $s(x)/x\to 0$ sufficiently slowly. Consequently, there exists a $p(x)$ such that

$$\mathbb{E}\{(g(x+\xi(x))-g(x))e^{\beta\xi(x)};\ \xi(x)>s(x)\} = o(p(x)). \tag{10.36}$$

Combining (10.34)–(10.36), we conclude that

$$\mathbb{E}(g(x+\xi(x))-g(x))e^{\beta\xi(x)} = -p(x)(\mathbb{E}\xi e^{\beta\xi}+o(1)).$$

Substituting this relation and (10.33) into (10.32), we obtain (10.31).

Consider, as usual, the transition kernel

$$Q(x,dy) = \frac{U_p(y)}{U_p(x)}P(x,dy), \quad y\ge\widehat{x}.$$

It follows from (10.31) that, for all sufficiently large $\widehat{x}$,

$$Q(x,\mathbb{R}) = \frac{1}{U_p(x)}\mathbb{E}\{U_p(x+\xi(x));\ x+\xi(x)\ge\widehat{x}\}$$
$$\le \frac{1}{U_p(x)}\mathbb{E}U_p(x+\xi(x)) \le 1 \quad \text{for all } x\ge\widehat{x}. \tag{10.37}$$

In other words, Q is a substochastic kernel. Furthermore, it follows from the asymptotic homogeneity that

$$Q(x,\mathbb{R}) \ge \mathbb{P}\{\xi(x)\ge 0\} \ge \mathbb{P}\{\xi\ge 0\}/2 \quad \text{for all } x\ge\widehat{x}, \tag{10.38}$$

if $\widehat{x}$ is chosen sufficiently large. Using (10.31) once again, we conclude that

$$q(x) := -\log Q(x,\mathbb{R}) = O(p(x)) \quad \text{as } x\to\infty. \tag{10.39}$$

Let $\{\widehat{X}_n\}$ be a Markov chain on $(\widehat{x},\infty)$ with transition kernel

$$\widehat{P}(x,dy) = \frac{Q(x,dy)}{Q(x,\mathbb{R})}$$

and let $\widehat{\xi}(x)$ denotes its jump from state x. It is immediate from the definition of U_p that $\widehat{\xi}(x)$ converges weakly to the distribution $e^{\beta y}\mathbb{P}\{\xi\in dy\}$ as $x\to\infty$. Furthermore, the assumption that $\mathbb{E}\Xi e^{\beta\Xi}<\infty$ and (10.38) together imply that the family of jumps $|\widehat{\xi}(x)|$ possesses an integrable majorant. Therefore, there exists an $\widehat{x}$ such that the family of jumps $\{\widehat{\xi}(x);\ x>\widehat{x}\}$ possesses a stochastic minorant with positive expectation. Thus, Corollary 10.6 applies to the chain $\{\widehat{X}_n\}$, and this in turn allows us to apply Theorem 10.2. If ξ is non-lattice then, for all $h>0$,

$$\widehat{H}(x,x+h] \to \frac{h}{\mathbb{E}\xi e^{\beta\xi}} \quad \text{as } x\to\infty.$$

If $\{X_n\}$ is an integer-valued Markov chain and $\mathbb{Z}$ is the minimal lattice for ξ then the previous relation is valid for h and x restricted to integers.

Combining (10.39) with the upper bound $\sup_x \widehat{H}(x,x+h]<\infty$, we conclude as in Lemma 8.9 that

$$\sum_{k=0}^{\infty}\mathbb{E}q(\widehat{X}_k) < \infty.$$

Thus, by Lemma 4.5,

$$\widehat{H}^{(q)}(x,x+h] \to \frac{h}{\mathbb{E}\xi e^{\beta\xi}}\mathbb{E}e^{-\sum_{k=0}^{\infty}q(\widehat{X}_k)} \quad \text{as } x\to\infty. \tag{10.40}$$

Here again h is an arbitrary positive number in the case when ξ is non-lattice, and h is an integer in the lattice case.

For the invariant distribution π we have the following representation, see (7.47):

$$\pi(dy) = c_* \frac{\widehat{H}^{(q)}(dy)}{U_p(y)}.$$

If ξ is lattice then

$$\pi\{n\} = c_* \frac{\widehat{H}^{(q)}\{n\}}{U_p(n)},$$

and the result follows from (10.40) and the fact that $U_p(x) \sim e^{\beta x}$.

In the non-lattice case, for any fixed $h > 0$,

$$c_* \frac{\widehat{H}^{(q)}(x, x+h]}{\max_{x \le y \le x+h} U_p(y)} \le \pi(x, x+h] \le c_* \frac{\widehat{H}^{(q)}(x, x+h]}{\min_{x \le y \le x+h} U_p(y)}.$$

Again using (10.40), we obtain lower and upper bounds:

$$che^{-\beta x - \beta h}(1 + o(1)) \le \pi(x, x+h] \le che^{-\beta x}(1 + o(1)).$$

Choosing h small and summing the bounds for $\pi(x + kh, x + (k+1)h]$ we obtain the required lower and upper bounds for $\pi(x, \infty)$, which completes the proof of the theorem. □

We now turn to the case where $\mathbb{E}e^{\beta\xi(x)}$ converges to 1 in a non-summable way. Our next result describes the behaviour of π in terms of a non-uniform exponential change of measure.

Theorem 10.10 *Suppose that the asymptotic homogeneity condition* (10.1) *and Cramér's condition* (10.16) *hold and, for some* $\varepsilon > 0$,

$$\sup_{x \in \mathbb{R}} \mathbb{E}e^{(\beta+\varepsilon)\xi(x)} < \infty. \tag{10.41}$$

Assume also that there exists a differentiable function $\beta(x) > 0$ *such that*

$$|\mathbb{E}e^{\beta(x)\xi(x)} - 1| \le \gamma(x) \tag{10.42}$$

and $|\beta'(x)| \le \gamma(x)$, *where* $\gamma(x)$ *is a bounded decreasing function that is integrable at infinity. Then, for some* $c > 0$,

$$\pi(x, \infty) \sim ce^{-\int_0^x \beta(y)dy} \quad \text{as } x \to \infty,$$

where x *runs through the integers in the lattice case.*

Proof The proof is quite similar to that of Theorem 10.8; the only alteration is a slightly trickier choice of the Lyapunov function U_p. Instead of $(1+g(x))e^{\beta x}$ we now define

$$U_p(x) := (1 + g(x))e^{\int_0^x \beta(y)dy}.$$

Let $\delta < \varepsilon$ and $c > 1/(\varepsilon - \delta)$. Observe that, with necessity, $\beta(x) \to \beta$ so that, by the condition (10.41), for all sufficiently large x,

$$\begin{aligned}
&\mathbb{E}\{e^{\beta(x)\xi(x)};\ |\xi(x)| > c\log x\} \\
&\quad \le \mathbb{E}\{e^{(\beta+\delta)\xi(x)};\ \xi(x) > c\log x\} + \mathbb{E}\{e^{(\beta-\delta)\xi(x)};\ \xi(x) < -c\log x\} \\
&\quad = O(e^{-c(\varepsilon-\delta)\log x} + e^{-c(\beta-\delta)\log x}) \ =\ O(1/x^{c(\varepsilon-\delta)}) \qquad (10.43)
\end{aligned}$$

as $x \to \infty$, where without loss of generality we can assume that $\varepsilon < \beta$. Similarly,

$$\mathbb{E}\{e^{\int_x^{x+\xi(x)}\beta(y)dy};\ |\xi(x)| > c\log x\} = O(1/x^{c(\varepsilon-\delta)}) \quad \text{as } x \to \infty.$$

Further, by the mean-value theorem, for some $\theta = \theta(x,\xi) \in (0,1)$, we have

$$\begin{aligned}
&\left|\mathbb{E}\{e^{\int_x^{x+\xi(x)}\beta(y)dy};\ |\xi(x)| \le c\log x\} - \mathbb{E}\{e^{\beta(x)\xi(x)};\ |\xi(x)| \le c\log x\}\right| \\
&\quad = \left|\mathbb{E}\{e^{\beta(x)\xi(x)}(\exp(\int_x^{x+\xi(x)}(\beta(y)-\beta(x))dy) - 1);\ |\xi(x)| \le c\log x\}\right| \\
&\quad = \Bigg|\mathbb{E}\Bigg\{\int_x^{x+\xi(x)}(\beta(y)-\beta(x))dy \\
&\qquad \times \exp\left(\beta(x)\xi(x) + \theta\int_x^{x+\xi(x)}(\beta(y)-\beta(x))dy\right);|\xi(x)| \le c\log x\Bigg\}\Bigg| \\
&\quad \le \frac{\gamma(x - c\log x)}{2}\mathbb{E}\{\xi^2(x)\exp\left(\beta(x)\xi(x) + \int_x^{x+\xi(x)}|\beta(y)-\beta(x)|dy\right);\ |\xi(x)| \le c\log x\},
\end{aligned} \qquad (10.44)$$

because, by the condition $|\beta'(x)| \le \gamma(x)$ on the derivative of $\beta(y)$, for $|\xi(x)| \le c\log x$ it holds that

$$\begin{aligned}
\left|\int_x^{x+\xi(x)}(\beta(y)-\beta(x))dy\right| &\le \int_x^{x+\xi(x)}|\beta(y)-\beta(x)|dy \\
&\le \sup_{|z|\le c\log x}|\beta'(x+z)|\xi^2(x)/2 \\
&\le \gamma(x - c\log x)\xi^2(x)/2.
\end{aligned}$$

Uniformly on the event $|\xi(x)| \le c\log x$, we have

$$\gamma(x - c\log x)\xi^2(x) \ \le\ c^2\gamma(x - c\log x)\log^2 x \to 0 \quad \text{as } x \to \infty,$$

since the function $\gamma(x)$ is decreasing and integrable at infinity. Therefore, for all sufficiently large x, the right-hand side of (10.44) is not greater than

$$\gamma(x - c\log x)\mathbb{E}\{\xi^2(x)e^{\beta(x)\xi(x)};\ |\xi(x)| \le c\log x\} = O(\gamma(x - c\log x))$$

as $x \to \infty$, owing to the condition (10.41). Hence, as $x \to \infty$,

$$\mathbb{E}e^{\int_x^{x+\xi(x)}\beta(y)dy} = \mathbb{E}e^{\beta(x)\xi(x)} + O(\gamma(x - c\log x) + 1/x^{c(\varepsilon-\delta)}).$$

Taking into account (10.42) and $c > 1/(\varepsilon - \delta)$, we conclude that there exists a decreasing function $p_1(x)$ that is integrable at infinity and is such that

$$\mathbb{E}e^{\int_x^{x+\xi(x)}\beta(y)dy} = 1 + O(p_1(x)) \quad \text{as } x \to \infty. \tag{10.45}$$

We have the equality

$$\mathbb{E}U_p(x + \xi(x)) - U_p(x) = U_p(x)\big(\mathbb{E}e^{\int_x^{x+\xi(x)}\beta(y)dy} - 1\big) + e^{\int_0^x \beta(y)dy}\mathbb{E}(g(x + \xi(x)) - g(x))e^{\int_x^{x+\xi(x)}\beta(y)dy}.$$

Using (10.45) and recalling that $g(x)$ is bounded, we get

$$\mathbb{E}U_p(x + \xi(x)) - U_p(x) = O(p_1(x)U_p(x)) + e^{\int_0^x \beta(y)dy}\mathbb{E}(g(x + \xi(x)) - g(x))e^{\beta(x)\xi(x)}.$$

Repeating the corresponding arguments from the proof of Theorem 10.8 and using (10.41), we obtain

$$\mathbb{E}\big\{(g(x + \xi(x)) - g(x))e^{\beta(x)\xi(x)}; |\xi(x)| > c\log x\big\} = o(1/x^{c(\varepsilon-\delta)})$$

and

$$\mathbb{E}\big\{(g(x + \xi(x)) - g(x))e^{\beta(x)\xi(x)}; |\xi(x)| \le c\log x\big\} \sim -p(x)\mathbb{E}\xi(x)e^{\beta(x)\xi(x)}.$$

Therefore, taking $p(x) \gg p_1(x)$, we get

$$\mathbb{E}U_p(x + \xi(x)) - U_p(x) \sim -p(x)U_p(x)\mathbb{E}\xi(x)e^{\beta(x)\xi(x)}.$$

Using (10.41) once again, we deduce the convergence $\mathbb{E}\xi(x)e^{\beta(x)\xi(x)} \to \mathbb{E}\xi e^{\beta\xi}$. Consequently, as $x \to \infty$,

$$\mathbb{E}U_p(x + \xi(x)) - U_p(x) = -p(x)U_p(x)\mathbb{E}\xi e^{\beta\xi}(1 + o(1)). \tag{10.46}$$

This means that U_p is an appropriate Lyapunov function, and the remaining part of the proof repeats that of Theorem 10.8. □

Since $\beta(x)$ is not given in a closed form, Theorem 10.10 cannot be seen as a final statement. For that reason we describe below two cases where $\beta(x)$ can be computed, given regular behaviour of the difference $\mathbb{E}e^{\beta\xi(x)} - 1$.

Corollary 10.11 *Assume the condition* (10.41) *and that there exists a differentiable function $\alpha(x)$ such that*

$$\alpha(x) = O(1/x^{1/2+\varepsilon}), \tag{10.47}$$

$$\alpha'(x) = O(\gamma(x)) \quad \textit{as } x \to \infty, \tag{10.48}$$

and

$$\mathbb{E}e^{\beta\xi(x)} - 1 = \alpha(x) + O(\gamma(x)) \quad \text{as } x \to \infty, \tag{10.49}$$

where $\gamma(x)$ is a decreasing function that is integrable at infinity. Suppose also that

$$\mathbb{E}\xi(x)e^{\beta\xi(x)} = m + O(\gamma(x)/\alpha(x)) \quad \text{as } x \to \infty, \tag{10.50}$$

where $m := \mathbb{E}\xi e^{\beta\xi}$. Then

$$\pi(x,\infty) \sim ce^{-\beta x + A(x)/m} \quad \text{as } x \to \infty, \tag{10.51}$$

where $c > 0$ and $A(x) := \int_0^x \alpha(y)dy$.

Proof Take $\beta(x) := \beta - \alpha(x)/m$. By Taylor's theorem, uniformly on the event $|\xi(x)| \le 1/\alpha(x)$,

$$e^{-\alpha(x)\xi(x)/m} = 1 - \alpha(x)\xi(x)/m + O(\alpha^2(x)\xi^2(x)) \quad \text{as } x \to \infty.$$

Similarly to (10.43), it follows from the condition (10.41) that

$$\mathbb{E}\{e^{\beta(x)\xi(x)};\ |\xi(x)| > 1/\alpha(x)\} = O(e^{-\varepsilon/2\alpha(x)}) \quad \text{as } x \to \infty.$$

Altogether this yields, by (10.50),

$$\begin{aligned}\mathbb{E}e^{\beta(x)\xi(x)} &= \mathbb{E}e^{\beta\xi(x)} - \alpha(x)\mathbb{E}\xi(x)e^{\beta\xi(x)}/m + O(\alpha^2(x) + e^{-\varepsilon/2\alpha(x)})\\ &= \mathbb{E}e^{\beta\xi(x)} - \alpha(x) + O(\gamma(x) + \alpha^2(x) + e^{-\varepsilon/2\alpha(x)})\\ &= 1 + O(\gamma(x) + 1/x^{1+2\varepsilon} + e^{-\varepsilon/2\alpha(x)}) \quad \text{as } x \to \infty.\end{aligned}$$

Thus, the function $\beta(x)$ satisfies all the conditions of Theorem 10.10 and the proof is complete. □

Notice that the key condition on the rate of convergence of $\mathbb{E}e^{\beta\xi(x)}$ to 1 that implies the asymptotics (10.51) in Corollary 10.11 is that the function $\alpha^2(x)$ is integrable at infinity. If this condition fails then the asymptotic behaviour of $\pi(x,\infty)$ is different from (10.51) and requires higher-moment assumptions, which are specified in the following corollary.

Corollary 10.12 *Assume the condition* (10.41) *and that there exists a differentiable function $\alpha(x)$ such that*

$$|\alpha(x)| = O\left(1/x^{\frac{1}{K+1}+\varepsilon}\right) \quad \text{as } x \to \infty$$

for some $K \in \mathbb{N}$ and $\varepsilon > 0$,

$$|\alpha'(x)| \le \gamma(x), \tag{10.52}$$

and

$$\mathbb{E}e^{\beta\xi(x)} - 1 = \alpha(x) + O(\gamma(x))$$

for some decreasing $\gamma(x)$ that is integrable at infinity. Assume also that, for all $k = 1, 2, \ldots, K$,

$$M_k(x) = M_k + \sum_{j=1}^{K-k} D_{k,j}\alpha^j(x) + O(\gamma(x)/\alpha^k(x)), \tag{10.53}$$

where $M_k(x) := \mathbb{E}\xi^k(x)e^{\beta\xi(x)}$ and $M_k := \mathbb{E}\xi^k e^{\beta\xi}$. Then there exist real numbers $c > 0$ and $R_1, R_2, \ldots, R_K$ such that

$$\pi(x,\infty) \sim c\exp\left\{-\beta x + \sum_{k=1}^{K} R_k \int_0^x \alpha^k(y)dy\right\} \quad \text{as } x \to \infty. \tag{10.54}$$

Proof Define

$$\Delta(x) := \sum_{k=1}^{K} R_k\alpha^k(x).$$

In view of Theorem 10.10 it suffices to show that there exist $R_1, R_2, \ldots, R_K$ such that

$$\left|\mathbb{E}e^{(\beta-\Delta(x))\xi(x)} - 1\right| \le q(x) \tag{10.55}$$

for some decreasing integrable function $q(x)$. Indeed, $\Delta(x)$ is differentiable and $|\Delta'(x)| \le C|\alpha'(x)|$. Therefore, we may apply Theorem 10.10 with $\beta(x) = \beta - \Delta(x)$.

Using a Taylor expansion, uniformly on the event $|\xi(x)| \le 1/\alpha(x)$ we have

$$\begin{aligned} e^{-\Delta(x)\xi(x)} &= 1 + \sum_{k=1}^{K} \frac{(-\Delta(x))^k\xi^k(x)}{k!} + O(\Delta^{K+1}(x)\xi^{K+1}(x)) \\ &= 1 + \sum_{k=1}^{K} \frac{(-\Delta(x))^k\xi^k(x)}{k!} + O(\alpha^{K+1}(x)\xi^{K+1}(x)) \quad \text{as } x \to \infty. \end{aligned}$$

Similarly to (10.43), it follows from the condition (10.41) that

$$\mathbb{E}\{e^{\beta(x)\xi(x)};\ |\xi(x)| > 1/\alpha(x)\} = O(e^{-\varepsilon/2\alpha(x)}) \quad \text{as } x \to \infty.$$

Therefore, as $x \to \infty$,

$$\begin{aligned}
&\mathbb{E}e^{(\beta-\Delta(x))\xi(x)}\\
&\quad = \mathbb{E}e^{\beta\xi(x)} + \sum_{k=1}^{K} \frac{M_k(x)}{k!}(-\Delta(x))^k + O(\alpha^{K+1}(x) + e^{-\varepsilon/2\alpha(x)})\\
&\quad = 1 + \alpha(x) + \sum_{k=1}^{K} \frac{M_k(x)}{k!}(-\Delta(x))^k + O(\gamma(x) + \alpha^{K+1}(x) + e^{-\varepsilon/2\alpha(x)}).
\end{aligned}$$

So, we need to identify constants $R_1, R_2, \ldots, R_K$ such that

$$\alpha(x) + \sum_{k=1}^{K} \frac{M_k(x)}{k!}(-\Delta(x))^k = O(\alpha^{K+1}(x)). \tag{10.56}$$

It follows from the assumption (10.53) and the bound $\Delta(x) = O(\alpha(x))$ that (10.56) is equivalent to

$$z + \sum_{k=1}^{K} \frac{1}{k!}\left(M_k + \sum_{j=1}^{K-k} D_{k,j} z^j\right)\left(-\sum_{j=1}^{K} R_j z^j\right)^k = O(z^{K+1}) \quad \text{as } z \to 0.$$

Consequently, the coefficients of z^k must be zero for all $k \le K$, and we can determine all the R_k recursively. For example, the coefficient of z equals $1 - m_1 R_1$. Thus, $R_1 = 1/m_1$. Further, the coefficient of z^2 is $-D_{1,1}R_1 - m_1 R_2 + m_2 R_1^2/2$ and, consequently,

$$R_2 = \frac{-D_{1,1}R_1 + m_2 R_1^2/2}{m_1}.$$

All further coefficients can be found recursively. □

If $\alpha(x)$ from Corollary 10.12 decreases more slowly than any power of x but (10.52) and (10.53) remain valid then we have, by the same arguments,

$$\pi(x, \infty) = \exp\left\{-\beta x + \sum_{k=1}^{K} R_k \int_0^x \alpha^k(y)dy + O\left(\int_0^x \alpha^{K+1}(y)dy\right)\right\},$$

which can be seen as a corrected logarithmic asymptotic for π. To obtain precise asymptotics one needs more information on the moments $M_k(x)$.

Corollary 10.13 *Assume the condition* (10.41) *and that there exists a differentiable function* $\alpha(x)$ *such that* (10.52) *holds,*

$$\mathbb{E}e^{\beta\xi(x)} - 1 = \alpha(x), \quad x \ge 0, \tag{10.57}$$

and

$$M_k(x) = M_k + \sum_{j=1}^{\infty} D_{k,j}\alpha^j(x) \quad \text{for all } k \geq 1. \tag{10.58}$$

Assume furthermore that

$$\sup_{k\geq 1} \sum_{j=1}^{\infty} D_{k,j} r^j \ < \ \infty \quad \text{for some } r > 0.$$

Then there exist real numbers $R_1, R_2, \ldots,$ *such that*

$$\pi(x,\infty) \ \sim \ c \exp\left\{-\beta x + \sum_{k=1}^{\infty} R_k \int_0^x \alpha^k(y)dy\right\} \quad \text{as } x \to \infty.$$

Proof For all sufficiently large x there is a positive solution $\beta(x)$ to the equation

$$\mathbb{E}e^{\beta(x)\xi(x)} \ = \ 1.$$

Since $\mathbb{E}e^{\gamma\xi(x)}$ is finite for all $\gamma \leq \beta+\varepsilon$, we may rewrite the last equation using a Taylor series:

$$\mathbb{E}e^{\beta\xi(x)} + \sum_{k=1}^{\infty} \frac{(-\Delta(x))^k}{k!} \mathbb{E}\xi^k(x)e^{\beta\xi(x)} \ = \ 1,$$

where $\Delta(x) \ = \ \beta - \beta(x)$. Taking into account (10.57) and (10.58), we then obtain

$$\alpha(x) + \sum_{k=1}^{\infty} \frac{(-\Delta(x))^k}{k!}\left(M_k + \sum_{j=1}^{\infty} D_{k,j}\alpha^j(x)\right) = 0. \tag{10.59}$$

Define

$$F(z,w) := z + \sum_{k\geq 1} \frac{M_k}{k!}(-w)^k + \sum_{k,j\geq 1} \frac{D_{k,j}}{k!} z^j (-w)^k.$$

Then (10.59) can be written as $F(\alpha(x),\Delta(x)) \ = \ 0$. In other words, we are looking for a function $w(z)$ satisfying $F(z,w(z)) \ = \ 0$. Since $F(0,0) \ = \ 0$ and $\frac{\partial}{\partial w}F(0,0) \ = \ -M_1 \ < \ 0$, we may apply Theorem B.4 from Flajolet and Sedgewick [65], which says that $w(z)$ is analytic in a vicinity of zero, that is, there exists a $\rho > 0$ such that

$$w(z) \ = \ \sum_{n=1}^{\infty} R_n z^n, \quad |z| < \rho.$$

Consequently,

$$\Delta(x) = \sum_{n=1}^{\infty} R_n \alpha^n(x)$$

for all x such that $|\alpha(x)| < \rho$.

Applying Theorem 10.10 with $\beta(x) = \beta - \Delta(x)$, we get

$$\pi(x,\infty) \sim ce^{-\beta x + \int_0^x \Delta(y)dy} \quad \text{as } x \to \infty.$$

Integrating $\Delta(y)$ termwise, we complete the proof. □

We finish this section with the following remark. In the proof of Corollary 10.13 we adapted the derivation of the Cramér series in large deviations for sums of independent random variables; see e.g. Petrov [132]. There is just one difference: we need the analyticity of an implicit function instead of the analyticity of the inverse of the moment generating function.

10.4 Local Central Limit Theorem

We first state a version of the central limit theorem for Markov chains on $\mathbb{R}$ with asymptotically constant drift.

Theorem 10.14 *Let the family of jumps $|\xi(x)|$ possess a square integrable majorant. Let $m_1(x) = \mu + o(1/\sqrt{x})$, $\mu > 0$, let $m_2(x) \to b > 0$ as $x \to \infty$, and let*

$$\limsup_{n\to\infty} X_n = \infty \quad \textit{with probability } 1. \tag{10.60}$$

Then the strong law of large numbers holds:

$$X_n/n \stackrel{a.s.}{\to} \mu \quad \textit{as } n \to \infty. \tag{10.61}$$

Furthermore,

$$\frac{X_n - \mu n}{\sqrt{bn}} \Rightarrow N_{0,1} \quad \textit{as } n \to \infty$$

and

$$\frac{\max_{k\le n} X_k - \mu n}{\sqrt{bn}} \Rightarrow N_{0,1} \quad \textit{as } n \to \infty.$$

These statements are immediate from Corollary 5.3 and Theorems 5.7 and 5.9 respectively, with $v(x) \equiv \mu$, so that $\beta = 0$. In this special case there is a shorter proof based on the characteristic functions method; see Korshunov [103, Theorem 5].

Theorem 10.15 *Let the family of jumps $\{\xi(x), x \in \mathbb{R}\}$ possess a stochastic square integrable minorant with positive mean (so that the condition (10.60) holds true) and a square integrable stochastic majorant. Assume the weak convergence $\xi(x) \Rightarrow \xi$, the relation $m_1(x) = \mu + o(1/\sqrt{x})$, and the upper bound $\mathbb{P}\{X_0 < -x\} = o(1/\sqrt{x})$ as $x \to \infty$.*

If ξ has a non-lattice distribution and, for all $A > 0$,

$$\sup_{|\lambda|\le A} \left|\mathbb{E}e^{i\lambda\xi(x)} - \mathbb{E}e^{i\lambda\xi}\right| = o(1/x) \quad \text{as } x \to \infty \tag{10.62}$$

then, for all $h > 0$,

$$\sup_{x\in\mathbb{R}} \left|\sqrt{2\pi bn}\,\mathbb{P}\{X_n \in (x, x+h]\} - he^{-(x-n\mu)^2/2bn}\right| \to 0 \quad \text{as } n \to \infty.$$

If ξ is integer-valued, $\mathbb{Z}$ is the minimal lattice for ξ, and

$$\sup_{|\lambda|\le\pi} \left|\mathbb{E}e^{i\lambda\xi(x)} - \mathbb{E}e^{i\lambda\xi}\right| = o(1/x) \quad \text{as } x \to \infty \tag{10.63}$$

then

$$\sup_{x\in\mathbb{Z}} \left|\sqrt{2\pi bn}\,\mathbb{P}\{X_n = x\} - e^{-(x-n\mu)^2/2bn}\right| \to 0 \quad \text{as } n \to \infty.$$

Proof Let η be a square integrable minorant with positive expectation for the family $\{\xi(x), x \in \mathbb{R}\}$. Let $\{\eta_k\}$ be independent copies of η and set $S_n := \eta_1 + \cdots + \eta_n$. Then, by the minorisation assumption, for all n we have

$$\begin{aligned}
&\mathbb{P}\{X_k < k\mathbb{E}\eta/2 \text{ for some } k \ge n\}\\
&\quad \le \mathbb{P}\{X_0 < -n\mathbb{E}\eta/4\} + \mathbb{P}\{S_k - k\mathbb{E}\eta/4 < k\mathbb{E}\eta/2 \text{ for some } k \ge n\}\\
&\quad \le o(1/\sqrt{n}) + \mathbb{P}\left\{\sup_{k\ge n} \frac{S_k - k\mathbb{E}\eta}{k} < -\mathbb{E}\eta/4\right\}
\end{aligned}$$

The sequence $(S_k - k\mathbb{E}\eta)/k$ constitutes a reverse martingale and hence it follows from the Kolmogorov inequality that

$$\mathbb{P}\left\{\sup_{k\ge n} \frac{S_k - k\mathbb{E}\eta}{k} < -\mathbb{E}\eta/4\right\} \le \frac{\mathbb{E}(S_n - n\mathbb{E}\eta)^2}{(\mathbb{E}\eta/4)^2 n^2} = O(1/n) = o(1/\sqrt{n}).$$

Therefore

$$\mathbb{P}\{X_k < k\mathbb{E}\eta/2 \text{ for some } k \ge n\} \le o(1/\sqrt{n}) \quad \text{as } n \to \infty. \tag{10.64}$$

We will proceed with the proof for the lattice case only; the non-lattice case can be treated similarly. By the inversion formula for lattice distributions,

$$\sqrt{n}\mathbb{P}\{X_n = x\} = \frac{1}{2\pi} \int_{-\pi\sqrt{n}}^{\pi\sqrt{n}} e^{-i\lambda((x-n\mu)/\sqrt{n})} \mathbb{E}e^{i\lambda((X_n-n\mu)/\sqrt{n})} d\lambda.$$

Therefore, using standard arguments,

$$\begin{aligned}
\sup_x \left| \sqrt{n}\mathbb{P}\{X_n = x\} - \frac{1}{\sqrt{2\pi b}} e^{-(x-n\mu)^2/2b} \right| \\
\le \frac{1}{2\pi} \int_{-A}^{A} \left| \mathbb{E}e^{i\lambda((X_n-n\mu)/\sqrt{n})} - e^{-\lambda^2 b/2} \right| d\lambda \\
+ \int_{|\lambda|\in(A,\pi\sqrt{n}]} \left| \mathbb{E}e^{i\lambda((X_n-n\mu)/\sqrt{n})} \right| d\lambda + \int_{|\lambda|>A} e^{-\lambda^2 b/2} d\lambda. \qquad (10.65)
\end{aligned}$$

It follows from the weak convergence to the normal law that $\mathbb{E}e^{i\lambda((X_n-n\mu/\sqrt{n})} \to e^{-\lambda^2 b/2}$ uniformly on compact λ-sets. Therefore, the first integral on the right-hand side of (10.65) converges to zero as $n \to \infty$, for any fixed A. Choosing A sufficiently large we can make the integral $\int_{|\lambda|>A} e^{-\lambda^2 b/2} d\lambda$ as small as we please. Thus, it remains to prove that the second integral in (10.65) is also small.

In order to prove this we need to show that the modulus of the characteristic function in the second integral is sufficiently small. Let us introduce an auxiliary time-inhomogeneous Markov chain $\{\widetilde{X}_n\}$ with jumps at time n:

$$\widetilde{\xi}_n(x) = \begin{cases} \xi(x) & \text{for } x > n\mathbb{E}\eta/4, \\ \xi(x_n) & \text{for } x \le n\mathbb{E}\eta/4, \end{cases}$$

where $x_n > n\mathbb{E}\eta/4$. Consider for simplicity n even and define $\widetilde{X}_k$ for $k \ge n/2$ only; set $\widetilde{X}_{n/2} = X_{n/2}$. Then it follows from the construction that, for all $u \in \mathbb{R}$,

$$\begin{aligned}
\left|\mathbb{E}e^{iuX_n}\right| &\le \left|\mathbb{E}e^{iu\widetilde{X}_n}\right| + \mathbb{P}\{X_k \ne \widetilde{X}_k \text{ for some } k > n/2\} \\
&\le \left|\mathbb{E}e^{iu\widetilde{X}_n}\right| + \mathbb{P}\{X_k \le k\mathbb{E}\eta/4 \text{ for some } k \ge n/2\}.
\end{aligned}$$

From this estimate and (10.64) we obtain

$$\left|\mathbb{E}e^{iuX_n}\right| \le \left|\mathbb{E}e^{iu\widetilde{X}_n}\right| + o(1/\sqrt{n}) \quad \text{as } n \to \infty \text{ uniformly for all } u \in \mathbb{R}. \qquad (10.66)$$

By the construction of $\{\widetilde{X}_k\}$, we have

$$\begin{aligned}
\left|\mathbb{E}e^{iu\widetilde{X}_{k+1}} - \mathbb{E}e^{iu\xi}\mathbb{E}e^{iu\widetilde{X}_k}\right| &= \left|\mathbb{E}e^{iu\widetilde{X}_k}\left(\mathbb{E}\{e^{iu\widetilde{\xi}_k(\widetilde{X}_k)} | \widetilde{X}_k\} - \mathbb{E}e^{iu\xi}\right)\right| \\
&\le \left|\mathbb{E}\{e^{iu\widetilde{\xi}_k(\widetilde{X}_k)} | \widetilde{X}_k\} - \mathbb{E}e^{iu\xi}\right| \\
&\le \sup_{x>k\mathbb{E}\eta/4} \left|\mathbb{E}e^{iu\xi(x)} - \mathbb{E}e^{iu\xi}\right|.
\end{aligned}$$

Then, for all $k \geq n/2$,

$$\left|\mathbb{E}e^{iu\widetilde{X}_{k+1}} - \mathbb{E}e^{iu\xi}\mathbb{E}e^{iu\widetilde{X}_k}\right| \leq \sup_{x>n\mathbb{E}\eta/8} |\mathbb{E}e^{iu\xi(x)} - \mathbb{E}e^{iu\xi}|$$
$$\leq \sup_{x>n\mathbb{E}\eta/8} \sup_{|u|\leq\pi} |\mathbb{E}e^{iu\xi(x)} - \mathbb{E}e^{iu\xi}|$$
$$=: \delta_n = o(1/n),$$

by the assumption of the theorem. Consequently, for $m = n/2$ we have

$$\left|\mathbb{E}e^{iu\widetilde{X}_n} - (\mathbb{E}e^{iu\xi})^{n-m}\mathbb{E}e^{iu\widetilde{X}_m}\right|$$
$$= \left|\sum_{k=m}^{n-1}(\mathbb{E}e^{iu\xi})^{n-k-1}\left(\mathbb{E}e^{iu\widetilde{X}_{k+1}} - \mathbb{E}e^{iu\xi}\mathbb{E}e^{iu\widetilde{X}_k}\right)\right|$$
$$\leq \delta_n \sum_{j=0}^{n-m-1} |\mathbb{E}e^{iu\xi}|^j,$$

hence

$$\left|\mathbb{E}e^{iu\widetilde{X}_n}\right| \leq |\mathbb{E}e^{iu\xi}|^{n/2} + \delta_n \sum_{j=0}^{n/2-1} |\mathbb{E}e^{iu\xi}|^j.$$

Since $\mathbb{Z}$ is the minimal lattice for ξ, there exists an $\varepsilon > 0$ such that $|\mathbb{E}e^{iu\xi}| \leq e^{-\varepsilon u^2}$ for all $u \in [-\pi, \pi]$. This implies that

$$\left|\mathbb{E}e^{iu\widetilde{X}_n}\right| \leq e^{-n\varepsilon u^2/2} + \delta_n \sum_{j=0}^{n/2\ 1} e^{-j\varepsilon u^2}.$$

Substituting this into (10.66) we obtain, uniformly for all $u \in \mathbb{R}$,

$$\left|\mathbb{E}e^{iuX_n}\right| \leq e^{-n\varepsilon u^2/2} + o(1/\sqrt{n}) + o(1/n)\sum_{j=1}^{n/2} e^{-j\varepsilon u^2} \quad \text{as } n \to \infty.$$

Hence the second term in (10.65) possesses the following upper bound:

$$\int_{|\lambda|\in(A,\pi\sqrt{n}]} \left|\mathbb{E}e^{i\lambda((X_n-n\mu)/\sqrt{n})}\right| d\lambda$$
$$\leq \int_{|\lambda|\in(A,\pi\sqrt{n}]} e^{-\varepsilon\lambda^2/2}d\lambda + o(1) + o(1/n)\sum_{j=1}^{n/2}\int_{|\lambda|\in(A,\pi\sqrt{n}]} e^{-\varepsilon\lambda^2 j/n}d\lambda$$
$$\leq 2\int_A^\infty e^{-\varepsilon\lambda^2/2}d\lambda + o(1) + o(1/n)\sum_{j=1}^{n/2}\int_{-\infty}^\infty e^{-\varepsilon\lambda^2 j/n}d\lambda$$
$$= 2\int_A^\infty e^{-\varepsilon\lambda^2/2}d\lambda + o(1) + o(1/n)\sum_{j=1}^{n/2}\sqrt{\frac{\pi n}{\varepsilon j}} \quad \text{as } n \to \infty.$$

Therefore,

$$\limsup_{n\to\infty}\int_{|\lambda|\in(A,\pi\sqrt{n}]}\left|\mathbb{E}e^{i\lambda((X_n-n\mu)/\sqrt{n})}\right|d\lambda\le 2\int_A^\infty e^{-\varepsilon\lambda^2/2}d\lambda.$$

Letting $A\to\infty$, we obtain the desired result. □

Theorem 10.16 *Assume that all the conditions of Theorem 10.15 hold.*
If ξ is a non-lattice random variable then, for all $h>0$,

$$\sum_{k=0}^n \mathbb{P}\{X_k\in(x,x+h]\} = \frac{h}{\mu}\Phi\Big(\frac{n\mu-x}{\sqrt{xb/\mu}}\Big)+o(1)$$

as $x\to\infty$ uniformly for all $n\ge 0$.
If ξ is a lattice random variable and $\mathbb{Z}$ is the minimal lattice for ξ then

$$\sum_{k=0}^n \mathbb{P}\{X_k=x\}=\frac{1}{\mu}\Phi\Big(\frac{n\mu-x}{\sqrt{xb/\mu}}\Big)+o(1)\quad \text{as } x\to\infty,$$

as $x\to\infty$ uniformly for all $n\ge 0$.

Proof We again consider the lattice case only. By the local limit theorem, for any fixed A, $B\in\mathbb{R}$, $A<B$, we have

$$\begin{aligned}\sum_{k=x/\mu+A\sqrt{x}}^{x/\mu+B\sqrt{x}}\mathbb{P}\{X_k=x\}&=\sum_{k=x/\mu+A\sqrt{x}}^{x/\mu+B\sqrt{x}}\frac{1}{\sqrt{2\pi bk}}e^{-(x-\mu k)^2/2bk}+o(1)\\&=\sum_{k=x/\mu+A\sqrt{x}}^{x/\mu+B\sqrt{x}}\frac{1}{\sqrt{2\pi bx/\mu}}e^{-(x-\mu k)^2\mu/2bx}+o(1).\end{aligned}$$

Thus, as $x\to\infty$,

$$\begin{aligned}\sum_{k=x/\mu+A\sqrt{x}}^{x/\mu+B\sqrt{x}}\mathbb{P}\{X_k=x\}&=\sum_{k=A\sqrt{x}}^{B\sqrt{x}}\frac{1}{\sqrt{2\pi xb/\mu}}e^{-(k/\sqrt{x})^2\mu^3/2b}+o(1)\\&=\int_A^B\frac{1}{\sqrt{2\pi b/\mu}}e^{-y^2\mu^3/2b}dy+o(1)\\&=\frac{1}{\mu}\big(\Phi(\mu^{3/2}B/\sqrt{b})-\Phi(\mu^{3/2}A/\sqrt{b})\big)+o(1).\end{aligned}\tag{10.67}$$

Together with Theorem 10.2 this implies that for any $\varepsilon>0$ there exist A and B such that, for all sufficiently large x,

$$\sum_{k=0}^{x/\mu+A\sqrt{x}} \mathbb{P}\{X_k = x\} + \sum_{k=x/\mu+B\sqrt{x}}^{\infty} \mathbb{P}\{X_k = x\} \le \varepsilon.$$

Therefore

$$\sum_{k=0}^{x/\mu+A\sqrt{x}} \mathbb{P}\{X_k{=}x\} \to 0$$

as $A \to -\infty$ uniformly for all x. Combining this with (10.67), we get the desired relation. □

10.5 Pre-Stationary Distributions

Theorem 10.17 *Let the distribution of X_n converge towards a stationary distribution π in the total-variation norm. Assume that the conditions of Theorem 10.8 are valid and that the majorant Ξ also satisfies the condition*

$$\mathbb{E}\Xi^2 e^{\beta\Xi} < \infty. \tag{10.68}$$

Assume also that

$$\mathbb{E}\xi(x)e^{\beta\xi(x)} = \mathbb{E}\xi e^{\beta\xi} + o(1/\sqrt{x}) \quad \text{as } x \to \infty. \tag{10.69}$$

If the limiting variable ξ is non-lattice, we assume that, for any $A > 0$,

$$\sup_{|\lambda|\le A} \left|\mathbb{E}e^{(\beta+i\lambda)\xi(x)} - \mathbb{E}e^{(\beta+i\lambda)\xi}\right| = o(1/x) \quad \text{as } x \to \infty. \tag{10.70}$$

If ξ is a lattice distribution and $\mathbb{Z}$ is the minimal lattice for ξ, we assume that

$$\sup_{|\lambda|\le \pi} \left|\mathbb{E}e^{(\beta+i\lambda)\xi(x)} - \mathbb{E}e^{(\beta+i\lambda)\xi}\right| = o(1/x) \quad \text{as } x \to \infty. \tag{10.71}$$

Then, uniformly for all $n \ge 1$,

$$\frac{\mathbb{P}\{X_n > x\}}{\pi(x,\infty)} = \Phi_{\sigma^2}\left(\frac{n\mathbb{E}\xi e^{\beta\xi} - x}{\sqrt{x/\mathbb{E}\xi e^{\beta\xi}}}\right) + o(1) \quad \text{as } x \to \infty, \tag{10.72}$$

where $\sigma^2 = \mathbb{E}\xi^2 e^{\beta\xi} - (\mathbb{E}\xi e^{\beta\xi})^2$.

Proof Let $\{\widehat{X}_n\}$ be the chain constructed in the proof of Theorem 10.8. We showed there that the family of its jumps $\widehat{\xi}(x)$ possesses a stochastic minorant with positive mean and finite second moment and a stochastic majorant with finite mean. Assumption (10.68) implies that there is a majorant with finite second moment.

We now turn to the asymptotic behaviour of $\mathbb{E}\widehat{\xi}(x)$. As we showed in the proof of Theorem 10.8, $\mathbb{E}\widehat{\xi}(x) \to \mathbb{E}\xi e^{\beta\xi}$. But, in order to apply Theorem 10.16, we also have to show that

$$\mathbb{E}\widehat{\xi}(x) = \mathbb{E}\xi e^{\beta\xi} + o(1/\sqrt{x}) \quad \text{as } x \to \infty. \tag{10.73}$$

It follows from (10.39) that

$$\mathbb{E}\widehat{\xi}(x) = \frac{\mathbb{E}\xi(x)U_p(x+\xi(x))}{U_p(x)}(1+o(1/x)) \quad \text{as } x \to \infty. \tag{10.74}$$

It is immediate from the definition (10.30) of U_p that

$$\mathbb{E}\{\xi(x)U_p(x+\xi(x));\ \xi(x) > s(x)\} \le U_p(x)\mathbb{E}\{\xi(x)e^{\beta\xi(x)};\ \xi(x) > s(x)\}.$$

Thus, owing to (10.68), for any $s(x) = o(x)$,

$$\frac{\mathbb{E}\{\xi(x)U_p(x+\xi(x));\ \xi(x) > s(x)\}}{U_p(x)} = o(1/s(x)) \quad \text{as } x \to \infty. \tag{10.75}$$

Furthermore, we have the upper bound

$$\frac{\mathbb{E}\{\xi(x)U_p(x+\xi(x));\ \xi(x) < -s(x)\}}{U_p(x)} = o(e^{-\beta s(x)/2}) \quad \text{as } x \to \infty. \tag{10.76}$$

Uniformly on the set $\{|\xi(x)| \le s(x)\}$ we have $g(x+\xi(x)) - g(x) \sim -p(x)\xi(x)$; see (10.29). Therefore

$$\begin{aligned}
&\mathbb{E}\{\xi(x)U_p(x+\xi(x));\ |\xi(x)| < s(x)\}\\
&\quad = e^{\beta x}\mathbb{E}\{\xi(x)(1+g(x+\xi(x)))e^{\beta\xi(x)};\ |\xi(x)| \le s(x)\}\\
&\quad = U_p(x)\mathbb{E}\{\xi(x)e^{\beta\xi(x)};\ |\xi(x)| \le s(x)\}\\
&\qquad - p(x)(1+o(1))e^{\beta x}\mathbb{E}\{\xi^2(x)e^{\beta\xi(x)};\ |\xi(x)| \le s(x)\}.
\end{aligned}$$

Again using (10.68), we obtain

$$\frac{\mathbb{E}\{\xi(x)U_p(x+\xi(x));\ |\xi(x)| < s(x)\}}{U_p(x)} = \mathbb{E}\xi(x)e^{\beta\xi(x)} + O(p(x) + 1/s(x)).$$

Combining this estimate with (10.75) and (10.76), and choosing $s(x)$ such that $s(x)/\sqrt{x} \to \infty$, we conclude that

$$\frac{\mathbb{E}\{\xi(x)U_p(x+\xi(x))\}}{U_p(x)} = \mathbb{E}\xi(x)e^{\beta\xi(x)} + o(1/\sqrt{x}).$$

The relation (10.73) now follows from the assumption (10.69). The same arguments show that (10.62) and (10.63) follow from (10.70) and (10.71) respectively. Thus, $\{\widehat{X}_n\}$ satisfies all the conditions of Theorem 10.16.

It follows from the conditions on jumps that $\mathbb{E}e^{\beta X_n} < \infty$ for all n, which implies that $\mathbb{P}\{X_n > x\} = o(e^{-\beta x})$ for any fixed n, and hence (10.72). So it remains to consider the case where $n \to \infty$.

Fix an $h > 0$. Applying (7.49) with $U = U_p$ we deduce that, for $x > \widehat{x}$,

$$\begin{aligned}&\mathbb{P}\{X_n \in (x, x+h]\}\\&\quad = \sum_{j=1}^{n} \int_B \mathbb{P}\{X_{n-j} \in dz\} \int_{\widehat{x}}^{\infty} P(z, du) U_p(u) \mathbb{E}_u \left\{ \frac{e^{-\sum_{k=0}^{j-2} q(\widehat{X}_k)}}{U_p(\widehat{X}_{j-1})}; \widehat{X}_{j-1} \in (x, x+h] \right\}.\end{aligned}$$

By the conditions (10.26) and (10.68) we have

$$P(z, (u, \infty)) \le \mathbb{P}\{\Xi > u - \widehat{x}\} \le c_2 e^{-\beta u}/u^2 \quad \text{for all } z \le \widehat{x} \text{ and } u > \widehat{x}. \tag{10.77}$$

The function U_p is increasing. Hence, for any $N_n = o(\sqrt{n})$,

$$\begin{aligned}&\sum_{j=n-N_n+1}^{n} \int_B \mathbb{P}\{X_{n-j} \in dz\}\\&\qquad \times \int_{\widehat{x}}^{\infty} P(z, du) U_p(u) \mathbb{E}_u \left\{ \frac{e^{-\sum_{k=0}^{j-2} q(\widehat{X}_k)}}{U_p(\widehat{X}_{j-1})}; \widehat{X}_{j-1} \in (x, x+h] \right\}\\&\quad \le \frac{1}{U_p(x)} \sum_{j=n-N_n+1}^{n} \int_B \mathbb{P}\{X_{n-j} \in dz\}\\&\qquad\qquad \times \int_{\widehat{x}}^{\infty} P(z, du) U_p(u) \mathbb{P}_u\{\widehat{X}_{j-1} \in (x, x+h]\}\\&\quad \le \frac{c_3}{U_p(x)\sqrt{n}} \sum_{j=n-N_n+1}^{n} \int_B \mathbb{P}\{X_{n-j} \in dz\} \int_x^{\infty} P(z, du) U_p(u)\\&\quad \le c_4 N_n / U_p(x)\sqrt{n} = o(1/U_p(x)) \quad \text{as } n \to \infty,\end{aligned} \tag{10.78}$$

where the second inequality follows by Theorem 10.15 applied to $\{\widehat{X}_n\}$. Since the distribution of X_{n-j} converges in total variation to π, for any $N_n \to \infty$ we have

$$\begin{aligned}&\sum_{j=1}^{n-N_n} \int_B \mathbb{P}\{X_{n-j} \in dz\} \int_{\widehat{x}}^{\infty} P(z, du) U_p(u) \mathbb{E}_u \left\{ \frac{e^{-\sum_{k=0}^{j-2} q(\widehat{X}_k)}}{U_p(\widehat{X}_{j-1})}; \widehat{X}_{j-1} \in (x, x+h] \right\}\\&\quad = (1+o(1)) \sum_{j=1}^{n-N_n} \int_B \pi(dz) \int_{\widehat{x}}^{\infty} P(z, du) U_p(u)\\&\qquad\qquad \times \mathbb{E}_u \left\{ \frac{e^{-\sum_{k=0}^{j-2} q(\widehat{X}_k)}}{U_p(\widehat{X}_{j-1})}; \widehat{X}_{j-1} \in (x, x+h] \right\}.\end{aligned} \tag{10.79}$$

Similarly to (10.78), we find that

$$\sum_{j=n-N_n+1}^{n} \int_B \pi(dz) \int_{\widehat{x}}^{\infty} P(z,du) U_p(u) \mathbb{E}_u \left\{ \frac{e^{-\sum_{k=0}^{j-2} q(\widehat{X}_k)}}{U_p(\widehat{X}_{j-1})}; \widehat{X}_{j-1} \in (x, x+h] \right\}$$
$$= o\Big(\frac{1}{U_p(x)}\Big).$$

Combining this with (10.78) and (10.79), we obtain

$$\begin{aligned}
&\mathbb{P}\{X_n \in (x, x+h]\} \\
&\quad = (1+o(1)) \sum_{j=1}^{n} \int_B \pi(dz) \int_{\widehat{x}}^{\infty} P(z,du) U_p(u) \\
&\qquad\qquad \times \mathbb{E}_u \left\{ \frac{e^{-\sum_{k=0}^{j-2} q(\widehat{X}_k)}}{U_p(\widehat{X}_{j-1})}; \widehat{X}_{j-1} \in (x, x+h] \right\} + o\Big(\frac{1}{U_p(x)}\Big) \\
&\quad = (1+o(1)) \int_{\widehat{x}}^{\infty} \mu(du) U_p(u) \sum_{j=1}^{n} \mathbb{E}_u \left\{ \frac{e^{-\sum_{k=0}^{j-2} q(\widehat{X}_k)}}{U_p(\widehat{X}_{j-1})}; \widehat{X}_{j-1} \in (x, x+h] \right\} \\
&\qquad + o\Big(\frac{1}{U_p(x)}\Big)
\end{aligned}$$

as $x \to \infty$, where

$$\mu(du) = \int_B \pi(dz) P(z, du)$$

is a measure on $(\widehat{x}, \infty)$; see (7.43). Therefore, as $x \to \infty$,

$$\mathbb{P}\{X_n \in (x, x+h]\} = (1+o(1)) \int_{\widehat{x}}^{\infty} \mu(du) U_p(u) \int_x^{x+h} e^{-\beta y} \widehat{H}_{u,n}^{(q)}(dy) + o(e^{-\beta x}),$$

where

$$\widehat{H}_{u,n}^{(q)}(dy) = \sum_{j=1}^{n} \mathbb{E}_u \Big\{ e^{-\sum_{k=0}^{j-2} q(\widehat{X}_k)}; \widehat{X}_{j-1} \in dy \Big\}.$$

In the non-lattice case, by Lemma 4.5, for any fixed $\Delta > 0$, we have

$$\widehat{H}_{u,n}^{(q)}(y, y+\Delta] \sim \mathbb{E}_u e^{-\sum_{k=0}^{\infty} q(\widehat{X}_k)} \sum_{j=1}^{n} \mathbb{P}_u\{\widehat{X}_{j-1} \in (y, y+\Delta\} \quad \text{as } y \to \infty;$$

hence

$$\begin{aligned}
&\mathbb{P}\{X_n \in (x, x+h]\} \\
&\quad = (1+o(1)) \int_{\widehat{x}}^{\infty} \mu(du) U_p(u) \mathbb{E}_u e^{-\sum_{k=0}^{\infty} q(\widehat{X}_k)} \int_x^{x+h} e^{-\beta y} \widehat{H}_{u,n}(dy) + o(e^{-\beta x}),
\end{aligned}$$

where the partial renewal measure of $\{\widehat{X}_n\}$,

$$\widehat{H}_{u,n}(dy) = \sum_{j=1}^{n} \mathbb{P}_u\{\widehat{X}_{j-1} \in dy\},$$

converges weakly to the Lebesgue measure on the interval $[x, x+h]$, with coefficient $\frac{1}{\mu}\Phi_{\sigma^2}\left(\frac{n\mu - x}{\sqrt{x/\mu}}\right)$, for any fixed $u > \widehat{x}$, by Theorem 10.16; here $\mu := \mathbb{E}\xi e^{\beta\xi}$. Then, for any fixed $u > \widehat{x}$, as $x \to \infty$,

$$\int_x^{x+h} e^{-\beta y}\widehat{H}_{u,n}(dy) = \frac{1}{\mu}\Phi_{\sigma^2}\left(\frac{n\mu - x}{\sqrt{x/\mu}}\right)\frac{1-e^{-\beta h}}{\beta}e^{-\beta x} + o(e^{-\beta x}).$$

Furthermore,

$$\int_x^{x+h} e^{-\beta y}\widehat{H}_{u,n}(dy) \le e^{-\beta x}\widehat{H}_{u,n}(x, x+h] \le c_5 e^{-\beta x};$$

hence the dominated convergence theorem is applicable, owing to (10.77). So, as $x \to \infty$, we have

$$\begin{aligned}\mathbb{P}\{X_n > x\} \\ &= \frac{1}{\beta\mu}\Phi_{\sigma^2}\left(\frac{n\mu - x}{\sqrt{x/\mu}}\right)\int_{\widehat{x}}^{\infty} \mu(du)U_p(u)\mathbb{E}_u e^{-\sum_{k=0}^{\infty} q(\widehat{X}_k)} e^{-\beta x} + o(e^{-\beta x}).\end{aligned}$$

Together with Theorem 10.8 this yields the required result (10.72). The lattice case can be concluded in a similar way. □

We can also determine the asymptotic behaviour of pre-stationary distributions in the case when (10.27) fails.

Theorem 10.18 *Assume that the conditions of Theorem 10.10 are valid. Assume also that*

$$\mathbb{E}\xi(x)e^{\beta(x)\xi(x)} = \mathbb{E}\xi e^{\beta\xi} + o(1/\sqrt{x}).$$

If the limiting variable ξ is non-lattice, we assume that, for any $A > 0$,

$$\sup_{|\lambda| \le A}\left|\mathbb{E}e^{(\beta(x)+i\lambda)\xi(x)} - \mathbb{E}e^{(\beta+i\lambda)\xi}\right| = o(1/x).$$

If $\mathbb{Z}$ is the minimal lattice for ξ, we assume that

$$\sup_{|\lambda| \le \pi}\left|\mathbb{E}e^{(\beta(x)+i\lambda)\xi(x)} - \mathbb{E}e^{(\beta+i\lambda)\xi}\right| = o(1/x).$$

Then, uniformly for all $n \ge 1$,

$$\frac{\mathbb{P}\{X_n > x\}}{\pi(x, \infty)} = \Phi_{\sigma^2}\left(\frac{n\mathbb{E}\xi e^{\beta\xi} - x}{\sqrt{x/\mathbb{E}\xi e^{\beta\xi}}}\right) + o(1) \quad \textit{as } x \to \infty,$$

where $\sigma^2 = \mathbb{E}\xi^2 e^{\beta\xi} - (\mathbb{E}\xi e^{\beta\xi})^2$.

The proof of this theorem is identical to that of Theorem 10.17 and for that reason we omit it.

10.6 Comments for Chapter 10

Theorem 10.2 specifies Theorem 1 from Korshunov [105] for transient Markov chains on $\mathbb{R}$.

Borovkov and Korshunov [24], [23, Section 27] proved exponential asymptotics for π under the condition

$$\int_0^\infty dx \int_{-\infty}^\infty e^{\beta y} \left|\mathbb{P}\{\xi(x) < y\} - \mathbb{P}\{\xi < y\}\right| dy < \infty, \qquad (10.80)$$

without assuming a domination condition such as (10.26).

On the other hand, it is worth mentioning that (10.27) is weaker than the conditions we found in the literature. First, (10.80) is definitely stronger than (10.27) and implies, in particular, that the expectations $\mathbb{E}\xi(x)e^{\beta\xi(x)}$ also converge at a summable rate. Furthermore, to show that the constant c in front of $e^{-\beta x}$ is positive, the following condition is introduced in [24]:

$$\int_0^\infty \left(\mathbb{E}e^{\beta\xi(x)} - 1\right)^- x \log x \, dx < \infty.$$

Second, for chains on $\mathbb{Z}^+$, Foley and McDonald [66] used an assumption which can be rewritten in our notation as follows:

$$\sum_{i=0}^\infty \sum_{j\in\mathbb{Z}} e^{\beta j} \left|\mathbb{P}\{\xi(i) = j\} - \mathbb{P}\{\xi = j\}\right| < \infty.$$

Theorems 10.8 and 10.17 were proven for the first time in [104] via the so-called evolution of masses, that is, via the analysis of non-stochastic kernels.

A lattice version of Theorem 10.10 was proven by Denisov, Korshunov, and Wachtel [43] following a different approach, which is based on a useful method of construction of harmonic functions for Markov kernels on $\mathbb{Z}^+$.

11

Applications

The main goal of this chapter is to demonstrate how the theory developed in the previous chapters can be useful in the study of various Markov models that give rise to Markov chains with asymptotically zero drift. Some of those models are popular in stochastic modelling: random walks conditioned to stay positive, state-dependent branching processes or branching processes with migration, stochastic difference equations. In contrast with the general approach discussed here, the methods available in the literature for investigation of these models are mostly model-tailored.

We also introduce some new models to which our approach is applicable. For example, in Section 11.4 we introduce a risk process with a surplus-dependent premium rate, which converges to the critical threshold in the nett-profit condition. Furthermore, we introduce a new class of branching processes with migration and state-dependent reproduction.

11.1 Random Walk Conditioned to Stay Positive

Let $\{S_n\}$ be a random walk with independent and identically distributed increments ξ_k, that is, $S_n = \xi_1 + \xi_2 + \cdots + \xi_n$, $n \ge 1$. Let $\tau(x)$ be the first time epoch when $\{S_n\}$, starting at x, is non-positive:

$$\tau(x) := \min\{n \ge 1 : x + S_n \le 0\}.$$

We shall assume that the random walk $\{S_n\}$ is oscillating, that is,

$$\liminf_{n\to\infty} X_n = -\infty, \quad \limsup_{n\to\infty} X_n = \infty \quad \text{with probability 1.}$$

In particular, $\mathbb{P}\{\tau(x) < \infty\} = 1$ for all starting points x. Let χ^- denote the first weak descending ladder height of $\{S_n\}$, that is, $\chi^- = -S_{\tau(0)}$. Let $V(x)$

denote the renewal function generated by the weak descending ladder heights of the random walk:

$$\begin{aligned} V(x) &:= 1 + \sum_{k=1}^{\infty} \mathbb{P}\{\chi_1^- + \chi_2^- + \cdots + \chi_k^- < x\} \\ &= \mathbb{E}\theta(x), \end{aligned} \tag{11.1}$$

where the χ_k^- are independent copies of χ^- and

$$\theta(x) \; := \; \min\{k \geq 1 : \chi_1^- + \chi_2^- + \cdots + \chi_k^- \geq x\}.$$

In particular, $V(0) = 1$.

It is well known – see e.g. Kozlov [109] – that $V(x)$ is a harmonic function for the process $\{S_n\}$ killed at leaving $(0, \infty)$. More precisely,

$$V(x) = \mathbb{E}\{V(x + S_1); \tau(x) > 1\} \quad \text{for all } x > 0.$$

This implies that Doob's h-transform

$$P(x, dy) := \frac{V(y)}{V(x)} \mathbb{P}\{x + S_1 \in dy, \tau(x) > 1\} \tag{11.2}$$

defines a stochastic transition kernel on $(0, \infty)$. Let $\{X_n\}$ be the corresponding Markov chain. It is usually called a *random walk conditioned to stay positive*. This definition via Doob's h-transform is equivalent to the construction of a random walk conditioned to stay positive via the weak limit of conditional distributions, see Bertoin and Doney [13]:

$$P(x, B) \; = \; \lim_{n \to \infty} \mathbb{P}\{x + S_1 \in B \mid \tau(x) > n\}.$$

We now show that if $\mathbb{E}\xi_1 = 0$ and $\mathbb{E}\xi_1^2 =: \sigma^2 \in (0, \infty)$ then $\{X_n\}$ has asymptotically zero drift. We first observe that these moment conditions allow us to apply Lemma 2.33 with $\gamma = 2$, $\alpha = \beta = 1$ and to conclude that, for some increasing $s(x) = o(x)$ and decreasing function $p(x) = o(1/x)$ that is integrable at infinity, we have

$$\mathbb{E}\{|\xi_1|;\ |\xi_1| > s(x)\} = o(p(x)) \quad \text{as } x \to \infty, \tag{11.3}$$

in particular, $\mathbb{E}\{\xi_1;\ \xi_1 > -x\} = o(1/x)$ since $\mathbb{E}\xi_1 = 0$. Then it follows from the definition (11.2) of the kernel P that

$$\begin{aligned} m_1(x) &:= \frac{1}{V(x)} \mathbb{E}\{V(x + \xi_1)\xi_1;\ \xi_1 > -x\} \\ &= \frac{1}{V(x)} \mathbb{E}\{(V(x + \xi_1) - V(x))\xi_1;\ \xi_1 > -x\} + \mathbb{E}\{\xi_1;\ \xi_1 > -x\} \\ &= \frac{1}{V(x)} \mathbb{E}\{(V(x + \xi_1) - V(x))\xi_1;\ \xi_1 > -x\} + o(1/x). \end{aligned}$$

The finiteness of the second moment also implies that the ladder heights have finite expectation, so by Blackwell's renewal theorem (see e.g. Durrett [56, Theorem 2.6.4]), for any fixed $y > 0$,

$$V(x+y) - V(x) \to \frac{y}{\mathbb{E}\chi^-} \quad \text{as } x \to \infty, \tag{11.4}$$

in the non-lattice case; in the lattice case both x and y are restricted to the lattice. Hence $(V(x+\xi_1) - V(x))\xi_1$ converges to $\xi_1^2/\mathbb{E}\chi^-$ as $x \to \infty$. By (11.4),

$$c_V := \sup_x (V(x+1) - V(x)) < \infty,$$

which yields

$$|V(x+y) - V(x)| \le c_V(|y|+1). \tag{11.5}$$

This allows us to apply the dominated convergence theorem to infer that

$$\mathbb{E}\{(V(x+\xi_1) - V(x))\xi_1;\ \xi_1 > -x\} \to \frac{\mathbb{E}\xi_1^2}{\mathbb{E}\chi^-} = \frac{\sigma^2}{\mathbb{E}\chi^-} \quad \text{as } x \to \infty.$$

By the elementary renewal theorem (see e.g. Durrett [56, Theorem 2.6.3]), it holds that $V(x) \sim x/\mathbb{E}\chi^-$ and hence

$$m_1(x) \sim \frac{\sigma^2}{x} \quad \text{as } x \to \infty. \tag{11.6}$$

For the second moment of jumps we have

$$\begin{aligned} m_2(x) &:= \frac{1}{V(x)}\mathbb{E}\{V(x+\xi_1)\xi_1^2;\ \xi_1 > -x\} \\ &= \frac{1}{V(x)}\mathbb{E}\{(V(x+\xi_1) - V(x))\xi_1^2;\ \xi_1 > -x\} + \mathbb{E}\{\xi_1^2;\ \xi_1 > -x\} \\ &= \frac{1}{V(x)}\mathbb{E}\{(V(x+\xi_1) - V(x))\xi_1^2;\ \xi_1 > -x\} + \sigma^2 + o(1). \end{aligned}$$

It follows from (11.5) that

$$|V(x+\xi_1) - V(x)|\xi_1^2 \le c_V(1+|\xi_1|)\xi_1^2 \le c_V(1+x)\xi_1^2 \quad \text{for all } |\xi_1| \le x,$$

hence

$$\frac{|V(x+\xi_1) - V(x)|}{V(x)}\xi_1^2 \to 0 \quad \text{as } x \to \infty$$

and, again by the dominated convergence theorem

$$\frac{1}{V(x)}\mathbb{E}\{(V(x+\xi_1) - V(x))\xi_1^2;\ |\xi_1| \le x\} \to 0 \quad \text{as } x \to \infty.$$

Therefore

$$m_2(x) = \frac{1}{V(x)}\mathbb{E}\{(V(x+\xi_1) - V(x))\xi_1^2;\ \xi_1 > x\} + \sigma^2 + o(1).$$

If $\mathbb{E}\{\xi_1^3;\ \xi_1 > 0\}$ is finite then we may also apply the dominated convergence theorem to the expectation over the event $\{\xi_1 > x\}$ and obtain $m_2(x) \to \sigma^2$ as $x \to \infty$. But if $\mathbb{E}\{\xi_1^3;\ \xi_1 > 0\} = \infty$ then $\mathbb{E}\{(V(x+\xi_1) - V(x))\xi_1^2;\ \xi_1 > x\}$ is infinite for all $x \geq 0$. Therefore, $m_2(x) \equiv \infty$ for any random walk with infinite $\mathbb{E}\{\xi_1^3;\ \xi_1 > 0\}$.

Clearly, one can show directly that any random walk conditioned to stay positive is transient while the classical Lamperti criterion for transience – according to which at least the second moment of jumps is assumed to be finite – is applicable only to a random walk conditioned to stay positive in the case of finite $\mathbb{E}\{\xi_1^3;\ \xi_1 > 0\}$.

Moreover, to the best of our knowledge, all known results on the convergence towards a Γ-distribution for Markov chains, see Klebaner [99], Kersting [94], or Denisov, Korshunov, and Wachtel [42], assume the finiteness of $m_2(x)$. However, it is well known that the finiteness of σ^2 for a random walk is sufficient for the convergence of X_n^2/n towards a Γ-distribution if X_n is a random walk conditioned to stay positive.

Random walks conditioned to stay positive represent an important class of Markov chains with asymptotically zero drift. So it would be very useful if the general limit theorems for Markov chains with asymptotically zero drift covered the well-known results for random walks conditioned to stay positive. This observation motivated us to state the conditions for Γ-convergence in Chapter 4 in terms of truncated moments and tail probabilities.

Repeating the arguments used above for truncation at the level $-x$, we conclude that

$$m_1^{[s(x)]}(x) \sim \frac{\sigma^2}{x} \quad \text{and} \quad m_2^{[s(x)]}(x) \to \sigma^2 \quad \text{as } x \to \infty, \tag{11.7}$$

where $s(x) = o(x)$ is defined in (11.3). Hence, for any $\varepsilon > 0$,

$$\frac{2m_1^{[s(x)]}(x)}{m_2^{[s(x)]}(x)} \geq \frac{2-\varepsilon}{x} \quad \text{for all sufficiently large } x.$$

Thus, in order to apply the criterion for transience, Theorem 2.21, it remains to show that

$$\mathbb{P}\{\xi(x) < -s(x)\} \leq \frac{p(x)}{x}, \tag{11.8}$$

for some decreasing integrable function p. According to the construction of $\{X_n\}$, this is equivalent to the following upper bound:

$$\frac{1}{V(x)}\mathbb{E}\{V(x+\xi_1);\ \xi_1 < -s(x)\} \le \frac{p(x)}{x}.$$

The function V is increasing; hence it suffices to show that

$$\mathbb{P}\{|\xi_1| > s(x)\} \le \frac{p(x)}{x}, \tag{11.9}$$

which in turn follows from Lemma 2.33 with $\gamma = 2$, $\beta = 0$, and $\alpha = 1$. Thus $\{X_n\}$ is transient, by Theorem 2.21.

To apply Theorem 4.8 on convergence to a Γ-distribution, we additionally need to check that

$$\mathbb{P}\{|\xi(x)| > s(x)\} \le \frac{p(x)}{x}, \tag{11.10}$$

which, due to (11.9), is equivalent to

$$\frac{1}{V(x)}\mathbb{E}\{V(x+\xi_1);\ \xi_1 > s(x)\} \le \frac{p(x)}{x}.$$

Since V has asymptotically linear growth, we may reduce the previous condition to

$$\mathbb{E}\{x+\xi_1;\ \xi_1 > s(x)\} \le p(x),$$

which follows from (11.3) and (11.9). Therefore, by Theorem 4.8,

$$\frac{X_n^2}{n} \Rightarrow \Gamma_{3/2,2\sigma^2} \quad \text{as } n \to \infty, \tag{11.11}$$

and, by Theorem 4.11, the sequence of processes

$$\frac{X_{[nt]}}{\sqrt{n\sigma^2}}, \quad t \in [0,1],$$

converges weakly in $D[0,1]$ to a Bessel process with drift coefficient $1/x$, that is, a three-dimensional Bessel process. In addition, convergence to a Γ-distribution is also accompanied by asymptotics for its integral renewal function; by Theorem 4.12,

$$H(0,x] := \sum_{n=1}^{\infty}\mathbb{P}\{X_n \le x\} \sim \frac{x^2}{\sigma^2} \quad \text{as } x \to \infty.$$

That a random walk conditioned to stay positive converges weakly to a limit was shown by Iglehart [85], with further improvements by Bolthausen [20].

A random walk conditioned to stay positive is a special example of a Markov chain with asymptotically zero drift. Its close connection to an ordinary random walk allows us to obtain a number of further results. More precisely, by the definition of the transition kernel of X, we have

$$\mathbb{P}_z\{X_n \in dx\} = \frac{V(x)}{V(z)}\mathbb{P}\{z + S_n \in dx, \tau(z) > n\}. \tag{11.12}$$

This allows us to use the fluctuation theory for random walks in order to derive results for a random walk conditioned to stay positive. For example, Caravenna and Chaumont [32] proved a functional limit theorem for X and Bryn-Jones and Doney [30] proved a local limit theorem for X. Using the results of Doney [50] one can also derive the asymptotics for local probabilities of small deviations of $\{X_n\}$. Finally, the results of Doney and Jones [51] can be transferred to the asymptotics of large-deviation probabilities for a random walk conditioned to stay positive.

We demonstrate the advantage of this connection to the fluctuation theory of ordinary random walks by the following version of Blackwell's theorem for random walks conditioned to stay positive.

Proposition 11.1 *Assume that $\mathbb{E}\xi_1 = 0$ and $\sigma^2 := \mathbb{E}\xi_1^2 \in (0,\infty)$. Then, for every fixed $\Delta > 0$,*

$$h(x) := H(x+\Delta) - H(x) \sim \frac{2\Delta}{\sigma^2}x \quad \textit{as } x \to \infty$$

if the distribution of ξ_1 is non-lattice, and

$$h(\Delta x) \sim \frac{2\Delta}{\sigma^2}x \quad \textit{as } x \to \infty,\ x \in \mathbb{Z},$$

if $\Delta\mathbb{Z}$ is the minimal lattice for ξ_1.

Proof Consider the non-lattice case. Define

$$u(x) := \mathbb{E}\sum_{n=1}^{\tau_0 - 1} \mathbb{I}\{S_n \in (x, x+\Delta]\} = \sum_{n=1}^{\infty} \mathbb{P}\{S_n \in (x, x+\Delta], \tau(0) > n\}.$$

Let χ_k^+ be independent copies of the first strictly ascending ladder height $\chi^+ := S_{\eta_+}$, where $\eta_+ = \min\{n \ge 1 : S_n > 0\}$. Then, by the classical duality lemma, see e.g. Feller [63, Section XII.2], it follows that

$$\sum_{n=1}^{\infty} \mathbb{P}\{S_n \in (x, x+\Delta], \tau(0) > n\} = \sum_{k=1}^{\infty} \mathbb{P}\{\chi_1^+ + \chi_2^+ + \cdots + \chi_k^+ \in (x, x+\Delta]\}.$$

Applying Blackwell's theorem (see e.g. Durrett [56, Theorem 2.6.4]), we conclude in the non-lattice case that

$$u(x) \to \frac{\Delta}{\mathbb{E}\chi^+} \quad \text{as } x \to \infty. \tag{11.13}$$

This gives us the asymptotics for h in the case where the initial value $X_0 = 0$. Indeed, by (11.12) with $z = 0$ where $V(0) = 0$,

$$\begin{aligned}\sum_{n=1}^{\infty} \mathbb{P}_0\{X_n \in (x, x+\Delta]\} &= \sum_{n=1}^{\infty} \int_x^{x+\Delta} \frac{V(y)}{V(0)} \mathbb{P}\{S_n \in dy, \tau(0) > n\} \\ &\sim V(x) \sum_{n=1}^{\infty} \mathbb{P}\{S_n \in (x, x+\Delta], \tau(0) > n\} \\ &= V(x)u(x) \quad \text{as } x \to \infty,\end{aligned}$$

by (11.4), which implies the long-tailedness of the function V, $V(x) \sim V(x+\Delta)$ as $x \to \infty$. Recalling that $V(x) \sim x/\mathbb{E}\chi^-$ and using (11.13), we obtain

$$\sum_{n=1}^{\infty} \mathbb{P}_0\{X_n \in (x, x+\Delta]\} \sim \frac{\Delta}{\mathbb{E}\chi^- \mathbb{E}\chi^+} x \quad \text{as } x \to \infty.$$

Then it only remains to apply the following identity, which holds true for any zero-drifted random walk with finite variance; see e.g. Feller [63, Section XVIII.5, Theorem 1, or Section XII.10, Problem 10],

$$\mathbb{E}\chi^- \mathbb{E}\chi^+ = \sigma^2/2. \tag{11.14}$$

Now let us consider an arbitrary initial value $X_0 = z$. In view of (11.12) and $V(x) \sim V(x+\Delta)$ as $x \to \infty$,

$$\begin{aligned}\sum_{n=1}^{\infty} \mathbb{P}_z\{X_n \in (x, x+\Delta]\} &\sim \frac{V(x)}{V(z)} \sum_{n=1}^{\infty} \mathbb{P}\{z + S_n \in (x, x+\Delta], \tau(z) > n\} \\ &= \frac{V(x)}{V(z)} \mathbb{E} \sum_{n=1}^{\tau(z)-1} \mathbb{I}\{S_n \in (x-z, x-z+\Delta]\}.\end{aligned}$$

Splitting the trajectory of $\{S_n\}$ into independent cycles by descending ladder epochs and recalling the definition of $u(x)$, we obtain

$$\mathbb{E} \sum_{n=1}^{\tau(z)-1} \mathbb{I}\{S_n \in (x-z, x-z+\Delta]\} = u(x-z) + \mathbb{E} \sum_{k=1}^{\theta(z)-1} u(x-z+\chi_1^- + \cdots + \chi_k^-), \tag{11.15}$$

where $\theta(z)$ was defined in (11.1). By (11.13),

$$\mathbb{E} \sum_{n=1}^{\tau(z)-1} \mathbb{I}\{S_n \in (x, x+\Delta]\} \sim \frac{\Delta}{\mathbb{E}\chi^+} \mathbb{E}\theta_z \quad \text{as } x \to \infty.$$

Recalling that $\mathbb{E}\theta(z) = V(z)$, see (11.1), and that $V(x) \sim x/\mathbb{E}\chi^-$ as $x \to \infty$, we finally get

$$\sum_{n=1}^{\infty} \mathbb{P}_z\{X_n \in (x, x+\Delta]\} \sim \frac{\Delta}{\mathbb{E}\chi^+ \mathbb{E}\chi^-} x = \frac{2\Delta}{\sigma^2} x$$

for all fixed z, owing to (11.14).

In order to derive the same asymptotics for any initial distribution of the chain it suffices to show that

$$\sup_{z \ge 0,\ x \ge 1} \frac{1}{x} \sum_{n=1}^{\infty} \mathbb{P}_z\{X_n \in (x, x+\Delta]\} < \infty, \tag{11.16}$$

which allows us to apply the dominated convergence theorem. It follows from (11.12) and (11.15) that

$$\begin{aligned} &\sum_{n=1}^{\infty} \mathbb{P}_z\{X_n \in (x, x+\Delta]\} \\ &\qquad \le \frac{V(x+\Delta)}{V(z)} \Big(u(x-z) + \mathbb{E} \sum_{k=1}^{\theta(z)-1} u(x - z + \chi_1^- + \cdots + \chi_k^-) \Big). \end{aligned}$$

Since $u_0 := \sup_x u(x) < \infty$,

$$\sum_{n=1}^{\infty} \mathbb{P}_z\{X_n \in (x, x+\Delta]\} \le \frac{V(x+\Delta)}{V(z)} u_0 \mathbb{E}\theta(z) = V(x+\Delta) u_0.$$

Now (11.16) follows from the asymptotic linearity of V, and the proof in the non-lattice case is complete. The lattice case is similar. □

We will demonstrate an alternative proof based on Corollary 6.12.

Proof Let us show that, under the conditions stated, a random walk conditioned to stay positive satisfies all the conditions of Corollary 6.12. First, the condition (6.2) holds with $\mu = \sigma^2$ and $b = \sigma^2$ as shown above in (11.7). Second, the condition (6.3) follows from (11.10).

Third, we also need to check the conditions (6.4), (6.5), and (6.53). To check the first, we note that

$$c_1 := \sup_x \frac{V(x + s(x))}{V(x)} < \infty,$$

hence, for $t \le s(x) = o(x)$,

$$\begin{aligned} \mathbb{P}\{|\xi(x)| > t, |\xi(x)| \le s(x)\} &= \Big(\int_{-s(x)}^{-t} + \int_t^{s(x)} \Big) \frac{V(x+u)}{V(x)} \mathbb{P}\{\xi_1 \in du\} \\ &\le c_1 \mathbb{P}\{|\xi_1| > t\}, \end{aligned}$$

and (6.4), (6.5) follow if we take $\widehat{\xi}$ to be defined by its tail:

$$\mathbb{P}\{\widehat{\xi} > t\} = \min\{1, c_1\mathbb{P}\{|\xi_1| > t\}\},$$

which is square integrable because ξ_1 is square integrable.

Next, using (11.5) once again we obtain

$$\begin{aligned}
&\mathbb{P}\{|\xi(x)| > t\}\\
&\quad = \Big(\int_{-x}^{-t} + \int_t^\infty\Big)\frac{V(x+u)}{V(x)}\mathbb{P}\{\xi_1 \in du\}\\
&\quad \le \mathbb{P}\{\xi_1 < -t\} + \int_t^\infty \Big(1 + c_V\frac{u+1}{V(x)}\Big)\mathbb{P}\{\xi_1 \in du\}\\
&\quad \le \mathbb{P}\{\xi_1 < -t\} + \Big(1 + \frac{c_V}{V(x)}\Big)\mathbb{P}\{\xi_1 > t\} + \frac{c_V}{V(x)}\mathbb{E}\{|\xi_1|; |\xi_1| > t\})\\
&\quad \le c_2(\mathbb{P}\{|\xi_1| > t\} + \mathbb{E}\{|\xi_1|; |\xi_1| > t\}) \quad \text{for all } x,\ t > 0.
\end{aligned}$$

The right-hand side is integrable since $\mathbb{E}\xi_1^2 < \infty$, so the condition (6.53) is satisfied too.

Finally, the asymptotic homogeneity (6.52) is immediate from (11.2) with $\xi = \xi_1$, because, for any fixed $u \in \mathbb{R}$, $V(x+u)/V(x) \to 1$ as $x \to \infty$; the proof is complete. □

11.2 Reflected Random Walk with Zero Drift

Let $\eta_n, n \ge 1$, be a sequence of independent and identically distributed random variables with zero mean and finite variance. The chain defined by

$$X_{n+1} = |X_n + \eta_{n+1}|, \quad n \ge 0, \tag{11.17}$$

is usually called a *reflected random walk*. It follows from (11.17) that

$$\begin{aligned}
\xi(x) &= (x+\eta)\mathbb{I}\{x+\eta \ge 0\} - (x+\eta)\mathbb{I}\{x+\eta < 0\} - x\\
&= \eta - 2(x+\eta)\mathbb{I}\{x+\eta < 0\} = \eta + 2(x+\eta)^-.
\end{aligned}$$

This representation implies that, for any function $s(x) < x$,

$$\begin{aligned}
&m_1^{[s(x)]}(x)\\
&\quad = \mathbb{E}\{\eta; |\eta| \le s(x)\} + \mathbb{E}\{\eta + 2(x+\eta)^-; |\eta + 2(x+\eta)^-| \le s(x),\ \eta < -x\}\\
&\quad = \mathbb{E}\{\eta; |\eta| \le s(x)\} - \mathbb{E}\{2x+\eta;\ |2x+\eta| \le s(x)\}.
\end{aligned}$$

From this equality and the assumption $\mathbb{E}\eta = 0$ we infer that

$$\begin{aligned}
|m_1^{[s(x)]}(x)| &\le \mathbb{E}\{|\eta|; |\eta| > s(x)\} + s(x)\mathbb{P}\{\eta \le -2x + s(x)\}\\
&\le 2\mathbb{E}\{|\eta|; |\eta| > s(x)\}.
\end{aligned}$$

The assumption $\mathbb{E}\eta^2 < \infty$ implies that there exists a function $s(x) = o(x)$ such that $\mathbb{E}[|\eta|; |\eta| > s(x)]$ is integrable; see Lemma 2.33 with $\gamma = 2$, $\alpha = \beta = 1$. Consequently, $|m_1^{[s(x)]}(x)|$ is also integrable. Taking into account that

$$m_2^{[s(x)]}(x) \to \mathbb{E}\eta^2 \in (0,\infty),$$

we finally obtain

$$\frac{m_1^{[s(x)]}(x)}{m_2^{[s(x)]}(x)} = o(p(x))$$

for some decreasing integrable function $p(x)$ satisfying $p'(x) = o(1/x^2)$. Therefore, the reflected random walk X_n satisfies (8.7) with $r(x) \equiv 0$. This implies that $U(x) = x$ in this case. Furthermore, the validity of (8.12)–(8.14) easily follows from the assumption $\mathbb{E}\eta^2 < \infty$. Consequently, we may apply Theorems 8.2, 8.18, and 8.24 to the invariant measure π of the reflected random walk $\{X_n\}$,

$$\pi(ax,x] \sim c(1-a)x \quad \text{as } x \to \infty, \tag{11.18}$$

to the down-crossing probability, which is, for a sufficiently large $\widehat{x}$,

$$\mathbb{P}_x\{\tau_{\widehat{x}} > n\} \sim \frac{V(x)}{\Gamma(3/2)\sqrt{2\mathbb{E}\eta^2}} n^{-1/2} \quad \text{as } n \to \infty, \tag{11.19}$$

and to the conditional distribution

$$\mathbb{P}\{X_n > u\sqrt{n} \mid \tau_{\widehat{x}} > n\} \to e^{-u^2/2\mathbb{E}\eta^2} \quad \text{as } n \to \infty. \tag{11.20}$$

In addition, we can apply Theorem 8.14 to obtain the local asymptotics for the invariant measure π of the reflected random walk,

$$\pi(x,x+h] \to ch \quad \text{as } x \to \infty, \tag{11.21}$$

for all $h > 0$ in the non-lattice case; in the lattice case both x and h should be restricted to the lattice.

The asymptotics in (11.19) and (11.20) coincide with that for an ordinary random walk, except that the function $V(x)$ can be different. This difference comes from the fact that reflection at zero can happen in such a way that the position after the reflection is larger than $\widehat{x}$.

One can also obtain the asymptotics (11.19) using the asymptotics for the first visit of a bounded set by a one-dimensional random walk. Namely, one can interpret $\tau_{\widehat{x}}$ as the first time the random walk visits a compact interval $[-\hat{x},\hat{x}]$. Then, for arithmetic random walks the asymptotics (11.19) follows from the results of Kesten and Spitzer [97] and that for general random walks from Vysotsky [150]; see also the references therein.

Relation (11.18) implies that $\{X_n\}$ is null recurrent. The recurrence of a reflected random walk with finite second moments of increments was shown by Kemperman [91]. Non-positivity in the case of zero mean is immediate from the fact that any ordinary driftless random walk is null recurrent.

The local asymptotics (11.21) was proven by Brofferio and Buraczewski in [28, Theorem 1.3] under the assumption that $\mathbb{E}|\eta^-|^{3/2} < \infty$ and $\mathbb{E}(\eta^+)^2 < \infty$.

11.3 State-Dependent Branching Processes with Migration

In this section we consider branching processes with reproduction law depending on the number of particles in the population: if there are k particles in the population then the number of offspring of every particle is an independent copy of a random variable $\zeta(k) \ge 0$. Furthermore, we assume that there is a migration of particles. This will be modelled by η: given k particles in the system, the number of migrants at time n is an independent copy of a random variable $\eta(k)$ – which may take both positive and negative values. As a consequence we have the following Markov chain:

$$Z_{n+1} := \left(\sum_{i=1}^{Z_n} \zeta_{n+1,i}(Z_n) + \eta_{n+1}(Z_n)\right)^+, \quad n \ge 0, \tag{11.22}$$

where $\{\zeta_{n,i}(k), n \ge 0, i \ge 1\}$ are independent copies of $\zeta(k)$ and $\{\eta_n(k), n \ge 1\}$ are independent copies of $\eta(k)$. Then $\{Z_n\}$ is a Markov chain on $\mathbb{Z}^+$.

There is an alternative way to introduce the migration of particles:

$$Y_{n+1} := \sum_{i=1}^{(Y_n+\eta_n(Y_n))^+} \zeta_{n+1,i}(Y_n), \quad n \ge 0. \tag{11.23}$$

The only difference between these two models is the order of branching and migration at every time step. In (11.22) one performs first branching and then migration, and in (11.23) these two mechanisms appear in the reverse order.

We shall assume that the offspring random variables $\zeta(k)$ are such that

$$k(\mathbb{E}\zeta(k) - 1) \to a_\zeta \in \mathbb{R} \quad \text{as } k \to \infty \tag{11.24}$$

and

$$\sigma^2(k) := \mathbb{V}\mathrm{ar}\,\zeta(k) \to \sigma^2 \in (0, \infty) \tag{11.25}$$

and that the expectation of the migration quantity $\eta(k)$ converges:

$$\mathbb{E}\eta(k) \to a_\eta \in \mathbb{R} \quad \text{as } k \to \infty. \tag{11.26}$$

Under these assumptions the asymptotic behaviour of the first two moments of jumps is as follows:

$$\begin{aligned}\mathbb{E}\{Z_{n+1} - Z_n \mid Z_n = k\} &\to a_\zeta + a_\eta, \\ \mathbb{E}\{(Z_{n+1} - Z_n)^2 \mid Z_n = k\} &\sim \sigma^2 k \quad \text{as } k \to \infty;\end{aligned}$$

for the second relation we need to assume that $\mathbb{E}\eta^2(k) = o(k)$.

The linear growth of the variances significantly complicates analysis of the Markov chain $\{Z_n\}$. In order to get bounded variances we consider a chain

$$X_n := \sqrt{Z_n}, \quad n \ge 0, \tag{11.27}$$

whose jumps are

$$\begin{aligned}\xi(\sqrt{k}) &\stackrel{d}{=} \sqrt{\left(\sum_{i=1}^{k} \zeta_{1,i}(k) + \eta_1(k)\right)^+} - \sqrt{k} \\ &= \sqrt{(S(k) + \eta_1(k))^+} - \sqrt{k},\end{aligned}$$

where $S(k) := \zeta_{1,1}(k) + \cdots + \zeta_{1,k}(k)$. It follows from the proof of the first result in the next subsection that this Markov chain has asymptotically zero drift and bounded second moment of jumps.

11.3.1 Classification of Near-Critical Branching Processes

We start with the classification of branching processes satisfying (11.24)–(11.26). Under some mild conditions on $\zeta(k)$ and $\eta(k)$ we will show that:

(i) if $a_\zeta + a_\eta > \sigma^2/2$ then $\{Z_n\}$ is transient;
(ii) if $0 < a_\zeta + a_\eta < \sigma^2/2$ then $\{Z_n\}$ is null recurrent;
(iii) if $a_\zeta + a_\eta < 0$ then $\{Z_n\}$ is positive recurrent.

We start by evaluating the first two truncated moments of the jumps $\xi(\sqrt{k})$ of the chain $\{X_n\}$ defined in (11.27) and of their left tails.

Proposition 11.2 *Let the moment conditions* (11.24)–(11.26) *hold and let the family of random variables* $\{|\eta(k)|, k \ge 0\}$ *possess an integrable majorant* η, *that is,*

$$|\eta(k)| \le_{\rm st} \eta \quad \text{for all } k \ge 0 \quad \text{and } \mathbb{E}\eta < \infty. \tag{11.28}$$

Then, there exists an increasing function $s(x) = o(x)$ *such that*

$$\mathbb{P}\{\xi(\sqrt{k}) < -s(\sqrt{k})\} \le p(\sqrt{k})/\sqrt{k} \quad \text{for all } k \ge 0, \tag{11.29}$$

where the decreasing function $p(x) > 0$ is integrable at infinity. If, in addition, for some increasing function $t(x) = o(x)$,

$$\mathbb{E}\{\zeta^2(k);\ \zeta(k) > t(k)\} \to 0 \quad \text{as } k \to \infty \tag{11.30}$$

then there exists an increasing function $\widetilde{s}(x) = o(x)$ such that, for all $s(x) \geq \widetilde{s}(x)$, as $k \to \infty$,

$$m_1^{[s(\sqrt{k})]}(\sqrt{k}) \sim \frac{a_\zeta + a_\eta - \sigma^2/4}{2\sqrt{k}} \quad \text{and} \quad m_2^{[s(\sqrt{k})]}(\sqrt{k}) \to \frac{\sigma^2}{4}. \tag{11.31}$$

Proof Let us introduce the events

$$\begin{aligned} A_k &:= \{|\xi(\sqrt{k})| \leq s(\sqrt{k})\} \\ &= \big\{\big|\sqrt{(S(k)+\eta(k))^+} - \sqrt{k}\big| \leq s(\sqrt{k})\big\}. \end{aligned} \tag{11.32}$$

Provided that $s(x) = o(x)$, an equivalent way to define A_k for all sufficiently large k is as

$$(\sqrt{k} - s(\sqrt{k}))^2 \ \leq\ S(k) + \eta(k) \ \leq\ (\sqrt{k} + s(\sqrt{k}))^2,$$

that is,

$$-2\sqrt{k}s(\sqrt{k}) + s^2(\sqrt{k}) \ \leq\ S(k) - k + \eta(k) \ \leq\ 2\sqrt{k}s(\sqrt{k}) + s^2(\sqrt{k}).$$

Therefore, again owing to $s(x) = o(x)$, for all sufficiently large k we have

$$\begin{aligned} &\{\xi^\pm(\sqrt{k}) > s(\sqrt{k})\} \\ &\quad \subseteq \{(S(k)-k)^\pm > \tfrac{3}{2}\sqrt{k}s(\sqrt{k})\} \cup \{\eta^\pm(k) > s^2(\sqrt{k})\}. \end{aligned} \tag{11.33}$$

The condition (11.24) may be rewritten as $\mathbb{E}S(k) - k \to a_\zeta$ as $k \to \infty$; hence, for all sufficiently large k,

$$\begin{aligned} &\mathbb{P}\{\xi^\pm(\sqrt{k}) > s(\sqrt{k})\} \\ &\quad \leq \mathbb{P}\{(S(k) - \mathbb{E}S(k))^\pm > \sqrt{k}s(\sqrt{k})\} + \mathbb{P}\{\eta > s^2(\sqrt{k})\}, \end{aligned} \tag{11.34}$$

owing to the majorisation condition (11.28).

By the exponential Chebyshev inequality,

$$\mathbb{P}\{S(k) - \mathbb{E}S(k) < -\sqrt{k}s(\sqrt{k})\} \leq \Big(\mathbb{E}\exp\Big(\frac{\mathbb{E}\zeta(k) - \zeta(k)}{\sqrt{k}}\Big)\Big)^k e^{-s(\sqrt{k})}.$$

Using a Taylor expansion, for some $\theta \in [0,1]$,

$$\begin{aligned} \mathbb{E}\exp\left(\frac{\mathbb{E}\zeta(k) - \zeta(k)}{\sqrt{k}}\right) &= 1 + \frac{1}{2k}\mathbb{E}(\mathbb{E}\zeta(k) - \zeta(k))^2 \exp\left(\theta\frac{\mathbb{E}\zeta(k) - \zeta(k)}{\sqrt{k}}\right) \\ &\leq 1 + \frac{1}{2k}\mathbb{V}\mathrm{ar}\,\zeta(k)\exp\left(\mathbb{E}\zeta(k)/\sqrt{k}\right), \end{aligned}$$

because the $\zeta(k)$ are non-negative. Therefore, for some $c < \infty$,

$$\mathbb{P}\{S(k) - \mathbb{E}S(k) < -\sqrt{k}s(\sqrt{k})\} \le ce^{-s(\sqrt{k})}. \tag{11.35}$$

Further, since $\mathbb{E}(\sqrt{\eta})^2 = \mathbb{E}\eta < \infty$, by Lemma 2.33 there exists an $s(x) = o(x)$ such that

$$\mathbb{P}\{\sqrt{\eta} > s(x)\} \le p_1(x)/x,$$

for some decreasing function $p_1(x)$ that is integrable at infinity. Therefore

$$\mathbb{P}\{\eta > s^2(\sqrt{k})\} \le p_1(\sqrt{k})/\sqrt{k}. \tag{11.36}$$

Substituting (11.35) and (11.36) into (11.34) we obtain

$$\mathbb{P}\{\xi(\sqrt{k}) < -s(\sqrt{k})\} \le ce^{-s(\sqrt{k})} + p_1(\sqrt{k})/\sqrt{k},$$

so (11.29) follows provided that $s(x) \ge 3\log x$.

Let us now show the relations (11.31) for the truncated moments. We start by showing that, as $k \to \infty$,

$$\mathbb{E}\{(S(k) - k)^2;\ |S(k) - k| > \sqrt{k}s(\sqrt{k})\} = o(k). \tag{11.37}$$

Since the variance of $\zeta(k)$, $k \ge 1$, is bounded, taking $y = x/2$ in (2.127) we conclude that

$$\begin{aligned}
&\mathbb{E}\{(S(k) - \mathbb{E}S(k))^2;\ S(k) - \mathbb{E}S(k) > x\} \\
&\quad \le 2C(2)(k/x)^2 + k\mathbb{E}\{(\zeta(k) - \mathbb{E}\zeta(k))^2;\ \zeta(k) - \mathbb{E}\zeta(k) > x/2\} \\
&\qquad + k^2\mathbb{P}\{\zeta(k) - \mathbb{E}\zeta(k) > x/2\}.
\end{aligned} \tag{11.38}$$

For $x = \sqrt{k}s(\sqrt{k})$, which is greater than $3t(k)$ provided $s(x) \ge 3t(x^2)/x = o(x)$, we obtain

$$\begin{aligned}
&\mathbb{E}\{(S(k) - \mathbb{E}S(k))^2;\ |S(k) - \mathbb{E}S(k)| > \sqrt{k}s(\sqrt{k})\} \\
&\quad \le Ck\Big(\frac{1}{s^2(\sqrt{k})} + \mathbb{E}\{\zeta^2(k);\ \zeta(k) > t(k)\}\Big),
\end{aligned} \tag{11.39}$$

and, by the condition (11.30),

$$\mathbb{E}\{(S(k) - \mathbb{E}S(k))^2;\ |S(k) - \mathbb{E}S(k)| > \sqrt{k}s(\sqrt{k})\} = o(k) \quad \text{as } k \to \infty,$$

which implies (11.37) because $|\mathbb{E}S(k) - k|$ is bounded owing to the condition (11.24).

It follows from (11.33) and then from (11.37) that

$$\begin{aligned}
&\mathbb{E}\{(S(k)-k)^2;\ A_k^c\}\\
&\quad \le \mathbb{E}\{(S(k)-k)^2;\ |S(k)-k| \ge \sqrt{k}s(\sqrt{k})\} + \mathbb{E}\{(S(k)-k)^2;\ \eta(k) \ge s^2(\sqrt{k})\}\\
&\quad = o(k) + (\mathbb{V}\mathrm{ar}\, S(k) + (\mathbb{E}S(k)-k)^2)\mathbb{P}\{\eta(k) \ge s^2(\sqrt{k})\}\\
&\quad = o(k) \quad \text{as } k \to \infty,
\end{aligned}$$

because the second moments of both ζ's and of the sequence $|\mathbb{E}S(k)-k|$ are bounded by the conditions (11.24) and (11.25). Hence the condition (11.25) allows us to conclude that

$$\frac{1}{k}\mathbb{E}\{(S(k)-k)^2;\ A_k\} = \sigma^2(k) + o(1) \ \to\ \sigma^2 \quad \text{as } k \to \infty. \tag{11.40}$$

Note also that

$$\begin{aligned}
\frac{1}{k}\mathbb{E}\{|\eta(k)(S(k)-k)|;\ A_k\} &\le \frac{1}{k}\mathbb{E}|\eta(k)(S(k)-k)|\\
&= \frac{1}{k}\mathbb{E}|\eta(k)|\mathbb{E}|S(k)-k|\\
&\le \frac{1}{k}\mathbb{E}\eta\sqrt{\mathbb{E}(S(k)-k)^2}\\
&= O(1/\sqrt{k}),
\end{aligned} \tag{11.41}$$

by the independence of $\eta(k)$ and $S(k)$ and by the majorisation condition (11.28). Moreover, since

$$A_k = \{k - 2\sqrt{k}s(\sqrt{k}) + s^2(\sqrt{k}) \le S(k) + \eta(k) \le k + 2\sqrt{k}s(\sqrt{k}) + s^2(\sqrt{k})\},$$

for all sufficiently large k, we have

$$A_k \subseteq \{|\eta(k)| \le 3\sqrt{k}s(\sqrt{k}) + |S(k)-k|\}; \tag{11.42}$$

hence

$$\mathbb{E}\{\eta^2(k);\ A_k\} \le \mathbb{E}\{\eta^2(k);\ |\eta(k)| \le 3\sqrt{k}s(\sqrt{k}) + |S(k)-k|\}.$$

By the assumption $\mathbb{E}\eta < \infty$ on the majorant for the $\eta(k)$, it follows from Lemma 2.26 with $V(x) = x$ and $p = 2$ that

$$\sup_k \mathbb{E}\{\eta^2(k);\ |\eta(k)| \le x\} = o(x) \quad \text{as } x \to \infty.$$

Thus, as $k \to \infty$,

$$\mathbb{E}\{\eta^2(k);\ A_k\} \le o(\sqrt{k}s(\sqrt{k}) + \mathbb{E}|S(k)-k|) \ = \ o(k). \tag{11.43}$$

Combining (11.40), (11.41), and (11.43), we conclude the convergence

$$\frac{1}{k}\mathbb{E}\{(S(k)-k+\eta(k))^2;\ A_k\} \to \sigma^2 \quad \text{as } k\to\infty. \tag{11.44}$$

The upper bound (11.37) implies that, for any fixed $n \ge 1$,

$$\begin{aligned}\mathbb{E}\{|S(k)-k|;\ |S(k)-k|>k/n\} &\le \frac{n}{k}\mathbb{E}\{(S(k)-k)^2;\ |S(k)-k|>k/n\}\\ &\to 0 \quad \text{as } k\to\infty.\end{aligned} \tag{11.45}$$

Therefore, for some $s(x)=o(x)$,

$$\mathbb{E}\{|S(k)-k|;\ |S(k)-k|>\sqrt{k}s(\sqrt{k})\}\to 0 \quad \text{as } k\to\infty. \tag{11.46}$$

It follows from (11.33) that

$$\begin{aligned}&\mathbb{E}\{|S(k)-k+\eta(k)|;\ |\xi(\sqrt{k})|>s(\sqrt{k})\}\\ &\quad\le \mathbb{E}\{|S(k)-k+\eta(k)|;\ |S(k)-k|>\sqrt{k}s(\sqrt{k})\}\\ &\qquad+\mathbb{E}\{|S(k)-k+\eta(k)|;\ |\eta(k)|>s^2(\sqrt{k})\}\\ &\quad\le \mathbb{E}\{|S(k)-k|;\ |S(k)-k|>\sqrt{k}s(\sqrt{k})\}+\mathbb{E}\eta\mathbb{P}\{|S(k)-k|>\sqrt{k}s(\sqrt{k})\}\\ &\qquad+\mathbb{E}|S(k)-k|\mathbb{P}\{\eta>s^2(\sqrt{k})\}+\mathbb{E}\{\eta;\ \eta>s^2(\sqrt{k})\},\end{aligned} \tag{11.47}$$

by the independence of $S(k)$ and $\eta(k)$ and by the majorisation condition (11.28). Now applying (11.46) and (11.36) we obtain

$$\mathbb{E}\{|S(k)-k+\eta(k)|;\ A_k^c\}\to 0 \quad \text{as } k\to\infty. \tag{11.48}$$

In turn, this implies that

$$\begin{aligned}\mathbb{E}\{S(k)-k+\eta(k);\ A_k\} &= \mathbb{E}(S(k)-k+\eta(k))+o(1)\\ &\to a_\zeta+a_\eta,\end{aligned} \tag{11.49}$$

by the conditions (11.24) and (11.26).

Using a Taylor expansion we conclude that

$$\begin{aligned}\xi(\sqrt{k}) &= \sqrt{k}\left(\sqrt{1+\frac{S(k)-k+\eta(k)}{k}}-1\right)\\ &= \frac{S(k)-k+\eta(k)}{2\sqrt{k}}-\frac{1}{8}\frac{(S(k)-k+\eta(k))^2}{k\sqrt{k}}(1+o(1))\end{aligned}$$

as $k\to\infty$ uniformly on the event A_k, where the uniformity of $o(1)$ on this event follows from the relation $s(x)=o(x)$. Then, using (11.49) and (11.44), we obtain

$$\begin{aligned}\sqrt{k}m_1^{[s(\sqrt{k})]}(\sqrt{k}) &= \sqrt{k}\mathbb{E}\{\xi(\sqrt{k});\ A_k\}\\ &\to \frac{a_\zeta+a_\eta-\sigma^2/4}{2} \quad \text{as } k\to\infty.\end{aligned}$$

To determine the asymptotic behaviour of the second truncated moment we note that, uniformly on the event A_k,

$$\xi^2(\sqrt{k}) = \left(\frac{S(k) - k + \eta}{\sqrt{S(k) + \eta(k)} + \sqrt{k}}\right)^2$$
$$= \frac{(S(k) - k + \eta(k))^2}{4k}(1 + o(1)) \quad \text{as } k \to \infty.$$

Using (11.44) once again we conclude that

$$m_2^{[s(\sqrt{k})]}(\sqrt{k}) \;=\; \mathbb{E}\{\xi^2(\sqrt{k});\ A_k\} \;\to\; \sigma^2/4 \quad \text{as } k \to \infty.$$

So, both convergences in (11.31) hold true and the proof is complete. □

Theorem 11.3 *Assume that* (11.24)–(11.28) *and* (11.30) *hold, and that*

$$\limsup_{n\to\infty} Z_n = \infty \quad \textit{with probability } 1. \tag{11.50}$$

If $a_\zeta + a_\eta > \sigma^2/2$, *then* $\{Z_n\}$ *is transient.*

Since $\{Z_n\}$ lives on the non-negative integers, the assumption (11.50) corresponds, modulo some periodicity issues, to the irreducibility of the state space of the branching process. This assumption obviously excludes the existence of absorbing states. So, standard non-degenerate critical Galton–Watson processes do not satisfy this condition, but if one adds a non-trivial immigration at zero then (11.50) follows. The same is true in the case of space-homogeneous immigration.

Probably the simplest sufficient condition for (11.50) is the following:

$$\mathbb{P}\{\eta(k) > 0\} > 0 \quad \text{for all } k \ge 0.$$

In this case no any further restriction on the offspring numbers $\zeta(k)$ is needed to guarantee (11.50). For a near-critical process satisfying (11.24) and (11.25), one can relax the restriction on migration mentioned above. Indeed, (11.24) and (11.25) imply that

$$\mathbb{E}\zeta(k)(\zeta(k) - 1) \to \sigma^2 > 0 \quad \text{as } k \to \infty.$$

Consequently, there exists a k_0 such that

$$\inf_{k>k_0} \mathbb{P}\{\zeta(k) \ge 2\} > 0.$$

Therefore, the desired irreducibility then follows from the conditions

$$\mathbb{P}\{\eta(k) > 0\} > 0 \quad \text{for all } k \le k_0$$

and

$$\mathbb{P}\{\eta(k) \geq -2k+1\} > 0 \quad \text{for all } k > k_0.$$

Proof of Theorem 11.3. It is sufficient to check that the chain $\{X_n\} = \{\sqrt{Z_n}\}$ satisfies all the conditions of Theorem 2.21. Fulfilment of the condition (2.108) of Theorem 2.21 follows from (11.31) and assumption $a_\zeta + a_\eta > \sigma^2/2$, while (2.109) follows from (11.29), which completes the proof. □

Theorem 11.4 *Assume that* (11.24) *and* (11.26) *hold and, for some* $\varepsilon > 0$,

$$\mathbb{P}\{\zeta(k) = 0\} \leq 1 - 2\varepsilon, \tag{11.51}$$

$$\mathbb{E}\{\eta(k);\ \eta(k) < -k\varepsilon\} \to 0 \quad \text{as } k \to \infty. \tag{11.52}$$

If $a_\zeta + a_\eta < 0$ *then the chain* $\{Z_n\}$ *is positive recurrent.*

If, in addition, (11.25), (11.28), *and* (11.30) *hold and* $a_\zeta + a_\eta < \sigma^2/2$ *then the chain* $\{Z_n\}$ *is recurrent.*

Proof For positive recurrence we show that the drift of $\{Z_n\}$,

$$\begin{aligned}\mathbb{E}\{Z_{n+1} - Z_n \mid Z_n = k\} &= \mathbb{E}(S(k) + \eta(k))^+ - k \\ &= \mathbb{E}(S(k) + \eta(k)) - k + \mathbb{E}(S(k) + \eta(k))^-,\end{aligned}$$

is negative and bounded away from zero for all sufficiently large k if $a_\zeta + a_\eta < 0$, because

$$\begin{aligned}\mathbb{E}(S(k) + \eta(k))^- &= \mathbb{E}\{(S(k) + \eta(k))^-;\ \eta(k) < -k\varepsilon\} \\ &\quad + \mathbb{E}\{(S(k) + \eta(k))^-;\ \eta(k) \geq -k\varepsilon\} \\ &\leq \mathbb{E}\{\eta^-(k);\ \eta(k) < -k\varepsilon\} + \mathbb{E}(S(k) - \varepsilon k)^-.\end{aligned}$$

The first expectation on the right-hand side tends to zero as $k \to \infty$ owing to the condition (11.52). The second expectation tends to zero too, because, by the condition (11.51), the $\zeta(k)$ stochastically dominate a Bernoulli random variable ζ with success probability 2ε, so that

$$\begin{aligned}\mathbb{E}(S(k) - \varepsilon k)^- &\leq \mathbb{E}((\zeta_1 - \varepsilon) + \cdots + (\zeta_k - \varepsilon))^- \\ &\leq \varepsilon k \mathbb{P}((\zeta_1 - \varepsilon) + \cdots + (\zeta_k - \varepsilon) < 0) \\ &\leq \varepsilon k (1 - \delta)^k \quad \text{for some } \delta > 0,\end{aligned}$$

since $\mathbb{E}(\zeta - \varepsilon) = \varepsilon > 0$; here the ζ_i are independent copies of ζ.

Let us now check that the chain $\{\sqrt{Z_n}\}$ satisfies all the conditions of Corollary 2.14 in the case $a_\zeta + a_\eta < \sigma^2/2$. In view of the condition (11.30),

$$\mathbb{P}\{\zeta(k) > t(k)\} \leq \frac{\mathbb{E}\{\zeta^2(k);\ \zeta(k) > t(k)\}}{t^2(k)} = o(1/k^2) \tag{11.53}$$

as $k \to \infty$, possibly with a faster increasing level $t(x)$. It follows from (11.46) and Chebyshev's inequality that

$$\mathbb{P}\{S(k) - \mathbb{E}S(k) > \sqrt{k}s(\sqrt{k})\} = o(1/k) \quad \text{as } k \to \infty,$$

possibly with a faster increasing $s(x)$. Together with (11.36) and (11.33) this yields an upper bound

$$\mathbb{P}\{\xi(\sqrt{k}) > s(\sqrt{k})\} = o(1/k) \quad \text{as } k \to \infty. \tag{11.54}$$

The condition (2.84) follows from upper bound

$$\begin{aligned}
&\mathbb{E}\{\xi^3(\sqrt{k});\ \xi(\sqrt{k}) \in [0, \sqrt{k}]\}\\
&\quad \le \mathbb{E}\{\xi^3(\sqrt{k});\ \xi(\sqrt{k}) \in [0, s(\sqrt{k})]\} + \mathbb{E}\{\xi^3(\sqrt{k});\ \xi(\sqrt{k}) \in (s(\sqrt{k}), \sqrt{k}]\}\\
&\quad \le s(\sqrt{k})\mathbb{E}\{\xi^2(\sqrt{k});\ \xi(\sqrt{k}) \in [0, s(\sqrt{k})]\} + k^{3/2}\mathbb{P}\{\xi(\sqrt{k}) > s(\sqrt{k})\}\\
&\quad = o(\sqrt{k}) \quad \text{as } k \to \infty,
\end{aligned}$$

because the first term on the right-hand side is of order $O(s(\sqrt{k})) = o(\sqrt{k})$ owing to the second convergence in (11.31) while the second term is of the same order by (11.54).

In order to show the validity of (2.85) we first note that, by the concavity of $\sqrt{y}$,

$$\xi(\sqrt{k}) \le \sqrt{S(k) - k + \eta(k)} \quad \text{on the event } \xi(\sqrt{k}) > 0$$

and

$$\{\xi(\sqrt{k}) \ge \sqrt{k}\} = \{S(k) - k + \eta \ge 3k\}.$$

Then, by the Markov inequality,

$$\begin{aligned}
&\mathbb{E}\{\xi^{\varepsilon/2}(\sqrt{k});\ \xi(\sqrt{k}) \ge \sqrt{k}\}\\
&\quad \le \mathbb{E}\{(S(k) - k + \eta(k))^{\varepsilon/4};\ S(k) - k + \eta(k) \ge 3k\}\\
&\quad \le \frac{1}{(\sqrt{k})^{2-\varepsilon/2}}\mathbb{E}\{S(k) - k + \eta(k);\ S(k) - k + \eta(k) \ge 3k\};
\end{aligned}$$

hence (2.85) follows because the expectation on the right-hand side tends to zero, as shown in (11.48).

For recurrence, it remains to prove that

$$\frac{2m_1^{[\sqrt{k}]}(\sqrt{k})}{m_2^{[\sqrt{k}]}(\sqrt{k})} \le \frac{1-\varepsilon}{\sqrt{k}}$$

for all sufficiently large k. Since $s(x) < x$, we have

$$\frac{2m_1^{[\sqrt{k}]}(\sqrt{k})}{m_2^{[\sqrt{k}]}(\sqrt{k})} \le \frac{2m_1^{[\sqrt{k}]}(\sqrt{k})}{m_2^{[s(\sqrt{k})]}(\sqrt{k})}$$

$$\le \frac{2m_1^{[s(\sqrt{k})]}(\sqrt{k})}{m_2^{[s(\sqrt{k})]}(\sqrt{k})} + \frac{2\mathbb{E}\{\xi(\sqrt{k});\ \xi(\sqrt{k}) \in (s(\sqrt{k}), \sqrt{k}]\}}{m_2^{[s(\sqrt{k})]}(\sqrt{k})}.$$

It follows from (11.31) that

$$\frac{2m_1^{[s(\sqrt{k})]}(\sqrt{k})}{m_2^{[s(\sqrt{k})]}(\sqrt{k})} \sim \frac{a_\zeta + a_\eta - \sigma^2/4}{\sigma^2/4}\frac{1}{\sqrt{k}}.$$

By (11.54),

$$\mathbb{E}\{\xi(\sqrt{k});\ \xi(\sqrt{k}) \in (s(\sqrt{k}), \sqrt{k}]\} \le \sqrt{k}\mathbb{P}\{\xi(\sqrt{k}) > s(\sqrt{k})\} = o(1/\sqrt{k});$$

hence the desired inequality follows because $a_\zeta + a_\eta < \sigma^2/2$. □

Theorem 11.5 *Assume that* (11.24)–(11.28) *and* (11.50) *hold. Let the family of random variables* $\{\zeta^2(k),\ k \ge 1\}$ *be uniformly integrable, that is,*

$$\sup_{k\ge 1} \mathbb{E}\{\zeta^2(k);\ \zeta(k) > t\} \to 0 \quad as\ t \to \infty. \tag{11.55}$$

If $a_\zeta + a_\eta \in (0, \sigma^2/2)$ *then the chain* $\{Z_n\}$ *is null-recurrent.*

Proof We apply Corollary 2.16. Note that the condition (11.55) implies the fulfilment of (11.30), so the first two truncated moments of jumps $\xi(\sqrt{k})$ satisfy the asymptotic relations (11.31).

Now let us show that the family of squares $\xi^2(\sqrt{k})$ is uniformly integrable. It follows from the definition of $\xi(\sqrt{k})$ that, for all $y > 0$,

$$\begin{aligned}\mathbb{P}\{\xi(\sqrt{k}) > y\} &= \mathbb{P}\{\sqrt{S(k) + \eta(k)} > \sqrt{k} + y\}\\ &= \mathbb{P}\{S(k) - k + \eta(k) > 2\sqrt{k}y + y^2\}\\ &\le \mathbb{P}\{S(k) - k > \sqrt{k}y\} + \mathbb{P}\{\eta > y^2\}.\end{aligned} \tag{11.56}$$

For the left tail, we have

$$\begin{aligned}\mathbb{P}\{\xi(\sqrt{k}) < -\,y\} &= \mathbb{P}\{\sqrt{(S(k) + \eta(k))^+} < \sqrt{k} - y\}\\ &\le \mathbb{P}\{S(k) - k + \eta(k) < -2\sqrt{k}y + y^2\}\\ &\le \mathbb{P}\{S(k) - k < (-2\sqrt{k}y + y^2)/2\} + \mathbb{P}\{\eta > (2\sqrt{k}y - y^2)/2\}.\end{aligned}$$

Since $\xi(\sqrt{k}) \ge -\sqrt{k}$, we have to consider only the values of $y \le \sqrt{k}$ in the last formula. However, for such values of y we have $-2\sqrt{k}y + y^2 \le -\sqrt{k}y$ and $-2\sqrt{k}y + y^2 \le -y^2$; therefore

$$\mathbb{P}\{\xi(\sqrt{k}) \le -y\} \le \mathbb{P}\{S(k) - k < -\sqrt{k}y/2\} + \mathbb{P}\{\eta > y^2/2\}.$$

Combining this estimate with (11.56), we obtain

$$\mathbb{P}\{|\xi(\sqrt{k})| \ge y\} \le \mathbb{P}\Big\{\Big|\frac{S(k)-k}{\sqrt{k}}\Big| > y/2\Big\} + \mathbb{P}\{\sqrt{\eta} > y/2\}. \quad (11.57)$$

By the conditions (11.24) and (11.25) and by the uniform integrability of $\zeta^2(k)$, the family of random variables $(S(k)-k)^2/k$ is uniformly integrable too. The random variable $\sqrt{\eta}$ is square integrable. Altogether this implies the uniform integrability of the family of squares $\{\xi^2(\sqrt{k}),\ k \ge 1\}$.

The condition (2.60) follows from (11.29) and (11.31), since $\xi(\sqrt{k}) \ge -\sqrt{k}$. Then uniform integrability and the asymptotics (11.31) allow us to apply Corollary 2.16 in the case $a_\zeta + a_\eta \in (0, \sigma^2/2)$ and to conclude the null recurrence of $\{X_n\}$, and hence of $\{Z_n\}$. □

11.3.2 Convergence to Γ-Distribution

Theorem 11.6 *Assume that* (11.24)–(11.28), (11.30), *and* (11.50) *hold, and that*

$$\mathbb{P}\{\zeta(k) > t(k)\} \le q(k)/k \quad (11.58)$$

for some increasing $t(x) = o(x)$ *and a decreasing integrable function* $q(x)$ *such that* $q(x)\sqrt{x}$ *decreases. If* $a_\zeta + a_\eta > \sigma^2/2$ *then*

$$\frac{Z_n}{n\sigma^2/4} = \frac{X_n^2}{n\sigma^2/4}$$

converges weakly as $n \to \infty$ *to a* Γ*-distribution with mean* $4(a_\zeta + a_\eta)/\sigma^2$ *and variance* $8(a_\zeta + a_\eta)/\sigma^2$*. In addition, the sequence of processes*

$$\sqrt{\frac{Z_{[nt]}}{n\sigma^2/4}}, \quad t \in [0,1],$$

converges weakly in the space $D[0,1]$ *to a Bessel process with drift coefficient* $(2(a_\zeta + a_\eta)/\sigma^2 - 1/2)/x$ *and unit diffusion coefficient.*

A sufficient condition for (11.58) is the existence of a square integrable majorant Ξ for the family of random variables $\{\zeta(k), k \ge 1\}$; see Lemma 2.33.

Proof It is sufficient to check that the chain $\{X_n\}$ satisfies all the conditions of Theorems 4.8 and 4.11. By Theorem 11.3, the chain $\{X_n\}$ is transient and by Proposition 11.2 the truncated moments of its jumps $\xi(\sqrt{k})$ satisfy (11.31), so the condition (4.45) follows with $\mu = (a_\zeta + a_\eta - \sigma^2/4)/2$ and $b = \sigma^2/4$. Then it remains to show that, for all k,

$$\mathbb{P}\{|\xi(\sqrt{k})| > s(\sqrt{k})\} \le p(\sqrt{k})/\sqrt{k}, \tag{11.59}$$

which in particular implies, since $\xi(\sqrt{k}) \ge -\sqrt{k}$, that

$$\mathbb{E}\{|\xi(\sqrt{k})|;\ \xi(\sqrt{k}) < -s(\sqrt{k})\} \le p(\sqrt{k}),$$

where the decreasing function $p(x) > 0$ is integrable at infinity. It follows from the Fuk–Nagaev inequality (2.126) with $x = \sqrt{k}s(\sqrt{k})$ and $y = x/2$ that

$$\mathbb{P}\{|S(k) - \mathbb{E}S(k)| > \sqrt{k}s(\sqrt{k})\} \le \frac{C}{s^4(\sqrt{k})} + k\mathbb{P}\{\zeta(k) > \sqrt{k}s(\sqrt{k})/2\}.$$

Let us choose $s(x) = o(x)$ such that $s(x) \ge x^{3/4}$ and $xs(x) \ge 2t(x^2)$, which is possible because $t(x) = o(x)$. Then, by the condition (11.58),

$$\begin{aligned}\mathbb{P}\{|S(k) - \mathbb{E}S(k)| > \sqrt{k}s(\sqrt{k})\} &\le \frac{C}{(\sqrt{k})^3} + k\mathbb{P}\{\zeta(k) > t(k)\}\\ &\le \frac{1}{\sqrt{k}}\left(\frac{C}{(\sqrt{k})^2} + q(k)\sqrt{k}\right).\end{aligned} \tag{11.60}$$

Together with (11.34), the upper bounds (11.60) and (11.36) imply that

$$\mathbb{P}\{|\xi(\sqrt{k})| \ge s(\sqrt{k})\} \le \frac{1}{\sqrt{k}}\left(\frac{C}{(\sqrt{k})^2} + \widetilde{q}(k)\sqrt{k}\right),$$

where $\widetilde{q}(x)$ is a decreasing integrable function. Since

$$\int_1^\infty \left(\frac{C}{x^2} + \widetilde{q}(x^2)x\right)dx = C + \frac{1}{2}\int_1^\infty \widetilde{q}(y)dy \ < \ \infty,$$

the chain $\{X_n\}$ satisfies the condition (11.59) and the proof is complete. □

Assume that all the conditions of Theorem 11.6 apart from (11.50) are valid but $\mathbb{P}\{\eta(k) \le 0\} = 1$, so the state 0 is absorbing and the extinction probability is positive. Denote

$$q \ := \ \mathbb{P}\{Z_n \to \infty\} \in (0,1).$$

In parallel, let us introduce a branching process $\{\widetilde{Z}_n\}$ governed by the same stochastic mechanism as $\{Z_n\}$ but with just one alteration: we add a transition at zero. If $\widetilde{Z}_n = 0$ we put $\widetilde{Z}_{n+1} = 1$; this alternating chain is transient provided that it is irreducible and Theorem 11.6 is applicable to it. Since $\{\widetilde{Z}_n\}$ visits 0 finitely many times, we conclude that the distribution of $Z_n/n\sigma^2$ conditioned on the event $\{Z_n \to \infty\}$ converges to the same Γ-distribution as that to which $\widetilde{Z}_n/n\sigma^2$ is converging, which implies that

$$\mathbb{P}\{Z_n/n\sigma^2 \le x\} \to (1-q) + q\Gamma(x) \quad \text{as } n \to \infty. \tag{11.61}$$

The next result is aimed at covering the null recurrent case.

Theorem 11.7 *Assume that* (11.24)–(11.28) *and* (11.50) *hold,* $a_\zeta + a_\eta > 0$, *and there exists a decreasing function* $\varepsilon(y) \to 0$ *such that*

$$\mathbb{P}\{\zeta(k) > y\} \le \frac{\varepsilon(y)}{y^2} \quad \textit{for all } k \ge 1,\ y > 0, \tag{11.62}$$

and

$$\int_1^\infty \frac{\varepsilon(y)}{y} dy < \infty. \tag{11.63}$$

Then $4Z_n/n\sigma^2$ *converges weakly as* $n \to \infty$ *to a* Γ*-distribution with mean* $4(a_\zeta + a_\eta)/\sigma^2$ *and variance* $8(a_\zeta + a_\eta)/\sigma^2$. *In addition, the sequence of processes*

$$2\sqrt{\frac{Z_{[nt]}}{n\sigma^2}}, \quad t \in [0, 1],$$

converges weakly in the space $D[0, 1]$ *to a Bessel process with drift coefficient* $(2(a_\zeta + a_\eta)/\sigma^2 - 1/2)/x$ *and unit diffusion coefficient.*

The conditions (11.62) and (11.63) imply the existence of a square integrable majorant Ξ for the family of random variables $\{\zeta(k), k \ge 1\}$, but not the other way around. A sufficient condition for (11.62) and (11.63) is the existence of a majorant Ξ such that $\mathbb{E}\Xi^2 \log^{1+\varepsilon}(1 + \Xi) < \infty$ for some $\varepsilon > 0$. Note that we use the monotonicity of the function $\varepsilon(y)$ when we justify (11.64) below.

Also, instead of the conditions (11.62) and (11.63) we could assume the existence of a majorant Ξ for the family $\{\zeta(k)\}$ such that $\Xi^2 \log(1 + \Xi)$ is integrable, because then the function $\mathbb{E}\{\Xi^2;\ \Xi > y\}/y^2$ would be integrable at infinity.

Proof Note that the condition (11.62) implies that the family $\{\zeta^2(k), k \ge 1\}$ is uniformly integrable, hence (11.30) holds, so the first two truncated moments of jumps $\xi(\sqrt{k})$ satisfy the asymptotic relations (11.31). Also, by Theorem 11.5, the chain $\{X_n\}$ is either null recurrent or transient.

To prove convergence to a Γ-distribution, let us check the conditions of Theorem 4.10. First, the null recurrence or transience of $\{X_n\}$ implies the convergence $X_n \to \infty$ in probability as $n \to \infty$. Second, the sequence $|\xi(\sqrt{k})|$ possesses a square integrable majorant Ξ. Indeed, using (2.126) with $x = \sqrt{k}y/2$ and $y = x/2$ we get from (11.57) that

$$\mathbb{P}\{|\xi(\sqrt{k})| \ge y\} \le c/y^4 + k\mathbb{P}\{|\zeta(k) - \mathbb{E}\zeta(k)| > \sqrt{k}y/4\} + \mathbb{P}\{\sqrt{\eta} > y/2\}.$$

Since $\zeta(k) \ge 0$ and the sequence $\mathbb{E}\zeta(k)$ is bounded, there exists an y_0 such that, for $y \ge y_0$,

$$\mathbb{P}\{|\xi(\sqrt{k})| \ge y\} \le c/y^4 + k\mathbb{P}\{\zeta(k) > \sqrt{k}y/2\} + \mathbb{P}\{\sqrt{\eta} > y/2\}.$$

Owing to (11.62) and the monotonicity of the function $\varepsilon(y)$ we have

$$\mathbb{P}\{\zeta(k) > \sqrt{k}y/2\} \le 4\frac{\varepsilon(\sqrt{k}y/2)}{ky^2} \le 4\frac{\varepsilon(y/2)}{ky^2}; \tag{11.64}$$

hence

$$\mathbb{P}\{|\xi(\sqrt{k})| \ge y\} \le c/y^4 + 4\varepsilon(y/2)/y^2 + \mathbb{P}\{\sqrt{\eta} > y/2\} \quad \text{for } y \ge y_0.$$

Let Ξ be a random variable taking values in $[y_0, \infty)$ such that

$$\mathbb{P}\{\Xi > y\} = \min\{1,\ c/y^4 + 4\varepsilon(y/2)/y^2 + \mathbb{P}\{\sqrt{\eta} > y/2\}\} \quad \text{for } y \ge y_0.$$

Clearly, Ξ is a stochastic majorant for the sequence $\xi(\sqrt{k})$. The finiteness of $\mathbb{E}\Xi^2$ follows from the condition (11.63) and the assumption $\mathbb{E}\eta < \infty$.

So, it only remains to determine the asymptotic behaviour of the first two full moments of jumps, that is, of $m_1(\sqrt{k})$ and $m_2(\sqrt{k})$. We know from the proof of Theorem 11.6 that

$$m_1^{[s(\sqrt{k})]}(\sqrt{k}) \sim \frac{a_\zeta + a_\eta - \sigma^2/4}{2\sqrt{k}} \quad \text{as } k \to \infty,$$

for any $s(x)$ such that $s(x)/x \to 0$ sufficiently slowly. From the existence of the majorant we infer that

$$\mathbb{E}\{|\xi(\sqrt{k})|;\ |\xi(\sqrt{k})| \ge s(\sqrt{k})\} \le \frac{1}{s(\sqrt{k})}\mathbb{E}\{\Xi^2;\ \Xi \ge s(\sqrt{k})\}.$$

Consequently, we can choose $s(x) = o(x)$ such that

$$\mathbb{E}\{|\xi(\sqrt{k})|;\ |\xi(\sqrt{k})| > s(\sqrt{k})\} = o(1/\sqrt{k}) \quad \text{as } k \to \infty.$$

This yields the asymptotics

$$m_1(\sqrt{k}) \sim \frac{a_\zeta + a_\eta - \sigma^2/4}{2\sqrt{k}} \quad \text{as } k \to \infty.$$

The existence of a square integrable majorant also implies that $m_2(\sqrt{k}) \to \sigma^2/4$ as $k \to \infty$. Thus, the weak convergence of $Z_n/n\sigma^2$ to a Γ-distribution now follows from Theorem 4.10 and the functional convergence follows from Theorem 4.11. □

11.3.3 Tail Asymptotics for Non-Extinction Probability of Recurrent Branching Processes

Basic topics in the theory of critical and near-critical recurrent branching processes are the asymptotic behaviour of the non-extinction probability and the limiting behaviour of the process conditioned on non-extinction. Let us demonstrate that the corresponding results for general Markov chains – Theorem 8.18 and Corollary 8.22 – may be applied to near-critical branching processes. For that, we have to find restrictions on $\zeta(k)$, $\eta(k)$, and η which guarantee the fulfilment of (8.4)–(8.6) and (8.12), (8.13).

The hardest task, from the technical point of view, consists of finding a regular function $r(x)$ such that (8.5) holds. In what follows we concentrate on the case when one can take $r(x) = c/x$.

We first prove a refined version of Proposition 11.2 where we assume refined versions of the conditions (11.24)–(11.28) on the moments of the $\zeta(k)$ and $\eta(k)$. From now on we take $s(x) = x/\log^{1+\varepsilon} x$.

Proposition 11.8 *For some $\varepsilon > 0$, let*

$$\sup_{k\ge 1} \mathbb{E}\zeta^2(k)\log^{3+3\varepsilon}(1+\zeta(k)) < \infty, \tag{11.65}$$

let the majorisation condition (11.28) *hold with η satisfying*

$$\mathbb{E}\eta\log^{1+2\varepsilon}(1+\eta) < \infty, \tag{11.66}$$

and let there exist a decreasing function $v(x)$ that is integrable at infinity and is such that $xv(x^2)$ is decreasing too, and for which, as $k \to \infty$, the following hold:

$$\begin{aligned}
\mathbb{E}\zeta(k) &= 1 + a_\zeta/k + o(v(k)), &(11.67)\\
\mathbb{E}\eta(k) &= a_\eta + o(kv(k)), &(11.68)\\
\mathbb{V}\mathrm{ar}\,\zeta^2(k) &= \sigma^2 + o(kv(k)), &(11.69)\\
\mathbb{E}\{\zeta^3(k);\ \zeta(k) \le k\} &\le k^2 v(k). &(11.70)
\end{aligned}$$

Then, for $s(x) = x/\log^{1+\varepsilon} x$, there exists a decreasing integrable function $p(x)$ such that

$$\begin{aligned}
m_1^{[s(\sqrt{k})]}(\sqrt{k}) &= \frac{a_\zeta + a_\eta - \sigma^2/4}{2\sqrt{k}} + o(p(\sqrt{k})), &(11.71)\\
m_2^{[s(\sqrt{k})]}(\sqrt{k}) &= \sigma^2/4 + o(\sqrt{k}\,p(\sqrt{k})) \quad \textit{as } k \to \infty. &(11.72)
\end{aligned}$$

Proof Owing to the condition (11.65), the condition (11.30) is valid for any $t(k) \to \infty$. Take $t(k) = \sqrt{k}$. Then it follows from the upper bound (11.38) with $s(x) = x/\log^{1+\varepsilon} x$ and the condition (11.65) that

$$\begin{aligned}
&\mathbb{E}\{(S(k)-k)^2;\ |S(k)-k|>\sqrt{k}s(\sqrt{k})\}\\
&\quad\le C\Big(\log^{2+2\varepsilon}k+k\mathbb{E}\{\zeta^2(k);\ \zeta(k)>\sqrt{k}s(\sqrt{k})\}\Big)\\
&\quad\le C\Big(\log^{2+2\varepsilon}k+k\frac{\mathbb{E}\zeta^2(k)\log^{3+3\varepsilon}(1+\zeta(k))}{\log^{3+3\varepsilon}(1+\sqrt{k})}\Big)\\
&\quad= O(k/\log^{3+3\varepsilon}k)\quad\text{as } k\to\infty. \qquad (11.73)
\end{aligned}$$

Therefore

$$\begin{aligned}
&\mathbb{E}\{|S(k)-k|;\ |S(k)-k|>\sqrt{k}s(\sqrt{k})\}\\
&\quad\le\frac{\mathbb{E}\{(S(k)-k)^2;\ |S(k)-k|>\sqrt{k}s(\sqrt{k})\}}{\sqrt{k}s(\sqrt{k})}\\
&\quad= O(1/\log^{2+2\varepsilon}k)\text{ as } k\to\infty. \qquad (11.74)
\end{aligned}$$

Using a Taylor expansion we obtain

$$\xi(\sqrt{k})=\frac{S(k)-k+\eta(k)}{2\sqrt{k}}-\frac{1}{8}\frac{(S(k)-k+\eta(k))^2}{k\sqrt{k}}+\theta\frac{(S(k)-k+\eta(k))^3}{k^2\sqrt{k}}, \qquad (11.75)$$

where $\theta=\theta((S(k)-k+\eta(k))/\sqrt{k})$ is bounded on the event A_k defined in (11.32). Let us estimate the expectation of every term in (11.75).

Recalling from (11.33) that $A_k^c\subseteq\{|S(k)-k+\eta(k)|>\sqrt{k}s(\sqrt{k})\}$ for all sufficiently large k, we obtain

$$\begin{aligned}
&\left|\mathbb{E}\Big\{\frac{S(k)-k+\eta(k)}{\sqrt{k}};\ A_k\Big\}-\frac{a_\zeta+a_\eta}{\sqrt{k}}\right|\\
&\quad\le\left|\frac{\mathbb{E}S(k)-k-a_\zeta}{\sqrt{k}}\right|+\left|\frac{\mathbb{E}\eta(k)-a_\eta}{\sqrt{k}}\right|\\
&\qquad+\frac{1}{\sqrt{k}}\mathbb{E}\{|S(k)-k+\eta(k)|;\ |S(k)-k+\eta(k)|>\sqrt{k}s(\sqrt{k})\}.
\end{aligned}$$

The first two terms on the right-hand side are of order $o(\sqrt{k}v(k))$ by the conditions (11.67) and (11.68). Also taking into account the upper bounds (11.47) and (11.74) we derive

$$\begin{aligned}
&\left|\mathbb{E}\Big\{\frac{S(k)-k+\eta(k)}{\sqrt{k}};\ A_k\Big\}-\frac{a_\zeta+a_\eta}{\sqrt{k}}\right|\\
&\quad\le o(\sqrt{k}v(k))+O(1/\sqrt{k}\log^{2+2\varepsilon}k)\\
&\qquad+\frac{1}{\sqrt{k}}\Big(\mathbb{E}|S(k)-k|\mathbb{P}\{\eta>s^2(\sqrt{k})\}+\mathbb{E}\{\eta;\ \eta>s^2(\sqrt{k})\}\Big).
\end{aligned}$$

The assumption (11.66) implies that

$$\frac{1}{\sqrt{k}}\mathbb{E}\{\eta;\ \eta > s^2(\sqrt{k})\} = \frac{1}{\sqrt{k}}\mathbb{E}\{\eta;\ \eta > k/\log^{2+2\varepsilon}\sqrt{k}\}$$
$$= o(1/\sqrt{k}\log^{1+\varepsilon} k) \quad \text{as } k \to \infty,$$

hence, by the Markov inequality,

$$\mathbb{P}\{\eta > s^2(\sqrt{k})\} \le \mathbb{E}\{\eta;\ \eta > s^2(\sqrt{k})\}/s^2(\sqrt{k})$$
$$= o((\log^{1+\varepsilon} k)/k) \quad \text{as } k \to \infty,$$

Combining this with the upper bound $\mathbb{E}|S(k)-k| = O(\sqrt{k})$, we conclude that

$$\left|\mathbb{E}\left\{\frac{S(k)-k+\eta(k)}{\sqrt{k}};\ A_k\right\} - \frac{a_\zeta + a_\eta}{\sqrt{k}}\right| \le o(\sqrt{k}v(k)) + O(1/\sqrt{k}\log^{1+\varepsilon} k)$$
$$= o(p(\sqrt{k})), \tag{11.76}$$

where the function

$$p(x) := xv(x^2) + 1/x\log^{1+\varepsilon/2} x \tag{11.77}$$

is decreasing and integrable at infinity because $\varepsilon > 0$ and

$$\int_1^\infty xv(x^2)dx \;=\; \frac{1}{2}\int_1^\infty v(y)dy \;<\; \infty.$$

For the second term on the right-hand side of (11.75), we have

$$\left|\mathbb{E}\left\{\frac{(S(k)-k+\eta(k))^2}{k};\ A_k\right\} - \sigma^2\right|$$
$$\le \left|\frac{\mathbb{E}(S(k)-k)^2}{k} - \sigma^2\right| + \frac{\mathbb{E}\{(S(k)-k)^2;\ A_k^c\}}{k}$$
$$+2\left|\frac{\mathbb{E}\{(S(k)-k)\eta(k);\ A_k\}}{k}\right| + \frac{\mathbb{E}\{\eta^2(k);\ A_k\}}{k}$$
$$=: E_1 + E_2 + E_3 + E_4. \tag{11.78}$$

The first term on the right-hand side of (11.75) may be bounded as follows:

$$E_1 = \left|\mathbb{V}\mathrm{ar}\,\zeta(k) - \sigma^2 + \frac{(\mathbb{E}S(k)-k)^2}{k}\right| \le o(kv(k)) + O(1/k), \tag{11.79}$$

by the conditions (11.67) and (11.69). Using (11.33), we obtain

$$E_2 \le \frac{\mathbb{E}\{(S(k)-k)^2;\ |S(k)-k| > \sqrt{k}s(\sqrt{k})\}}{k}$$
$$+\frac{\mathbb{E}\{(S(k)-k)^2;\ |\eta(k)| > s^2(\sqrt{k})\}}{k}$$
$$\le O(1/\log^{3+3\varepsilon} k) + c_2\mathbb{P}\{\eta > s^2(\sqrt{k})\} \quad \text{as } k \to \infty,$$

by the upper bound (11.73) and the independence of $S(k)$ and $\eta(k)$. Therefore, by the condition (11.66),

$$E_2 = O(1/\log^{3+3\varepsilon} k). \tag{11.80}$$

By (11.41),

$$E_3 = O(1/\sqrt{k}) \quad \text{as } k \to \infty. \tag{11.81}$$

Finally, owing to the condition (11.66) we deduce, similarly to (11.43), that

$$E_4 = o(1/\log^{1+\varepsilon} k) \quad \text{as } k \to \infty. \tag{11.82}$$

Combining (11.78)–(11.82), we obtain

$$\frac{\mathbb{E}\{(S(k) - k + \eta(k))^2;\ A_k\}}{k\sqrt{k}} = \frac{\sigma^2}{\sqrt{k}} + o(p(\sqrt{k})), \tag{11.83}$$

where $p(x)$ is defined in (11.77).

As follows from the definition of A_k, see (11.32),

$$|S(k) - k + \eta(k)| \le 3\sqrt{k}s(\sqrt{k}) \quad \text{on the event } A_k.$$

Hence the remainder term in (11.75) possesses the following upper bound:

$$\begin{aligned} \mathbb{E}\{|S(k) - k + \eta(k)|^3;\ A_k\} &\le 3\sqrt{k}s(\sqrt{k})\mathbb{E}\{(S(k) - k + \eta(k))^2;\ A_k\} \\ &= O(k\sqrt{k}s(\sqrt{k})) \quad \text{as } k \to \infty, \end{aligned} \tag{11.84}$$

as follows from (11.83).

Combining (11.75), (11.76), (11.83), and (11.84), we conclude that

$$\mathbb{E}\{\xi(\sqrt{k});\ A_k\} = \frac{a_\zeta + a_\eta - \sigma^2/4}{2\sqrt{k}} + o(p(\sqrt{k})), \tag{11.85}$$

where $p(x)$ is defined in (11.77); so (11.71) is proven.

In order to prove (11.72) we first use a Taylor expansion for the function

$$(\sqrt{1+u} - 1)^2 = \frac{u^2}{4} - \frac{u^3}{6}\frac{3}{4(1+\theta_1 u)^{5/2}}, \quad \theta_1 \in (0,1),$$

to conclude that

$$\xi^2(\sqrt{k}) = \frac{(S(k) - k + \eta(k))^2}{4k} + \widetilde{\theta}\frac{(S(k) - k + \eta(k))^3}{k^2},$$

where $\widetilde{\theta} = \widetilde{\theta}(S(k), \eta(k))$ is bounded on the event A_k. Then we apply (11.83) and (11.84) to obtain (11.72). □

Under the conditions of Proposition 11.8 we have that, with $s(x) = x/\log^{1+\varepsilon} x$,

$$\frac{2m_1^{[s(x)]}(x)}{m_2^{[s(x)]}(x)} = -\frac{\sigma^2/4 - a_\zeta - a_\eta}{\sigma^2/4}\frac{1}{x} + o(p(x)) \quad \text{as } x \to \infty.$$

This means that (8.5) holds with

$$r(x) = \frac{\rho - 1}{1 + x},$$

where

$$\rho = \frac{\sigma^2/2 - a_\zeta - a_\eta}{\sigma^2/4}.$$

Theorem 11.9 *Assume that all the conditions of Proposition 11.8 are valid and that* $a_\zeta + a_\eta < \sigma^2/2$.

Assume that $\mathbb{E}\eta \log^{3+3\varepsilon}(1+\eta) < \infty$ *and* $\mathbb{E}\eta^{\rho/2} < \infty$ *and that*

$$\sup_{k\ge1} \mathbb{E}\zeta^{\rho/2}(k) < \infty. \tag{11.86}$$

Then, for each starting state z,

$$\mathbb{P}_z\{Z_k > z_* \text{ for all } k \le n\} \sim \frac{c(z)}{n^{\rho/2}} \quad \text{as } n \to \infty \tag{11.87}$$

and, for all $u > 0$,

$$\mathbb{P}_z\left\{\frac{2Z_n}{n\sigma^2} > u \;\middle|\; Z_k > z_* \text{ for all } k \le n\right\} \to e^{-u} \quad \text{as } n \to \infty, \tag{11.88}$$

where z_* *is the minimal accessible state of* $\{Z_n\}$.

It is easy to see that if $\mathbb{P}\{\zeta(k) = 0\} > 0$ and $\mathbb{P}\{\eta(k) \le 0\} > 0$ for all k then $z_* = 0$. Furthermore, if $\mathbb{P}\{\eta(k) \le 0\} = 1$ then 0 is an absorbing state and we have the following statements, typical for branching processes:

$$\mathbb{P}_z\{Z_n > 0\} \sim c(z)/n^{\rho/2} \quad \text{as } n \to \infty$$

and, for all $u > 0$,

$$\mathbb{P}_z\left\{\frac{2Z_n}{n\sigma^2} > u \;\middle|\; Z_n > 0\right\} \to e^{-u} \quad \text{as } n \to \infty.$$

Proof We again put $s(x) = x/\log^{1+\varepsilon} x$ and check the sufficient conditions for the results from Section 8.6. We start with the following auxiliary upper bound, for all $\rho > 0$:

$$\mathbb{E}\{(S(k)-\mathbb{E}S(k))^{\rho/2};\ S(k)-\mathbb{E}S(k)>\sqrt{k}s(\sqrt{k})\} = (\sqrt{k})^{\rho-1}o(q(\sqrt{k})) \tag{11.89}$$

as $k \to \infty$, for some decreasing function $q(x)$ that is integrable at infinity. By Lemma 2.38, for all $\rho > 0$, $\mathbb{E}|S(k)-\mathbb{E}S(k)|^{\rho/2} = O(k^{\rho/4})$ and hence $\mathbb{E}|S(k)-k|^{\rho/2} = O(k^{\rho/4})$. Then, by the condition (11.86),

$$\begin{aligned}
&\mathbb{E}\{(S(k)-\mathbb{E}S(k))^{\rho/2};\ S(k)-\mathbb{E}S(k)>\sqrt{k}s(\sqrt{k})\}\\
&\quad \le \mathbb{E}|S(k)-\mathbb{E}S(k)|^{\rho/2}\\
&\quad = (\sqrt{k})^{\rho-1}O(1/(\sqrt{k})^{\rho/2-1})\\
&\quad = (\sqrt{k})^{\rho-1}o(q_1(\sqrt{k})) \quad \text{as } k\to\infty,
\end{aligned}$$

for some decreasing function $q_1(x)$ that is integrable at infinity; the above relation holds provided that $\rho > 4$. If $\rho \in (0,4]$ then, by a Chebyshev-type inequality and by the upper bound (11.38),

$$\begin{aligned}
&\mathbb{E}\{(S(k)-\mathbb{E}S(k))^{\rho/2};\ S(k)-\mathbb{E}S(k)>\sqrt{k}s(\sqrt{k})\}\\
&\quad \le \frac{\mathbb{E}\{(S(k)-\mathbb{E}S(k))^{2};\ S(k)-\mathbb{E}S(k)>\sqrt{k}s(\sqrt{k})\}}{(\sqrt{k}s(\sqrt{k}))^{2-\rho/2}}\\
&\quad \le \frac{C}{(\sqrt{k}s(\sqrt{k}))^{2-\rho/2}}\left[\frac{1}{s^2(\sqrt{k})}+k\mathbb{E}\{\zeta^2(k);\ \zeta(k)>\sqrt{k}s(\sqrt{k})/2\}\right].
\end{aligned}$$

Applying the condition (11.65) we conclude that

$$\begin{aligned}
&\mathbb{E}\{(S(k)-\mathbb{E}S(k))^{\rho/2};\ S(k)-\mathbb{E}S(k)>\sqrt{k}s(\sqrt{k})\}\\
&\quad \le \frac{C_1}{(\sqrt{k}s(\sqrt{k}))^{2-\rho/2}}\frac{k}{\log^{3+3\varepsilon}\sqrt{k}}\\
&\quad = \frac{C_1}{k^{1-\rho/2}\log^{(1+\rho/2)(1+\varepsilon)}\sqrt{k}} \qquad (11.90)\\
&\quad = (\sqrt{k})^{\rho-1}o(q_2(\sqrt{k})) \quad \text{as } k\to\infty,
\end{aligned}$$

which completes the proof of (11.89) for all $\rho > 0$.

In Proposition 11.8 we checked the condition (8.5) for the chain $\{\sqrt{Z_n}\}$. The fulfilment of the condition (8.12) for the left tail was proven in Proposition 11.2. For the right tail, it is enough to notice that, by the Chebyshev inequality and by the upper bound (11.90) with $\rho = 2$, we have

$$\begin{aligned}
&\mathbb{P}\{S(k)-\mathbb{E}S(k)>\sqrt{k}s(\sqrt{k})\}\\
&\quad \le \frac{\mathbb{E}\{(S(k)-\mathbb{E}S(k))^{2};\ S(k)-\mathbb{E}S(k)>\sqrt{k}s(\sqrt{k})\}}{(\sqrt{k}s(\sqrt{k}))^{2}}\\
&\quad = O(1/k\log^{1+\varepsilon}k) \quad \text{as } k\to\infty.
\end{aligned}$$

It only remains to validate the conditions (8.11), (8.13), and (8.14) under the assumptions of Theorem 11.9.

Since $r(x) = \frac{\rho-1}{1+x}$, the function $U(x)$ is asymptotically equivalent to cx^ρ with some positive constant c. Thus, we can replace $U(x)$ by x^ρ in (8.11) and (8.14). In particular, (8.11) then follows from (8.14).

We start with (8.13). From the upper bound

$$|\xi(\sqrt{k})| \le \frac{|S(k) - k + \eta(k)|}{\sqrt{k}}$$

and (11.84) we conclude that

$$\mathbb{E}\{|\xi(\sqrt{k})|^3;\ |\xi(\sqrt{k})| \le s(\sqrt{k})\} = O(s(\sqrt{k})) \ = \ O((\sqrt{k})^2/\sqrt{k}\log^{1+\varepsilon} k).$$

This implies (8.13) with, say, $p(x) = 1/x \log^{1+\varepsilon/2} x$.

Let us now check the fulfilment of (8.14). First we note that, by the concavity of the root function,

$$\begin{aligned}
&\mathbb{E}\{\xi^\rho(\sqrt{k});\ \xi(\sqrt{k}) > s(\sqrt{k})\}\\
&\quad \le \mathbb{E}\{(S(k) - k + \eta(k))^{\rho/2};\ \xi(\sqrt{k}) > s(\sqrt{k})\}\\
&\quad \le \mathbb{E}\{(S(k) - k + \eta(k))^{\rho/2};\ S(k) + \eta(k) > (\sqrt{k} + s(\sqrt{k}))^2\}\\
&\quad \le \mathbb{E}\{(S(k) - k + \eta(k))^{\rho/2};\ S(k) - k > \sqrt{k}s(\sqrt{k})\}\\
&\qquad + \mathbb{E}\{(S(k) - k + \eta(k))^{\rho/2};\ \eta(k) > s^2(\sqrt{k})\}.
\end{aligned} \tag{11.91}$$

Owing to the independence of $S(n)$ and $\eta(k)$, the first expectation on the right-hand side is not greater, up to a constant factor, than the sum

$$\begin{aligned}
&\mathbb{E}\eta^{\rho/2}\mathbb{P}\{S(k) - \mathbb{E}S(k) > \sqrt{k}s(\sqrt{k})\}\\
&\quad + \mathbb{E}\{(S(k) - \mathbb{E}S(k))^{\rho/2};\ S(k) - \mathbb{E}S(k) > \sqrt{k}s(\sqrt{k})\}\\
&\qquad \le c\mathbb{E}\{(S(k) - \mathbb{E}S(k))^{\rho/2};\ S(k) - \mathbb{E}S(k) > \sqrt{k}s(\sqrt{k})\}, \quad c < \infty,
\end{aligned}$$

because of the condition $\mathbb{E}\eta^{\rho/2} < \infty$. Then it follows from (11.89) that, as $k \to \infty$,

$$\mathbb{E}\{(S(k) - k + \eta(k))^{\rho/2};\ S(k) - k > \sqrt{k}s(\sqrt{k})\} = (\sqrt{k})^{\rho-1}o(q(\sqrt{k})). \tag{11.92}$$

The second expectation on the right-hand side of (11.91) is not greater, up to a constant factor, than the sum

$$\mathbb{E}(S(k) - k)^{\rho/2}\mathbb{P}\{\eta > s^2(\sqrt{k})\} + \mathbb{E}\{\eta^{\rho/2};\ \eta > s^2(\sqrt{k})\},$$

owing to the independence of $S(n)$ and η. Again by Lemma 2.38,

$$\mathbb{E}|S(k) - k|^{\rho/2} = O((\sqrt{k})^{\rho/2}) \ = \ O((\sqrt{k})^{\rho-1}/(\sqrt{k})^{\rho/2-1}) \quad \text{for all } \rho > 0.$$

Also for all $\rho > 0$ we have

$$\mathbb{P}\{\eta > s^2(\sqrt{k})\} \le \frac{(\mathbb{E}\eta)\log^{2+2\varepsilon}\sqrt{k}}{k},$$

so that

$$\begin{aligned}\mathbb{E}(S(k)-k)^{\rho/2}\mathbb{P}\{\eta > s^2(\sqrt{k})\} &= (\sqrt{k})^{\rho-1}o\Big(\frac{\log^{2+2\varepsilon}\sqrt{k}}{(\sqrt{k})^{\rho/2+1}}\Big)\\ &= (\sqrt{k})^{\rho-1}o(q_3(\sqrt{k})) \quad \text{as } k\to\infty.\end{aligned}$$

If $\rho > 1$ then, owing to the condition $\mathbb{E}\eta^{\rho/2} < \infty$,

$$\begin{aligned}\mathbb{E}\{\eta^{\rho/2};\ \eta > s^2(\sqrt{k})\} = o(1) &= (\sqrt{k})^{\rho-1}o(1/(\sqrt{k})^{\rho})\\ &= (\sqrt{k})^{\rho-1}o(q_4(\sqrt{k})) \quad \text{as } k\to\infty.\end{aligned}$$

If $\rho \in (0,1]$ then, owing to the condition $\mathbb{E}\eta\log^{3+3\varepsilon}(1+\eta) < \infty$,

$$\begin{aligned}\mathbb{E}\{\eta^{\rho/2};\ \eta > s^2(\sqrt{k})\} &\le \frac{\mathbb{E}\{\eta\log^{3+3\varepsilon}\eta;\ \eta > s^2(\sqrt{k})\}}{(s^2(\sqrt{k}))^{1-\rho/2}\log^{3+3\varepsilon}s^2(\sqrt{k})}\\ &= (\sqrt{k})^{\rho-1}o\Big(\frac{\log^{(2-\rho)(1+\varepsilon)}\sqrt{k}}{\sqrt{k}\log^{3+3\varepsilon}(\sqrt{k})}\Big)\\ &= (\sqrt{k})^{\rho-1}o(q_5(\sqrt{k})) \quad \text{as } k\to\infty.\end{aligned}$$

Altogether this implies that

$$\mathbb{E}\{(S(k)+\eta(k))^{\rho/2};\ \eta(k) > s^2(\sqrt{k})\} = (\sqrt{k})^{\rho-1}o(q(\sqrt{k})). \tag{11.93}$$

Substituting (11.92) and (11.93) into (11.91) we arrive at (8.14). Relations (11.87) and (11.88) now follow from Corollaries 8.22 and 8.25. □

The processes $\{Z_n\}$ and $\{Y_n\}$ – defined in (11.22) and (11.23) respectively – are formally different. But it is intuitively clear that the difference in their definitions should have no influence on their asymptotic behaviour. Let us show how, in the case of an identically distributed $\zeta(k)$ and a non-positive η, one can transfer the asymptotics for one process into corresponding asymptotics for the other. Indeed, if we define

$$W_{2k+1} := (W_{2k}+\eta_{k+1})^+, \quad W_{2k+2} = \sum_{i=1}^{W_{2k+1}} \zeta_{k+1,i}, \quad k \ge 0,$$

then $Y_0 = W_0 = m$ implies that $Y_n = W_{2n}$ and $Z_n = W_{2n+1}$ with $Z_0 = (m+\eta_1)^+$. In the case of an emigration process – where $\mathbb{P}\{\eta \le 0\} = 1$ – we have that the sequence of events $\{W_k = 0\}$ is increasing. If (11.87) is valid for every fixed starting point Z_0 then it is also valid for $Z_0 = (m+\eta)^+$. As a result, we have

$$\mathbb{P}\{Y_n > 0 \mid Y_0 = m\} \sim \sum_{j=1}^{m} \mathbb{P}\{m+\eta = j\}\mathbb{P}\{Z_n > 0 \mid Z_0 = j\}$$
$$\sim c(m)n^{-\rho/2}.$$

Furthermore, let $\{Z_n = W_{2n+1}\}$ satisfy the conditions of Theorem 11.9. Then it follows that

$$\mathbb{P}\{Z_n \le k \mid Z_n > 0\} \to 0 \quad \text{as } n \to \infty, \text{ for all } k > 0. \tag{11.94}$$

Recalling that $Y_n = \sum_{i=1}^{Z_{n-1}} \zeta_{n,i}$, (11.94) implies the following version of the weak law of large numbers: for all $\varepsilon > 0$,

$$\mathbb{P}\left\{\left|\frac{Y_n}{Z_{n-1}} - 1\right| > \varepsilon \;\middle|\; Z_{n-1} > 0\right\} \to 0 \quad \text{as } n \to \infty. \tag{11.95}$$

This yields, owing to (11.88),

$$\mathbb{P}\left\{\frac{2Y_n}{n\sigma^2} > u \;\middle|\; Z_{n-1} > 0\right\} \sim \mathbb{P}_z\left\{\frac{2Z_n}{n\sigma^2} > u \;\middle|\; Z_{n-1} > 0\right\}, \quad u > 0.$$

We also have the inequalities

$$\mathbb{P}\{Z_{n-1} > 0\} \ge \mathbb{P}\{Y_n > 0\} \ge \mathbb{E}[1 - \mathbb{P}^{Z_{n-1}}\{\zeta = 0\}].$$

Combining them with (11.95) we conclude that

$$\mathbb{P}\{Z_{n-1} > 0\} \sim \mathbb{P}\{Y_n > 0\} \quad \text{as } n \to \infty.$$

Therefore

$$\mathbb{P}\left\{\frac{2Y_n}{n\sigma^2} > u \;\middle|\; Y_n > 0\right\} \to e^{-u}, \quad u > 0.$$

If $\inf_k \mathbb{P}\{\eta(k) > 0\} > 0$ then 0 is non-absorbing and, consequently, $\{Z_n\}$ is irreducible. Then we can apply Theorem 8.2 to $\sqrt{Z_n}$ and derive the tail behaviour of the stationary measure of $\{Z_n\}$: for any constants $a < b$ we have

$$\pi_Z(ak, bk) \sim C \int_{\sqrt{ak}}^{\sqrt{bk}} y^{1-\rho} dy \quad \text{as } k \to \infty.$$

It follows from Theorem 11.9 that $\{Z_n\}$ is positive recurrent when $\rho > 2$. In this case we may also apply Theorem 8.17 and obtain tail asymptotics for Z_n.

If $\rho \in (0,2)$ then the pre-limiting behaviour of Z_n is described in Theorem 4.10. If $\rho = 2$ then, by (11.87), $\{Z_n\}$ is also null recurrent but its behaviour is not covered by Theorem 4.10. Here we can apply Theorem 8.29. Since $G(x) \sim \log x$ under the assumptions of Theorem 11.9, we conclude that

$$\lim_{n\to\infty} \mathbb{P}\left\{\frac{\log Z_n}{\log n} \le x\right\} = x, \quad x \in [0,1]. \tag{11.96}$$

11.4 Cramér–Lundberg Risk Processes with Level-Dependent Premium Rate

In the context of the collective theory of risk, the classical *Cramér–Lundberg model* (or *Sparre Andersen model*) is defined as follows. An insurance company receives a constant inflow of premium at rate v, that is, the premium income is assumed to be linear in time with the rate v. It is also assumed that the claims incurred by the insurance company arrive according to a homogeneous renewal process $N(t)$ with intensity λ and that the sizes (amounts) $\xi_n \ge 0$ of the claims are independent copies of a random variable ξ with finite mean. The ξ are assumed to be independent of the process $N(t)$. The company has an initial risk reserve $x = R(0) \ge 0$. Then the risk reserve $R(t)$ at time t is equal to

$$R(t) = x + vt - \sum_{i=1}^{N(t)} \xi_i.$$

The quantity

$$\mathbb{P}\{R(t) \ge 0 \text{ for all } t \ge 0\} = \mathbb{P}\Big\{\min_{t\ge 0} R(t) \ge 0\Big\}$$

gives the probability of ultimate survival and

$$\begin{aligned}\psi(x) &:= \mathbb{P}\{R(t) < 0 \text{ for some } t \ge 0\}\\ &= \mathbb{P}\Big\{\min_{t\ge 0} R(t) < 0\Big\}\end{aligned}$$

is the probability of ruin. We have

$$\psi(x) = \mathbb{P}\Big\{\sum_{i=1}^{N(t)} \xi_i - vt > x \text{ for some } t \ge 0\Big\}.$$

Since $v > 0$, ruin can only occur during a claim epoch. Therefore,

$$\psi(x) = \mathbb{P}\Big\{\sum_{i=1}^{n} \xi_i - vT_n > x \text{ for some } n \ge 1\Big\},$$

where T_n is the nth claim epoch, so that $T_n = \tau_1 + \cdots + \tau_n$ where the τ_k are independent copies of a random variable τ with finite mean $1/\lambda$, so that $N(t) := \max\{n \geq 1 : T_n \leq t\}$. Denote $X_i := \xi_i - v\tau_i$ and $S_n := X_1 + \cdots + X_n$; then

$$\psi(x) = \mathbb{P}\Big\{\sup_{n\geq 1} S_n > x\Big\}.$$

This relation represents the ruin-probability problem as a tail-probability problem for the maximum of the associated random walk $\{S_n\}$. Let the *net-profit condition*

$$v > v_c := \mathbb{E}\xi/\mathbb{E}\tau = \lambda\mathbb{E}\xi$$

hold; thus $\{S_n\}$ has a negative drift, $\mathbb{E}S_1 = \mathbb{E}\xi_1 - v\mathbb{E}\tau < 0$. Hence, by the strong law of large numbers, $S_n \to -\infty$ a.s. so $\psi(x) \downarrow 0$ as $x \to \infty$.

If $v \leq v_c$ then $\psi(x) = 1$ for all x.

The classic case occurs when the distribution of X_1 satisfies the following well-known Cramér condition: there exists a $\beta > 0$ such that

$$\mathbb{E}e^{\beta X_1} = 1. \tag{11.97}$$

Under this condition, the sequence $e^{\beta S_n}$ is a martingale and, by the Doob maximal inequality, the following Lundberg inequality holds true:

$$\psi(x) = \mathbb{P}\left\{\sup_{n\geq 1} e^{\beta S_n} > e^{\beta x}\right\} \leq e^{-\beta x}, \quad x > 0. \tag{11.98}$$

If we additionally assume that

$$\mathbb{E}X_1 e^{\beta X_1} < \infty$$

and that the distribution of X_1 is non-lattice then the Cramér–Lundberg approximation holds; that is, there exists a constant $c_0 \in (0, 1)$ such that

$$\psi(x) \sim c_0 e^{-\beta x} \quad \text{as } x \to \infty, \tag{11.99}$$

see e.g. Theorem VI.3.2 in Albrecher and Asmussen [2]; in the lattice case x must be taken as a multiple of the lattice step. The most important feature of these results is that the upper bound (11.98) depends on the distribution of X_1 only via the parameter β. If the moment condition (11.97) on the distribution of X_1 does not hold then the tail asymptotics for $\psi(x)$ is typically determined by the tail of the claim size ξ. The most prominent situation occurs when the distribution of ξ is of subexponential type; see the discussion on the maximum of random walk in Section 1.3.

Risk models with non-constant premium rates have also become rather popular in the collective risk literature. There are two main approaches.

One approach leads to a Markovian model when the premium rate is a function of the current level of the risk reserve $R(t)$; see e.g. Albrecher and Asmussen [2, Chapter VIII], Albrecher, Constantinescu, Palmowski, Regensburger, and Rosenkranz [3], Boxma and Mandjes [26], Czarna, Pérez, Rolski, and Yamazaki [40], and Marciniak and Palmowski [120]. The second approach considers a premium rate that depends on the whole claims history; see e.g. Li, Ni, and Constantinescu [117].

In this section we follow the first approach and consider a risk process where the premium rate $v(y)$ depends on the current level of risk reserve $R(t) = y$, so that $R(t)$ satisfies the equality

$$R(t) = x + \int_0^t v(R(s))ds - \sum_{j=1}^{N(t)} \xi_j; \tag{11.100}$$

$v(y)$ is assumed to be a bounded càdlàg function, which is bounded away from zero on each interval; given that v is a càdlàd function, there are at the most countably many discontinuity points of v, which together with the boundedness of $1/v$ on any interval implies by the Lebesgue–Vitali theorem that the function $1/v$ is Riemann integrable. The probability of ruin given an initial risk reserve x is again denoted by $\psi(x)$; it is a decreasing function of x as in the classic case.

Since ruin can occur only at a claim epoch, the ruin probability may be reduced to that for the embedded Markov chain $R_n := R(T_n), n \geq 1, R_0 := x$; that is,

$$\psi(x) = \mathbb{P}\{R_n < 0 \text{ for some } n \geq 0\}.$$

In this section we consider the case where $v(y)$ approaches the critical value v_c at infinity, that is,

$$v(y) \to v_c \quad \text{as } y \to \infty. \tag{11.101}$$

Then the Markov chain $\{R_n\}$ has asymptotically zero drift and, as follows from Theorem 3.1, the ruin probability decays more slowly than any exponential function; that is, for any $\lambda > 0$,

$$e^{\lambda x}\psi(x) \to \infty \quad \text{as } x \to \infty.$$

The main goal in this section is to investigate how the rate of convergence in (11.101) is reflected in how quickly the ruin probability $\psi(x)$ becomes vanishing for large x. Let us get some intuition on what kind of phenomena we could expect here by considering a model where $\psi(x)$ is known in closed form.

To the best of our knowledge, the only case where $\psi(x)$ is explicitly calculable is the case of exponentially distributed τ and ξ, say, with parameters λ and μ respectively, so that $v_c = \lambda/\mu$. In this case, for some $c_0 \in (0, 1)$,

$$\psi(x) = c_0 \int_x^\infty \frac{1}{v(y)} \exp\Big\{-\mu y + \lambda \int_0^y \frac{dz}{v(z)}\Big\} dy$$
$$= c_0 \int_x^\infty \frac{1}{v(y)} \exp\Big\{\lambda \int_0^y \Big(\frac{1}{v(z)} - \frac{1}{v_c}\Big) dz\Big\} dy, \qquad (11.102)$$

provided that the outer integral is convergent from 0 to infinity; see e.g. Corollary 1.9 in Albrecher and Asmussen [2, Chapter VIII]. Then, by (11.101),

$$\psi(x) \sim \frac{c_0}{v_c} \int_x^\infty \exp\Big\{\lambda \int_0^y \Big(\frac{1}{v(z)} - \frac{1}{v_c}\Big) dz\Big\} dy \quad \text{as } x \to \infty.$$

If the premium rate $v(z) \geq v_c$ approaches v_c at the rate θ/z, $\theta > 0$, more precisely, if

$$\Big|v(z) - v_c - \frac{\theta}{z}\Big| \leq p(z) \quad \text{for all } z > 1, \qquad (11.103)$$

where $p(z) > 0$ is a decreasing function that is integrable at infinity, then we get

$$\frac{1}{v(z)} = \frac{1}{v_c} - \frac{\theta}{v_c^2 z} + O(p(z) + z^{-2})$$

and consequently

$$\lambda \int_0^y \Big(\frac{1}{v(z)} - \frac{1}{v_c}\Big) dz = -\frac{\theta \mu^2}{\lambda} \log y + c_1 + o(1) \quad \text{as } y \to \infty,$$

where c_1 is a finite real number. Let $\theta > \lambda/\mu^2$. Then, for $C := c_0 e^{c_1} / (\theta\mu - \lambda/\mu) > 0$, we have

$$\psi(x) \sim \frac{C}{x^{\theta\mu^2/\lambda - 1}} \quad \text{as } x \to \infty. \qquad (11.104)$$

A similar asymptotic expression can be obtained in the case where the Laplace transforms of variables ξ_1 and τ_1 are rational functions; see Albrecher, Constantinescu, Palmowski, Regensburger, and Rosenkranz [3].

If the premium rate $v(z)$ approaches v_c at the rate θ/z^α, $\theta > 0$, and $\alpha \in (0, 1)$, or, more precisely, if

$$\Big|v(z) - v_c - \frac{\theta}{z^\alpha}\Big| \leq p(z) \quad \text{for all } z > 1, \qquad (11.105)$$

where $p(z) > 0$ is decreasing function that is integrable at infinity, then we get

$$\frac{1}{v(z)} = \frac{1}{v_c} \sum_{j=0}^\infty \Big(-\frac{\theta}{v_c}\Big)^j \frac{1}{z^{\alpha j}} + O(p(z)).$$

Let $\gamma := \min\{k \in \mathbb{N} : k\alpha > 1\}$. Then

$$\frac{1}{v(z)} = \frac{1}{v_c}\sum_{j=0}^{\gamma-1}\Big(-\frac{\theta}{v_c}\Big)^j \frac{1}{z^{\alpha j}} + O(p_1(z)),$$

where $p_1(z) = p(z) + z^{-\gamma\alpha}$ is integrable at infinity. Consequently, if $1/\alpha$ is not an integer then

$$\begin{aligned}\lambda\int_0^y \Big(\frac{1}{v(z)} - \frac{1}{v_c}\Big)dz &= \frac{\lambda}{v_c}\int_1^y \sum_{j=1}^{\gamma-1}\Big(-\frac{\theta}{v_c}\Big)^j \frac{1}{z^{\alpha j}}dz + c_2 + o(1)\\ &= \frac{\lambda}{v_c}\sum_{j=1}^{\gamma-1}\Big(-\frac{\theta}{v_c}\Big)^j \frac{y^{1-\alpha j}}{1-\alpha j} + c_3 + o(1) \quad \text{as } y \to \infty,\end{aligned}$$

where c_3 is a finite number because $p_1(x)$ is integrable. In the case where $1/\alpha$ is an integer, we have

$$\begin{aligned}&\lambda\int_0^y \Big(\frac{1}{v(z)} - \frac{1}{v_c}\Big)dz\\ &= \frac{\lambda}{v_c}\sum_{j=1}^{\gamma-2}\Big(-\frac{\theta}{v_c}\Big)^j \frac{y^{1-\alpha j}}{1-\alpha j} + \frac{\lambda}{v_c}\Big(-\frac{\theta}{v_c}\Big)^{\gamma-1}\log y + c_4 + o(1) \quad \text{as } y \to \infty.\end{aligned}$$

For example, let $\alpha \in (1/2, 1)$. Then

$$\lambda\int_0^y \Big(\frac{1}{v(z)} - \frac{1}{v_c}\Big)dz = -\frac{\theta\mu^2}{\lambda(1-\alpha)}y^{1-\alpha} + c_3 + o(1) \quad \text{as } y \to \infty.$$

Therefore, for $C_1 := c_0 e^{c_3}/\theta\mu > 0$ and $C_2 := \theta\mu^2/\lambda(1-\alpha) > 0$, it follows that

$$\psi(x) \sim C_1 x^\alpha e^{-C_2 x^{1-\alpha}} \quad \text{as } x \to \infty. \tag{11.106}$$

Let us extend these results to distributions that are not necessarily exponential, for which there are no closed-form expressions like (11.102) available for $\psi(x)$. In that case we can derive only lower and upper bounds for $\psi(x)$ that have the same decay rate at infinity.

11.4.1 Approaching Critical Premium Rate at Rate of θ/x

Denote the jumps of the embedded Markov chain $\{R_n = R(T_n)\}$ by $\xi(x)$ and its kth truncated moment by $m_k^{[s(x)]}(x)$.

To avoid the trivial case where $\psi(x) = 0$ for all sufficiently large x, we assume that

$$\psi(x) > 0 \quad \text{for all } x. \tag{11.107}$$

A sufficient condition for this is that, for all $x_0 > 0$ there exists an $\varepsilon = \varepsilon(x_0) > 0$ such that

$$\mathbb{P}\{\xi(x) \le -\varepsilon\} > \varepsilon \quad \text{for all } x \in [0, x_0].$$

In turn, for the above inequality to hold it suffices to assume that the random variable ζ is unbounded owing to the inequality $\xi(y) \le \overline{v}\tau - \zeta$, which is valid for all y, where $\overline{v} := \sup_{z>0} v(z)$.

Theorem 11.10 *Assume* (11.107) *and that both* $\mathbb{E}\xi^2$ *and* $\mathbb{E}\tau^2$ *are finite. If*

$$\theta > \frac{\operatorname{Var}\xi + v_c^2 \operatorname{Var}\tau}{2\mathbb{E}\tau},$$

then R_n *is transient in the sense that* $\psi(x) < 1$ *for all sufficiently large* x. *Set*

$$\rho = \theta \frac{2\mathbb{E}\tau}{\operatorname{Var}\xi + v_c^2 \operatorname{Var}\tau} - 1 > 0.$$

If both $\mathbb{E}\tau^2 \log(1+\tau)$ *and* $\mathbb{E}\xi^{\rho+2}$ *are finite then there exist positive constants* c_1 *and* c_2 *such that*

$$\frac{c_1}{(1+x)^\rho} \le \psi(x) \le \frac{c_2}{(1+x)^\rho} \quad \text{for all } x > 0.$$

To prove this result we first need to establish some truncated-moment relations.

Proposition 11.11 *Assume the rate of convergence* (11.103) *and that both* $\mathbb{E}\tau_1^2$ *and* $\mathbb{E}\xi_1^2$ *are finite. Then there exists an increasing function* $s(x) = o(x)$ *such that*

$$\frac{2m_1^{[s(x)]}(x)}{m_2^{[s(x)]}(x)} \ge \frac{\rho+1}{x} + o(p_1(x)) \quad \text{as } x \to \infty,$$

for some decreasing integrable function $p_1(x)$, *and where*

$$\rho := \frac{2\theta\mathbb{E}\tau}{\operatorname{Var}\xi + v_c^2 \operatorname{Var}\tau} - 1.$$

If, in addition, both $\mathbb{E}\tau^2 \log(1+\tau)$ *and* $\mathbb{E}\xi^2 \log(1+\xi)$ *are finite, then there exists an increasing function* $s(x) = o(x)$ *such that*

$$\frac{2m_1^{[s(x)]}(x)}{m_2^{[s(x)]}(x)} = \frac{\rho+1}{x} + o(p_2(x)) \quad \text{as } x \to \infty,$$

for some decreasing integrable function $p_2(x)$.

Proof The dynamics of the risk reserve between two consecutive claims is governed by the differential equation $R'(t) = v(R(t))$ where by R' we mean the right derivative of V. This equation is solvable in R because the function $1/v$ is Riemann integrable owing to the boundedness of $1/v$ and its right continuity; let $V_x(t)$ denote a solution with initial value x, so that

$$V_x(t) = x + \int_0^t v(V_x(s))ds.$$

By (11.103),

$$\begin{aligned} v(y) &\le v_c + \theta/y + p(y) \\ &\le v_c + \theta/x + p(x) \quad \text{for all } y \ge x; \end{aligned}$$

therefore

$$V_x(t) - x \le v_c t + \theta t/x + p(x)t, \quad t > 0. \tag{11.108}$$

On the other hand, again by (11.103),

$$\begin{aligned} v(y) &\ge v_c + \theta/y - p(y) \\ &\ge v_c + \theta/y - p(x) \quad \text{for all } y \ge x, \end{aligned}$$

Hence

$$\begin{aligned} V_x(t) - x &\ge v_c t + \theta \int_0^t \frac{ds}{V_x(s)} - p(x)t \\ &\ge v_c t + \theta \int_0^t \frac{ds}{x + (v_c + \theta/x + p(x))s} - p(x)t \\ &= v_c t + \frac{\theta}{v_c + \theta/x + p(x)} \log\big(1 + (v_c + \theta/x + p(x))t/x\big) - p(x)t, \end{aligned}$$

where the second inequality follows from the upper bound (11.108). Therefore,

$$V_x(t) - x \ge v_c t + \frac{\theta}{v_c + \theta/x + p(x)} \log\big(1 + v_c t/x\big) - p(x)t. \tag{11.109}$$

Since $\xi(x) = V_x(\tau) - x - \xi$, it follows from (11.108) and (11.109) that

$$\begin{aligned} v_c\tau - \xi + \frac{\theta}{v_c + \theta/x + p(x)} \log\left(1 + \frac{v_c\tau}{x}\right) - p(x)\tau \\ \le \xi(x) \le v_c\tau - \xi + \frac{\theta\tau}{x} + p(x)\tau. \end{aligned} \tag{11.110}$$

Recalling that $v_c = \mathbb{E}\xi/\mathbb{E}\tau$, we get

$$\frac{\theta}{v_c + \theta/x + p(x)} \mathbb{E} \log\left(1 + \frac{v_c\tau}{x}\right) - p(x)\mathbb{E}\tau \le m_1(x) \le \frac{\theta}{x}\mathbb{E}\tau + p(x)\mathbb{E}\tau.$$

By the inequality $\log(1+z) \geq z - z^2/2$ for $z \geq 0$, we have

$$\mathbb{E}\log\Big(1+\frac{v_c\tau}{x}\Big) \geq \frac{v_c\mathbb{E}\tau}{x} - \frac{v_c^2\mathbb{E}\tau^2}{2x^2}.$$

Therefore

$$m_1(x) = \frac{\theta\mathbb{E}\tau}{x} + O(p(x)+1/x^2) \quad \text{as } x\to\infty. \tag{11.111}$$

From this expression we have

$$\begin{aligned} m_2(x) &= \mathbb{V}\mathrm{ar}\,\xi(x) + m_1^2(x) \\ &= \mathbb{V}\mathrm{ar}\,(V_x(\tau)-x-\xi) + O(p^2(x)+1/x^2) \\ &= \mathbb{V}\mathrm{ar}\,(V_x(\tau)-x) + \mathbb{V}\mathrm{ar}\,\xi + O(p^2(x)+1/x^2) \quad \text{as } x\to\infty. \end{aligned}$$

Recalling that

$$v_c t - p(x)t \leq V_x(t) - x \;\leq\; v_c t + \frac{\theta}{x}t + p(x)t,$$

we get

$$(v_c - p(x))\mathbb{E}\tau \leq \mathbb{E}(V_x(\tau)-x) \;\leq\; (v_c + \theta/x + p(x))\mathbb{E}\tau$$

and

$$(v_c - p(x))^2\mathbb{E}\tau^2 \leq \mathbb{E}(V_x(\tau)-x)^2 \;\leq\; (v_c + \theta/x + p(x))^2\mathbb{E}\tau^2.$$

Hence

$$\mathbb{V}\mathrm{ar}\,(V_x(\tau)-x) = v_c^2\mathbb{V}\mathrm{ar}\,\tau + O(1/x) \quad \text{as } x\to\infty,$$

which in its turn implies

$$m_2(x) = \mathbb{V}\mathrm{ar}\,\xi + v_c^2\mathbb{V}\mathrm{ar}\,\tau + O(1/x) \quad \text{as } x\to\infty. \tag{11.112}$$

Together with (11.111) this yields

$$\frac{2m_1(x)}{m_2(x)} = \frac{2\theta\mathbb{E}\tau}{\mathbb{V}\mathrm{ar}\,\xi + v_c^2\mathbb{V}\mathrm{ar}\,\tau}\,\frac{1}{x} + O(p(x)+1/x^2) \quad \text{as } x\to\infty.$$

Recall that we need such an expansion for the truncated moments. For any truncation level $s(x)$ we have

$$\begin{aligned} &|V_x(\tau)-x-\xi|\mathbb{I}\{|V_x(\tau)-x-\xi|>s(x)\} \\ &\quad\leq (V_x(\tau)-x+\xi)\mathbb{I}\{V_x(\tau)-x>s(x) \text{ or } \xi>s(x)\} \\ &\quad\leq (V_x(\tau)-x)\mathbb{I}\{V_x(\tau)-x>s(x)\} + \xi\mathbb{I}\{\xi>s(x)\} \\ &\qquad + \xi\mathbb{I}\{V_x(\tau)-x>s(x)\} + (V_x(\tau)-x)\mathbb{I}\{\xi>s(x)\}. \end{aligned} \tag{11.113}$$

Since $V_x(t) - x \le \overline{v}t$, where $\overline{v} = \sup_z v(z) < \infty$, we get

$$\begin{aligned}|m_1(x) - m_1^{[s(x)]}| &\le \mathbb{E}\{|V_x(\tau) - x - \xi|;\ |V_x(\tau) - x - \xi| > s(x)\}\\ &\le \overline{v}\,\mathbb{E}\{\tau;\ \tau > s(x)/\overline{v}\} + \mathbb{E}\{\xi; \xi > s(x)\}\\ &\quad + \mathbb{E}\xi\,\mathbb{P}\{\tau > s(x)/\overline{v}\} + \overline{v}\,\mathbb{E}\tau\mathbb{P}\{\xi > s(x)\}.\end{aligned}$$

It follows from the finiteness of $\mathbb{E}\tau^2$ and $\mathbb{E}\xi^2$ that there exists an increasing function $s_1(x) = o(x)$ such that both $\mathbb{E}\{\tau;\ \tau > s_1(x)/\overline{v}\}$ and $\mathbb{E}\{\xi;\ \xi > s_1(x)\}$ are integrable; see Lemma 2.33. Consequently, $|m_1(x) - m_1^{[s_1(x)]}(x)|$ is bounded by a decreasing integrable function. Combining this with (11.111), we conclude that

$$m_1^{[s_1(x)]}(x) = \frac{\theta\mathbb{E}\tau}{x} + o(p_2(x)) \quad \text{as } x \to \infty, \tag{11.114}$$

where p_2 is a decreasing integrable function.

It follows from (11.112) and (11.114) that

$$\frac{2m_1^{[s(x)]}(x)}{m_2^{[s(x)]}(x)} \ge \frac{2m_1^{[s(x)]}(x)}{m_2(x)} \ge \frac{1+\rho}{x} + o(p_3(x)) \quad \text{as } x \to \infty,$$

and the first result follows.

Similarly to (11.113),

$$\begin{aligned}&(V_x(\tau) - x - \xi)^2\mathbb{I}\{|V_x(\tau) - x - \xi| > s(x)\}\\ &\quad \le 2[(V_x(\tau) - x)^2 + \xi^2]\mathbb{I}\{V_x(\tau) - x > s(x) \text{ or } \xi > s(x)\}\\ &\quad \le 2(V_x(\tau) - x)^2\mathbb{I}\{V_x(\tau) - x > s(x)\} + 2\xi^2\mathbb{I}\{\xi > s(x)\}\\ &\qquad + 2\xi^2\mathbb{I}\{V_x(\tau) - x > s(x)\} + 2(V_x(\tau) - x)^2\mathbb{I}\{\xi > s(x)\}.\end{aligned}$$

Then, owing to the upper bound $V_x(t) - x \le \overline{v}t$, for some $c_2 < \infty$ we have

$$\begin{aligned}0 \le m_2(x) - m_2^{[s(x)]}(x) &= \mathbb{E}\{(V_x(\tau) - x - \xi)^2;\ |V_x(\tau) - x - \xi| > s(x)\}\\ &\le c_2\Big(\mathbb{E}\{\tau^2;\ \tau > s(x)/\overline{v}\} + \mathbb{E}\{\xi^2;\ \xi > s(x)\}\\ &\quad + \mathbb{E}\xi^2\mathbb{P}\{\tau > s(x)/\overline{v}\} + \mathbb{E}\tau^2\mathbb{P}\{\xi > s(x)\}\Big).\end{aligned}$$

It follows from the finiteness of $\mathbb{E}\xi^2\log(1+\xi)$ and $\mathbb{E}\tau^2\log(1+\tau)$ that there exists an increasing function $s_2(x) = o(x)$ such that both $x^{-1}\mathbb{E}\{\tau^2;\ \tau > s_2(x)/\overline{v}\}$ and $x^{-1}\mathbb{E}\{\xi^2;\ \xi > s_2(x)\}$ are integrable at infinity; see Lemma 2.36. Then $(m_2(x) - m_2^{[s_2(x)]}(x))/x$ is integrable too. From this fact and (11.112) we get

$$m_2^{[s_2(x)]}(x) = \mathbb{V}\mathrm{ar}\,\xi + v_c^2\,\mathbb{V}\mathrm{ar}\,\tau + o(xp_4(x)) \quad \text{as } x \to \infty, \tag{11.115}$$

for some decreasing function $p_4(x)$ that is integrable at infinity. Now taking $s(x) = \max(s_1(x), s_2(x)) = o(x)$, we conclude the desired result from (11.114) and (11.115). □

Proof of Theorem 11.10. By Proposition 11.11,

$$\frac{2m_1^{[s(x)]}(x)}{m_2^{[s(x)]}(x)} \geq \frac{1+\varepsilon}{x}$$

for some small ε and for all $x \geq x_0(\varepsilon)$. Furthermore, from the elementary bound $\mathbb{P}\{\xi(x) < -s(x)\} \leq \mathbb{P}\{\xi > s(x)\}$ and the finiteness of $\mathbb{E}\xi^2$, we infer that, for some increasing function $s(x) = o(x)$,

$$\mathbb{P}\{\xi(x) < -s(x)\} \leq p(x)/x,$$

where $p(x)$ is a decreasing function that is integrable at infinity; see Lemma 2.33. In addition, there exists an x_0 sufficiently large that the Markov chain $\{R_n\}$ dominates, above the level x_0, a similar Markov chain generated by a risk process with constant premium rate v_c. The latter represents a zero-drift random walk which is null recurrent and hence satisfies the condition (2.100). Thus, all the conditions of Theorem 2.21 are valid and, consequently,

$$\mathbb{P}_x\{R_n > x_0 \text{ for all } n\} \to 1 \quad \text{as } x \to \infty,$$

which implies the first claim of the theorem.

To prove the second claim of the theorem, let us first show that all the conditions of Theorem 3.2 hold true. The conditions (3.9)–(3.11) are valid for $\{R_n\}$ with $r(x) = (\rho+1)/(x+1)$, as follows from Proposition 11.11. For this $r(x)$ we have $U(x) = 1/\rho(x+1)^\rho$ for $x > 0$ and $U(x) = 1/\rho$ for $x \leq 0$. The condition (3.14) on the right tail of $\xi(x)$ holds because

$$\begin{aligned}\mathbb{P}\{\xi(x) > s(x)\} &\leq \mathbb{P}\{V_x(\tau) - x > s(x)\} \\ &\leq \mathbb{P}\{\tau > s(x)/\overline{v}\} = o(p(x)/x) \quad \text{as } x \to \infty,\end{aligned}$$

owing to the assumption that $\mathbb{E}\tau^2 < \infty$, see Lemma 2.33, and to the relation $U(x) \sim xe^{-R(x)}/\rho$. By the same argument, the condition (3.15) holds because

$$\begin{aligned}\mathbb{E}\{U(x+\xi(x));\ \xi(x) < -s(x)\} &\leq c\mathbb{P}\{\xi(x) < -s(x)\} \\ &\leq c\mathbb{P}\{\xi > s(x)\} \\ &= o(p(x)/x^{\rho+1}) \quad \text{as } x \to \infty,\end{aligned}$$

owing to the assumption $\mathbb{E}\xi^{\rho+2} < \infty$, again by Lemma 2.33. Obviously,

$$|\xi(x)| \leq V_x(\tau) - x + \xi \leq c_1\tau + \xi =: \Xi,$$

where Ξ is square integrable, so, by Lemma 2.27 with $\alpha = 1$ and $\gamma = 2$, the condition (3.13) on the third truncated moment is also met for $\{R_n\}$.

Hence, Theorem 3.2 applies; thus we conclude the following lower bound, for some $\widehat{x} \ge 0$ and $c > 0$:

$$\mathbb{P}_x\{R_n \le \widehat{x} \text{ for some } n\} \ge cU(x) \quad \text{for all } x \ge \widehat{x}.$$

Hence the second claim of the theorem follows, because by the strong Markov property, for all $x > 0$,

$$\begin{aligned} \psi(x) &= \mathbb{P}_x\{R_n < 0 \text{ for some } n\} \ge \mathbb{P}_x\{R_n \le \widehat{x} \text{ for some } n\} \inf_{y\in[0,\widehat{x}]} \psi(y) \\ &\ge cU(x)\psi(\widehat{x}), \end{aligned} \tag{11.116}$$

since $\psi(x)$ is decreasing; here $\psi(x_\lambda) > 0$ owing to the condition (11.107). □

11.4.2 Approaching Critical Premium Rate at the Rate of θ/x^α

In this subsection we consider the case (11.105) with some $\alpha \in (0, 1)$. Define

$$\gamma := \min\{k \ge 1 : \alpha k > 1\}.$$

The main result in this subsection is as follows.

Theorem 11.12 *Assume* (11.107) *and the rate of convergence* (11.105). *Let* $\mathbb{E}\tau^{\gamma+1} < \infty$ *and* $\mathbb{E}e^{r\xi^{1-\alpha}} < \infty$ *for some*

$$r > \frac{r_1}{1-\alpha},$$

where

$$r_1 := \frac{2\theta\mathbb{E}\tau}{\mathbb{V}\mathrm{ar}\,\xi + v_c^2\,\mathbb{V}\mathrm{ar}\,\tau}. \tag{11.117}$$

Then there exist constants $r_2, r_3, \ldots, r_{\gamma-1} \in \mathbb{R}$, *defined recursively in the proof below, and* $0 < C_1 < C_2 < \infty$ *such that:*

(i) *If* $\alpha = 1/(\gamma - 1)$ *for an integer* $\gamma \ge 2$ *then, for all* $x > 1$,

$$\begin{aligned} \frac{C_1 x^\alpha}{x^{r_{\gamma-1}}} \exp\left\{-\sum_{j=1}^{\gamma-2} \frac{r_j}{1-\alpha j} x^{1-\alpha j}\right\} &\le \psi(x) \\ &\le \frac{C_2 x^\alpha}{x^{r_{\gamma-1}}} \exp\left\{-\sum_{j=1}^{\gamma-2} \frac{r_j}{1-\alpha j} x^{1-\alpha j}\right\}. \end{aligned} \tag{11.118}$$

(ii) *If* $\alpha < 1/(\gamma - 1)$ *then*

$$C_1 x^\alpha \exp\left\{-\sum_{j=1}^{\gamma-1} \frac{r_j}{1-\alpha j} x^{1-\alpha j}\right\} \le \psi(x)$$
$$\le C_2 x^\alpha \exp\left\{-\sum_{j=1}^{\gamma-1} \frac{r_j}{1-\alpha j} x^{1-\alpha j}\right\}. \tag{11.119}$$

As can be seen from these bounds, roughly speaking the ruin probability decays as a Weibullian distribution with shape parameter $1 - \alpha$. However, further terms in the exponent are needed to make the lower and upper bounds precise up to a constant multiplier.

To prove these bounds for the ruin probability under the rate (11.105) of approaching the critical value v_c, we first derive asymptotic estimates for the moments of $V_x(\tau) - x$.

Lemma 11.13 *Let* $\mathbb{E}\tau^\gamma < \infty$ *and*

$$v_-(x) \le v(x) \le v_+(x) \quad \textit{for all } x, \tag{11.120}$$

where both $v_-(x)$ *and* $v_+(x)$ *are decreasing functions. Then, for all* $k \le \gamma$,

$$\mathbb{E}\tau^k v_-(x + \tau v_+(x)) \le \mathbb{E}(V_x(\tau) - x)^k \le v_+^k(x)\mathbb{E}\tau^k. \tag{11.121}$$

If, in addition, $\mathbb{E}\tau^{\gamma+1-\alpha} < \infty$ *and* (11.105) *holds true then there exists an integrable decreasing function* $p_1(x)$ *such that, for all* $k \le \gamma$,

$$\mathbb{E}(V_x(\tau) - x)^k = (v_c + \theta/x^\alpha)^k \mathbb{E}\tau^k + O(p_1(x)) \quad \textit{as } x \to \infty. \tag{11.122}$$

Proof By (11.120), we have $v(z) \le v_+(x)$ for all $z \ge x$. Hence

$$V_x(t) = x + \int_0^t v(V_x(s))ds$$
$$\le x + \int_0^t v_+(x)ds = x + tv_+(x), \tag{11.123}$$

and the second inequality on the right-hand side of (11.121) follows. It follows from the left-hand inequality in (11.120) and from the last upper bound for $V_x(t)$ that

$$V_x(t) - x \ge \int_0^t v_-(V_x(t))ds \ge tv_-(x + tv_+(x)), \tag{11.124}$$

and the left-hand side bound in (11.121) is proven.

Owing to (11.105), $v(z)$ is sandwiched between the two eventually decreasing functions $v_\pm(z) := v_c + \theta/z^\alpha \pm p(z)$. Therefore, applying the right-hand bound in (11.121) we get

$$\begin{aligned}\mathbb{E}(V_x(\tau)-x)^k &\le (v_c+\theta/x^\alpha+p(x))^k\mathbb{E}\tau^k \\ &= (v_c+\theta/x^\alpha)^k\mathbb{E}\tau^k + O(p(x)) \quad \text{as } x\to\infty. \end{aligned} \tag{11.125}$$

From the lower bound in (11.121) we deduce that, for all $k\le\gamma$,

$$\begin{aligned}\mathbb{E}(V_x(\tau)-x)^k &\ge \mathbb{E}\tau^k\Big(v_c+\frac{\theta}{(x+\tau v_+(x))^\alpha}-p(x)\Big)^k \\ &\ge \mathbb{E}\tau^k\Big(v_c+\frac{\theta}{(x+\overline{v}\tau)^\alpha}\Big)^k + O(p(x)), \quad \overline{v}=\sup_z v(z)<\infty.\end{aligned}$$

By the inequality $1/(1+y)^\alpha \ge 1-\alpha y\wedge 1$, we infer that, for $c_2=\alpha\,\overline{v}$,

$$\frac{1}{(x+\overline{v}\,t)^\alpha} \ge \frac{1}{x^\alpha}\Big(1-\frac{c_2 t}{x}\wedge 1\Big).$$

Therefore, for all $k\le\gamma$,

$$\begin{aligned}&\mathbb{E}(V_x(\tau)-x)^k \\ &\ge \mathbb{E}\tau^k\Big(v_c+\frac{\theta}{x^\alpha}-\frac{c_2\theta\tau}{x^{\alpha+1}}\mathbb{I}\{\tau\le x/c_2\}-\frac{1}{x^\alpha}\mathbb{I}\{\tau>x/c_2\}\Big)^k + O(p(x)) \\ &\ge \Big(v_c+\frac{\theta}{x^\alpha}\Big)^k\mathbb{E}\tau^k - \frac{c_3}{x^\alpha}\mathbb{E}\{\tau^k;\ \tau>x/c_2\} \\ &\quad - c_3\sum_{j=1}^k \frac{1}{x^{j(\alpha+1)}}\mathbb{E}\{\tau^{k+j};\ \tau\le x/c_2\} - c_3 p(x),\end{aligned} \tag{11.126}$$

for some $c_3<\infty$. Then, owing to the integrability of $p(x)$, in order to prove that

$$\mathbb{E}(V_x(\tau)-x)^k \ge (v_c+\theta/x^\alpha)^k\mathbb{E}\tau^k - p_1(x) \tag{11.127}$$

holds for some decreasing integrable function $p_1(x)$, it suffices to show that

$$x^{-\alpha}\mathbb{E}\{\tau^\gamma;\ \tau>x\}$$

and

$$x^{-j(\alpha+1)}\mathbb{E}\{\tau^{\gamma+j};\ \tau\le x\}$$

are bounded by decreasing functions that are integrable at infinity. Indeed, the integral of the first function – which decreases itself – is finite owing to the finiteness of the $(\gamma+1-\alpha)$th moment of τ. Concerning the second function, first notice that

$$x^{-j(\alpha+1)}\mathbb{E}\{\tau^{\gamma+j};\ \tau\le x\} \le \frac{\mathbb{E}\{\tau^{\gamma+1};\ \tau\le x\}}{x^{1+\alpha}}, \quad j\ge 1.$$

Owing to the moment condition on τ and Lemma 2.27, the right-hand side is bounded by a decreasing function that is integrable at infinity. So (11.127) is proven, which, together with (11.125), completes the proof of the lemma. □

Proposition 11.14 *Assume the rate of convergence* (11.105). *If both* $\mathbb{E}\tau^{1+\gamma}$ *and* $\mathbb{E}\xi^{1+\gamma}$ *are finite then there exists* $s(x) = o(x^\alpha)$ *such that, for all* $k \le \gamma$,

$$m_k^{[s(x)]}(x) = \sum_{j=0}^{k} \frac{a_{k,j}}{x^{\alpha j}} + O(x^{\alpha(k-1)} p_2(x)) \quad \text{as } x \to \infty,$$

where $p_2(x)$ *is a decreasing function that is integrable at infinity and*

$$a_{k,j} := \binom{k}{j} \theta^j \mathbb{E}\tau^j (v_c\tau - \xi)^{k-j}, \quad j \le k \le \gamma.$$

Proof It follows from the definition of $\xi(x)$ that

$$\mathbb{E}\xi^k(x) = \mathbb{E}(V_x(\tau) - x - \xi)^k = \sum_{i=0}^{k} \binom{k}{i} \mathbb{E}(V_x(\tau) - x)^i \mathbb{E}(-\xi)^{k-i}.$$

Applying Lemma 11.13, we then obtain

$$\begin{aligned} m_k(x) &:= \mathbb{E}\xi^k(x) \\ &= \sum_{i=0}^{k} \binom{k}{i} \left(v_c + \frac{\theta}{x^\alpha}\right)^i \mathbb{E}\tau^i \mathbb{E}(-\xi)^{k-i} + O(p_1(x)) \\ &= \sum_{i=0}^{k} \binom{k}{i} \mathbb{E}\tau^i \mathbb{E}(-\xi)^{k-i} \sum_{j=0}^{i} \binom{i}{j} v_c^{i-j} \left(\frac{\theta}{x^\alpha}\right)^j + O(p_1(x)) \\ &=: \sum_{j=0}^{k} \frac{a_{k,j}}{x^{\alpha j}} + O(p_1(x)) \quad \text{as } x \to \infty, \end{aligned}$$

where

$$\begin{aligned} a_{k,j} &:= \binom{k}{j} \theta^j \sum_{i=j}^{k} \binom{k-j}{i-j} \mathbb{E}\tau^i \mathbb{E}(-\xi)^{k-i} v_c^{i-j} \\ &= \binom{k}{j} \theta^j \mathbb{E} \sum_{i=0}^{k-j} \binom{k-j}{i} \tau^{i+j} (-\xi)^{k-j-i} v_c^i \\ &= \binom{k}{j} \theta^j \mathbb{E}\tau^j (v_c\tau - \xi)^{k-j}. \end{aligned}$$

Now, in view of (11.113) we have

$$\begin{aligned}
&|m_k(x) - m_k^{[s(x)]}(x)| \\
&\quad = O\Big(\mathbb{E}\{(V_x(\tau) - x)^k;\ V_x(\tau) - x > s(x)\} + \mathbb{E}\{\xi^k; \xi > s(x)\} \\
&\qquad + \mathbb{E}\xi^k \mathbb{P}\{V_x(\tau) - x > s(x)\} + \mathbb{E}(V_x(\tau) - x)^k \mathbb{P}\{\xi > s(x)\}\Big) \\
&\quad = O\Big(\mathbb{E}\{\tau^k;\ \tau > s(x)/\overline{v}\} + \mathbb{E}\{\xi^k; \xi > s(x)\}\Big) \quad \text{as } x \to \infty.
\end{aligned}$$

Since $\mathbb{E}\tau^{\gamma+1} < \infty$, for all $k \le \gamma$ we have

$$\begin{aligned}
x^{-\alpha(k-1)}\mathbb{E}\{\tau^k;\ \tau > s(x)/\overline{v}\} &= o(1/x^{\alpha(k-1)} s^{\gamma+1-k}(x)) \\
&= o(1/s^{\gamma}(x)) \quad \text{as } x \to \infty,
\end{aligned}$$

for any $s(x) = o(x^\alpha)$. By the definition of γ, $\alpha\gamma > 1$. Therefore, for an increasing function $s(x) = x^\alpha/\log x$ of order $o(x^\alpha)$, the function $1/s^\gamma(x)$ is integrable at infinity. The same arguments work for ξ; hence the function $x^{-\alpha(k-1)}|m_k(x) - m_k^{[s(x)]}(x)|$ is dominated by a decreasing function that is integrable at infinity function, and the proof of the proposition is complete. □

Proof of Theorem 11.12. We first show that there exist constants $r_1, r_2, \ldots, r_{\gamma-1}$ such that the function

$$r(x) := \sum_{j=1}^{\gamma-1} \frac{r_j}{(b+x)^{\alpha j}}$$

satisfies (3.45); here b is a positive number. We can determine all these numbers recursively. Indeed, as proven in Proposition 11.14,

$$m_1^{[s(x)]}(x) = \frac{\theta \mathbb{E}\tau}{x^\alpha} + o(p_2(x)) \quad \text{as } x \to \infty$$

and

$$m_2^{[s(x)]}(x) = \mathbb{V}\mathrm{ar}\,\xi + v_c^2\,\mathbb{V}\mathrm{ar}\,\tau + O(x^{-\alpha}) \quad \text{as } x \to \infty.$$

For r_1, defined in (11.117), it holds that

$$-m_1^{[s(x)]}(x) + \sum_{j=2}^{\gamma} (-1)^j \frac{m_j^{[s(x)]}(x)}{j!} r^{j-1}(x) = O(x^{-2\alpha}) \quad \text{as } x \to \infty,$$

for any choice of $r_2, r_3, \ldots, r_{\gamma-1}$. Then we can choose r_2 such that the coefficient of $x^{-2\alpha}$ is also zero, giving

$$r_2 = \frac{\mathbb{E}(v_c\tau - \xi)^3 r_1^2/3 - 2\theta\mathbb{E}\tau(v_c\tau - \xi) r_1}{\mathbb{V}\mathrm{ar}\,\xi + v_c^2\,\mathbb{V}\mathrm{ar}\,\tau},$$

and so on. It is clear that the numbers $r_1, r_2, \ldots, r_{\gamma-1}$ do not depend on the parameter b. Therefore, we can take b large enough that the function $r(x)$ is decreasing on $[0, \infty)$. The conditions (3.44) and (3.46) are satisfied for $r(x)$.

We have

$$U(x) = \int_x^\infty \exp\Big\{-\int_0^y \sum_{j=1}^{\gamma-1} \frac{r_j}{(b+z)^{\alpha_j}} dz\Big\} dy.$$

The conditions (3.47), (3.48), and (3.49) are immediate from the moment assumptions on τ and ξ. Thus, the announced bounds for the ruin probability follow from Theorem 3.10, as in (11.116). □

11.5 Stochastic Difference Equations: Approach via Asymptotically Homogeneous Chains

Let (A_n, B_n) be a sequence of independent and identically distributed random vectors in $(\mathbb{R}^+)^2$ and let R_0 be independent of them. Consider a stochastic linear recursion

$$R_n = A_n R_{n-1} + B_n, \quad n \ge 1, \tag{11.128}$$

with starting point R_0. The sequence $\{R_n\}$ is a Markov chain. The assumption $B_n \ge 0$ is not at all standard but we chose it because non-negative stochastic difference equations allow a more straightforward analysis, via Markov chains on $\mathbb{R}^+$.

We also assume that $\mathbb{P}\{A_1 > 1\} > 0$, which guarantees that

$$\mathbb{P}\Big\{\limsup_{n\to\infty} R_n = \infty\Big\} = 1.$$

It is immediate from (11.128) that

$$R_n = R_0 \prod_{j=1}^n A_j + \sum_{k=1}^n B_k \prod_{j=k+1}^n A_j, \quad n \ge 1.$$

Then, for every $n \ge 1$, the distribution of the variable R_n coincides with that of

$$D_n := R_0 \prod_{j=1}^n A_j + \sum_{k=1}^n B_k \prod_{j=1}^{k-1} A_j, \tag{11.129}$$

which is called a *perpetuity*. The coincidence of their marginal distributions is not the only connection between the sequences $\{R_n\}$ and $\{D_n\}$. Vervaat [147] showed that the Markov chain $\{R_n\}$ is positive recurrent if and only if

$$D_\infty := \sum_{k=1}^{\infty} B_k \prod_{j=1}^{k-1} A_j < \infty \quad \text{a.s.}$$

In this case, the sequence $\{R_n\}$ converges weakly to the distribution of D_∞ and, furthermore, this distribution is the unique solution to the fixed-point equation

$$D_\infty \stackrel{d}{=} A_1 D'_\infty + B_1,$$

where D'_∞ is independent of (A_1, B_1) and D'_∞ and D_∞ are identically distributed.

We are going to show how one can determine the asymptotic behaviour of the invariant distribution of $\{R_n\}$ by using results from Chapter 10. First we notice that the chain $\{R_n\}$ is not asymptotically homogeneous in space. In order to transform it to an asymptotically homogeneous chain we define a function

$$f(x) := \begin{cases} \log x & \text{for } x \geq e, \\ x/e & \text{for } x \in [0, e], \end{cases} \tag{11.130}$$

so that $f(x) : \mathbb{R}^+ \to \mathbb{R}^+$ is a continuous strictly increasing function such that $f(x) \geq \log x$ for all $x \geq 0$. Since $f(x)$ is strictly increasing, the sequence

$$X_n = f(R_n), \quad n \geq 0, \tag{11.131}$$

is a Markov chain on the state space $\mathbb{R}^+$. Let $\xi(x)$ denote the jumps of this chain. It is immediate from the definition of $f(x)$ that, for all $x \geq 1$,

$$\xi(x) = \begin{cases} \log\left(A_1 + e^{-x} B_1\right), & \text{if } A_1 e^x + B_1 \geq e, \\ A_1 e^x + B_1/e - x, & \text{if } A_1 e^x + B_1 \in [0, e]. \end{cases} \tag{11.132}$$

Therefore,

$$\xi(x) \Rightarrow \log A_1 \in [-\infty, \infty),$$

that is, $\{X_n\}$ is asymptotically homogeneous. Furthermore,

$$\mathbb{P}\{R_n > x\} = \mathbb{P}\{D_n > x\} = \mathbb{P}\{X_n > \log x\}, \quad x \geq e. \tag{11.133}$$

11.5.1 Positive Recurrent Case

If $\mathbb{E} \log A_1 \in [-\infty, 0)$ then, according to Lemma 1.7 in [147], $D_\infty < \infty$ provided that $\mathbb{E} \log(1 + B_1) < \infty$. In the following theorem we describe the asymptotic behaviour of the distribution of D_∞, which is also a stationary distribution for the chain $\{R_n\}$.

Theorem 11.15 *Suppose that* $\mathbb{E}A_1^\beta = 1$ *for some* $\beta > 0$ *and* $\mathbb{E}(A_1+B_1)^\beta < \infty$. *Then*

$$\log \mathbb{P}\{D_\infty > x\} \sim -\beta \log x \quad \text{as } x \to \infty. \tag{11.134}$$

If, in addition,

$$\mathbb{E}(\log^+(A_1 + B_1))(A_1 + B_1)^\beta < \infty \tag{11.135}$$

and the distribution of $\log A_1$ *is non-lattice then, for some* $c > 0$,

$$\mathbb{P}\{D_\infty > x\} \sim \frac{c}{x^\beta} \quad \text{as } x \to \infty. \tag{11.136}$$

Proof The logarithmic asymptotics follow from the asymptotic homogeneity of the chain $\{X_n = f(D_n)\}$ and Theorem 10.7.

It follows from (11.132) that

$$\xi(x) \le \log^+(A_1 + B_1), \quad x \ge 1.$$

For $x \le 1$ we have

$$\xi(x) \le f(A_1 ex + B_1) \le 1 + \log^+(A_1 + B_1).$$

As a result,

$$\xi(x) \le \Xi := 1 + \log^+(A_1 + B_1) \quad \text{for all } x \ge 0, \tag{11.137}$$

and $\mathbb{E}\Xi e^{\beta\Xi} < \infty$. Thus, to apply Theorem 10.8, it is sufficient to check that $|\mathbb{E}e^{\beta\xi(x)} - 1|$ is dominated by a decreasing integrable function.

Using (11.132), we get the following lower bound, for all $x > 1$,

$$\begin{aligned}
\mathbb{E}e^{\beta\xi(x)} &\ge \mathbb{E}\{(A_1 + e^{-x}B_1)^\beta;\ A_1 + e^{-x}B_1 > e^{1-x}\}\\
&\ge \mathbb{E}A_1^\beta - \mathbb{E}\{(A_1 + e^{-x}B_1)^\beta;\ A_1 + e^{-x}B_1 \le e^{1-x}\}\\
&\ge 1 - e^{\beta-\beta x}.
\end{aligned}$$

To obtain an upper bound, we first notice that, for all $x > 1$,

$$\begin{aligned}
\mathbb{E}e^{\beta\xi(x)} &= \mathbb{E}\{e^{\beta\xi(x)};\ \xi(x) \le 1 - x\} + \mathbb{E}\{e^{\beta\xi(x)};\ \xi(x) > 1 - x\}\\
&\le e^{\beta-\beta x} + \mathbb{E}\{(A_1 + e^{-x}B_1)^\beta;\ A_1 + e^{-x}B_1 > e^{1-x}\}\\
&\le e^{\beta-\beta x} + \mathbb{E}(A_1 + e^{-x}B_1)^\beta.
\end{aligned}$$

If $\beta \le 1$ then $(u+v)^\beta \le u^\beta + v^\beta$ for all $u, v \ge 0$. Set $u = A_1$ and $v = e^{-x}B_1$; then

$$\mathbb{E}e^{\beta\xi(x)} \le \mathbb{E}A_1^\beta + (e^\beta + \mathbb{E}B_1^\beta)e^{-\beta x}.$$

If $\beta > 1$ then

$$\begin{aligned}(u+v)^\beta &\le u^\beta + \beta v(u+v)^{\beta-1}\\ &\le u^\beta + c_\beta v u^{\beta-1} + c_\beta v^\beta,\end{aligned}$$

where $c_\beta = \beta 2^{\beta-1}$. Therefore

$$\begin{aligned}\mathbb{E}e^{\beta\xi(x)} &\le \mathbb{E}A_1^\beta + (e^\beta + c_\beta \mathbb{E}B_1^\beta)e^{-\beta x} + c_\beta e^{-x}\mathbb{E}A_1^{\beta-1}B_1\\ &= 1 + (e^\beta + c_\beta \mathbb{E}B_1^\beta)e^{-\beta x} + c_\beta e^{-x}\mathbb{E}A_1^{\beta-1}B_1,\end{aligned}$$

where $\mathbb{E}A_1^{\beta-1}B_1 < \infty$, because

$$A_1^{\beta-1}B_1 \le (A_1+B_1)^{\beta-1}(A_1+B_1) \;=\; (A_1+B_1)^\beta.$$

As a result,

$$|\mathbb{E}e^{\beta\xi(x)} - 1| = O\big(e^{-(1\wedge\beta)x}\big),$$

which completes the proof. □

11.5.2 Null Recurrent Case

As mentioned above, the distribution of R_n is the same as that of D_n defined in (11.129). The sequence D_n dominates an increasing sequence

$$T_n := \sum_{k=1}^n B_k \prod_{j=1}^{k-1} A_j.$$

If $\mathbb{E}\log A_1 = 0$ then $S_n := \log A_1 + \cdots + \log A_n$ is an oscillating random walk, so $S_n > 0$ infinitely often with probability 1. Equivalently,

$$\prod_{j=1}^{k-1} A_j > 1 \quad \text{infinitely often with probability 1},$$

which implies the convergence $T_n \to \infty$ as $n \to \infty$ with probability 1. Hence, in the case $\mathbb{E}\log A_1 = 0$,

$$R_n \to \infty \quad \text{in probability as } n \to \infty. \tag{11.138}$$

Theorem 11.16 *Assume that* $\mathbb{E}\log A_1 = 0$, $\sigma^2 := \mathbb{E}\log^2 A_1 \in (0,\infty)$, *and* $\mathbb{E}\log^2 B_1 < \infty$. *Then*

$$\frac{\log R_n}{\sqrt{\sigma^2 n}} \Rightarrow |\eta| \quad \text{as } n \to \infty,$$

where η has a standard normal distribution. In addition, the process

$$\frac{\log R_{[tn]}}{\sqrt{\sigma^2 n}}, \quad t \in [0,1],$$

converges weakly in $D[0,1]$ to a Bessel process with drift 0 and diffusion coefficient 1 as $n \to \infty$, that is, to a reflected Brownian motion $|B(t)|$.

Proof Note that the weak convergence of $\log R_n/\sqrt{\sigma^2 n}$ to $|\eta|$ is equivalent to the weak convergence of $X_n^2/\sigma^2 n = f^2(R_n)/\sigma^2 n$ towards η^2. Since η^2 is Γ-distributed with parameters 1/2 and 1/2, the desired convergence would be proven if it could be shown that the conditions of Theorem 4.10 hold with $\mu = 0$. Then the functional convergence would follow automatically; see Theorem 4.11.

We start by constructing a square integrable majorant for the jumps $\xi(x)$. It follows from the definition of $f(x)$ that

$$\xi(x) = f(A_1 e^x + B_1) - x \geq \log(A_1 e^x + B_1) - x \geq \log A_1,$$

because $B_1 \geq 0$. Furthermore, according to (11.137),

$$\xi(x) \leq 1 + \log^+(A_1 + B_1).$$

From these two inequalities we infer that

$$|\xi(x)|^2 \leq 2\left(1 + \log^2 A_1 + \log^2(A_1 + B_1)\right).$$

Since the random variable on the right-hand side is integrable, we have constructed a suitable majorant.

Recalling that $\xi(x) \Rightarrow \log A_1$ and using the Lebesgue theorem, we infer that

$$m_2(x) = \mathbb{E}\xi^2(x) \ \to \ \mathbb{E}\log^2 A_1 = \sigma^2 \quad \text{as } x \to \infty.$$

Therefore, it remains to determine the asymptotic behaviour of $m_1(x) = \mathbb{E}\xi(x)$. We start with the following decomposition, for all $x > 1$:

$$\begin{aligned}\mathbb{E}\xi(x) &= \mathbb{E}f(A_1 e^x + B_1) - x \\ &= \mathbb{E}\{\log(A_1 + e^{-x} B_1);\ A_1 e^x + B_1 > e\} \\ &\quad + \mathbb{E}\left\{\frac{A_1 e^x + B_1}{e} - x;\ A_1 e^x + B_1 \leq e\right\}.\end{aligned}$$

Hence we deduce the following upper bound:

$$\begin{aligned}&\left|\mathbb{E}\xi(x) - \mathbb{E}\log(A_1 + e^{-x} B_1)\right| \\ &\quad \leq \left|\mathbb{E}\left\{\log(A_1 + e^{-x} B_1) + \frac{A_1 e^x + B_1}{e} - x;\ A_1 e^x + B_1 \leq e\right\}\right|.\end{aligned}$$

We now have, by the positivity of B_1,

$$\begin{aligned}
&\mathbb{E}\{|\log(A_1+e^{-x}B_1)|;\ A_1e^x+B_1\le e\}\\
&\quad=\mathbb{E}\{|\log(A_1+e^{-x}B_1)|;\ \log(A_1+e^{-x}B_1)\le 1-x\}\\
&\quad\le\mathbb{E}\{|\log A_1|;\ \log A_1\le 1-x\}\\
&\quad=o(p_1(x))
\end{aligned}$$

and, since $\frac{A_1e^x+B_1}{e}-x\in[-x,1-x]$ if $A_1e^x+B_1\le e$,

$$\begin{aligned}
\mathbb{E}\Big\{\Big|\frac{A_1e^x+B_1}{e}-x\Big|;\ A_1e^x+B_1\le e\Big\}&\le x\mathbb{P}\{\log A_1\le 1-x\}\\
&=o(p_1(x))
\end{aligned}$$

for some decreasing function $p_1(x)$ that is integrable at infinity; this follows from the assumption $\mathbb{E}\log^2 A_1<\infty$; see Lemma 2.33. Therefore

$$\big|\mathbb{E}\xi(x)-\mathbb{E}\log(A_1+e^{-x}B_1)\big|=o(p_1(x))\quad\text{as } x\to\infty. \tag{11.139}$$

By the assumption $\mathbb{E}\log A_1=0$,

$$\begin{aligned}
\mathbb{E}\log(A_1+e^{-x}B_1)&=\mathbb{E}\log A_1+\mathbb{E}\log(1+e^{-x}B_1/A_1)\\
&=\mathbb{E}\{\log(1+e^{-x}B_1/A_1);\ B_1/A_1\le e^{x/2}\}\\
&\quad+\mathbb{E}\{\log(1+e^{-x}B_1/A_1);\ B_1/A_1\in(e^{x/2},e^x]\}\\
&\quad+\mathbb{E}\{\log(1+e^{-x}B_1/A_1);\ B_1/A_1>e^x\}\\
&=:E_1+E_2+E_3.
\end{aligned}$$

Using the inequality $\log(1+u)\le u$ we derive $E_1\le\log(1+e^{-x/2})\le e^{-x/2}$. Next, we have

$$\begin{aligned}
E_2&\le(\log 2)\mathbb{P}\{B_1/A_1>e^{x/2}\}\\
&\le\mathbb{P}\{\log B_1-\log A_1>x/2\}\\
&\le\mathbb{P}\{\log B_1>x/4\}+\mathbb{P}\{-\log A_1>x/4\}\\
&=o(p_2(x))\quad\text{as } x\to\infty,
\end{aligned}$$

for some decreasing function $p_2(x)$ that is integrable at infinity; this follows from the assumptions $\mathbb{E}\log^2 A_1<\infty$ and $\mathbb{E}\log^2 B_1<\infty$; see Lemma 2.33. Finally, by the same moment conditions,

$$E_3\le\mathbb{E}\{\log(2B_1/A_1);\ \log(B_1/A_1)>x\}\ =\ o(p_3(x))\quad\text{as } x\to\infty,$$

for some decreasing function $p_3(x)$ that is integrable at infinity; see Lemma 2.33. Combining all this, we obtain

$$m_1(x)=o(p_4(x)). \tag{11.140}$$

for some decreasing function $p_4(x)$ that is integrable at infinity. Thus, all the moment conditions of Theorem 4.10 are met with $\mu = 0$. Together with the convergence to infinity (11.138) this completes the proof. □

Theorem 11.17 *Under the conditions of Theorem 11.16, the chain $\{R_n\}$ is null recurrent. In addition, if π_R is an invariant measure of $\{R_n\}$ satisfying $\pi_R[0,x] < \infty$ for all x then*

$$\pi_R(x_1,x_2] \sim c\log(x_2/x_1)$$

as x_1, $x_2 \to \infty$ in such a way that

$$1 < \liminf \frac{\log x_2}{\log x_1} \le \limsup \frac{\log x_2}{\log x_1} < \infty.$$

Proof We start by checking the moment condition of Corollary 2.16 for $\{X_n\}$. It follows from the existence of a square integrable majorant for the family of jumps that, for any $s(x) \to \infty$,

$$m_2^{[s(x)]}(x) \to \sigma^2 > 0 \quad \text{as } x \to \infty$$

and that there exists an $s(x) = o(x)$ such that

$$\mathbb{E}\{|\xi(x)|; |\xi(x)| \ge s(x)\} = o(p_5(x)) \quad \text{as } x \to \infty, \tag{11.141}$$

for some decreasing function $p_5(x)$ that is integrable at infinity; see Lemma 2.33. Together with (11.140) this implies that

$$m_1^{[s(x)]}(x) = o(p_4(x) + p_5(x)) \tag{11.142}$$

and, hence,

$$\frac{2m_1^{[s(x)]}(x)}{m_2^{[s(x)]}(x)} = o(p_4(x) + p_5(x)) = o(1/x) \quad \text{as } x \to \infty.$$

Thus, applying Corollary 2.16, we conclude that the chain $X_n = f(R_n)$ is null recurrent. Consequently, $\{R_n\}$ is null recurrent as well.

Furthermore, (11.142) and $m_2^{[s(x)]}(x) \to \sigma^2$ imply that the function $U(x)$ defined in (8.9) has asymptotically linear growth: $U(x) \sim Cx$ as $x \to \infty$. Notice that the chain $\{X_n\}$ satisfies the moment conditions (8.12), (8.13), and (8.14) from Theorem 8.2. Indeed, the condition (8.12) is immediate from the existence of a square integrable majorant. For the same reason, the condition (8.13) follows from Lemma 2.27 with $\alpha = 1$ and $\gamma = 2$. The condition (8.14) follows from (11.141) and from the fact that $U(x) \sim Cx$.

Then it follows from Theorem 8.2 that the stationary measure of $\{X_n\}$ has linear growth:

$$\pi_X(y_1, y_2) \sim c(y_2 - y_1)$$

provided that $y_1, y_2 \to \infty$ in such a way that

$$1 < \liminf \frac{y_2}{y_1} \leq \limsup \frac{y_2}{y_1} < \infty.$$

But it is clear that $\pi_R(x_1, x_2] = \pi_X(\log x_1, \log x_2]$ for all sufficiently large $x_1 < x_2$ and so the proof is complete. □

11.6 Application to the ALOHA Network

We can also illustrate the results with the Markov chain arising from the model of the original ALOHA packet-switching network, originally proposed by Abramson [1] and which was indeed a motivation for Borovkov, Fayolle, and Korshunov [25]. Let us first briefly recall the salient features of the system.

(a) A single error-free channel is shared among an infinite population of users (or stations), which retransmit messages of constant length (packets). Time is slotted and so may be considered discrete. Users are syncronised with respect to the slots, so that packets are transmitted at the beginning of slots only. Each slot is equal to the time required to transmit a packet.

(b) Each transmission is within reception range of every user. When more than one user transmits simultaneously, packets collide (interfere) and none is received correctly. These collisions are treated as transmission errors, the corresponding users (stations) become blocked, and each user must strive to retransmit its colliding packet until it is correctly received. The users all employ the same algorithm for this purpose and have to resolve the contention without the benefit of any other source of information on other users' activity apart from the common channel.

(c) Each user with a colliding packet will repeatedly transmit each successive time with a certain probability until it hits a free slot and thus succeeds.

The main drawback of the ALOHA protocol described above is that, left to their own devices, the nodes congest the channel, which, in the absence of additional control, is non-ergodic. The approach suggested by Lam and Kleinrock in [110] was to let the retransmission probabilities be a function of the number of blocked stations at time t. Such a retransmission control policy can stabilise the channel.

Let A_n be the number of new packets generated by stations which are not blocked during the nth slot. We shall assume that the A_n, $n \geq 1$, form a sequence of independent and identically distributed random variables, with $\mathbb{P}\{A_1 = k\} = p(k)$, $k \geq 0$, and finite expectation. Let X_n, $n \geq 0$, be the number of blocked stations at time n (i.e. as observed at the beginning of the nth slot) and $f(X_n)$ the probability that a blocked station retransmits during this nth slot. So, we consider a centralised ALOHA algorithm where information about the number of blocked stations is available to the stations. Given $X_n = k$, the random number of messages in the nth slot has a binomial distribution with success probability $f(k)$. Hence, $\{X_n\}$ forms a Markov chain with transition probabilities

$$P(0, j) = \begin{cases} p(0) + p(1) & \text{for } j = 0, \\ p(j) & \text{for } j \geq 2, \end{cases}$$

and, for $i \geq 1$,

$$P(i, j) = \begin{cases} p(0) i f(i)(1 - f(i))^{i-1} & \text{for } j = i - 1, \\ p(1)(1 - f(i))^i & \text{for } j = i, \\ p(1)(1 - (1 - f(i))^i) & \text{for } j = i + 1, \\ p(j - i) & \text{for } j \geq i + 2. \end{cases}$$

Define the quantity

$$q(k) = p(0) k f(k)(1 - f(k))^{k-1} + p(1)(1 - f(k))^k, \tag{11.143}$$

which represents the probability of successful transmission in the nth slot, given the event $X_n = k$. Clearly, if

$$\mathbb{E}A_1 < \liminf_{k\to\infty} q(k)$$

then $\{X_n\}$ is positive recurrent and possesses a probabilistic invariant measure. However, if

$$\mathbb{E}A_1 > \limsup_{k\to\infty} q(k)$$

then $\{X_n\}$ is transient.

Our goal is to describe the asymptotic behaviour of $\{X_n\}$ in the asymptotically zero-drift case. Assume that, for all sufficiently large k,

$$f(k) = f/k \quad \text{for some } f > 0. \tag{11.144}$$

Then (11.143) gives the following limiting probability of successful transmission:

$$q = \lim_{k\to\infty} q(k) = e^{-f}(f p(0) + p(1)). \tag{11.145}$$

Its maximal value $p(0)\exp(p(1)/p(0) - 1)$ is attained at $f = 1 - p(1)/p(0)$.

By direct computation, the first and second moments of the jumps of $\{X_n\}$ are given by

$$m_1(k) = \mathbb{E}A_1 - q + \mu/k + O(1/k^2), \tag{11.146}$$

$$m_2(k) = b + O(1/k) \quad \text{as } k \to \infty, \tag{11.147}$$

where

$$\mu = \frac{f^2 e^{-f}}{2}[p(1) + p_0(f-2)], \qquad b = \mathbb{E}A_1^2 + (p(0)f - p(1))e^{-f}.$$

It can never happen that $\mu \le -b/2$ because

$$\begin{aligned} e^f(2\mu + b) &= f^2[p(1) + p_0(f-2)] + e^f\mathbb{E}A_1^2 + (p(0)f - p(1)) \\ &\ge f^2[p(1) + p_0(f-2)] + (1+f)p(1) + (p(0)f - p(1)) \\ &= (f^2 + f)p(1) + p_0 f(f-1)^2 > 0. \end{aligned}$$

Theorem 11.18 *Let $\mathbb{E}A_1 = q$. Then the Markov chain $\{X_n\}$ of the ALOHA protocol is non-ergodic and the following main situations can occur:*

(i) *If $\mu > b/2$ then $\{X_n\}$ is transient;*
(ii) *If $-b/2 < \mu < b/2$ then $\{X_n\}$ is null recurrent.*

In addition, X_n^2/n converges weakly as $n \to \infty$ to a $\Gamma_{1/2+\mu/b,2b}$-distribution and, moreover, the process

$$\frac{\log X_{[tn]}}{\sqrt{bn}}, \quad t \in [0,1],$$

converges weakly in $D[0,1]$ to a Bessel process with drift μ/bx and diffusion coefficient 1 as $n \to \infty$.

Proof The claims are immediate from Corollaries 2.19 and 2.16 and Theorems 4.10 and 4.11. □

11.7 Comments for Chapter 11

11.7.1 Cramér–Lundberg Risk Processes with Level-Dependent Premium Rate

This subsection was based on Denisov, Gotthardt, Korshunov, and Wachtel [45].

11.7.2 Near-Critical Branching Processes

Apparently Lamperti [114] was the first to apply Markov chains to the study of branching processes and, in particular, Markov chains with asymptotically zero drift; see [115]. The use of a square root transform for critical Galton–Watson branching processes was suggested by Nagaev and Wachtel in [131].

Kersting [92] studied transience and recurrence criteria for sequences of the form $X_{n+1} = X_n + g(X_n) + \xi_{n+1}$, where $\{\xi_k\}$ are square integrable martingale differences. It is worth mentioning that state-dependent branching processes with migration – which were considered in Section 11.3 – can be represented in this form.

For state-dependent processes without migration, weak convergence to a Γ-distribution has been obtained in several papers. Klebaner [98] showed this convergence for processes satisfying $\max_{k\ge1} \mathbb{E}\zeta^m(k) < \infty$ for all $m \ge 1$. Höpfner [81] proved the same result under weaker moment assumptions. He showed that (11.61) holds for processes satisfying $\mathbb{E}\zeta(k) = 1 + a/k$, $|\sigma^2(k) - \sigma^2| = O(1)$, and $\max_{k\ge1} \mathbb{E}\zeta^2(k)\log(1+\zeta(k)) < \infty$. The restrictions in Theorem 11.6 are significantly weaker than those in the papers cited above.

The convergence of critical branching processes with immigration to a Γ-distribution was first proven by Seneta [140]. More precisely, he showed that if the $\zeta(k)$ are identically distributed with expectation 1 and variance σ^2 and if η is non-negative with finite expectation then Z_n/n converges weakly to a Γ-distribution. If $\mathbb{E}\eta > \sigma^2/2$ then this is a particular case of our Theorem 11.6. If $\mathbb{E}\eta \le \sigma^2/2$ then, in order to apply Theorem 11.7, we have to check the validity of (11.62) and (11.63). For identically distributed variables this condition is satisfied in particular if $\mathbb{E}\zeta^2 \log^{1+\varepsilon}(1+\zeta) < \infty$ for some $\varepsilon > 0$.

For size-dependent processes without migration the asymptotic behaviour of the non-extinction probability and the corresponding conditional distribution was studied earlier by Höpfner [82]. The assumptions in that paper are quite restrictive: $\mathbb{E}\zeta(k) = 1 + a/k$ with some $a \in (0, \sigma^2/2]$, $|\sigma^2(k) - \sigma^2| = O(1/k)$, and $\max_{k\ge1} \mathbb{E}\zeta^{2+\delta}(k) < \infty$ for some $\delta > 0$. If $a < \sigma^2/2$ then the results in [82] coincide with that in Theorem 11.9, but if $a = \sigma^2/2$ (this corresponds to $\rho = 0$) then (11.88) is still valid and $\mathbb{P}\{Z_n > 0\} \sim c/\log n$. This particular case is not covered by Theorem 11.9.

Zubkov [153] investigated the recurrence times to zero for branching processes with immigration. He showed that if $\mathbb{E}\eta < \sigma^2/2$ then there exists a slowly varying function L such that

$$\mathbb{P}\{\min_{k\le n} Z_k > 0\} \sim L(n) n^{2\mathbb{E}\eta/\sigma^2 - 1}.$$

It is also shown there that one can take $L(n) \equiv C > 0$ if and only if $\mathbb{E}\eta \log(1+\eta) < \infty$. Vatutin [146] showed that (11.88) holds under the same conditions. Zubkov's result shows that the restrictions $\mathbb{E}|\eta| \log(1 + |\eta|) < \infty$ and (11.65) in Theorem 11.9 are optimal purely for the power tail of the recurrence times.

Vatutin [145] initiated the study of branching processes with emigration. More precisely, he considered sequence $\{Y_n\}$ given by (11.23) with identically distributed $\zeta(k)$ with mean 1 and $\eta \equiv -1$. For $\sigma^2 = \mathbb{E}(\zeta - 1)^2 > 2$ he proved that $\mathbb{P}\{Y_n > 0 \mid Y_0 = m\} \sim L_m(n)n^{-1-2/\sigma^2}$ and that $L_m(n) \equiv c_m > 0$ if and only if $\mathbb{E}\zeta^2 \log(1 + \zeta) < \infty$. Moreover, for $\sigma^2 < 2$ he showed that $\mathbb{P}\{Y_n > 0|Y_0 = m\} \sim c_m n^{-1-2/\sigma^2}$ if and only if $\mathbb{E}\zeta^{1+2/\sigma^2} < \infty$. Finally, assuming that all moments of ζ are finite, he proved that $2Y_n/n\sigma^2$ conditioned on non-extinction converges weakly to a standard exponential distribution. Kaverin [89] generalized this result to all processes Y_n satisfying $\mathbb{E}(-\eta)^{[2+2/\sigma^2]} < \infty$, $\mathbb{E}\zeta^{1+2/\sigma^2} < \infty$ in the case $\sigma^2 < 2$ and $\mathbb{E}\zeta^2 \log(1 + \zeta) < \infty$ in the case $\sigma^2 = 2$. Specialising Theorem 11.9 to identically distributed $\zeta(k)$ and non-positive η, we conclude that (11.87) and (11.88) hold for all processes Z_n satisfying $\mathbb{E}(-\eta) < \infty$, $\mathbb{E}\zeta^2 \log(1 + \zeta) < \infty$, and $\mathbb{E}\zeta^{1+2/\sigma^2} < \infty$ in the case $\sigma^2 < 2$. We see that our restrictions on the emigration component η are much weaker than that in [89].

Kosygina and Mountford [108] proved (11.87) for a special model of branching processes with migration. This model appears in the description of excited random walks on integers.

The first result of this type was obtained by Foster [69] for a critical Galton–Watson process with immigration at zero. Formally, we cannot say that Foster's result follows from (11.96). But since all the calculations we made in the proof of Theorem 11.9 are valid for processes without migration, it is easy to see that adding immigration at zero does not change the asymptotic behaviour of truncated moments. Therefore, Theorem 8.29 is applicable to the process from [69] if the number of immigrating individuals has finite mean.

Nagaev and Khan [130] proved (11.96) for a critical process with migration. More precisely, they have considered the sequence Y_n defined in (11.23) with identically distributed $\zeta(k)$ with mean 1 and finite variance. Let us compare our moment assumptions with that in [130]. First we note that if the $\zeta(k)$ are identically distributed and have finite variance then (11.67)–(11.70) hold automatically. The assumption (11.65), which states $\mathbb{E}\zeta^2 \log^{1+\varepsilon}(1+|\zeta|) < \infty$, is a bit more restrictive than the second-moment assumption in [130]. Further, we have assumed that $\mathbb{E}|\eta| \log(1 + |\eta|)$ is finite, which is weaker than the corresponding condition in [130]. It is assumed there that $\mathbb{E}\eta^2 < \infty$ and $\mathbb{P}\{\eta > -m\} = 1$ for some $m \geq 1$.

Comparing our theorems with the results known in the literature for branching processes with migration, we conclude that the only weakness of the transformation $\sqrt{Z_n}$ is that it is not clear how to deal with the case when one has tail asymptotics with non-trivial slowly varying functions. Recall that the only obstacle is showing (8.5) in the case when $2m_1^{[s(x))]}(x)/m_2^{[s(x))]}(x) - c/x$ is not integrable for any constant c.

11.7.3 Stochastic Difference Equations

Theorem 11.15 is due to Kesten [95, Theorem 5]; for a complete proof and further related results see Goldie [73, Theorem 4.1]. In these papers a weaker moment condition $\mathbb{E}(\log A_1)A_1^\beta < \infty$ was used. We have imposed (11.135) since we have to construct a majorant Ξ for jumps $\xi(x)$ such that $\Xi e^{\beta\Xi} < \infty$. One can prove the Kesten–Goldie result by using results for asymptotically homogeneous chains under optimal moment assumptions. Such a proof can be found in Korshunov [107].

Theorem 11.16 was proved by Hitczenko and Wesolowski in [78].

The asymptotic behaviour of π_R in the null recurrent case discussed in Theorem 11.17 has been studied in the literature. The most general results were proven by Babillot, Bougerol, and Elie [11] and by Brofferio and Buraczewski [28]: if $\mathbb{E}\log A_1 = 0$ and $\mathbb{E}|\log A_1|^{2+\delta} + \mathbb{E}|\log B_1|^{2+\delta} < \infty$ then it was proven in [11] that there exists a slowly varying function $L(x)$ such that

$$\pi_R(ax, bx] \sim \log(b/a)L(x), \quad x \to \infty;$$

it was shown in [28, Theorem 1.1] that $L(x)$ is a constant.

Theorem 11.17 says nothing about $\pi_R(ax, bx]$, since $\log(bx)/\log(ax) \to 1$ as $x \to \infty$. But our result implies that a slowly varying function from the previous relation cannot converge to either zero or infinity. On the basis of our Theorem 11.17, it is plausible to expect that $L(x)$ is a constant under the assumption that the second moments of both A_1 and B_1 are finite.

For thorough discussion on stochastic difference equations see the book by Buraczewski, Damek, and Mikosch [31].

References

[1] N. Abramson. The ALOHA system – another alternative for computer communications. *Proc. Fall Joint Computer Conf., AFIPS Press*, 37:281–285, 1970.

[2] H. Albrecher and S. Asmussen. *Ruin Probabilities, 2nd Ed.* World Scientific Publishing, Hackensack, NJ, 2010.

[3] H. Albrecher, C. Constantinescu, Z. Palmowski, G. Regensburger, and M. Rosenkranz. Exact and asymptotic results for insurance risk models with surplus-dependent premiums. *SIAM J. Appl. Math.*, 73:47–66, 2013.

[4] D. Aldous. *Probability Approximations Via the Poisson Clumping Heuristic*. Springer, New York, 1989.

[5] K. S. Alexander. Excursions and local limit theorems for Bessel-like random walks. *Electron. J. Probab.*, 16:1–44, 2011.

[6] K. S. Alexander and N. Zygouras. Quenched and annealed critical points in polymer pinning models. *Commun. Math. Phys.*, 291:659–689, 2009.

[7] S. Asmussen. *Ruin Probabilities*. World Scientific, Singapore, 2000.

[8] S. Asmussen. *Applied Probability and Queues*. Springer, New York, 2003.

[9] S. Aspandiiarov and R. Iasnogorodski. Asymptotic behaviour of stationary distributions for countable Markov chains, with some applications. *Bernoulli*, 5:535–569, 1999.

[10] K. B. Athreya, D. McDonald, and P. Ney. Limit theorems for semi-Markov processes and renewal theory for Markov chains. *Ann. Probab.*, 6:788–797, 1978.

[11] M. Babillot, P. Bougerol, and L. Elie. The random difference equation $X_n = A_n X_{n-1} + B_n$ in the critical case. *Ann. Probab.*, 25:478–493, 1997.

[12] Q. Berger. Strong renewal theorems and local large deviations for multivariate random walks and renewals. *Electron. J. Probab.*, 24:47 pp., 2019.

[13] J. Bertoin and R. A. Doney. On conditioning a random walk to stay nonnegative. *Ann. Probab.*, 22:2152–2167, 1994.

[14] J. Bertoin and R. A. Doney. Some asymptotic results for transient random walks. *Adv. Appl. Probab.*, 28:207–226, 1996.

[15] J. Bertoin and I. Kortchemski. Self-similar scaling limits of Markov chains on the positive integers. *Ann. Appl. Probab.*, 26:2556–2595, 2016.

[16] P. Billingsley. *Convergence of Probability Measures*. J. Wiley & Sons, New York, 1968.

[17] N. Bingham, C. Goldie, and J. Teugels. *Regular Variation*. Cambridge University Press, Cambridge, 1987.

[18] D. Blackwell. A renewal theorem. *Duke Math. J.*, 15:145–150, 1948.

[19] D. Blackwell. Extension of a renewal theorem. *Pacific J. Math.*, 3:315–320, 1953.

[20] E. Bolthausen. On a functional central limit theorem for random walks conditioned to stay positive. *Ann. Probab.*, 3:480–485, 1976.

[21] A. N. Borodin and P. Salminen. *Handbook of Brownian Motion – Facts and Formulae, 2nd Ed.* Birkhäuser, Basel, 2002.

[22] A. Borovkov and K. Borovkov. *Asymptotic Analysis of Random Walks. Heavy-Tailed Distributions*. Cambridge University Press, Cambridge, 2008.

[23] A. A. Borovkov. *Ergodicity and Stability of Stochastic Processes*. John Wiley, Chichester, 1998.

[24] A. A. Borovkov and D. Korshunov. Large-deviation probabilities for one-dimensional Markov chains. Part 1: Stationary distributions. *Theory Probab. Appl.*, 41:1–24, 1997.

[25] A. A. Borovkov, G. Fayolle, and D. Korshunov. Transient phenomena for Markov chains and their applications. *Adv. Appl. Probab.*, 24:322–342, 1992.

[26] O. Boxma and M. Mandjes. Affine storage and insurance risk models. *Math. Oper. Res.*, 46:1282–1302, 2021.

[27] H. Brézis, W. Rosenkrantz, and B. Singer. An extension of Khintchine's estimate for large deviations to a class of Markov chains converging to a singular diffusion. *Comm. Pure Appl. Math.*, 24:705–726, 1971.

[28] S. Brofferio and D. Buraczewski. On unbounded invariant measures of stochastic dynamical systems. *Ann. Probab.*, 43:1456–1492, 2015.

[29] B. M. Brown. Martingale central limit theorems. *Ann. Math. Statist.*, 42:59–66, 1971.

[30] A. Bryn-Jones and R. A. Doney. A functional limit theorem for random walks conditioned to stay non-negative. *J. Lond. Math. Soc. (2)*, 74:244–258, 2006.

[31] D. Buraczewski, E. Damek, and T. Mikosch. *Stochastic Models with Power-Law Tails. The Equation $X = AX + B$*. Springer, Switzerland, 2016.

[32] F. Caravenna and L. Chaumont. Invariance principles for random walks conditioned to stay positive. *Ann. Inst. Henri Poincaré Probab. Statist.*, 44:170–190, 2008.

[33] F. Caravenna and R. Doney. Local large deviations and the strong renewal theorem. *Electron. J. Probab.*, 24:48 pp., 2019.

[34] A. S. Cherny and H.-J. Engelbert. *Singular Stochastic Differential Equations*. Springer, Berlin, 2005.

[35] G. Choquet and J. Deny. Sur l'équation de convolution $\mu = \mu * \sigma$. *C. R. Acad. Sci. Paris Série A*, 250:799–801, 1960.

[36] K. L. Chung and J. B. Walsh. *Markov Processes, Brownian Motion, and Time Symmetry*. Springer, 2005.

[37] D. R. Cox and W. L. Smith. A direct proof of a fundamental theorem of renewal theory. *Scand. Actuarial J.*, 36:139–150, 1953.

[38] H. Cramér. *Collective Risk Theory*. Esselte, Stockholm, 1955.

[39] E. Csáki, A. Földes, and P. Révész. Transient nearest neighbor random walk and Bessel process. *J. Theoret. Probab.*, 22:992–1009, 2009.

[40] I. Czarna, J.-L. Pérez, T. Rolski, and K. Yamazaki. Fluctuation theory for level-dependent Lévy risk processes. *Stochastic Process. Appl.*, 129:5406–5449, 2019.

[41] J. De Coninck, F. Dunlop, and T. Huilett. Random walk weakly attracted to a wall. *J. Stat. Phys.*, 133:271–280, 2008.

[42] D. Denisov, D. Korshunov, and V. Wachtel. Potential analysis for positive recurrent Markov chains with asymptotically zero drift: power-type asymptotics. *Stochastic Process. Appl.*, 123:3027–3051, 2013.

[43] D. Denisov, D. Korshunov, and V. Wachtel. Markov chains on $\mathbb{Z}^+$: analysis of stationary measure via harmonic functions approach. *Queueing Syst.*, 91: 265–295, 2019.

[44] D. Denisov, D. Korshunov, and V. Wachtel. Renewal theory for transient Markov chains with asymptotically zero drift. *Trans. Amer. Math. Soc.*, 373:7253–7286, 2020.

[45] D. Denisov, N. Gotthardt, D. Korshunov, and V. Wachtel. Probabilistic approach to risk processes with level-dependent premium rate. *Insur. Math. Econ.*, 118:142–156, 2024.

[46] D. E. Denisov. On the existence of a regularly varying majorant of an integrable monotone function. *Math. Notes*, 76:129–133, 2006.

[47] H. Dette. First return probabilities of birth and death chains and associated orthogonal polynomials. *Proc. Amer. Math. Soc.*, 129:1805–1815, 2001.

[48] S. W. Dharmadhikari and K. Jogdeo. Bounds on moments of certain random variables. *Ann. Math. Statist.*, 40:1506–1509, 1969.

[49] R. A. Doney. An analogue of the renewal theorem in higher dimensions. *Proc. London Math. Soc.*, 16:669–684, 1966.

[50] R. A. Doney. Local behaviour of first passage probabilities. *Probab. Theory Related Fields*, 152:559–588, 2012.

[51] R. A. Doney and E. M. Jones. Large deviation results for random walk conditioned to stay positive. *Electron. Commun. Probab.*, paper no. 38:11 pp., 2012.

[52] J. L. Doob. Conditional Brownian motion and the boundary limits of harmonic functions. *Bull. Soc. Math. France*, 85:431–458, 1957.

[53] J. L. Doob. *Classical Potential Theory and Its Probabilistic Counterpart.* Springer-Verlag, New York, 1984.

[54] J. Duraj and V. Wachtel. Invariance principles for random walks in cones. *Stochastic Process. Appl.*, 130:3920–3942, 2020.

[55] R. Durrett. Conditioned limit theorems for some null recurrent Markov chains. *Ann. Probab.*, 6:798–828, 1978.

[56] R. Durrett. *Probability – Theory and Examples, 5th Ed.* Cambridge University Press, Cambridge, 2019.

[57] P. Embrechts, C. Klüppelberg, and T. Mikosch. *Modelling Extremal Events for Insurance and Finance*. Springer, Berlin, 1997.

[58] P. Erdös, W. Feller, and H. Pollard. A property of power series with positive coefficients. *Bull. Amer. Math. Soc.*, 55:201–204, 1949.

[59] K. B. Erickson. Strong renewal theorems with infinite mean. *Trans. Amer. Math. Soc.*, 151:263–291, 1970.

[60] S. N. Ethier and T. G. Kurtz. *Markov Processes. Characterization and Convergence*. J. Wiley & Sons, New York, 1986.

[61] G. Fayolle, V. A. Malyshev, and M. V. Menshikov. *Topics in the Constructive Theory of Countable Markov Chains*. Cambridge University Press, Cambridge, 1995.

[62] W. Feller. On the integral equation of renewal theory. *Ann. Math. Statist.*, 12: 243–267, 1941.

[63] W. Feller. *An Introduction to Probability Theory and Its Applications, Vol. 2*. Wiley, New York, 1971.

[64] W. Feller and S. Orey. A renewal theorem. *J. Math. Mech.*, 10:619–624, 1961.

[65] P. Flajolet and R. Sedgewick. *Analytic Combinatorics*. Cambridge University Press, Cambridge, 2009.

[66] R. D. Foley and D. R. McDonald. Constructing a harmonic function for an irreducible nonnegative matrix with convergence parameter $R > 1$. *Bull. London Math. Soc.*, 44:533–544, 2012.

[67] S. Foss, D. Korshunov, and S. Zachary. *An Introduction to Heavy-Tailed and Subexponential Distributions, 2nd Ed.* Springer, New York, 2013.

[68] F. G. Foster. On the stochastic matrices associated with certain queueing processes. *Ann. Math. Statist.*, 24:355–360, 1953.

[69] J. Foster. A limit theorem for a branching process with state-dependent immigration. *Ann. Math. Statist.*, 42:1773–1776, 1971.

[70] D. X. Fuk and S. V. Nagaev. Probabilistic inequalities for sums of independent random variables. *Theor. Probab. Appl.*, 16:643–660, 1971.

[71] P. Gaenssler, J. Strobel, and W. Stute. On central limit theorems for martingale triangular arrays. *Acta Mathematica Acad. Sci. Hungaricae*, 31:205–216, 1978.

[72] A. Garsia and J. Lamperti. A discrete renewal theorem with infinite mean. *Comment. Math. Helv.*, 37:221–234, 1962.

[73] C. M. Goldie. Implicit renewal theory and tails of solutions of random equations. *Ann. Appl. Probab.*, 1:126–166, 1991.

[74] I. S. Gradshteyn and I. M. Ryzhik. *Table of Integrals, Series, and Products, 7th Ed.* Elsevier/Academic Press, Amsterdam, 2007.

[75] D. Guibourg and L. Hervé. Multidimensional renewal theory in the non-centered case. Application to strongly ergodic Markov chains. *Potential Anal.*, 38: 471–497, 2013.

[76] Y. Guivarc'h, M. Keane, and B. Roynette. *Marches Aléatoires sur les Groupes de Lie (in French)*. Springer, Berlin–New York, 1977.

[77] T. E. Harris. First passage and recurrence distributions. *Trans. Amer. Math. Soc.*, 73:471–486, 1952.

[78] P. Hitczenko and J. Wesolowski. Renorming divergent perpetuities. *Bernoulli*, 17:880–894, 2011.

[79] J. L. Hodges and M. Rosenblatt Jr. Recurrence-time moments in random walks. *Pacific J. Math.*, 3:127–136, 1953.

[80] J. Hoffman-Jørgensen and G. Pisier. The law of large numbers and the central limit theorem in Banach spaces. *Ann. Probab.*, 4:587–599, 1976.

[81] R. Höpfner. On some classes of population-size-dependent Galton–Watson processes. *J. Appl. Probab.*, 22:25–36, 1985.

[82] R. Höpfner. Some results on population-size-dependent Galton–Watson processes. *J. Appl. Probab.*, 23:297–306, 1986.

[83] O. Hryniv, M. V. Menshikov, and A. R. Wade. Excursions and path functionals for stochastic processes with asymptotically zero drift. *Stochastic Process. Appl.*, 123:1891–1921, 2013.

[84] T. Huillet. Random walk with long-range interaction with a barrier and its dual: exact results. *J. Comput. Appl. Math.*, 233:2449–2467, 2010.

[85] D. Iglehart. Functional central limit theorems for random walks conditioned to stay positive. *Ann. Probab.*, 2:608–619, 1974.

[86] S. Karlin and J. McGregor. Random walks. *Illinois J. Math.*, 3:66–81, 1959.

[87] S. Karlin and H. M. Taylor. *A First Course in Stochastic Processes, 2nd Ed.* Academic Press, New York, 1975.

[88] S. Karlin and H. M. Taylor. *A Second Course in Stochastic Processes*. Academic Press, New York, 1981.

[89] S. V. Kaverin. A refinement of limit theorems for critical branching processes with an emigration. *Theory Probab. Appl.*, 35:574–580, 1990.

[90] G. Keller, G. Kersting, and U. Rösler. On the asymptotic behaviour of discrete time stochastic growth processes. *Ann. Probab.*, 15:305–343, 1987.

[91] J. H. B. Kemperman. The oscillating random walk. *Stochastic Process. Appl.*, 2: 1–29, 1974.

[92] G. Kersting. On recurrence and transience of growth models. *J. Appl. Probab.*, 23:614–625, 1986.

[93] G. Kersting. A law of large numbers for stochastic difference equations. *Stochastic Process. Appl.*, 40:1–13, 1992.

[94] G. Kersting. Asymptotic Γ-distribution for stochastic difference equations. *Stochastic Process. Appl.*, 40:15–28, 1992.

[95] H. Kesten. Random difference equations and renewal theory for products of random matrices. *Acta Math.*, 131:207–248, 1973.

[96] H. Kesten. Renewal theory for functionals of a Markov chain with general state space. *Ann. Probab.*, 2:355–386, 1974.

[97] H. Kesten and F. Spitzer. Ratio theorems for random walks I. *J. Anal. Math.*, 11: 285–322, 1963.

[98] F. C. Klebaner. On population size dependent branching processes. *Adv. Appl. Probab.*, 16:30–55, 1984.

[99] F. C. Klebaner. Stochastic difference equations and generalized gamma distributions. *Ann. Probab.*, 17:178–188, 1989.

[100] C. Klüppelberg and S. Pergamenchtchikov. Renewal theory for functionals of a Markov chain with compact state space. *Ann. Probab.*, 31:2270–2300, 2003.

[101] D. A. Korshunov. Tightness and continuity of a family of invariant measures for Markov chains depending on a parameter. *Siberian Math. J.*, 37:730–746, 1996.

[102] D. A. Korshunov. On distribution tail of the maximum of a random walk. *Stochastic Process. Appl.*, 72:97–103, 1997.

[103] D. A. Korshunov. Limit theorems for general Markov chains. *Siberian Math. J.*, 42:301–316, 2001.

[104] D. A. Korshunov. One-dimensional asymptotically homogeneous Markov chains: Cramér transform and large deviation probabilities. *Siberian Adv. Math.*, 14(4):30–70, 2004.

[105] D. A. Korshunov. The key renewal theorem for a transient Markov chain. *J. Theoret. Probab.*, 21:234–245, 2008.

[106] D. A. Korshunov. Moments for stationary Markov chains with asymptotically zero drift. *Siberian Math. J.*, 52:655–664, 2011.

[107] D. A. Korshunov. A look at perpetuities via asymptotically homogeneous in space Markov chains. *arXiv: 1603.08410*, 2016.

[108] E. Kosygina and T. Mountford. Limit laws of transient excited random walks on integers. *Ann. Inst. Henri Poincaré Probab. Stat.*, 47:575–600, 2011.

[109] M. V. Kozlov. The asymptotic behavior of the probability of non-extinction of critical branching processes in a random environment. (In Russian. English summary). *Teor. Verojatnost. i Primenen.*, 21:813–825, 1976.

[110] S. S. Lam and L. Kleinrock. Packet switching in a multiaccess broadcast channel: dynamic control procedures. *IEEE Trans. Commun.*, 24:891–904, 1975.

[111] J. Lamperti. Criteria for the recurrence or transience of stochastic processes I. *J. Math. Anal. Appl.*, 1:314–330, 1960.

[112] J. Lamperti. A new class of probability limit theorems. *J. Math. Mech.*, 11: 749–772, 1962.

[113] J. Lamperti. Criteria for stochastic processes ii: passage time moments. *J. Math. Anal. Appl.*, 7:127–145, 1963.

[114] J. Lamperti. Maximal branching processes and 'long-range percolation'. *J. Appl. Probab.*, 7:89–98, 1970.

[115] J. Lamperti. Remarks on maximal branching processes. *Theory Probab. Appl.*, XVII:46–54, 1972.

[116] D. Levin and Y. Peres. *Markov Chains and Mixing Times, 2nd Ed.* American Mathematical Society, 2017.

[117] B. Li, W. Ni, and C. Constantinescu. Risk models with premiums adjusted to claims number. *Insurance Math. Econom.*, 65:94–102, 2015.

[118] M. Maejima. On local limit theorems and Blackwell's renewal theorem for independent random variables. *Ann. Inst. Stat. Math.*, 27:507–520, 1975.

[119] V. A. Malyshev and M. V. Menshikov. Ergodicity, continuity and analyticity of countable Markov chains. *Trans. Moscow Math. Soc.*, 1:1–48, 1982.

[120] E. Marciniak and Z. Palmowski. On the optimal dividend problem for insurance risk models with surplus-dependent premiums. *J. Optim. Theory Appl.*, 168: 723–742, 2016.

[121] M. Menshikov, S. Popov, and A. Wade. *Non-Homogeneous Random Walks. Lyapunov Function Methods for Near-Critical Stochastic Systems.* Cambridge University Press, Cambridge, 2017.

[122] M. V. Menshikov and S. Y. Popov. Exact power estimates for countable Markov chains. *Markov Proc. Relat. Fields*, 1:57–78, 1995.

[123] M. V. Menshikov and A. R. Wade. Rate of escape and central limit theorem for the supercritical Lamperti problem. *Stochastic Process. Appl.*, 120:2078–2099, 2010.

[124] M. V. Menshikov, I. M. Asymont, and R. Yasnogorodskii. Markov processes with asymptotically zero drifts. *Probl. Inform. Trans.*, 31:248–261, 1995.

[125] M. V. Menshikov, M. Vachkovskaia, and A. R. Wade. Asymptotic behaviour of randomly reflecting billiards in unbounded domains. *J. Stat. Phys.*, 132: 1097–1133, 2008.

[126] S. Meyn and R. Tweedie. *Markov Chains and Stochastic Stability.* Wiley, 1993.

[127] M. D. Moustafa. Input–output Markov processes. *Proc. Koninkl. Ned. Akad. Wetensch. Ser. A*, 60:112–118, 1957.

[128] A. V. Nagaev. Renewal theorems in $\mathbb{R}^d$. *Theory Probab. Appl.*, 24:572–581, 1980.

[129] S. V. Nagaev. Large deviations of sums of independent random variables. *Ann. Probab.*, 7:745–789, 1979.

[130] S. V. Nagaev and L. V. Khan. Limit theorems for a critical Galton–Watson branching process with migration. *Theory Probab. Appl.*, 24:514–525, 1980.

[131] S. V. Nagaev and V. I. Wachtel. Probability inequalities for a critical Galton–Watson process. *Theory Probab. Appl.*, 50:225–247, 2006.

[132] V. V. Petrov. *Sums of Independent Random Variables*. Springer, Berlin, 1975.

[133] J. Pitman. One-dimensional Brownian motion and the three-dimensional Bessel process. *Adv. Appl. Probab.*, 7:511–526, 1975.

[134] S. Popov. *Two-Dimensional Random Walk – from Path Counting to Random Interlacements*. Cambridge University Press, Cambridge, 2021.

[135] D. Revuz and M. Yor. *Continuous Martingales and Brownian Motion, 3rd Ed.* Springer, Berlin, 1999.

[136] B. Rogozin. Asymptotics of renewal functions. *Theory Probab. Appl.*, 21: 669–686, 1977.

[137] T. Rolski, H. Schmidli, V. Schmidt, and J. Teugels. *Stochastic Processes for Insurance and Finance*. Wiley, Chichester, 1998.

[138] W. A. Rosenkrantz. A local limit theorem for a certain class of random walks. *Ann. Math. Statist.*, 37:855–859, 1966.

[139] N. Sandrić. Recurrence and transience property for a class of Markov chains. *Bernoulli*, 19:2167–2199, 2013.

[140] E. Seneta. An explicit-limit theorem for the critical Galton–Watson process with immigration. *J. Royal Statist. Soc. Ser. B*, 32:149–152, 1970.

[141] V. Shurenkov. On Markov renewal theory. *Theory Probab. Appl.*, 29:247–265, 1984.

[142] W. L. Smith. On some general renewal theorems for nonidentically distributed variables. In *Proc. 4th Berkeley Sympos. Math. Statist. and Prob., Vol. 2*: Contributions to Probability Theory, 467–514, 1961.

[143] F. Spitzer. *Principles of Random Walk*. D. Van Nostrand, Princeton, N.J.–Toronto–London, 1964.

[144] V. A. Vatutin. The asymptotic probability of the first degeneration for branching processes with immigration. *Theory Probab. Appl.*, 19:25–34, 1974.

[145] V. A. Vatutin. A critical Galton–Watson branching process with emigration. *Theory Probab. Appl.*, 22:465–481, 1977.

[146] V. A. Vatutin. A conditional limit theorem for a critical branching process with immigration. *Math. Notes*, 21:405–411, 1977.

[147] W. Vervaat. On a stochastic difference equation and a representation of nonnegative infinitely divisible random variables. *Adv. Appl. Probab.*, 11:750–783, 1979.

[148] M. Voit. Strong laws of large numbers for random walks associated with a class of one-dimensional convolution structures. *Monatsh. Math.*, 113:59–74, 1992.

[149] M. Voit. Central limit theorems for Markov processes associated with Laguerre polynomials. *J. Math. Anal. Appl.*, 182:731–741, 1994.

[150] V. Vysotsky. Limit theorems for random walks that avoid bounded sets, with applications to the largest gap problem. *Stochastic Process. Appl.*, 125: 1886–1910, 2015.

[151] J. A. Williamson. Some renewal theorems for non-negative independent random variables. *Trans. Amer. Math. Soc.*, 114:417–445, 1965.

[152] W. Woess. *Denumerable Markov Chains*. European Mathematical Society, Zurich, 2009.

[153] A. M. Zubkov. The life spans of a branching process with immigration. *Theory Probab. Appl.*, 17:174–183, 1972.

Author Index

Subject Index

For EU product safety concerns, contact us at Calle de José Abascal, 56–1°,
28003 Madrid, Spain or eugpsr@cambridge.org.

www.ingramcontent.com/pod-product-compliance
Lightning Source LLC
LaVergne TN
LVHW011531200926
848484LV00001B/226